AutoCAD Plant 3D 2023 for Designers

(7th Edition)

CADCIM Technologies
525 St. Andrews Drive
Schererville, IN 46375, USA
(www.cadcim.com)

Contributing Author
Sham Tickoo
Professor
Purdue University Northwest
Department of Mechanical Engineering Technology
Hammond, Indiana, USA

CADCIM Technologies

AutoCAD Plant 3D 2023 for Designers 7th Edition
Sham Tickoo

CADCIM Technologies
525 St Andrews Drive
Schererville, Indiana 46375, USA
www.cadcim.com

ISBN 978-1-64057-155-6

www.cadcim.com

DEDICATION

To teachers, who make it possible to disseminate knowledge
to enlighten the young and curious minds
of our future generations

To students, who are dedicated to learning new technologies
and making the world a better place to live in

SPECIAL RECOGNITION

A special thanks to Mr. Denis Cadu and the ADN team of Autodesk Inc.
for their valuable support and professional guidance to
procure the software for writing this textbook

THANKS

To the faculty and students of the MET department of
Purdue University Northwest for their cooperation

To employees of CADCIM Technologies for their valuable help

Online Training Program Offered by CADCIM Technologies

CADCIM Technologies provides effective and affordable virtual online training on various software packages including Computer Aided Design, Manufacturing, and Engineering (CAD/CAM/CAE), computer programming languages, animation, architecture, and GIS. The training is delivered 'live' via Internet at any time, any place, and at any pace to individuals as well as the students of colleges, universities, and CAD/CAM training centers. The main features of this program are:

Training for Students and Companies in a Classroom Setting

Highly experienced instructors and qualified engineers at CADCIM Technologies conduct the classes under the guidance of Prof. Sham Tickoo of Purdue University Northwest, USA. This team has authored several textbooks that are rated "one of the best" in their categories and are used in various colleges, universities, and training centers in North America, Europe, and in other parts of the world.

Training for Individuals

CADCIM Technologies with its cost effective and time saving initiative strives to deliver the training in the comfort of your home or work place, thereby relieving you from the hassles of traveling to training centers.

Training Offered on Software Packages

CADCIM provides basic and advanced training on the following software packages:

***CAD/CAM/CAE**: CATIA, Pro/ENGINEER Wildfire, Creo Parametric, Creo Direct, SOLIDWORKS, Autodesk Inventor, Solid Edge, NX, AutoCAD, AutoCAD LT, AutoCAD Plant 3D, Customizing AutoCAD, EdgeCAM, and ANSYS*

***Architecture and GIS**: Autodesk Revit (Architecture, Structure, MEP), AutoCAD Civil 3D, AutoCAD Map 3D, Navisworks, Oracle Primavera, and Bentley STAAD Pro*

***Animation and Styling**: Autodesk 3ds Max, Autodesk Maya, Autodesk Alias, Foundry NukeX, and MAXON CINEMA 4D*

***Computer Programming**: C++, VB.NET, Oracle, AJAX, and Java*

For more information, please visit the following link: ***https://www.cadcim.com***

Note

If you are a faculty member, you can register by clicking on the following link to access the teaching resources: ***https://www.cadcim.com/Registration.aspx***. The student resources are available at ***https://www.cadcim.com***. We also provide **Live Virtual Online Training** on various software packages. For more information, write us at ***sales@cadcim.com***.

Table of Contents

Dedication iii
Preface xiii

Chapter 1: Introduction to AutoCAD Plant 3D

Introduction to AutoCAD Plant 3D 2023 1-2
Starting AutoCAD Plant 3D 1-2
  Start Tab Enhanced 1-2
Working on a Project 1-3
AutoCAD Plant 3D User Interface 1-4
  Drawing Area 1-5
  Command Window 1-5
  PROJECT MANAGER 1-5
  DATA MANAGER 1-6
  Navigation Bar 1-7
  ViewCube 1-7
  In-canvas Viewport Controls 1-8
  Status Bar 1-8
  PROPERTIES Palette 1-9
Different Workspaces in AutoCAD Plant 3D 1-10
Grips 1-11
Invoking Commands in AutoCAD Plant 3D 1-12
  Invoking Commands Using Command Prompt 1-12
  Invoking Commands Using Ribbon 1-12
  TOOL PALETTES 1-14
  Application Menu 1-15
  Menu Bar 1-16
  Shortcut Menu 1-16
AutoCAD Plant 3D Dialog Boxes 1-18
Creating Backup Files 1-19
  Converting Auto-saved and Backup Files into AutoCAD Format 1-19
  Using the Drawing Recovery Manager to Recover Files 1-20
Closing a Drawing 1-20
Opening a Project Drawing 1-21
Opening a Drawing that is not in the Project 1-21
  Opening an Existing Drawing Using the Select File Dialog Box 1-22
Quitting AutoCAD Plant 3D 1-24
AutoCAD Plant 3D Help 1-24
Autodesk App Store 1-25
Additional Help Resources 1-25
Self-Evaluation Test 1-26
Review Questions 1-27

Chapter 2: Creating Projects and P&IDs

Introduction 2-2
PROJECT MANAGER 2-2
Creating a New Project in AutoCAD Plant 3D 2-2
Creating a New Drawing 2-6
Grouping Project Files 2-6
Designing a P&ID 2-7
Adding Equipment to a P&ID 2-8
Adding Pipe Lines 2-9
Assigning Tags to a Line 2-10
Assigning Custom Service Category to a Line 2-11
Creating and Adding Custom Pipe Line Segments to TOOL PALETTES 2-13
Adding Custom Tags to Pipe Line Segments 2-14
Adding Valves 2-17
Adding Instruments and Instrumentation Lines 2-18
Adding Fittings 2-20
Adding the Off Page Connectors 2-20
Connecting the Off Page Connectors 2-20
Validating the Drawing 2-21
Checking for Errors 2-22
Editing the Drawing 2-24
Moving an Equipment 2-25
Moving a Valve 2-25
Moving a Line 2-25
Editing a Line 2-26
Grouping Lines 2-27
Editing a P&ID Symbol 2-28
Substituting Components 2-28
Converting AutoCAD Components into P&ID Symbols 2-28
Adding Intelligence to the Custom P&ID Symbols 2-29
Tutorial 1 2-35
Self-Evaluation Test 2-52
Review Questions 2-52
Exercise 1 2-53

Chapter 3: Creating Structures

Introduction 3-2
Creating a Grid 3-2
Editing Grids 3-3
Setting the Representation of the Structural Member 3-3
Adding Members 3-4
Creating Stairs 3-6
Editing Stairs 3-7
Creating Railings 3-7
Creating Ladders 3-9

Ladder Tab 3-10
Cage Tab 3-10
Creating a Plate/Grate 3-11
Creating Footings 3-12
Editing the Structural Members 3-12
Changing the Length of a Member 3-12
Restoring the Member to its Original Length 3-13
Cutting Member at Intersections 3-13
Creating Miter Joints 3-14
Trimming/Extending a Member 3-14
Exploding a Structure 3-15
Visibility Options 3-15
Hiding and Displaying Components 3-16
Exchanging Data with other Applications 3-16
Tutorial 1 3-16
Tutorial 2 3-21
Self-Evaluation Test 3-29
Review Questions 3-30
Exercise 1 3-30

Chapter 4: Creating Equipment

Introduction 4-2
Creating Equipment 4-2
Placing Equipment in a Drawing 4-3
Adding a Vessel 4-3
Adding a Heat Exchanger 4-4
Adding a Pump 4-4
Adding a Heater 4-5
Creating a Customized Equipment 4-6
Modifying Equipment 4-7
Converting Solid Models into Equipment 4-8
Converting Inventor Models into Equipment 4-9
Attaching Objects to an Equipment 4-9
Detaching Objects from an Equipment 4-10
Adding Nozzles to a Customized Equipment 4-10
Adding Nozzles to a Converted Equipment 4-13
Modifying Nozzles 4-13
Tutorial 1 4-14
Tutorial 2 4-25
Self-Evaluation Test 4-30
Review Questions 4-30

Chapter 5: Editing Specifications and Catalogs

Introduction 5-2
Getting Started with AutoCAD Plant 3D Spec Editor 5-2
Working with Spec Files 5-2

Creating a New Spec File from an Existing Spec 5-3
Adding Parts to the Spec Sheet 5-5
Editing the Parts Added to a Spec 5-5
Setting the Part Use Priority 5-7
Adding Notes to a Group 5-8
Editing the Long Description Styles 5-9
Assigning a Long Description Style to Multiple Specs 5-11
Assigning Operators (Actuators) to Valves 5-11
Working with the Catalog Editor 5-13
Creating a New Catalog from an Existing Catalog 5-15
Adding a New Part to a Catalog 5-15
Creating a New Component Using Parametric Graphics 5-16
Creating a New Component Using Block Based Graphics 5-17
Modifying the Branch Table 5-19
Creating Branch Table Legends 5-19
Assigning Legends to a Branch Table 5-20
Tutorial 1 5-21
Tutorial 2 5-26
Tutorial 3 5-31
Self-Evaluation Test 5-39
Review Questions 5-39

Chapter 6: Routing Pipes

Introduction 6-2
Selecting a Spec 6-2
Working with the Spec Viewer 6-2
Routing a Pipe 6-3
Routing a Pipe with a New Line Number 6-4
Setting the Route Line 6-5
Routing a Pipe from a Line 6-6
Routing a Pipe using a P&ID 6-6
Routing a Pipe from an Equipment 6-7
Working with the Compass 6-7
Connecting two Open Ports of Pipes 6-9
Changing the Pipe Size while Routing 6-11
Changing the Orientation Plane while Routing a Pipe 6-11
Creating a Cutback Elbow 6-12
Creating a Roll Elbow 6-13
Creating Bends 6-14
Changing the Elevation while Routing a Pipe 6-14
Routing Pipe at an Offset 6-15
Routing a Pipe at a Slope 6-16
Creating Branches 6-18
Creating a Tee Branch 6-18
Creating an O-let Branch 6-19
Creating a Stub-In Branch 6-20
Creating a Stub-In Branch at an Offset from the Center of the Header Pipe 6-20

Creating a Branch from an Elbow 6-21
Creating a Stub-In Branch at a Precise Location 6-21
Adding a Reinforced Pad to a Stub-In Branch 6-22
Creating a Weld Connection 6-22
Creating Autodesk Connection Point 6-23
Editing an Autodesk Connection Point 6-23
Routing a Pipe from an Autodesk Connection Point 6-24
Tutorial 1 6-24
Self-Evaluation Test 6-41
Review Questions 6-42

Chapter 7: Adding Valves, Fittings, and Pipe Supports

Introduction 7-2
Adding Valves and Fittings 7-2
Adding Valves and Fittings to a Pipe using the Spec Sheet 7-2
Adding Valves and Fittings Using a P&ID 7-3
Placing Valves and Fittings while Routing a Pipe 7-4
Placing Custom Parts 7-5
Mapping a P&ID Object to a Plant 3D Object 7-6
Adding Pipe Supports 7-8
Adding a Dummy Leg 7-10
Adding a Hanger and Connecting it to a Structural Member 7-10
Modifying the Pipe Supports 7-11
Copying and Moving a Pipe Support 7-13
Connecting Two Pipe Supports 7-14
Converting Solids into Pipe Supports 7-14
Attaching Objects to a Pipe Support 7-14
Detaching Objects from a Support 7-15
Insulating a Pipe 7-15
Modifying the Pipe Components Using Grips 7-15
Substituting a Pipe Component 7-15
Rotating a Pipe Component 7-15
Flipping a Pipe Component 7-15
Flipping a Component Inline with the Pipe 7-16
Changing the Elevation of the Pipe 7-16
Changing the Valve Operator 7-16
Validating a 3D Model 7-18
Tutorial 1 7-19
Self-Evaluation Test 7-27
Review Questions 7-28

Chapter 8: Creating Isometric Drawings

Introduction 8-2
Isometric Drawing Types 8-2
Creating a Quick Isometric Drawing 8-2
Creating a Production Isometric Drawing 8-5

Viewing Isometric Results 8-5
Placing Iso Messages and Annotations 8-6
Exporting a Pipe Component File 8-8
Creating an Iso from a Pipe Component File 8-8
Locking a Line Number 8-9
Configuring Isometric Drawing Settings 8-9
Configuring Iso Style Settings 8-9
Configuring Annotation Settings 8-10
Configuring Dimensional Settings 8-11
Configuring Themes 8-12
Configuring Sloped and Offset Piping Settings 8-13
Setting the Title Block and Display Properties 8-14
Tutorial 1 8-16
Tutorial 2 8-21
Tutorial 3 8-25
Self-Evaluation Test 8-32
Review Questions 8-33

Chapter 9: Creating Orthographic Drawings

Introduction 9-2
Creating Orthographic Drawings 9-2
Generating the First View 9-2
Creating the Adjacent View 9-6
Adding Annotations to the Drawing 9-7
Adding Dimensions to the Drawing 9-8
Locating a Component in the 3D Model 9-8
Editing a Drawing View 9-8
Updating a View 9-8
Adding Gaps to Pipes 9-8
Generating Bill of Material 9-8
Displaying Pipes as Single Lines New 9-9
Creating BOM Annotation 9-9
Tutorial 1 9-10
Self-Evaluation Test 9-17
Review Questions 9-17

Chapter 10: Managing Data and Creating Reports

Introduction 10-2
DATA MANAGER 10-2
Viewing data in the DATA MANAGER 10-3
Modifying the Display of Data 10-3
Displaying the Data by Object Type and Area 10-3
Zooming to Plant Objects 10-4
Locating an Object in the Drawing Area 10-4
Scrolling through the Data in the DATA MANAGER 10-5
Editing Data in the DATA MANAGER 10-5

Placing Annotations in a P&ID Using the DATA MANAGER 10-5
Filtering the Information in the Data Table 10-6
Viewing only the Selected Record in the Data Table 10-6
Viewing all the records in the Data Table except the Selected One 10-6
Viewing the Data of the Objects Selected in the Drawing Area 10-6
Exporting Data from the DATA MANAGER 10-7
Importing Data to the DATA MANAGER 10-7
Accepting or Rejecting Changes in the Imported Data 10-8
Viewing Reports in the DATA MANAGER 10-8
Exporting the Project Reports 10-9
Importing the Project Reports 10-9
Working with the Report Creator 10-10
Generating Reports using the AutoCAD Plant Report Creator 10-11
Tutorial 1 10-14
Tutorial 2 10-18
Self-Evaluation Test 10-20
Review Questions 10-20

Project: Thermal Power Plant

Index I-1

This page is intentionally left blank

Preface

AutoCAD Plant 3D

AutoCAD Plant 3D 2023, a product of Autodesk Inc, is a plant layout design software. It is extensively used in the industry to create and modify the P&IDs and 3D models of the process plants. Built on the AutoCAD platform, this software allows you to carry out the design process in a project based manner. The modern toolset available with this software allows designers and engineers, who model and document process plants, to generate and share isometric drawings, orthographic drawings, and materials reports. You can also integrate Plant 3D with Navisworks and can import designs created in Revit and Inventor.

The **AutoCAD Plant 3D 2023 for Designers** textbook explains the readers to effectively use the designing tools in AutoCAD Plant 3D. The accompanying tutorials and exercises, which relate to the real world projects, help you understand the usage and abilities of the tools available in AutoCAD Plant 3D 2023. You will learn how to setup a project, create and edit P&IDs, design a 3D Plant model, generate isometric/orthographic drawings, as well as how to publish and print drawings. The chapters in this textbook are structured in a pedagogical sequence that makes this textbook very effective in learning the features and capabilities of the software.

Since AutoCAD Plant 3D is based on AutoCAD platform, a user must have the basic knowledge of AutoCAD. In this textbook, the basic tools of AutoCAD are not explained while explaining the working of Plant 3D tools assuming that the user knows AutoCAD basics.

The salient features of this textbook are as follows:

- **Tutorial Approach**
  The author has adopted tutorial point-of-view and learn-by-doing theme throughout the textbook. This approach guides the users understand the concepts and processes easily.

- **Tips and Notes**
  Additional information related to various topics is provided to the users in the form of tips and notes.

- **Learning Objectives**
  The first page of every chapter summarizes the topics that are covered in that chapter.

- **Self-Evaluation Test, Review Questions, and Exercises**
  Every chapter ends with Self-Evaluation Test so that the users can assess their knowledge of the chapter. The answers to Self-Evaluation Test are given at the end of the chapter. Also, Review Questions and Exercises are given at the end of the chapters and they can be used by the instructor as test questions and exercises.

- **Heavily Illustrated Text**
  The text in this book is heavily illustrated with about 600 line diagrams and screen capture images.

Symbols Used in the Book

Note
The author has provided additional information related to various topics in the form of notes.

Tip
The author has provided a lot of useful information to the users about the topic being discussed in the form of tips.

Enhanced
This symbol indicates that the command or tool being discussed is enhanced.

New
This icon indicates that the command or tool being discussed is new.

Formatting Conventions Used in the Book

Refer to the following list for the formatting conventions used in this textbook.

- Command names are capitalized and written in boldface letters.
  Example: The **MOVE** command

- A key icon appears when you have to respond by pressing the ENTER or the RETURN key.
  Enter

- Command sequences are indented. The responses are indicated in boldface. The directions are indicated in italics and the comments are enclosed in parentheses.
  Command: **MOVE**
  Select object: **G**
  Enter group name: *Enter a group name (the group name is group1)*

- The methods of invoking a tool/option from the **Ribbon**, **Menu Bar**, **Quick Access Toolbar**, **Application menu**, Status Bar, and Command prompt are enclosed in a shaded box.
  Ribbon: Structure > Parts > Footing
  Menu Bar: File > Open
  Quick Access Toolbar: Open
  Application Menu: Open > Drawing
  Command: OPEN

Naming Conventions Used in the Book

Tool

If you click on an item in a panel of the **Ribbon** and a command is invoked to create/edit an object or perform some action, then that item is termed as **tool**.

For example:
To Create: Route Pipe tool, **Create Supports** tool, **Create Ortho View** tool
To Edit: Modify Equipment tool, **Attach Supports** tool, **Detach Supports** tool
Action: Zoom tool, **Move** tool, **Copy** tool

If you click on an item in a panel of the **Ribbon** and a dialog box is invoked wherein you can set the properties to create/edit an object, then that item is also termed as tool, refer to Figure 1.

For example:
To Create: Create Equipment tool, **Create Supports** tool
To Edit: Edit Attributes tool, **Block Editor** tool

*Figure 1 Various tools in the **Ribbon***

Button

If you click on an item in a toolbar or a panel of the **Ribbon** and the display of the corresponding object is toggled on/off, then that item is termed as **Button**. For example, **Grid** button, **Snap** button, **Ortho** button, **Route Pipe** button, **Line to Pipe** button, and so on; refer to Figure 2.

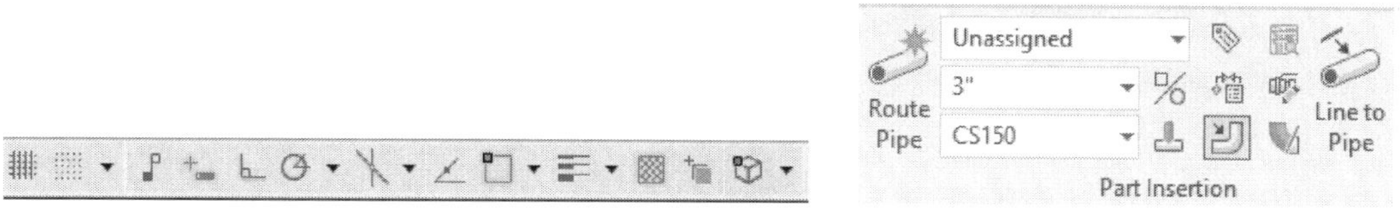

*Figure 2 Various buttons displayed in the Status Bar and **Ribbon***

The item in a dialog box that has a 3D shape like a button is also termed as **Button**. For example, **OK** button, **Cancel** button, **Apply** button, and so on.

Dialog Box

In this textbook, different terms are used for referring to the components of a dialog box. Refer to Figure 3 for the terminology used.

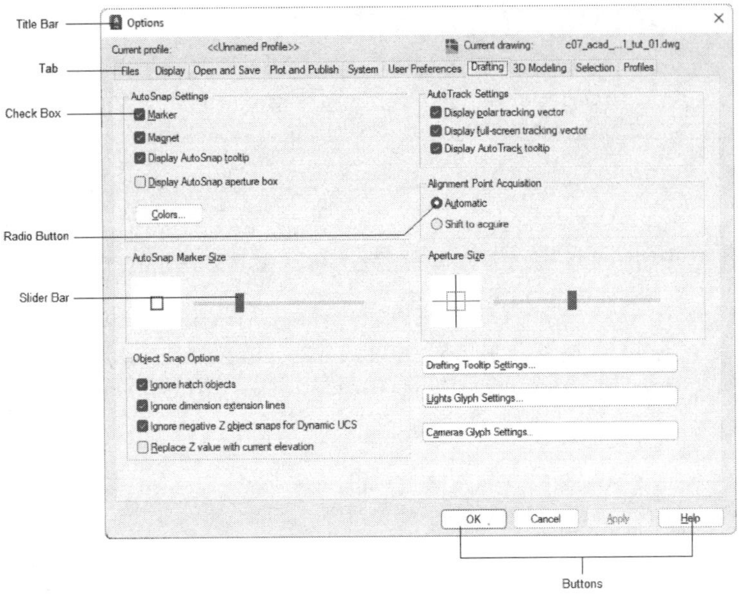

Figure 3 The components in a dialog box

Drop-down

A drop-down is the one in which a set of common tools are grouped together for performing an action. You can identify a drop-down with a down arrow on it. These drop-downs are given a name based on the tools grouped in them. For example, **Project** drop-down, **Settings** drop-down, **Create Light** drop-down, and so on; refer to Figure 4.

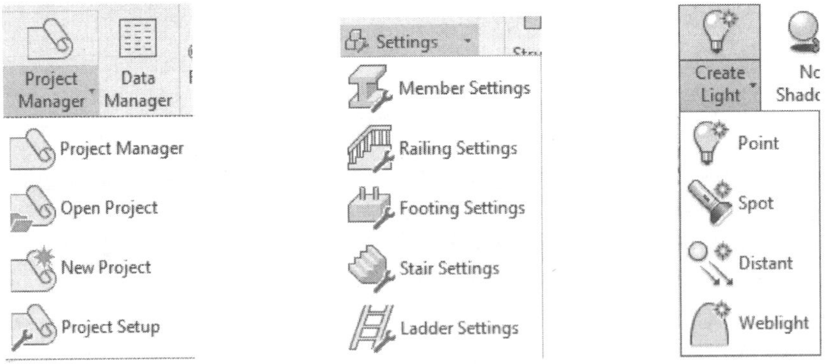

*Figure 4 The **Project**, **Settings**, and **Create Light** drop-downs*

Drop-down List

A drop-down list is the one in which a set of options are grouped together. You can set parameters using these options. You can identify a drop-down list with a down arrow on it. To know the name of a drop-down list, move the cursor over it; its name will be displayed as a tool tip. For example, **Line Number Selector** drop-down list, **Linetype** drop-down list, **Object Color** drop-down list, **Visual Styles** drop-down list, and so on, refer to Figure 5.

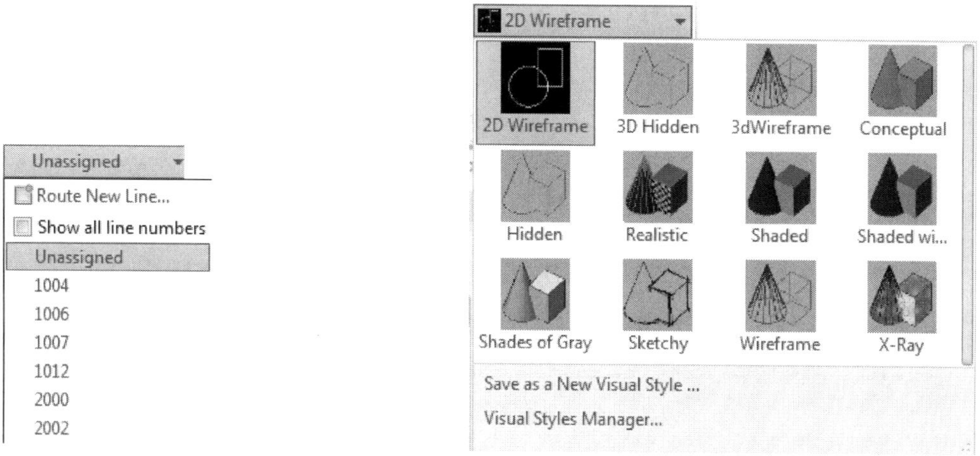

***Figure 5** The **Line Number Selector** and **Visual Styles** drop-down lists*

Options

Options are the items that are available in shortcut menu, drop-down list, Command prompt, **Properties** panel, and so on. For example, choose the **Properties** option from the shortcut menu displayed on right-clicking in the drawing area, refer to Figure 6.

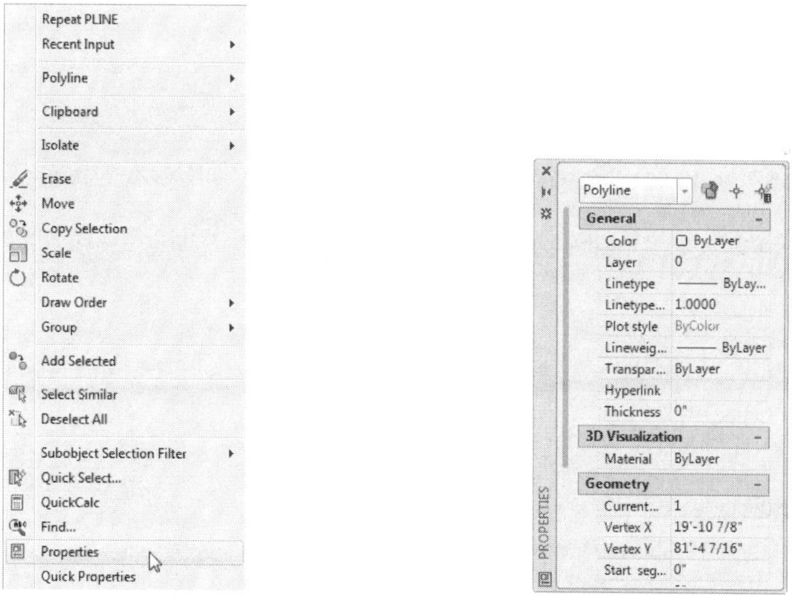

***Figure 6** Options in a shortcut menu and the **PROPERTIES** palette*

Free Companion Website

It has been our constant endeavor to provide you the best textbooks and services at affordable price. In this endeavor, we have come out with a free companion website that will facilitate the process of teaching and learning of AutoCAD Plant 3D. If you purchase this textbook, you will get access to the companion website.

The following resources are available for the faculty and students in this website:

Faculty Resources

- **Technical Support**
  The faculty can get online technical support by contacting ***techsupport@cadcim.com***.

- **Instructor Guide**
  Solutions to all review questions and exercises in the textbook are provided in this link to help faculty members test the skills of the students.

- **Part Files**
  The part files used in illustrations, tutorials, and exercises are available for free download.

- **Input Files**
  The input files used in tutorials and exercises are available for free download.

Student Resources

- **Technical Support**
  The students can get online technical support by contacting ***techsupport@cadcim.com***.

- **Part Files**
  The part files used in illustrations and tutorials are available for free download.

- **Input Files**
  The input files used in tutorials and exercises are available for free download.

You can access additional learning resources by visiting ***https://allaboutcadcam.blogspot.com***.

If you face any problem in accessing these files, please contact the publisher at ***sales@cadcim.com*** or the author at ***stickoo@pnw.edu*** or ***tickoo525@gmail.com***.

Video Courses

CADCIM offers video courses in CAD, CAE Simulation, BIM, Civil/GIS, and Animation domains on various e-Learning/Video platforms. To enroll for the video courses, please visit the CADCIM website using the link ***https://www.cadcim.com/video-courses.***

Stay Connected

You can now stay connected with us through Facebook and Twitter to get the latest information on textbooks, videos, and teaching/learning resources. To stay informed of such updates, follow us on Facebook (***www.facebook.com/cadcim***) and Twitter (***@cadcimtech***). You can also subscribe to our You Tube channel (***www.youtube/cadcimtech***) to get the information about our latest video tutorials.

Chapter 1

Introduction to AutoCAD Plant 3D

Learning Objectives

After completing this chapter, you will be able to:

- *Start AutoCAD Plant 3D*
- *Work with the components of the initial screen of AutoCAD Plant 3D*
- *Work with the PROJECT MANAGER*
- *Invoke AutoCAD commands from the keyboard, menus, shortcut menus, Tool Palettes, and Ribbon*
- *Work with the components of dialog boxes in AutoCAD Plant 3D*

INTRODUCTION TO AutoCAD Plant 3D 2023

AutoCAD Plant 3D is a purpose-built plant design software. This software is used to design and document process plants. AutoCAD Plant 3D contains various predefined shapes of plant components. These predefined shapes, which are solid models, carry the intelligence of AutoCAD Plant 3D drawings. While installing AutoCAD Plant 3D, the standard specifications and part catalogs related to piping are also installed.

AutoCAD Plant 3D comes along with AutoCAD P&ID which is used to create piping and instrumentation drawings. AutoCAD P&ID contains various piping and instrumentation symbols. These symbols, most of which are AutoCAD blocks with attributes, carry the intelligence of AutoCAD P&ID drawings.

STARTING AutoCAD Plant 3D

When you install AutoCAD Plant 3D 2023, its icon is created and displayed on the desktop. You can launch the application by double-clicking on that icon. The interface screen of AutoCAD Plant 3D 2023 is shown in Figure 1-1.

Start Tab

In the initial interface of Plant 3D, the **Start** tab remains chosen by default, refer to Figure 1-1. This tab consists of two areas namely **AutoCAD Plant 3D 2023** and **Announcements**, and these areas contain options to access drawing template files, recently opened drawings, sheet sets, and various other learning options. The areas and their respective options are discussed next.

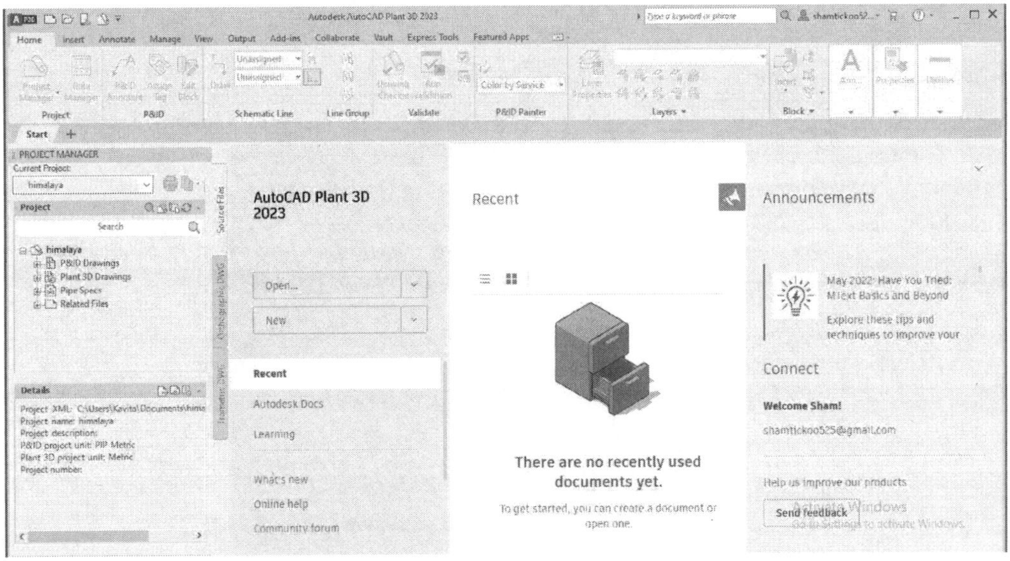

Figure 1-1 The interface of AutoCAD Plant 3D 2023

AutoCAD Plant 3D 2023

This area is displayed by default whenever you launch AutoCAD Plant 3D. The options contained in it are discussed next.

Open

When you choose this button, the **Select File** dialog box will be displayed. It allows you to open any existing file by either specifying its name in the **File Name** edit box or by browsing and selecting the file from the location where it is saved, and then choosing the Open button.

Note

*If you click on the down arrow next to the **Open** button, a list of options will be displayed. Using these options, you will be able to open any sheet sets, sample drawings, existing projects, collaboration projects, and projects from the vault.*

New

When you choose this button, a new drawing file will be created.

Note

*If you click on the down arrow next to the **New** button a list of options will be displayed. Using these options, you will be able to select template files, create sheet sets and new projects.*

Recent

When you click on this button, the **Recent** page gets displayed. This page lists the project files that you have recently worked on. You can access these files by simply clicking on the file name displayed in this page.

AutoDesk Docs

Autodesk Docs provides a way to open and save files to your connected drives. You will need to have Desktop Connector installed to access the connected drives on Autodesk Docs.

Learning

When you click on this button; the **Learning** page gets displayed. This page provides you the options that allow you to access to the learning resources related to AutoCAD Plant 3D.

Announcements

This area provides you with the options to connect with the Autodesk community. Note that you should be logged in to your A360 account to access the options in this area.

WORKING ON A PROJECT

A piping design project includes the drawing and other forms of data. These data sources are inter-related in a project. The drawing data includes P&IDs, 3D models, Isometric drawings, and Orthographic drawings. The other data forms include catalogs and specs for piping, process data, and so on. Figure 1-2 shows a project flow diagram.

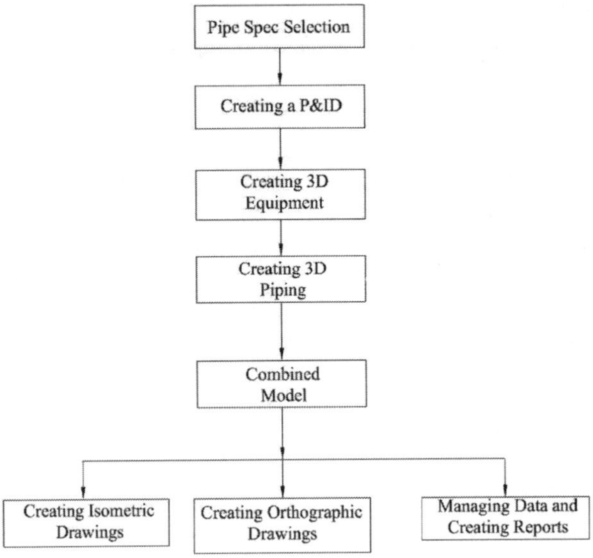

***Figure 1-2** A project flow diagram*

AutoCAD Plant 3D USER INTERFACE

The AutoCAD Plant 3D 2023 interface consists of the drawing area, command window, PROJECT MANAGER, menu bar, toolbars, Status Bar, and so on, refer to Figure 1-3. These components are discussed next.

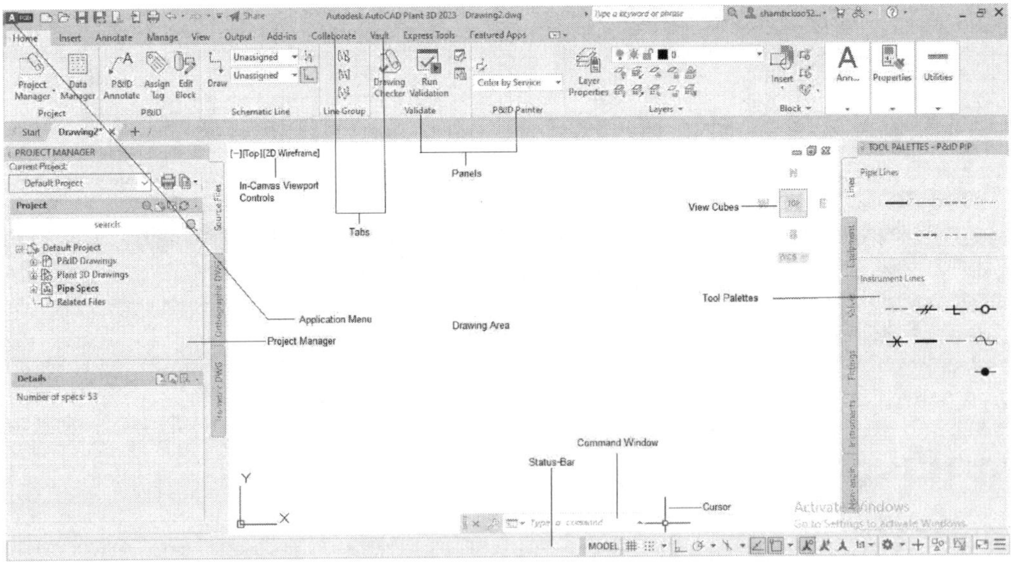

***Figure 1-3** AutoCAD Plant 3D 2023 user interface*

Drawing Area

The drawing area covers the major portion of the screen. In this area, you can draw objects and use the commands. To draw objects, you need to define the coordinate points which can be done by using the pointing device. The position of the pointing device is represented on the screen by the cursor. The drawing area also has the standard Windows buttons such as **Close**, **Minimize**, and **Restore Down** at the top right corner. These buttons have the same functions as in other standard window.

Command Window

The command window located at the bottom of the drawing area has the command prompt where you can enter the commands. It also displays the subsequent prompt sequences and the messages. You can change the size of the window by placing the cursor on the top edge (double line bar known as the grab bar) and then dragging it. This way you can increase its size to see all the previous commands you have used.

PROJECT MANAGER

Ribbon: Project > Project Manager drop-down > Project Manager
Command: PROJECTMANAGER

PROJECT MANAGER is used to access and manage all the drawings in a project. In addition to that, you can configure project settings, and export and import data from the **PROJECT MANAGER**. Figure 1-4 shows the **PROJECT MANAGER**. When you launch AutoCAD Plant 3D for the first time, the PROJECT MANAGER is displayed on the left in the interface, refer to Figure 1-3. There are three tabs in the **PROJECT MANAGER: Source Files**, **Orthographic DWG**, and **Isometric DWG**. You can create and access the P&ID and Plant 3D drawing files using the **Source Files** tab. The other two tabs can be used to create and access the orthographic and isometric drawings.

The areas in the **PROJECT MANAGER** and the options available in them are discussed next.

Current Project Area

When you open the **PROJECT MANAGER** for the first time, the **Source Files** tab is chosen by default and the **Default Project** is displayed in this area, refer to Figure 1-4. Using the options in the drop-down list in the **Current Project** area, you can open existing projects, collaboration projects, access files from Vault, start a new project, and access the sample project. The **Publish** button in this area enables you to publish the drawing set to a DWF, DWFx, or PDF format. The **Reports** drop-down next to the **Publish** button is used for displaying a list of reports that you can run. Using the options in this drop-down, you can view, modify, import, and export data related to a project.

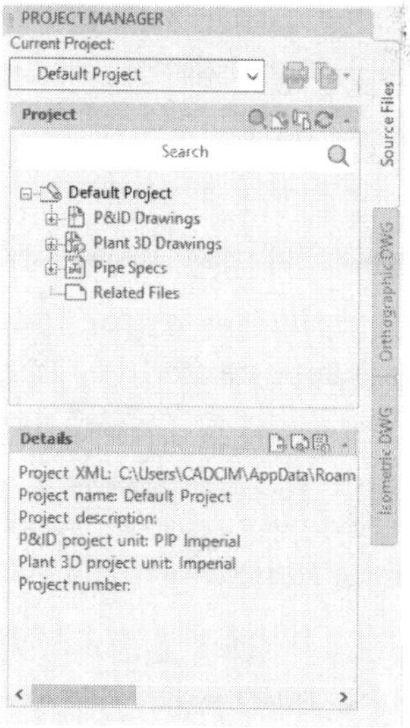

*Figure 1-4 The **PROJECT MANAGER***

Project Area

All the drawing files related to a project are organized under their respective nodes in this area. For instance, if you create a project with the name CADCIM, a node named CADCIM will be created as a project node along with four sub-nodes: **P&ID Drawings**, **Plant 3D Drawings**, **Pipe Specs**, and **Related Files**. These sub-nodes contain respective project files. For example, any P&ID file related to the CADCIM project will be displayed under the **P&ID Drawings** sub-node of the CADCIM node in the **Project** area. The piping specifications which are being used in the project will be displayed under the **Pipe Specs** sub-node of the CADCIM node.

The toolbar available in this area contains four buttons which are discussed next.

Hide/Show search box

This toggle button is used to show or hide a search box below the **Project Area**. This search box can be used to search a drawing in the project. To search a drawing, type its name in the search box to display the results under the linked node.

New Drawing

When you choose this button, the **New DWG** dialog box will be displayed. Using this dialog box, you can create a new drawing and add it to the project tree.

Copy Drawing to Project

When you choose this button, the **Select Drawings to Copy to Project** dialog box will be displayed. Using this dialog box, you can select one or more files to copy in the project.

Refresh DWG Status

This button enables you to know if any of the drawings or the files in the project are being edited by the other user.

Details Area

This area lists all the information related to the project such as project path, project name and its description.

The data used in a project is arranged in folders that are created while creating a project. For example, if you create a project with the name CADCIM, a folder named CADCIM will be created as the project folder. Also, a series of the sub-folders will be created in the project folder that contains data related to the project, refer to Figure 1-5.

DATA MANAGER

Ribbon: Project > Data Manager
Command: DATAMANAGER

When you create a P&ID or a Plant 3D model, each and every part of the drawing is assigned with some properties. These properties can be accessed through the **DATA MANAGER**. Using the **DATA MANAGER**, you can view, import, export, and create reports from the project data. Choose the **Data Manager** button from the **Project** panel to display the **DATA MANAGER**. Figure 1-6 shows a partial view of the **DATA MANAGER**.

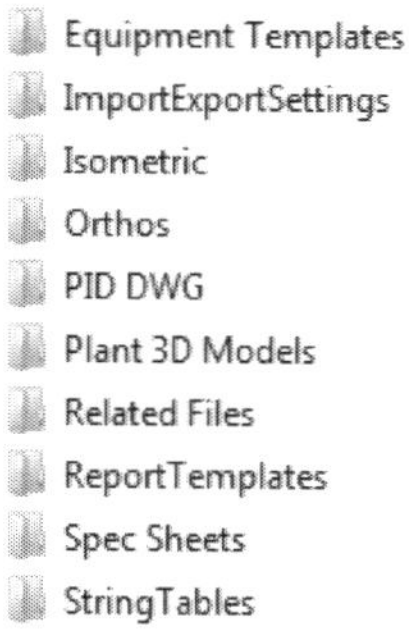

***Figure 1-5** Sub-folders of the project folder*

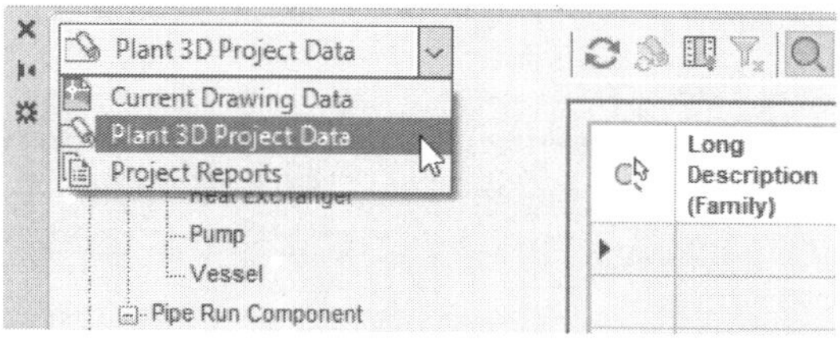

Figure 1-6** Partial view of the **DATA MANAGER

Navigation Bar

The Navigation Bar is displayed in the drawing area and contains various navigation tools which are grouped together, refer to Figure 1-7. You can toggle on/off the visibility of this bar by using the **Navigation Bar** toggle button available in the **Viewport Tools** palette of the **View** tab of the **Ribbon**. These tools are discussed next.

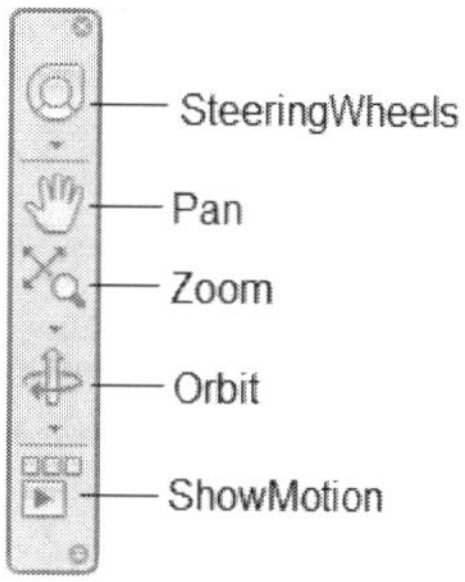

Figure 1-7** The tools in the **Navigation Bar

SteeringWheels

The SteeringWheels allow you to access different 2D and 3D navigation tools. SteeringWheels are divided into different sections known as wedges. Each wedge on a wheel represents a single navigation tool. You can pan, zoom, or rewind the current view of a scene in different ways using the SteeringWheels.

Pan Tool

This tool allows you to adjust the viewpoint of a drawing. To adjust the viewpoint, choose this tool, press and hold the left mouse button, and then drag it in the drawing area. You can exit this command by pressing ESC.

Zoom Tools

The zoom tools are used to adjust the view of the drawing on the screen without affecting the actual size of the object.

Orbit Tools

The orbit tools are used to rotate the view in the 3D space.

ShowMotion Tool

This tool is used to capture different views in a sequence and animate them when required.

ViewCube

ViewCube is available at the top right corner of the drawing area and is used to switch between the standard and isometric views or roll back to the current view.

In-canvas Viewport Controls

In-canvas Viewport Controls are available at the top left corner of the drawing screen. They enable you to change the view, the visual style, as well as the viewport.

Status Bar

The bar displayed at the bottom of the screen is called Status Bar. It contains useful information and buttons, refer to Figure 1-8, which makes it easy to change the status of some of the AutoCAD functions. You can toggle between the on and off states of most of these functions by clicking on them.

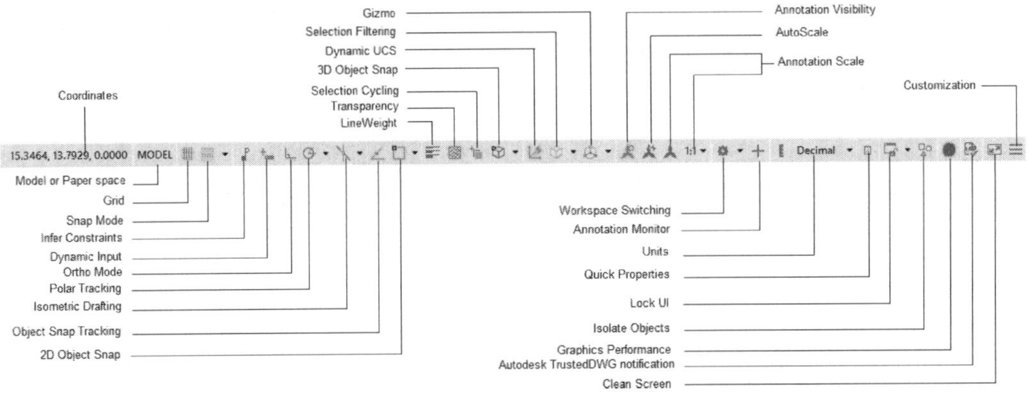

Figure 1-8 *The Status Bar*

Tray Settings

You can control the display of icons and notifications in the tray at the right end of the Status Bar by making appropriate settings in the **Tray Settings** dialog box. To invoke the **Tray Settings** dialog box, enter **TRAYSETTINGS** in the command prompt and press ENTER. The **Tray Settings** dialog box is shown in Figure 1-9.

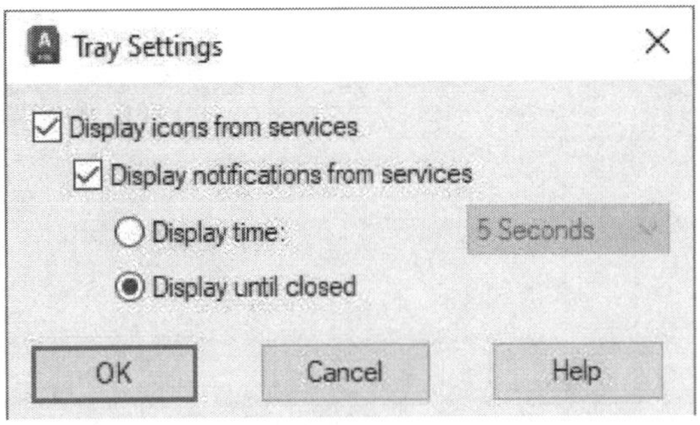

Figure 1-9 *The* ***Tray Settings*** *dialog box*

Clean Screen

The **Clean Screen** button is available at the lower right corner of the screen. This button, when chosen, displays an expanded view of the drawing area by hiding all the toolbars except the command window, Status Bar, and menu bar. The expanded view of the drawing area can also be displayed by choosing **View > Clean Screen** from the menu bar or by using the CTRL+0 keys. You can choose the **Clean Screen** button again to restore the previous display state.

Customization

You can hide the display of some of the buttons in the Status Bar. To do so, click on the **Customization** button available at the lower right corner of the screen; a menu will be displayed. In this menu, clear the check mark next to the names of the buttons you need to hide.

Plot/Publish Details Report Available

This icon is displayed when some plotting or publishing activity is performed in the background. When you click on this icon, the **Plot and Publish Details** dialog box, which provides the details about the plotting and publishing activities, will be displayed. You can copy this report to the clipboard by choosing the **Copy to Clipboard** button from the dialog box.

Manage Xrefs

The **Manage Xrefs** icon is displayed whenever an external reference drawing is attached to the selected drawing. This icon displays an alert message whenever the external referenced drawing needs to be reloaded. To find detailed information regarding the status of each Xref in the drawing and the relation between various Xrefs, click on the **Manage Xrefs** icon; the **EXTERNAL REFERENCES** palette will be displayed.

PROPERTIES Palette

The **PROPERTIES** palette is used to set the current properties and to change the general properties of the selected objects. The **PROPERTIES** palette is displayed on right-clicking on an object and then choosing the **Properties** option from the shortcut menu displayed. If you right-click on the **PROPERTIES** palette, a shortcut menu will be displayed. You can choose **Allow Docking** from this menu to dock or hide the palette.

In AutoCAD Plant 3D, the **PROPERTIES** palette displays an additional section for the properties specific to the selected object, refer to Figure 1-10. For example, if you select a Plant 3D object, the **Plant 3D** section will be displayed in the **PROPERTIES** palette with the 3D properties of the object. The list of properties displayed varies depending on the object selected. For example, if you select pipe support, the dimensions of the selected geometry will be displayed in the **Part Geometry** section, refer to Figure 1-11. In this section, you can edit the geometry of the selected pipe support. Similarly, if you select a valve from a Plant 3D model, properties of the valve will be displayed. You can change the valve actuator using the **PROPERTIES** palette.

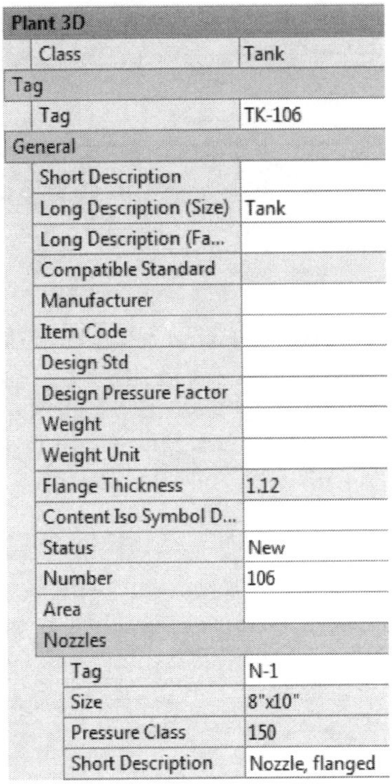

***Figure 1-10** The additional section displayed for a Plant 3D object*

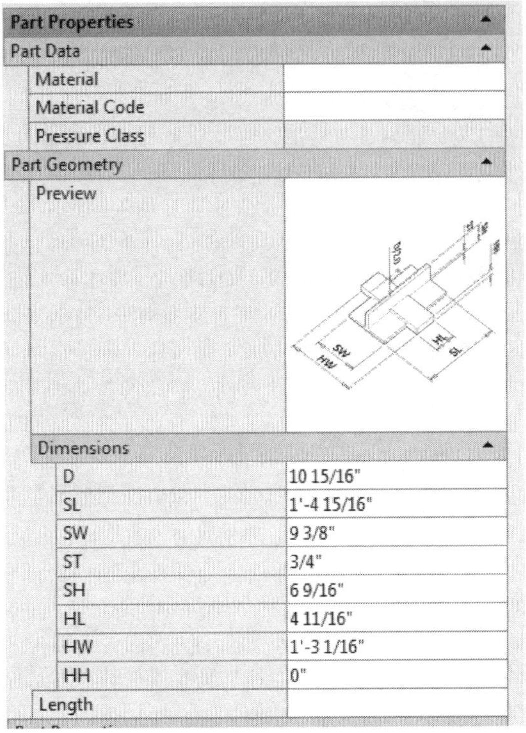

***Figure 1-11** The dimensions of the pipe support displayed in the **PART GEOMETRY** palette*

DIFFERENT WORKSPACES IN AutoCAD PLANT 3D

A workspace is defined as a customized arrangement of **Ribbon**, toolbars, menus, and window palettes in the AutoCAD environment. In AutoCAD Plant 3D, there are different workspaces to create a P&ID and Plant 3D model. Some of them are 3D Piping, P&ID PIP, P&ID ISO, P&ID ISA, P&ID DIN, and P&ID JIS-ISO.

You will notice that there are different workspaces for creating a P&ID. These workspaces are based on different standards. When you invoke a P&ID workspace of a particular standard, the tool palette will display the symbols related to that standard.

You can select any of the predefined workspaces from the **Workspace** drop-down list available in the title bar located next to the **Quick Access Toolbar**, refer to Figure 1-12. This drop-down list does not appear by default in the title bar. To make this drop-down list available, click on the down arrow in the title bar, a flyout will be displayed. Next, choose the **Workspace** option from the flyout displayed. You can also set the workspace from the flyout displayed on choosing the **Workspace Switching** button in the Status Bar, refer to Figure 1-13.

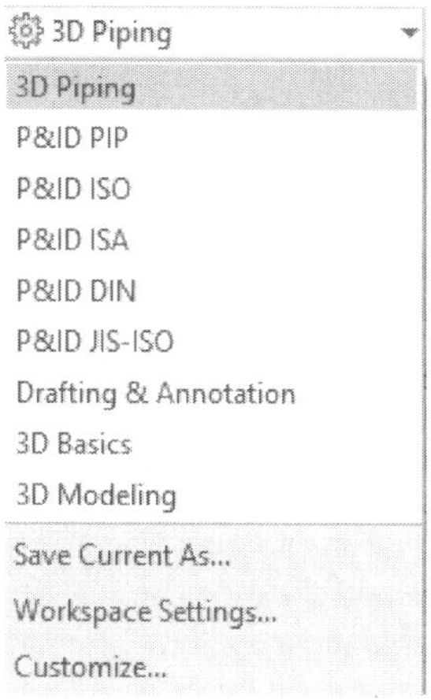

***Figure 1-12** The predefined workspaces*

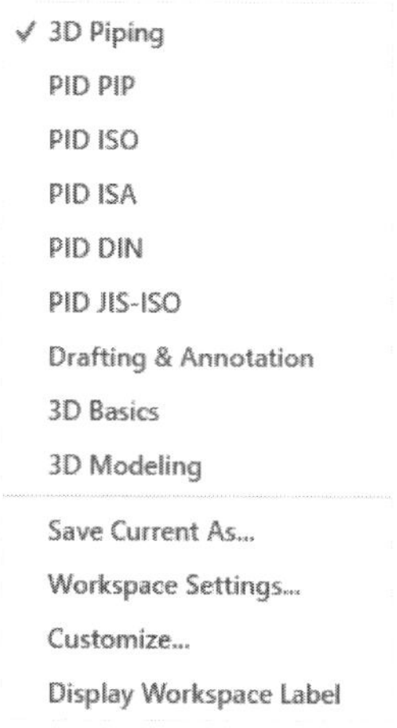

***Figure 1-13** The flyout displayed on choosing the **Workspace Switching** button*

GRIPS

Grips provide a convenient and quick means of editing objects. Grips are displayed on the key points of an object when the object is selected. There are different grips available in AutoCAD Plant 3D. The following table shows the grip symbols, their names, and their usage.

Grip Symbol	Grip Name	Usage
	Continuation grip	It is used to start or continue routing a pipe.
	Substitution grip	It is used to substitute a component.
	Add Nozzle grip	You can add a nozzle to an equipment using this grip.
	Elevation grip	You can change the elevation of a pipe using this grip.
	Edit nozzle grip	It is used to modify a nozzle.
	Rotation grip	It is used to rotate a valve or fitting.
	Flip grip	It is used to flip a valve or a fitting.

	Connection grip	It is displayed when a schematic line is connected to a component or an equipment.
	Stretch grip	It is displayed at the midpoint of a schematic line. You can move the schematic line using this grip.
	Endline grip	It is used to increase or decrease the length of a schematic line.
	Flip grip (in P&ID)	It is used to change the direction of the schematic line.
	Gap grip	It is displayed when a gap is added on the schematic line. You can increase or decrease the gap using this grip.

INVOKING COMMANDS IN AutoCAD Plant 3D

When you are in the drawing area, you need to invoke AutoCAD Plant 3D commands to perform any operation. For example, to draw a line, first you need to invoke the **LINE** command and then define the start point and the endpoint of the line. Similarly, if you want to erase objects, you must invoke the **ERASE** command and then select the objects for erasing. In AutoCAD Plant 3D, you can invoke the commands using different methods which are discussed next.

Invoking Commands Using Command Prompt

You can invoke any AutoCAD Plant 3D command using the keyboard by typing the command name at the command prompt and then pressing the ENTER key. As you type the first letter of a command, AutoCAD Plant 3D displays all available commands starting with the letter typed. You can also use the **Dynamic Input** button to directly enter the command in the **Pointer Input** box. The **Pointer Input** box is a small box displayed on the right of the cursor, as shown in Figure 1-14. However, if the cursor is currently placed on any toolbaror menu bar, or if the **Dynamic Input** is turned off, the command can be entered only through the command prompt. The following example shows how to invoke the **LINE** command using the keyboard:

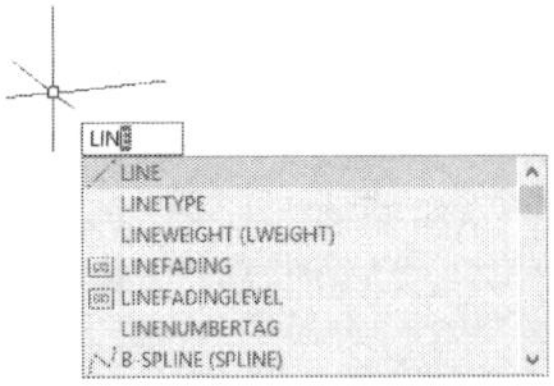

***Figure 1-14** The **Pointer Input** box displayed when the **Dynamic Input** is on*

Command: **LINE** or **L** Enter (L is command alias)

Invoking Commands Using Ribbon

In AutoCAD Plant 3D, you can also invoke a tool from the **Ribbon**. In the **Ribbon**, the tools for creating pipes, equipment, supports, and the other Plant 3D components are available in different panels instead of being spread out in the entire drawing area in different toolbars and menus, refer to Figure 1-15.

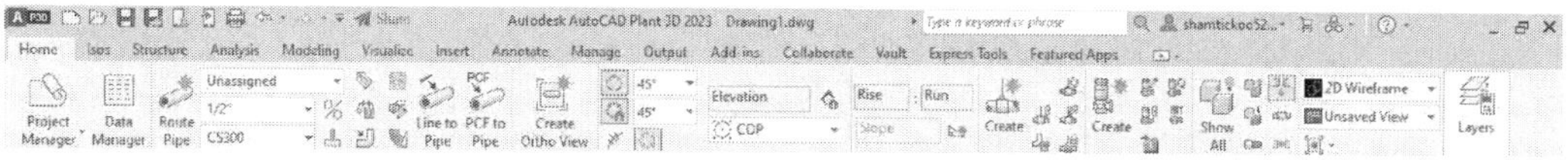

*Figure 1-15 The **Ribbon** for the **3D Piping** workspace*

In AutoCAD Plant 3D, there are different tabs and panels for performing different tasks such as creating a P&ID, isometric drawings, orthographic drawing, and so on. These are discussed next.

Home Tab of the P&ID Workspace

The **Home** tab of the P&ID workspace contains tools that are used to create a P&ID. The **Home** tab of **P&ID** workspace is shown in Figure 1-16.

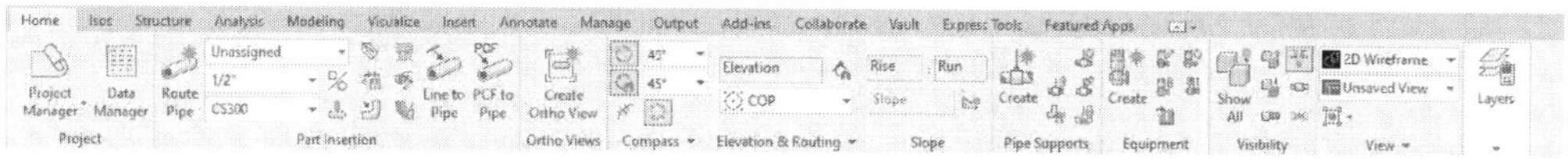

*Figure 1-16 The **Home** tab of the **P&ID** workspace*

Home Tab of the 3D Piping Workspace

This is one of the most important tabs provided in the 3D Piping workspace. This tab provides all the tools which are used to create 3D pipings, equipments, and pipe supports. The **Home** tab of the **3D Piping** workspace is shown in Figure 1-17.

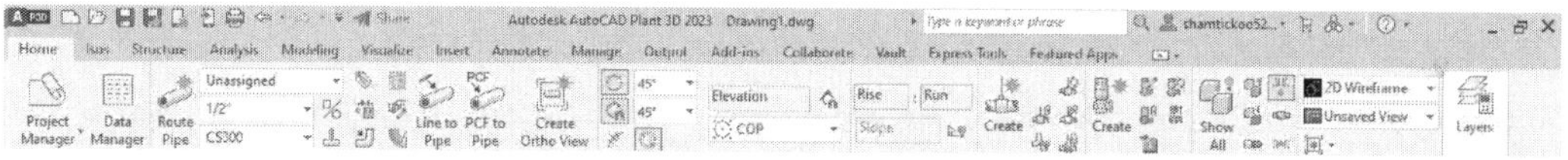

*Figure 1-17 The **Home** tab of the **3D Piping** workspace*

Isos Tab

The tools in the **Isos** tab are used to generate isometric drawings. The **Isos** tab is shown in Figure 1-18.

*Figure 1-18 The **Isos** tab*

Structure Tab

The tools in the **Structure** tab are used to create and modify structures. The **Structure** tab is shown in Figure 1-19.

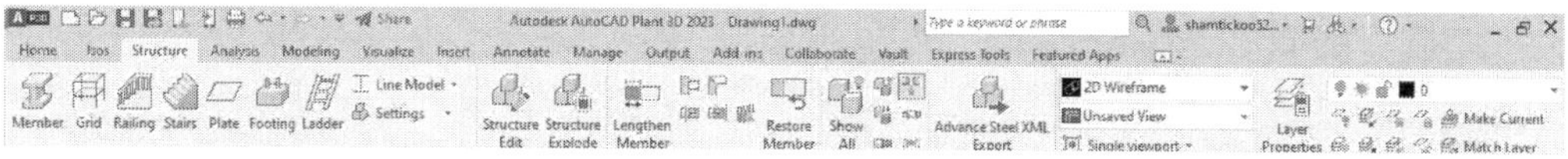

*Figure 1-19 The **Structure** tab*

Ortho View Tab

The tools in the **Ortho View** tab are used to annotate and dimension a view. Also, you can create adjacent views and locate the view components in a 3D Model. The **Ortho View** tab is shown in Figure 1-20. Note that the **Ortho View** tab will be displayed only when you create orthographic drawings. The method of creating orthographic drawings will be discussed in later chapters.

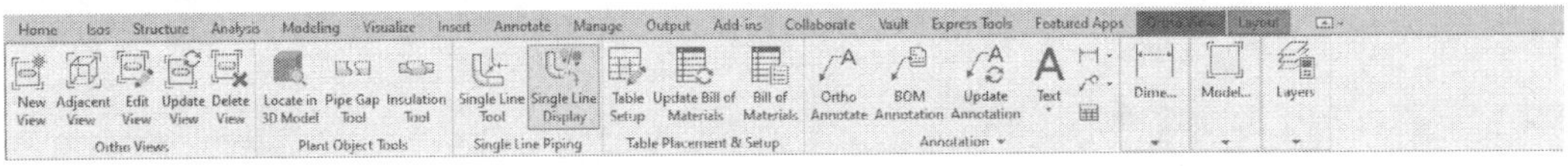

*Figure 1-20 The **Ortho View** tab*

Ortho Editor Tab

The tools in the **Ortho Editor** tab are used to create orthographic views. The **Ortho Editor** tab is shown in Figure 1-21. You will learn more about **Ortho Editor** tab in later chapters.

*Figure 1-21 The **Ortho Editor** tab*

TOOL PALETTES

AutoCAD Plant 3D has provided different tool palettes as an easy and convenient way of placing symbols and 3D parts in the current drawing. The **TOOL PALETTES** display items based on the workspace in which you are currently working. The **TOOL PALETTES** in different workspaces are discussed next.

TOOL PALETTES in P&ID PIP Workspace

The **TOOL PALETTE - P&ID PIP** contains various tabs such as **Lines**, **Equipment**, **Valves** and so on, refer to Figure 1-22. The symbols in each tab are grouped into different areas. You can also create more custom symbols and add them to the **TOOL PALETTES**. You can change the Tool Palette by choosing the **Properties** button and then selecting the required Tool Palette from the flyout displayed.

TOOL PALETTES in 3D Piping Workspace

In the **3D Piping** workspace, the **TOOL PALETTES** contains three tabs: **Dynamic Pipe Spec**, **Pipe Supports Spec**, and **Instrumentation Spec**, refer to Figure 1-23. The **Dynamic Pipe Spec** tab contains the piping components from the selected specification. You can add more components to the **Dynamic Pipe Spec** tab by invoking the **Spec Viewer**. You will learn more about the **Spec Viewer** in the later chapters. In addition, you can add a customized part to the Tool Palettes. Also, you can change the components displayed in this tab by selecting a different specification. The **Pipe Supports Spec** tab contains pipe supports. In this tab, you can dynamically select a support and place it in the 3D model. The **Instrumentation Spec** tab contains various measuring instruments which can be installed on the pipings as per the requirement.

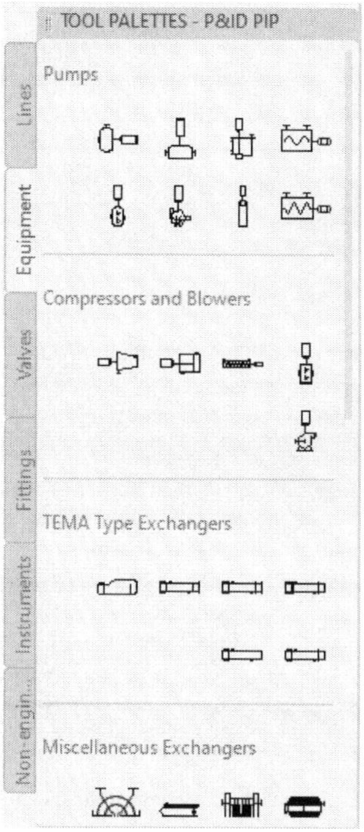

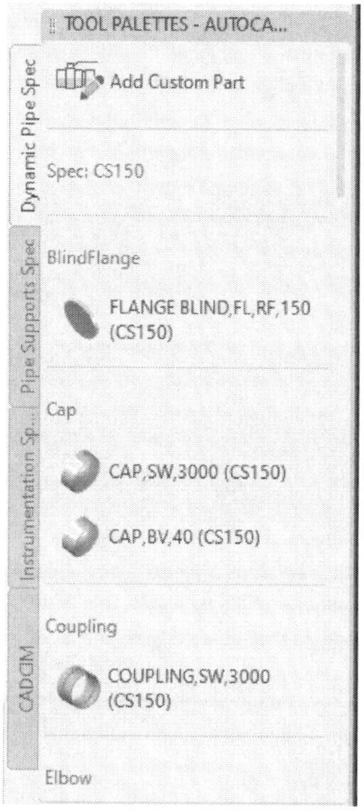

Figure 1-22 *P&ID* ***TOOL PALETTES*** ***Figure 1-23*** *3D Piping* ***TOOL PALETTES***

Application Menu

The **Application Menu** is available at the top left corner of the AutoCAD Plant 3D window. It contains some of the tools that are available in the **Standard** toolbar. Click on the **Application Menu** to display the tools, as shown in Figure 1-24.

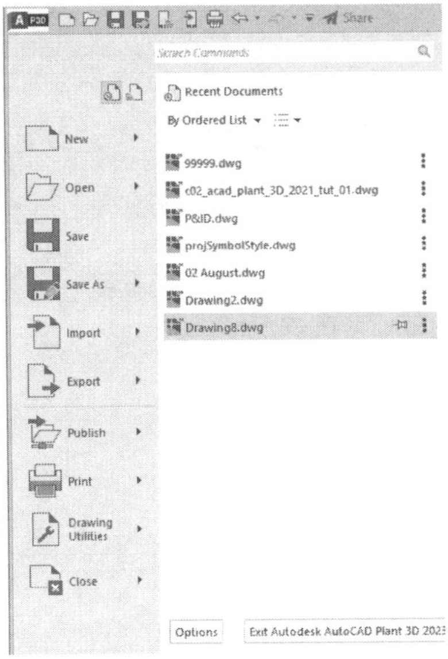

*Figure 1-24 The **Application Menu***

You can search a command using the search field on the top of the **Application Menu**. To search a tool, enter the complete or partial name of the command in the search field; the possible tool list will be displayed. If you click on a tool from the list, the corresponding command will get activated. By default, the **Recent Documents** button is chosen in the **Application Menu**. Therefore, the recently opened drawings will be listed. If you have opened multiple drawing files, choose the **Open Documents** button; the documents that are opened will be listed in the **Application Menu**. To set the preferences of the file, choose the **Options** button available at the bottom of the **Application Menu**. To exit AutoCAD Plant 3D, choose the **Exit Autodesk AutoCAD Plant 3D 2023** button next to the **Options** button.

Menu Bar

You can also invoke commands from the Menu bar. Menu bar is not displayed by default in AutoCAD Plant 3D. To display the Menu bar, click on the down arrow in the **Quick Access Toolbar**; a flyout will be displayed. Choose the **Show Menu Bar** option from it; the Menu bar will be displayed. As you move the cursor over the Menu bar, different titles get highlighted. You can click on the desired item to display a menu. Now, you can invoke a command by left-clicking on it in the menu. An arrow displayed on the right side of some of the menu items indicates that they have a cascading menu. The cascading menu provides various options to execute the same AutoCAD Plant 3D command.

Shortcut Menu

AutoCAD Plant 3D provides shortcut menus to invoke the recently used tools easily. These shortcut menus are context-specific which means that the tools available in them are dependent on the place/object for which they are displayed. A shortcut menu is invoked by right-clicking in the drawing area. It generally contains an option to select the previously invoked tool again apart from the commonly used tools, refer to Figure 1-25.

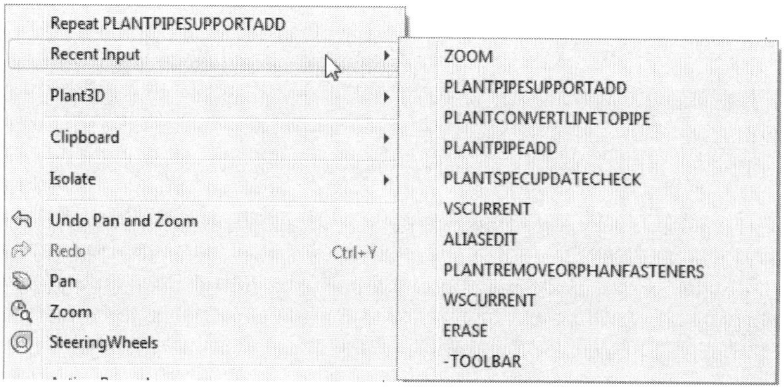

Figure 1-25 *Partial view of the shortcut menu with the recently used commands*

If you right-click in the drawing area while a command is active, a shortcut menu with the options of that particular command will be displayed. Figure 1-26, shows the shortcut menu displayed on right-clicking when the **Route Pipe** tool is active.

You can also right-click on the command window to display the shortcut menu. This menu displays the input setting options and also some of the Windows options such as **Copy** and **Paste**, refer to Figure 1-27. The commands and their prompt entries are displayed in the History window (previous command lines not visible) and can be selected, copied, and pasted in the command

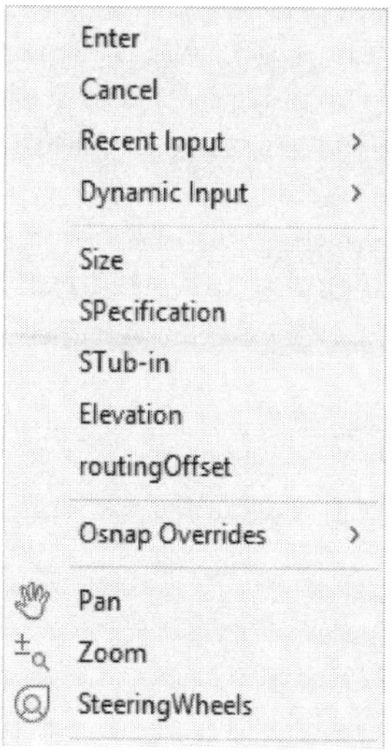

Figure 1-26 *Shortcut menu displayed on right-clicking when the* ***Route Pipe*** *tool is active*

line using the shortcut menu. As you press the up arrow key, the previously entered commands are displayed in the command window. Once the desired command is displayed at the command prompt, you can execute it by simply pressing the ENTER key.

You can also copy and edit any previously invoked command by locating it in the History window and then selecting the lines. After selecting the desired command lines from the History window, right-click to display a shortcut menu. Choose **Copy** from the menu and then paste the selected lines at the end of the command line.

You can right-click on the coordinate display area of the Status Bar to display the shortcut menu. This menu contains the options to modify the display of coordinates, as shown in Figure 1-28. You can also right-click on any of the toolbars to display the shortcut menu from where you can choose any toolbar to be displayed.

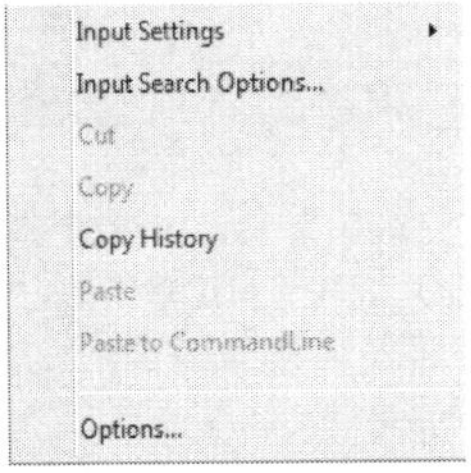

Figure 1-27 *Shortcut menu displayed on right-clicking on command window*

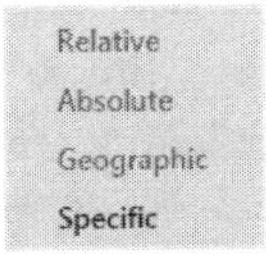

Figure 1-28 *Shortcut menu displayed on right-clicking in the coordinate display area*

AutoCAD Plant 3D DIALOG BOXES

On invoking certain commands in AutoCAD Plant 3D, the related dialog boxes are displayed. When you choose an item in the Menu bar which displays three dots along with its name [**...**], a dialog box is displayed. For example, when **Options...** option is chosen from the **Tools** menu, the **Options** dialog box is displayed. A dialog box contains a number of components like the dialog label, radio buttons, text or edit boxes, check boxes, slider, image boxes, and command buttons. Some of the components in a dialog box are shown in Figure 1-29.

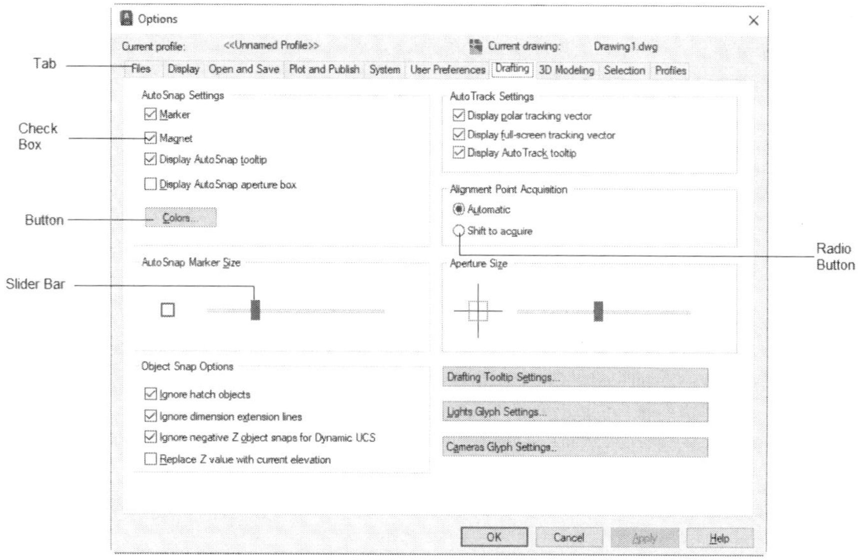

Figure 1-29 *Components of a dialog box*

The title bar displays the name of the dialog box. The tabs have different sections that contain various groups of related options under them. The check boxes are toggle options for making a particular option available or unavailable. When you click on an option which has a down arrow attached to its icon and a list of options is displayed, then it is termed as drop-down list. You can select options using the radio buttons. Note that only one button can be selected at a time from a section. The text box is an area where you can enter a text such as a file name. It is also called an edit box, because you can make any change to the text entered. In some dialog boxes, you will find the [...] button, which displays another related dialog box. There are certain buttons such as the **OK**, **Cancel**, and **Help** buttons that are also displayed at the bottom of the dialog box. The button with a dark border is the default button.

CREATING BACKUP FILES

If the drawing file already exists and you use **Save** or **Save As** tool to update the current drawing, AutoCAD Plant 3D creates a backup file. AutoCAD Plant 3D takes the previous copy of the drawing and changes it from the file type *.dwg* to *.bak*, and the updated drawing is saved as a drawing file with the *.dwg* extension. For example, if the name of the drawing is *myproj.dwg*, AutoCAD Plant 3D will change it to *myproj.bak* and save the current drawing as *myproj.dwg*.

Converting Auto-saved and Backup Files into AutoCAD Format

Sometimes, you may need to convert the auto-saved and backup files to AutoCAD format. To change the backup file into an AutoCAD format, open the folder in which you have saved the backup or the auto-saved drawing. Now, choose **Options** from the **View** tab; the **Folder Options** dialog box will be displayed. In this dialog box, choose the **View** tab and uncheck the **Hide extensions for known file types** check box under the **Advanced settings** area. Next, choose the **OK** button to exit the dialog box. Rename the auto-saved drawing or the backup file with a different name and also change the extension of the drawing from *.sv$* or *.bak* to *.dwg*. After you rename the drawing, you will notice that the icon of the auto-saved drawing or the backup file is replaced by the AutoCAD icon. This indicates that the auto-saved drawing or the backup file is converted to an AutoCAD Plant 3D drawing.

Using the Drawing Recovery Manager to Recover Files

The files that are saved automatically can also be retrieved by using the **DRAWING RECOVERY MANAGER**. You can open the **DRAWING RECOVERY MANAGER** by choosing **Drawing Utilities > Open the Drawing Recovery Manager** from the **Application Menu** or by entering **DRAWINGRECOVERY** at the command prompt.

If the automatic save operation is performed on a drawing and the system crashes accidentally, then next time when you run AutoCAD Plant 3D, the **Drawing Recovery** message box will be displayed, as shown in Figure 1-30. The message box informs you that the program unexpectedly crashed and you can open the most relevant backup file created by AutoCAD Plant 3D. Choose the **Close** button from the **Drawing Recovery** message box; the **DRAWING RECOVERY MANAGER** is displayed on the left in the drawing area, as shown in Figure 1-31.

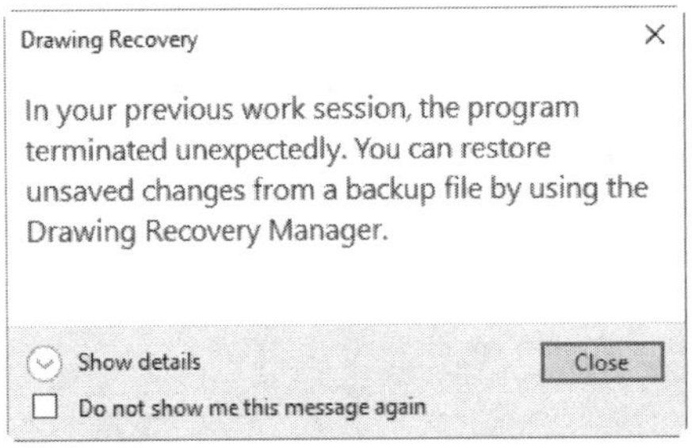

***Figure 1-30** The **Drawing Recovery** message box*

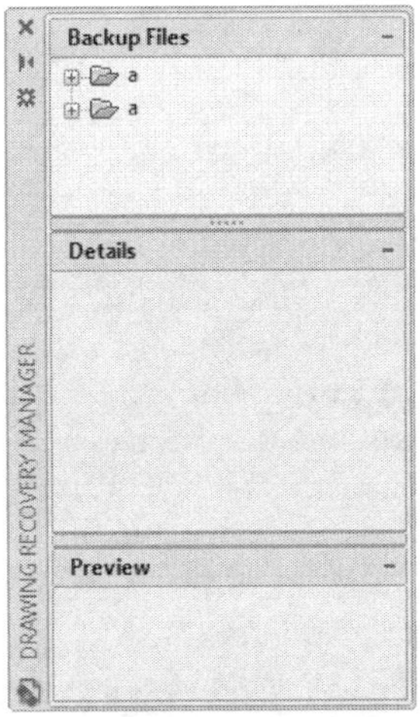

Figure 1-31** The **DRAWING RECOVERY MANAGER

The **Backup Files** rollout lists the original files, the backup files, and the automatically saved files. Select a file; its preview will be displayed in the **Preview** rollout. Also, the information corresponding to the selected file will be displayed in the **Details** rollout. To open a backup file, double-click on its name in the **Backup Files** rollout. Alternatively, right-click on the file name and then choose **Open** from the shortcut menu. It is recommended that you save the backup file at the desired location before you start working on it.

CLOSING A DRAWING

You can use the **CLOSE** command to close the current drawing file without actually quitting AutoCAD Plant 3D. If you choose **Close > Current Drawing** from the **Application Menu** or enter **CLOSE** at the command prompt, the current drawing file will be closed. If multiple drawing files are opened, choose **Close > All Drawings** from the **Application Menu**. If you have not saved the drawing after making the last changes to it and you invoke the **CLOSE** command, AutoCAD Plant 3D displays a message box that allows you to save the drawing before closing. This dialog box gives you an option to discard the current drawing or the changes made to it. It also gives you an option to cancel the command. After closing the drawing, you are still in AutoCAD Plant 3D from where you can open a new or an already saved drawing file. You can also use the close button (**X**) corresponding to the drawing area to close the drawing.

Note

You can close a drawing even if a command is active.

OPENING A PROJECT DRAWING

You can open an existing drawing file that has been saved previously in a project. The drawings are located under the **P&ID Drawings** and **Plant 3D Drawings** nodes in the **PROJECT MANAGER**. To open a drawing, expand the respective drawings node and right-click to display a shortcut menu. Next, choose the **Open** option from the shortcut menu, refer to Figure 1-32; the drawing will open. Alternatively, double-click on the drawing in the **PROJECT MANAGER** to open it.

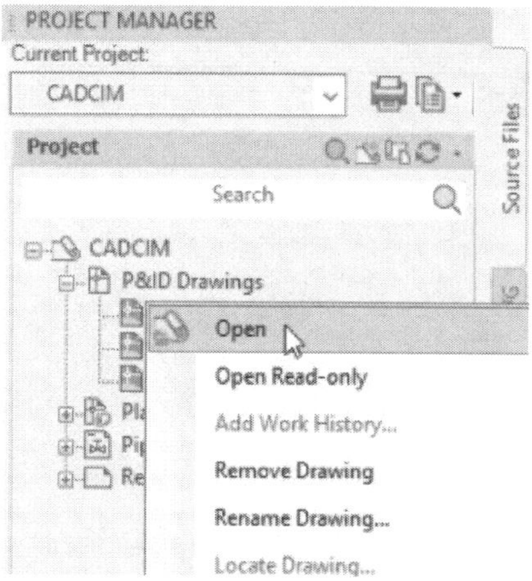

Figure 1-32** Opening a drawing from the **PROJECT MANAGER

To view a drawing without altering it, you must choose the **Open Read-Only** option from the shortcut menu. In other words, choosing the read-only option protects the drawing file from being changed. AutoCAD Plant 3D does not prevent you from editing the drawing. But if you try to save the opened drawing with the original file name, AutoCAD Plant 3D warns you that the drawing file is write-protected. However, you can save the edited drawing to a file with a different file name using the **SAVEAS** command. This way you can preserve your drawing.

OPENING A DRAWING THAT IS NOT IN THE PROJECT

Application Menu: Open > Drawing	**Quick Access Toolbar:** Open
Menu Bar: File > Open	**Command:** OPEN

You can open a drawing file that does not exist in the currently opened project using the **Select File** dialog box. This dialog box is discussed next.

Opening an Existing Drawing Using the Select File Dialog Box

If you are already in the drawing editor and you want to open a drawing file, choose the **Open** tool from the **Quick Access Toolbar**; the **Select File** dialog box will be displayed. Alternatively, invoke the **OPEN** command to display the **Select File** dialog box by using the command prompt, as shown in Figure 1-33. You can select the drawing to be opened using this dialog box. This dialog box is similar to the standard dialog boxes. You can choose the file you want to open from the folder in which it is stored. You can access the required folder by using the **Look in** drop-down list. You can then select the name of the drawing from the list box or you can enter the name of the drawing file you want to open in the **File name** edit box. After selecting the drawing file, you can choose the **Open** button to open the file.

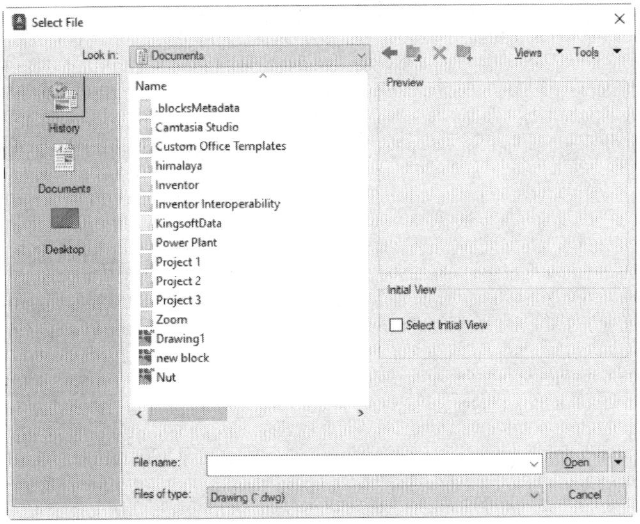

***Figure 1-33** The **Select File** dialog box*

When you select a file name, the preview of the selected drawing file is displayed in the **Preview** box. If you are not sure about the file name of a particular drawing but know the contents, you can still select the file and look for that drawing in the **Preview** box. You can also change the file type by selecting it in the **Files of type** drop-down list. Apart from the .dwg files, you can open the .dwt (template) files, .dws files or the .dxf files. The **Open** button also has a drop-down list, as shown in Figure 1-34. You can choose any of the methods given in this list for opening the file. Note that you cannot make any changes to the drawing file that is not in the current project. If you do so, the **Alert** message box will be displayed, as shown in Figure 1-35. This message box warns you that the object can only be inserted into a project drawing, and if you want to add this drawing to the current project, choose the **Yes** button; the message box will be closed and the drawing will be added to the currently opened project.

***Figure 1-34** The **Open** drop-down list*

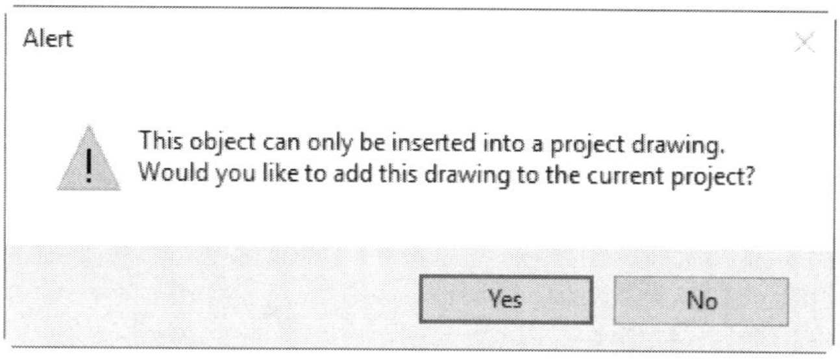

*Figure 1-35 The **Alert** message box*

Alternatively, to add files to the current drawing, right-click on the **P&ID Drawings** or **Plant 3D Drawings** node and choose the **Copy Drawing to Project** option from the shortcut menu displayed, refer to Figure 1-36; the **Select Drawings to Copy to Project** dialog box will be displayed, refer to Figure 1-37. Browse to the file location and double-click on the drawing file to be added to the project; the drawing file will be added to the current project.

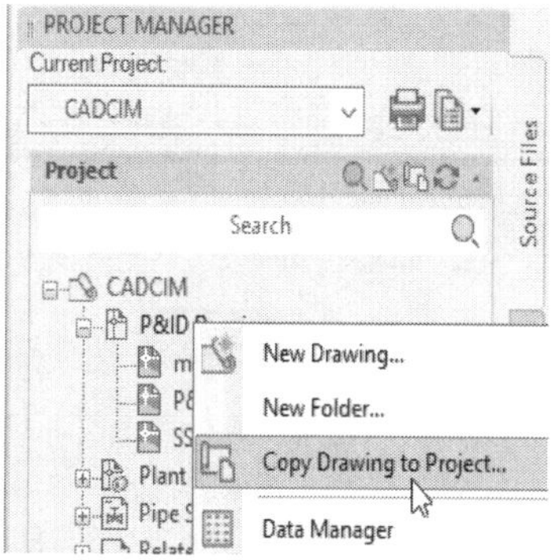

Figure 1-36 Copying a drawing to project

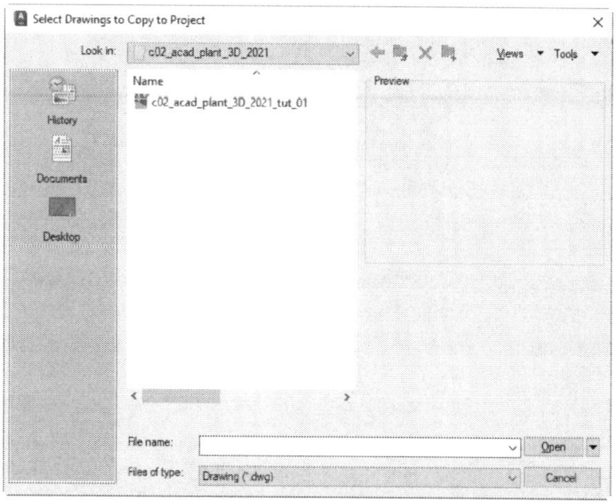

*Figure 1-37 The **Select Drawings to Copy to Project** dialog box*

QUITTING AutoCAD Plant 3D

You can exit the AutoCAD Plant 3D program by using the **EXIT** or **QUIT** command. Even if a command is active, you can choose **Exit Autodesk AutoCAD Plant 3D 2023** from the **Application Menu** to close the AutoCAD Plant 3D program. In case the drawing has not been saved, it allows you to first save the work through a dialog box. Note that if you choose **No** in this dialog box, all the changes made in the current file till the last save will be lost. You can also use the **Close** button (**X**) of the main AutoCAD Plant 3D window (present in the title bar) to end the AutoCAD Plant 3D session.

AutoCAD Plant 3D HELP

Titlebar: ? > Help **Shortcut Key:** F1 **Command:** HELP or ?

You can get online help and documentation on the working of AutoCAD Plant 3D 2023 commands from the **Help** menu in the InfoCenter located on the right in the title bar, refer to Figure 1-38. You can also access the **Help** menu by pressing the F1 function key. Some important options in the **Help** menu are discussed next.

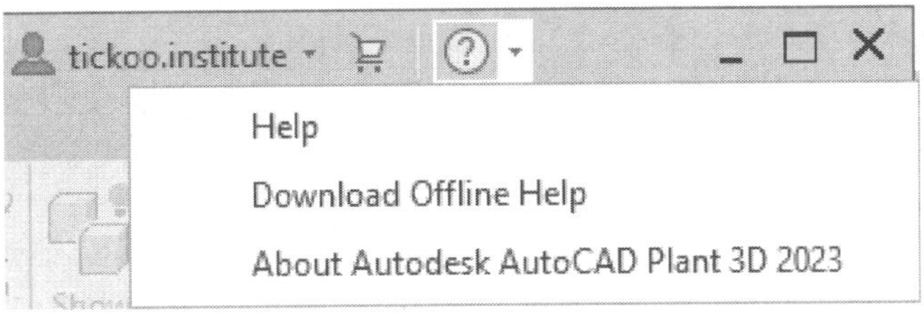

*Figure 1-38 The **Help** menu in the **InfoCenter***

Figure 1-39 shows the **Autodesk AutoCAD Plant 3D 2023 - Help** page after choosing the **Help** button from the InfoCenter bar.

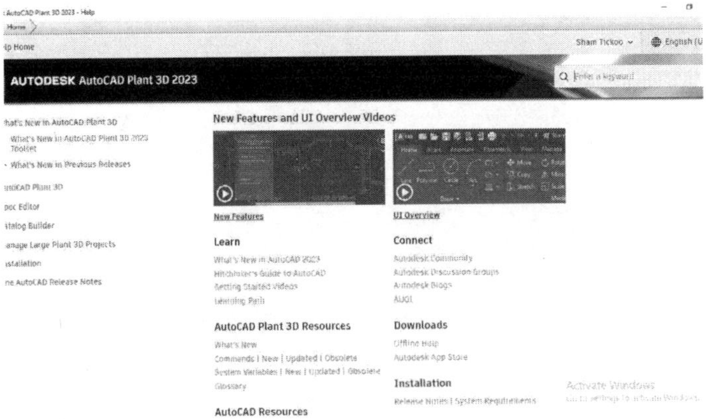

*Figure 1-39 The **Autodesk AutoCAD Plant 3D 2023 - Help page***

Customer Involvement Program

This option is used to share your software and system configuration information with Autodesk. The collected information is used by Autodesk for the improvement of Autodesk software.

Open Collaboration Project

This feature is available in the **Current Project** drop-down of the **PROJECT MANAGER**. Using this feature you can quickly filter projects when searching for collaboration and vault projects. You can collaborate with your clients/team members to discuss about a single project or multiple projects online.

About Autodesk AutoCAD Plant 3D 2023

This option gives you information about the release, serial number, licensed to, and also the legal description about AutoCAD Plant 3D.

Autodesk App Store

Autodesk App Store helps you to download various applications of AutoCAD, get connected to the AutoCAD network, share information and designs, and so on. On choosing the **Autodesk App Store** button from the title bar, the **AUTODESK APP STORE** page will open in the default browser, as shown in Figure 1-40.

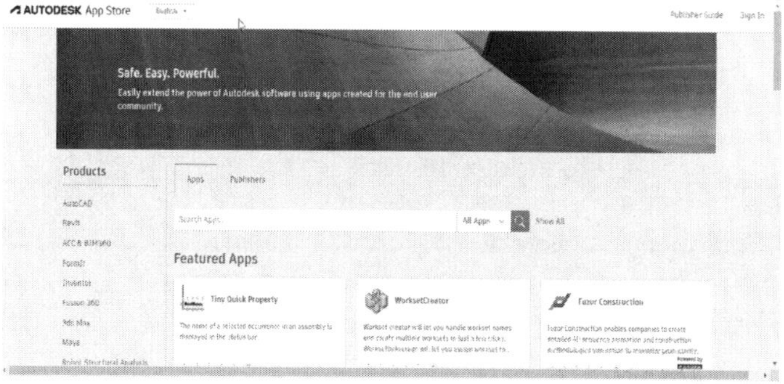

***Figure 1-40** The **AUTODESK App Store** page*

You can download various Autodesk apps such as PlantDataManager, Rename Project, and so on from this page. Some of them are free of cost. You can also publish your own Autodesk products for other users of Autodesk products.

Also, you can download applications for software other than AutoCAD family such as Autodesk Alias, Revit, Simulation, and so on. You can also search for the applications by entering the name of the app in the **Search Apps** text box

ADDITIONAL HELP RESOURCES

- You can get help for a command while working in AutoCAD Plant 3D by pressing the F1 key. If you press the F1 key when a command is active, the help window containing information about that command will be displayed. After using the help, you can exit the window and continue with the current project.

- You can get help about a dialog box by choosing the **Help** button from that dialog box.
- Autodesk provides several resources that can be used to get assistance on Plant 3D tools and functions. These resources can be accessed by visiting the following links:

   a. Autodesk community forum website *https://forums.autodesk.com/*

   b. AutoCAD Plant 3D Technical Assistance website *https://knowledge.autodesk.com/support/autocad-plant-3d#?sort=score*

- You can also get help by contacting the author, Prof. Sham Tickoo, at *stickoo@pnw.edu* and *tickoo525@gmail.com*.
- You can download AutoCAD Plant 3D drawings, programs, and special topics by registering yourself at the website by visiting: *https://cadcim.com/Registration.aspx*.

Self-Evaluation Test

Answer the following questions and then compare them to those given at the end of this chapter:

1. The items in the ________ tab in the **TOOL PALETTES** contain the piping components from the selected specification.
2. The ________ grip is used to modify a nozzle.
3. If you want to work on a drawing without altering the original drawing, you must select the __________ option.
4. The __________ is used to start or continue routing a pipe.
5. The___________ palette displays an additional section for the properties specific to the selected item.
6. You can use the __________ command to close the current drawing file without actually quitting AutoCAD Plant 3D.
7. You can press F3 key to invoke the **AutoCAD** text window which displays the previously used commands and prompts. (T/F)
8. If you do not have an internet connection, you cannot access the Help files. (T/F)
9. You can close a drawing in AutoCAD Plant 3D 2023 even if a command is active. (T/F)
10. If the current drawing is unnamed and you save the drawing for the first time, you will be prompted to specify the file name in the **Save Drawing As** dialog box. (T/F)

Review Questions

Answer the following questions:

1. Which of the following combination of keys should be pressed to turn on or off the display of the **TOOL PALETTES** window?

   (a) CTRL+3 (b) CTRL+0
   (c) CTRL+5 (d) CTRL+2

2. Which of the following commands is used to exit the AutoCAD Plant 3D program?

   (a) **QUIT** (b) **END**
   (c) **CLOSE** (d) None of these

3. Which of the following commands is invoked when you choose the **Save** option from the **File** menu or choose the **Save** tool in the **Quick Access Toolbar**?

   (a) **SAVE** (b) **LSAVE**
   (c) **QSAVE** (d) **SAVEAS**

4. By default, the angles are positive if measured in the __________ direction.

5. You can get help for a command while working in AutoCAD Plant 3D by pressing the __________ key.

6. The shortcut menu invoked by right-clicking in the command window displays the most recently used commands and some of the commonly used options such as **Copy**, **Paste**, and so on. (T/F)

7. Flip grip is used to flip a valve or a fitting. (T/F)

8. The **TOOL PALETTES** display items based on the workspace in which you are currently working. (T/F)

9. You cannot make any changes in the drawing file which is not in the current project. (T/F)

10. The file name that you enter to save a drawing in the **Save Drawing As** dialog box can be 255 characters long but cannot contain spaces and punctuation marks. (T/F)

Answers to Self-Evaluation Test

1. **Dynamic Pipe Spec**, 2. **Edit Nozzle**, 3. **Open Read-only**, 4. Continuation grip, 5. **PROPERTIES**, 6. **CLOSE**, 7. F, 8. T, 9. T, 10. T

Chapter 2

Creating Projects and P&IDs

Learning Objectives

After completing this chapter, you will be able to:

- *Create a new project*
- *Create a new drawing*
- *Create a P&ID drawing*
- *Add an equipment to a P&ID*
- *Assign a tag to a line*
- *Add valves*
- *Add instruments and instrumentation lines*
- *Add fittings*
- *Add off-page connectors*
- *Validate the P&ID drawing*
- *Edit the drawing*
- *Substitute components*
- *Convert AutoCAD components into P&ID Symbols*
- *Work with the Project Setup dialog box*

INTRODUCTION

In this chapter, you will learn how to create a new project and a new drawing. Also, you will be briefly introduced to the P&ID module.

PROJECT MANAGER

The **PROJECT MANAGER** is used to access existing projects, create new projects, add new drawings to a project, re-order drawing files, and modify the existing information in a project. It also allows you to export and import data, create project reports, include referenced drawings (xrefs), and link or copy files to the project folders. By default, the **PROJECT MANAGER** is docked on the left of your screen. Figure 2-1 shows the **PROJECT MANAGER**.

The **PROJECT MANAGER** contains three tabs: **Source Files**, **Orthographic DWG**, and **Isometric DWG**. The **Source Files** tab contains a project tree that displays P&ID drawings, Plant 3D drawings, Pipe Specs, and all related files of the project. The **Orthographic DWG** tab contains the list of all the orthographic drawing files. The **Isometric DWG** tab contains the isometric line diagrams of the plant layout.

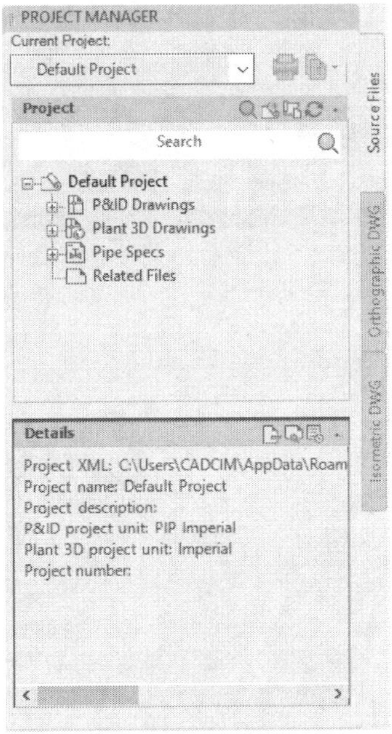

*Figure 2-1 The **PROJECT MANAGER***

CREATING A NEW PROJECT IN AutoCAD Plant 3D

To create a new project, choose the **New Project** option from the drop-down in the **Current Project** area of the **PROJECT MANAGER**. You can also create a new project by choosing the **New Project** option from the **Project Manager** drop-down list of the **Project** panel in the **Home** tab; the **Project Setup Wizard** with the **Specify general settings** page will be displayed, as shown in Figure 2-2. Alternatively, use the **NEWPROJECT** command to create a new project. Steps to create a new project are discussed next.

1. Enter a project name in the **Enter a name for this project** edit box. Next, enter a description of the project in the **Enter an optional description** edit box.

2. Choose the Browse button adjacent to the **Specify the directory where program-generated files are stored** edit box; the **Select Project Directory** dialog box will be displayed. Browse to the directory *C:\Users\user_name\Documents* and choose the **Open** button; the location of the project is automatically displayed in the edit box.

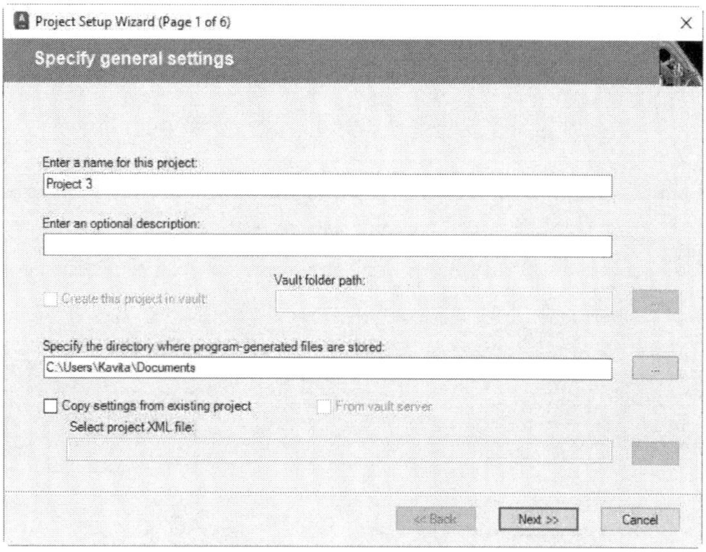

***Figure 2-2** The **Project Setup Wizard** with the **Specify general settings** page*

3. Clear the **Copy settings from existing project** check box, if selected, and then choose the **Next** button; the **Specify unit settings** page will be displayed, as shown in Figure 2-3.

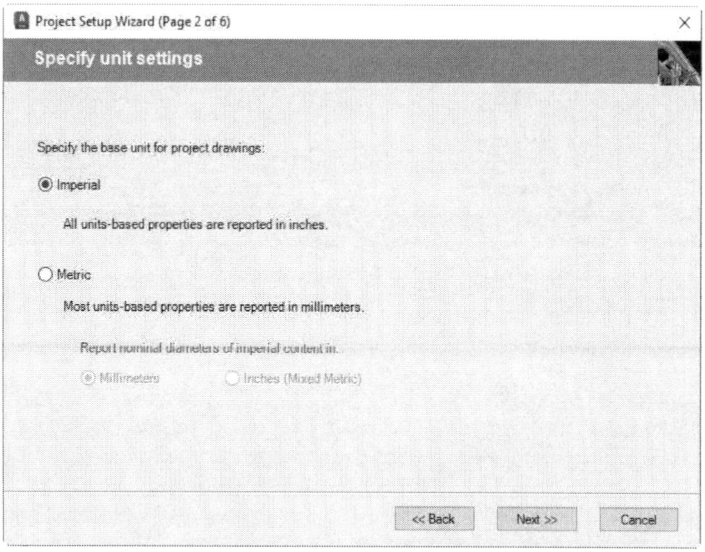

***Figure 2-3** The **Specify unit settings** page*

4. Choose the **Imperial** radio button, if not chosen by default, and then choose the **Next** button; the **Specify P&ID settings** page will be displayed, as shown in Figure 2-4.

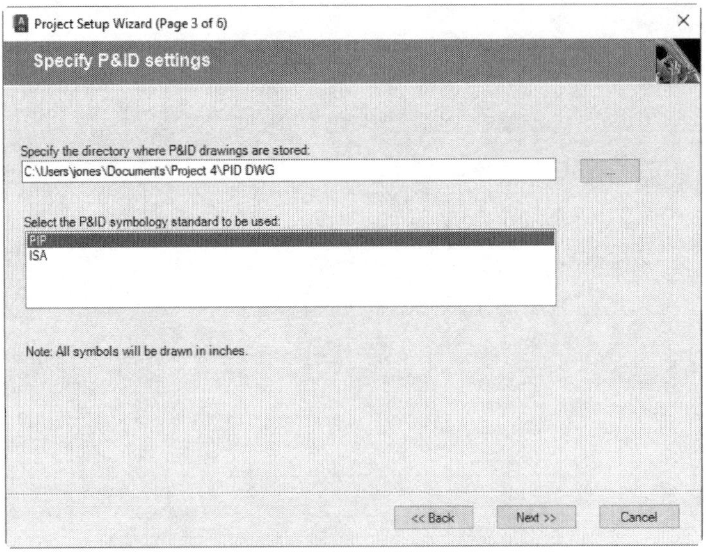

***Figure 2-4** The **Specify P&ID settings** page*

5. Specify the directory as *C:\Users\user_name\Documents\users_project\PID DWG* in the **Specify the directory where P&ID drawings are stored** edit box by using the Browse button available next to this edit box.

6. Select **PIP** as the P&ID standard from the **Select the P&ID symbology standard to be used** list box and choose the **Next** button; the **Specify Plant 3D directory settings** page will be displayed, as shown in Figure 2-5.

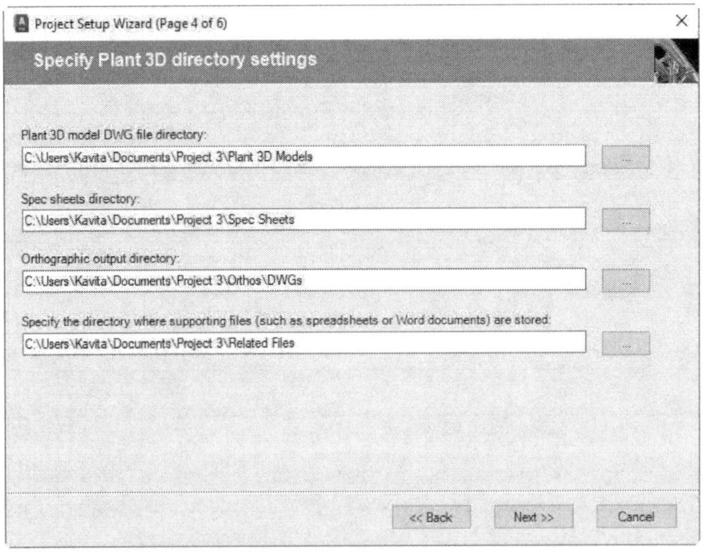

***Figure 2-5** The **Specify Plant 3D directory settings** page*

Note

In this book, the Imperial standard has been used throughout.

7. Accept the default settings in this page and choose the **Next** button; the **Specify database settings** page will be displayed, as shown in Figure 2-6.

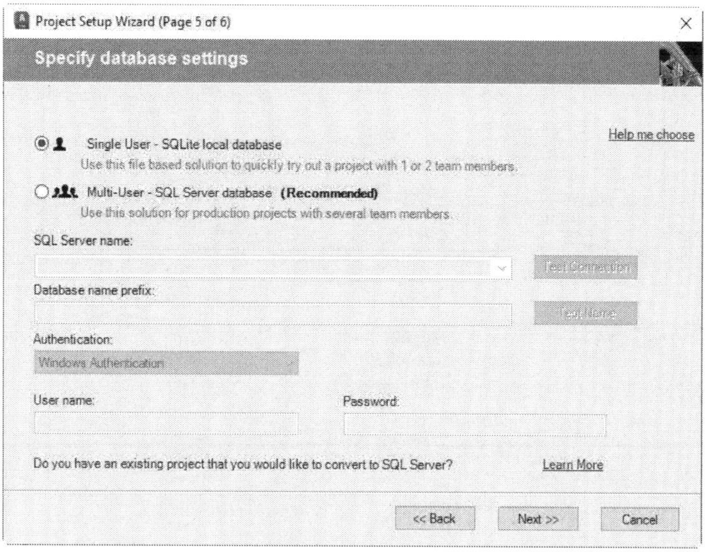

***Figure 2-6** The **Specify database settings** page*

8. In this page, select the option to specify the database settings. Select the **Single User - SQLite local database** radio button, if you are working on a stand-alone system.

   Note that you need to skip Steps 9 through 11, if you have selected the **Single User - SQLite local database** radio button.

9. If your system is connected to a server, select the **Multi User - SQL Server database** radio button. Next, you need to specify the server name in the **SQL Server name** edit box and then choose the **Test Connection** button next to it.

10. After connecting to the server, you need to enter a database prefix in the **Database name prefix** edit box. You can also choose the **Test Name** button to automatically generate a prefix.

11. Next, specify the authentication type by using the **Authentication** drop-down list. If you select **SQL Server Authentication**, you need to specify the user name and password in the **User name** and **Password** edit boxes, respectively.

12. After specifying the database settings, choose the **Next** button; the **Finish** page will be displayed.

13. Choose the **Finish** button; a new project will be created and listed in the **PROJECT MANAGER**.

Creating a New Drawing

To create a new drawing, first select the node (For example: **P&ID Drawings**) from the **PROJECT MANAGER**. Next, choose the **New Drawing** button from the **Project** toolbar, as shown in Figure 2-7; the **New DWG** dialog box will be displayed, as shown in Figure 2-8. In this dialog box, enter the file name, author, and then select a dwg template by choosing the Browse button next to the **DWG template** edit box. Choose the **OK** button; a drawing file with the specified name will be created.

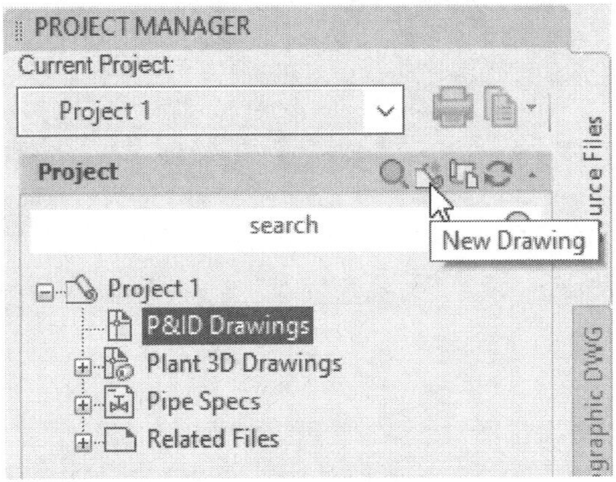

***Figure 2-7** Choosing the **New Drawing** button*

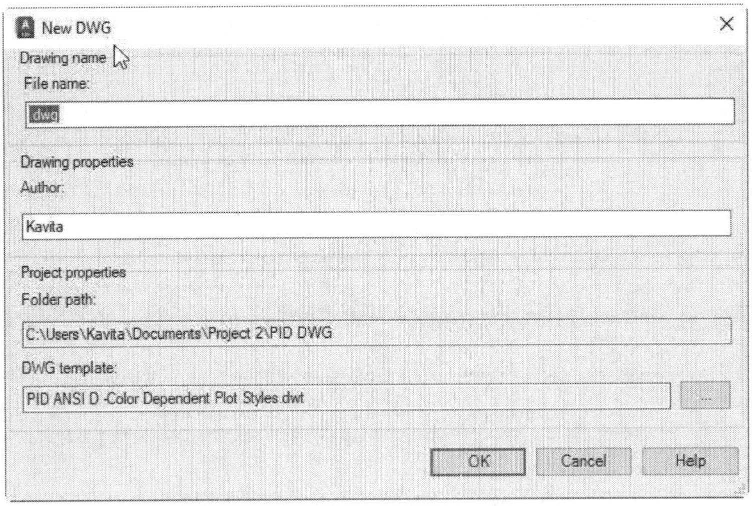

***Figure 2-8** The **New DWG** dialog box*

Grouping Project Files

You can group drawings into a folder. This ensures that the folder path retains the same folder hierarchy, even if the project files are moved to another computer. To do so, right-click on the required node (for example **P&ID Drawings** node) in the **PROJECT MANAGER** and choose the **New Folder** option from the shortcut menu displayed; the **New Folder** dialog box will be displayed, as shown in Figure 2-9. Next, enter a name in the **Folder name** edit box. Now, choose the **OK** button; the new folder will be added to the **P&ID Drawings** node in the project tree. Click and drag the drawing file to the newly created folder to move the file, refer to Figure 2-10.

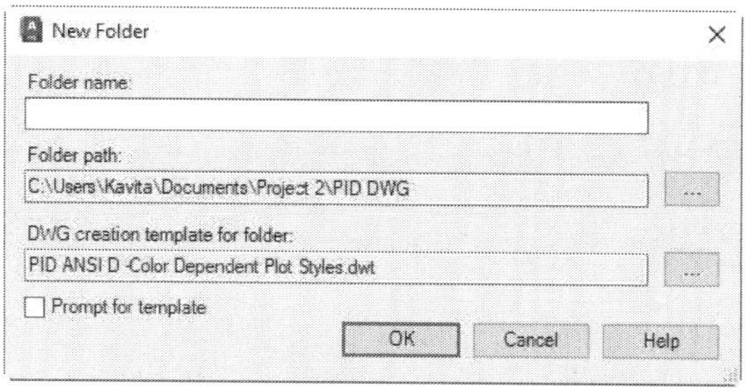

***Figure 2-9** The **New Folder** dialog box*

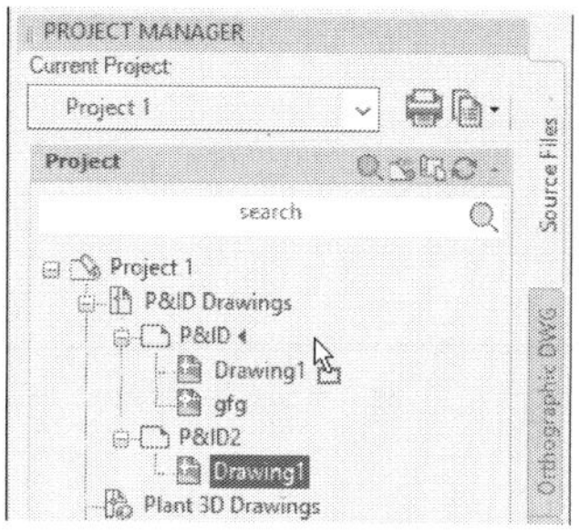

***Figure 2-10** Dragging the drawing file into the folder*

DESIGNING A P&ID

A P&ID is a graphical representation of a plant process and is created by combining various types of components, pipings, and instrumentations that are required to design, construct, and operate the plant. To design a P&ID, you need to create a new P&ID drawing file. The procedure to create a new drawing file is explained earlier. After creating a P&ID drawing file, the **3D Piping** environment will be displayed by default, refer to Figure 2-11.

The **TOOL PALETTES**, which is available at the right side of the window, is also loaded with modeling components. In order to display P&ID environment and P&ID components in the **TOOL PALETTES**, choose the **Workspace Switching** button available in the right side of the Status Bar; a flyout will be displayed. Choose the **PID PIP** option from the flyout, refer to Figure 2-12; the **P&ID PIP** environment will be displayed and the **TOOL PALETTES** will be loaded with P&ID components, as shown in Figure 2-13.

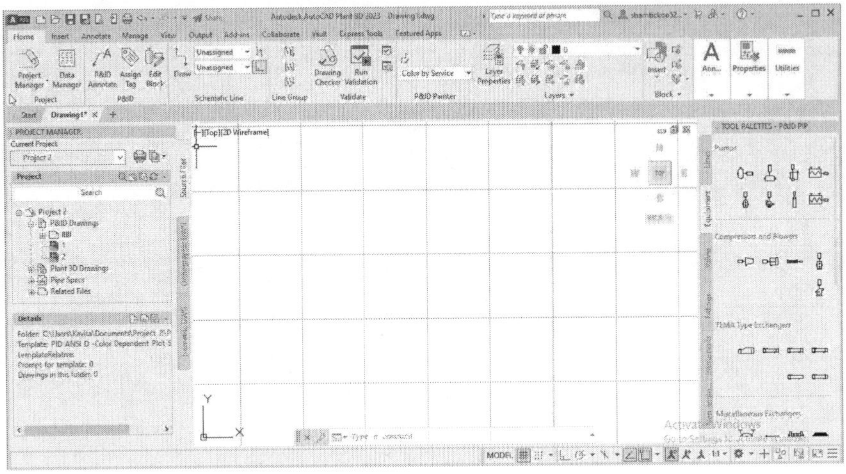

***Figure 2-11** The P&ID environment*

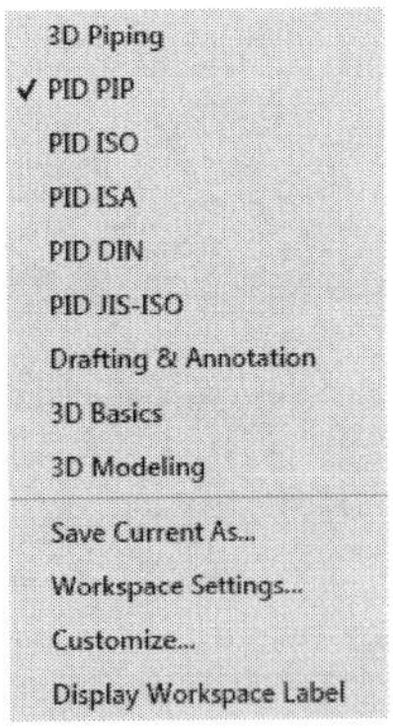

***Figure 2-12** Choosing the **PID PIP** option from the flyout*

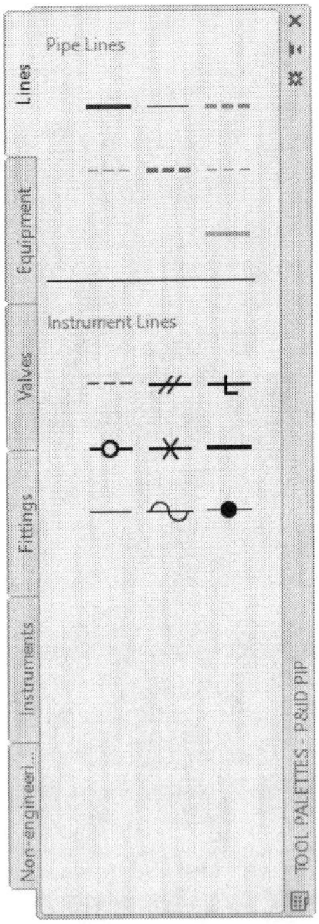

***Figure 2-13** The **TOOL PALETTES** loaded with P&ID components*

Adding Equipment to a P&ID

Equipment includes all the components that are associated with the process plant such as pumps, heat exchangers, compressors, blowers, tanks, vessels, and so on. To place an equipment in a P&ID, first you need to check whether the **TOOL PALETTES - P&ID PIP** is displayed or not. If it is not displayed, choose the **Tool Palettes** button from the **Palettes** panel in the **View** tab;

the **TOOL PALETTES - P&ID PIP** will be displayed on the right side of the window. In the **TOOL PALETTES - P&ID PIP**, choose the **Equipment** tab; the symbols of various equipments will be displayed in it. Choose the symbol of the desired equipment from the **TOOL PALETTES**; it will be attached to the cursor and you will be prompted to specify the insertion point. Click in the drawing area to place the symbol; the **Assign Tag** dialog box will be displayed, as shown in Figure 2-14.

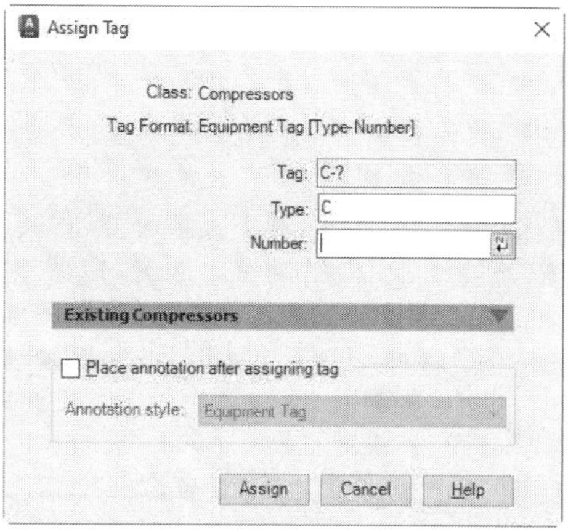

*Figure 2-14 The **Assign Tag** dialog box*

Note that in some equipments after clicking in the drawing area to place the symbol, you will be prompted to enter the XY scale factor. You can scale the equipment horizontally or vertically by dragging the cursor in the respective direction. Alternatively, you can enter a scale factor at the command prompt.

In the **Assign Tag** dialog box, enter a numeric value in the **Number** edit box and select the **Place annotation after assigning tag** check box. Next, choose the **Assign** button; the **Assign Tag** dialog box will be closed and you will be prompted to select the annotation position. Place the annotation below the equipment at an appropriate location.

Adding Pipe Lines

You can select a line from the **Pipe Lines** area in the **Lines** tab of the **TOOL PALETTES - P&ID PIP**. For example, to connect a vessel and a condenser with a primary line, choose the **Primary Line Segment** tool from the **Pipe Lines** area in the **Lines** tab of the **TOOL PALETTES - P&ID PIP**; you will be prompted to select the start point. Select a point on the top of the vessel. Next, connect the line to the top nozzle of the condenser, as shown in Figure 2-15; a nozzle will be created automatically on the vessel and an arrow will show the direction of flow. Note that the direction of flow is determined by the direction in which you move the cursor while creating the line.

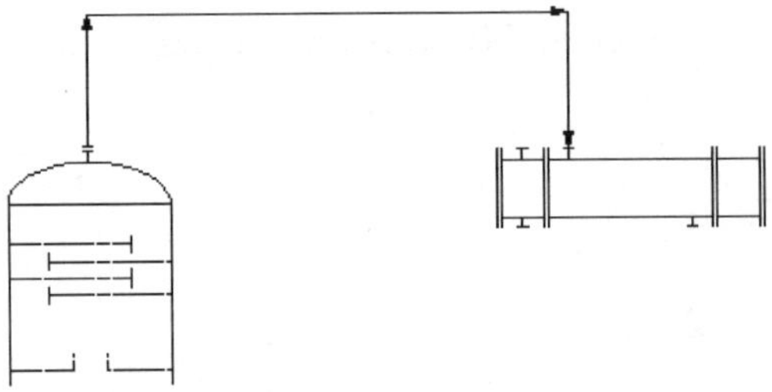

Figure 2-15 Schematic line connecting the vessel and the condenser

You can easily distinguish between a line connected to a component and a line which is not connected. On selecting the line that is connected to a component, a connection grip is displayed at the end of the line, as shown in Figure 2-16. But if the line is not connected to a component, an end grip is displayed at the end of the line, as shown in Figure 2-17.

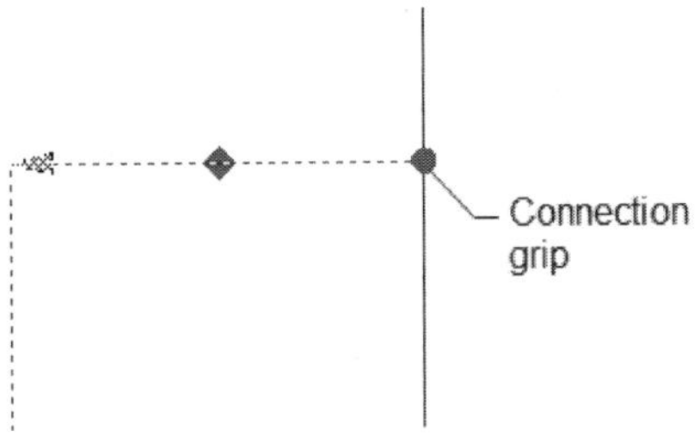

Figure 2-16 A connected line showing the connection grip

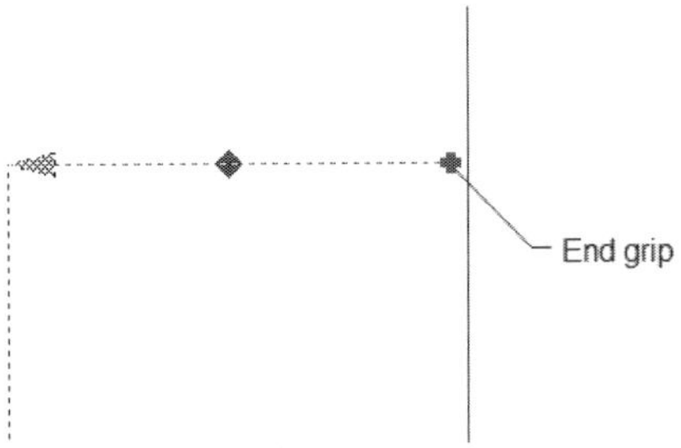

Figure 2-17 A line not connected to any line or a component

Note

After connecting two components with a pipe line, the 'From' and 'To' information is automatically added to the line. This information is displayed when you hover the cursor over the line. The 'From' field shows the component from which the line originates and the 'To' field indicates the destination component.

Assigning Tags to a Line

After creating a line, you need to assign a tag to it. The tag data consists of the information such as pipe size, specification, service, and line number; refer to Figure 2-18 for the representation of tag data on the line. In order to assign a tag to a line, choose the **Assign Tag** tool from the **P&ID** panel in the **Home** tab and select a line. Next, press ENTER; the **Assign Tag** dialog box will be displayed. Enter the data in the respective edit boxes and choose the **Assign** button; the tag data will be assigned to the line. Also, you can select the **Place annotation after assigning tag** check box from the **Assign Tag** dialog box in order to place the annotation near the line.

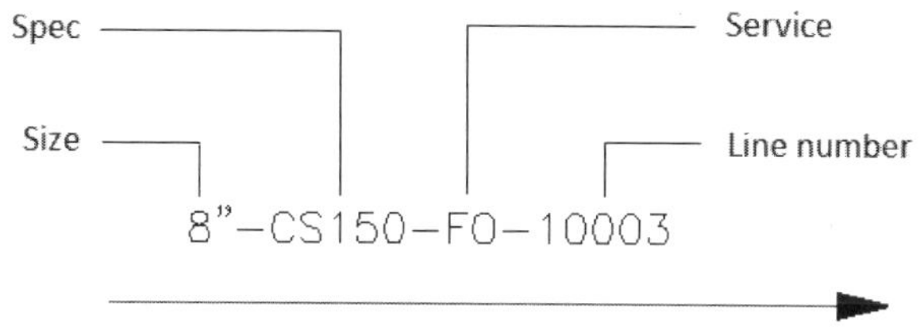

Figure 2-18 Annotation of a pipeline tag

Assigning Custom Service Category to a Line

As discussed earlier, after creating a line, you need to assign a tag to it. By default, you can assign pipe size, specification, service, and line number to a line. While assigning the service category to a line, the required service category might not be available in the **Pipe Line Group.Service** drop-down list of the **Assign Tag** dialog box. You can add the required service category to the **Pipe Line Group.Service** drop-down list using the **Project Setup** dialog box. Let us suppose you need to add Sludge Line service to the list. The procedure to add the Sludge Line service is given next.

Choose the **Project Setup** tool from the **Project Manager** drop-down of the **Project** panel in the **Home** tab; the **Project Setup** dialog box will be displayed. In this dialog box, expand the **P&ID DWG Settings** node. Next, expand the **P&ID Class Definitions** sub-node and then choose the **Pipe Line Group** option available under the **Non Engineering Items**; the corresponding options will be displayed in the **Properties** area of the dialog box, refer to Figure 2-18(a).

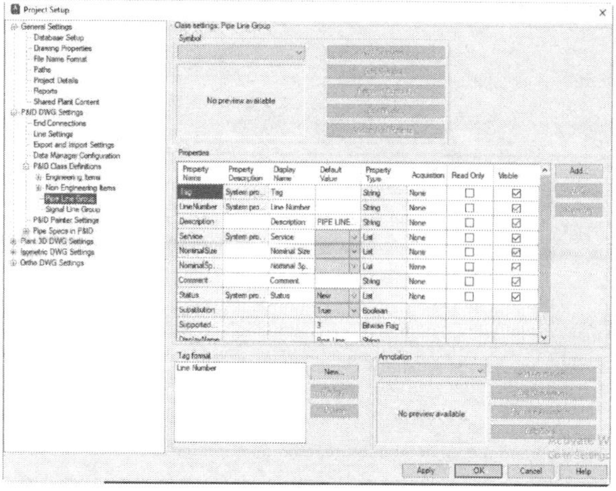

*Figure 2-18(a) The **Project Setup** dialog box with the **Pipe Line Group** option chosen*

In the **Properties** area, select the **Service** option available under the **Property Name** column. Next, choose the **Edit** button available at the right in the **Properties** area; the **Selection List Property** dialog box will be displayed, refer to Figure 2-18(b).

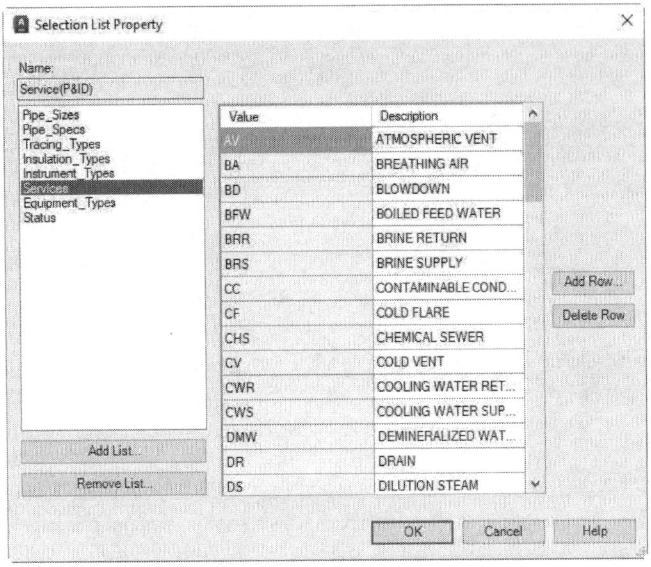

*Figure 2-18(b) The **Selection List Property** dialog box*

In this dialog box, the **Services** option is chosen by default, refer to Figure 2-18(b). Now, choose the **Add Row** button; the **Add Row** dialog box will be displayed. Enter **SL** in the **Value** edit box of the dialog box, refer to Figure 2-18(c) and then choosse the **OK** button; the new value **SL** will be added to the **Value** column of the **Selection List Property** dialog box. Select the newly added value **SL** from the **Value** column and then enter **SLUDGE LINE** in the respective **Description** edit box, refer to Figure 2-18(d) and then choose the **OK** button.

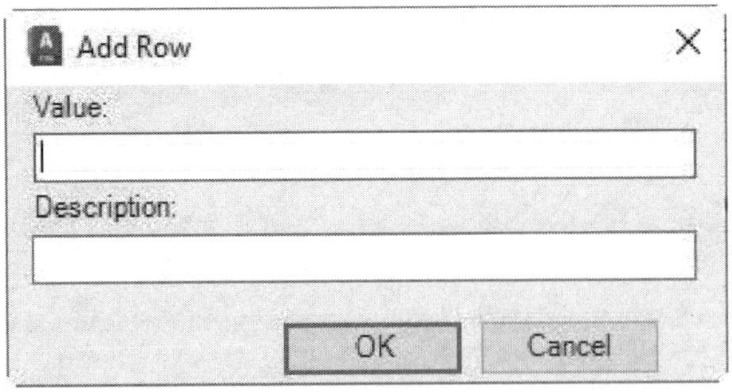

*Figure 2-18(c) The **Add Row** dialog box*

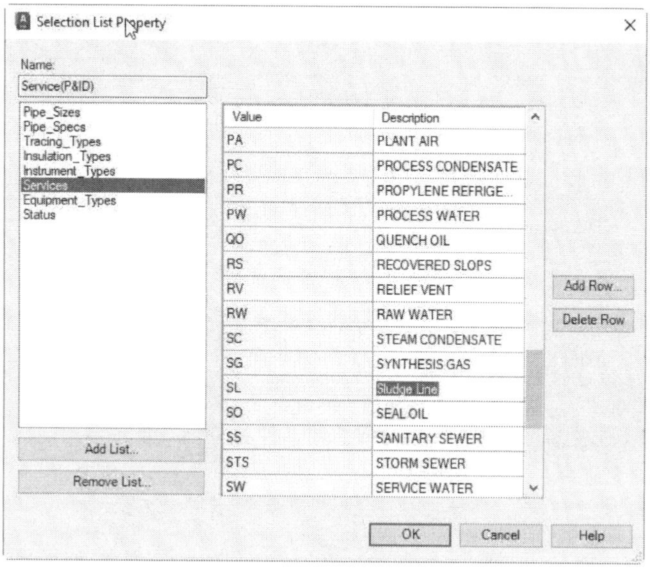

Figure 2-18(d) *The new value **SL** added to the dialog box*

Next, choose **Apply** and then the **OK** button to exit the **Project Setup** dialog box; the new service category **SL - SLUDGE LINE** will be added to the **Pipe Line Group.Service** drop-down list of the **Assign Tag** dialog box.

Creating and Adding Custom Pipe Line Segments to TOOL PALETTES

You can create custom pipe line segements and add them to **TOOL PALETTES**. To do so, choose the **Lines** tab in the **TOOL PALETTES**. Next, invoke the **Project Setup** dialog box. In this dialog box, expand the **P&ID DWG Settings** node. Next, expand **P&ID Class Definitions** > **Engineering Items** > **Lines** > **Pipe Line Segments** sub-nodes. Next, click on the **Pipe Line Segments** sub-node and right click on it; a shortcut menu will be displayed. Choose the New option from the shortcut menu; Create Class dialog box will be displayed, refer to Figure 2-18(e).

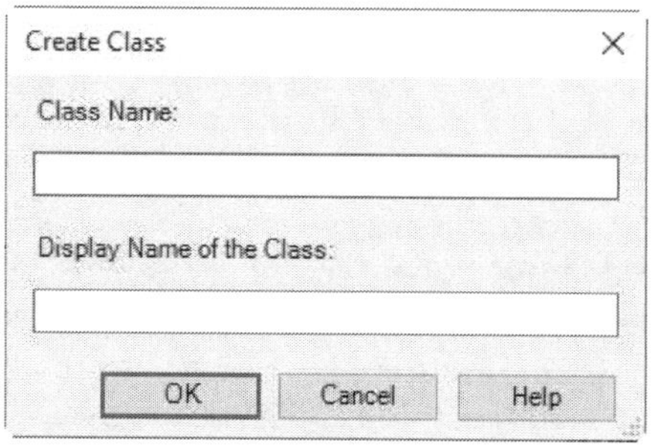

Figure 2-18(e) *The **Create Class** dialog box*

Enter **Sludgeline** in the **Class Name** edit box and **Sludge Line** in the **Display Name of the Class** edit box. Choose the OK button; the **Sludge Line** will be added and highlighted under the **Pipe**

Line Segments sub-node. Next, choose the **Edit Line** button available in the **Line** area of the **Project Setup** dialog box; the Line **Settings** dialog box will be displayed, refer to Figure 2-18(f).

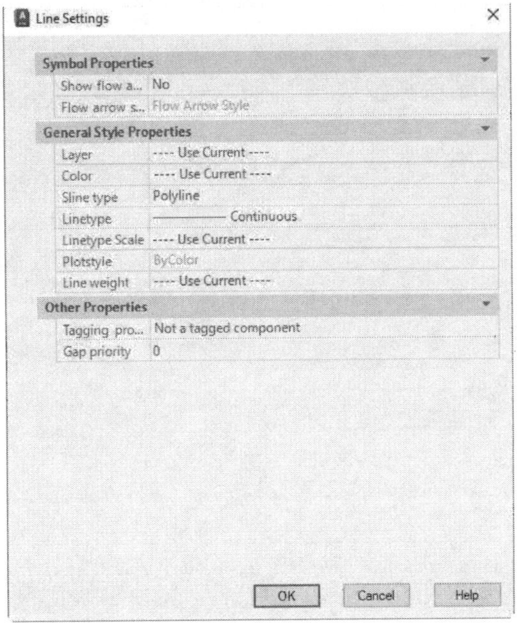

***Figure 2-18(f)** The **Line Settings** dialog box*

Choose the **Yes** option from the **Show flow arrows** drop-down. Next, assign the Layer, Linetype, Color, Lineweight, Tagging prompt, Gap priority value and so on from the respective drop-downs. Choose the **OK** button to exit the dialog box. Next, choose the **Add to Tool Palette** button from the **Line** area of the **Project Setup** dialog box; the **Create Tool** message box will appear. Choose the **OK** button from the message box; the **Sludgeline** will be added to the **Line** tab of the **TOOL PALETTES**. Next, choose **Apply** and then the **OK** button to exit the **Project Setup** dialog box.

Adding Custom Tags to Pipe Line Segments

As discussed earlier, after creating a line, you need to assign tag to it. By default, you can assign pipe size, specification, service, and line number to a line. In some cases, you might want to assign other tag properties to the pipe segment like zone number, material of the pipeline, slope, and so on.

Let us assign the size, zone number, material of the pipeline, and slope to the **Primary Line Segment** available in the **Pipe Lines** area of the **Lines** tab in the **TOOL PALETTES**. To do so, invoke the **Project Setup** dialog box. Expand the nodes **P&ID DWG Settings > P&ID Class Definitions > Engineering Items > Lines > Pipe Line Segments**. Next, selectthe **Primary Line Segment** option; it will be highlighted and corresponding options will be displayed in the **Properties** area of the dialog box. Click on the **Add** button in this area; the **Add Property** dialog box will be displayed, refer to Figure 2-18(g).

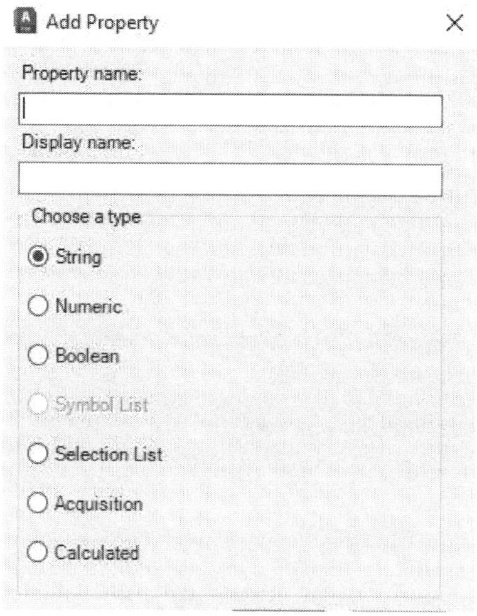

*Figure 2-18(g) The **Add Property** dialog box*

Enter **Zone_Number** in the **Property name** edit box and click on the **OK** button; the property will be added to the **Properties** area. Similarly add material and slope properties to the **Properties** area.

Choose the **New** button from the **Tag format** area of the dialog box; the **Tag Format Setup** dialog box will be displayed, refer to Figure 2-18(h).

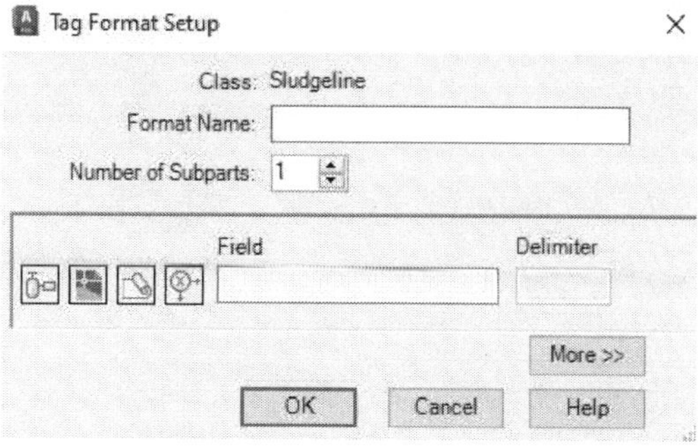

*Figure 2-18(h) The **Tag Format Setup** dialog box*

Enter **Pipeline Tag[Size- Zone Number- Material- Slope]** in the **Format Name** edit box and **4** in the **Number of Subparts** edit box; the **Tag Format** dialog box will be modified, refer to Figure 2-18(i).

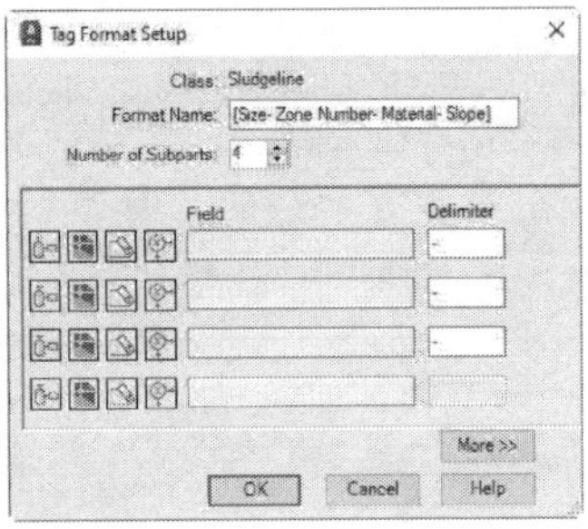

Figure 2-18(i) The modified ***Tag Format Setup*** *dialog box*

Click on the **Select Class Properties** button, the first button in the first row; the **Select Class Property** dialog box will be displayed, refer to Figure 2-18(j).

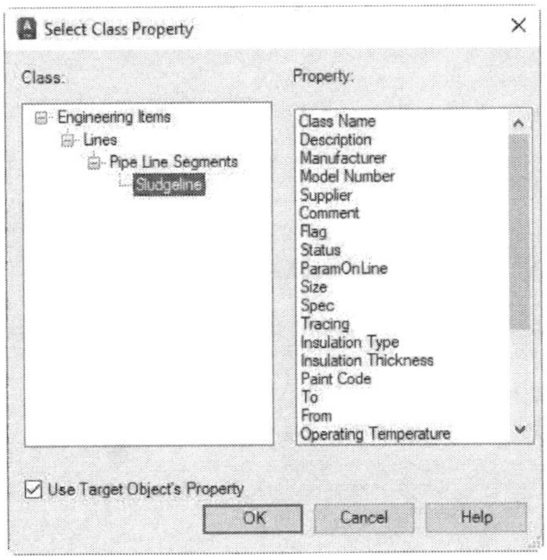

Figure 2-18(j) The ***Select Class Property*** *dialog box*

Choose the **Size** option from the **Property** list box and then choose the **OK** button. Similarly, add Zone_Number, Material, and Slope to the remaining rows of the **Tag Format Setup** dialog box, refer to Figure 2-18(k).

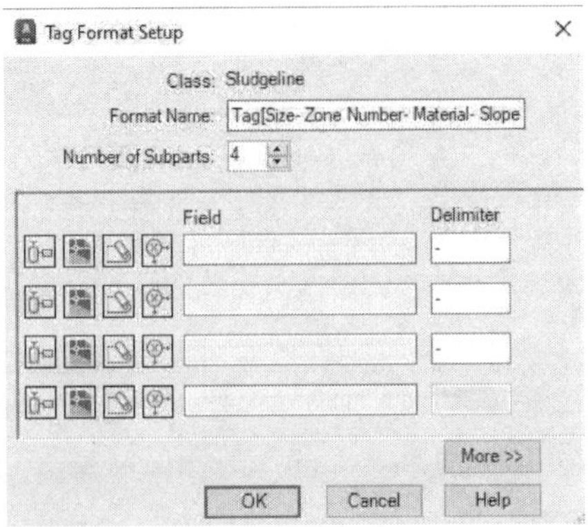

Figure 2-18(k) The ***Tag Format Setup*** *dialog box with class properties added*

Choose the **OK** button to exit the **Tag Format Setup** dialog box. Next, scroll down to the **TagFormatName** row of the **Properties** area and then choose the **Pipeline Tag[Size- Zone Number- Material- Slope]** option from the drop-down list in the **Default Value** column. Next, choose **Apply** and then the **OK** button to exit the **Project Setup** dialog box; the newly created tag format will be assigned to the Primary Line Segment.

Now, create a Primary Line Segment and then choose the **Assign Tag** tool to assign tag to it; the **Assign Tag** dialog box will be displayed with the new class properties listed in it, refer to Figure 2-18(l).

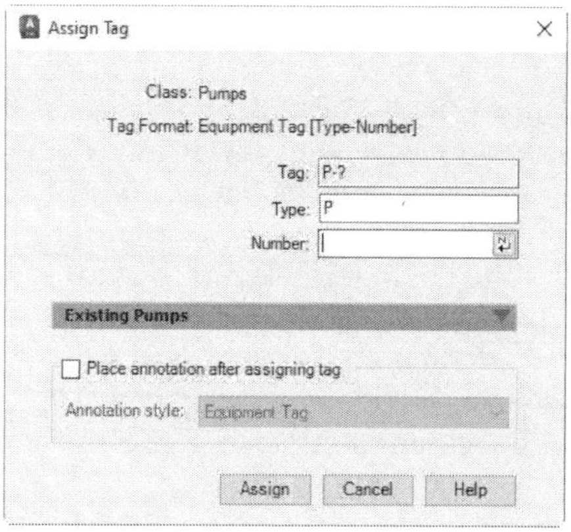

***Figure 2-18(l)** The **Assign Tag** dialog box with new tag format*

Adding Valves

The **Valves** tab in the **TOOL PALETTES - P&ID PIP** contains a group of valve symbols. You can add any type of valve to a schematic line. For example, to place a gate valve on the schematic line, you need to choose the **Valves** tab from **TOOL PALETTES - P&ID PIP** and then choose the **Gate Valve** tool from the **Valves** area; the valve symbol will be attached to the cursor and you will be prompted to select the insertion point. Select an insertion point on the line; the valve will be placed on the specified point, as shown in Figure 2-19.

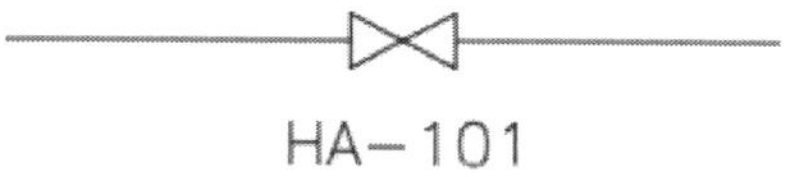

***Figure 2-19** A gate valve placed on the line*

Adding Instruments and Instrumentation Lines

Instruments play a major role in maintaining the safety and streamlining the process in the plant. The **Instruments** tab in the **TOOL PALETTES - P&ID PIP** contains **Control Valve**, **Relief Valves**, **Primary Element Symbols (Flow)**, **General Instruments**, and **Miscellaneous Instrument Symbols** areas. Figure 2-20 shows various instrument symbols available in the **Instruments** tab.

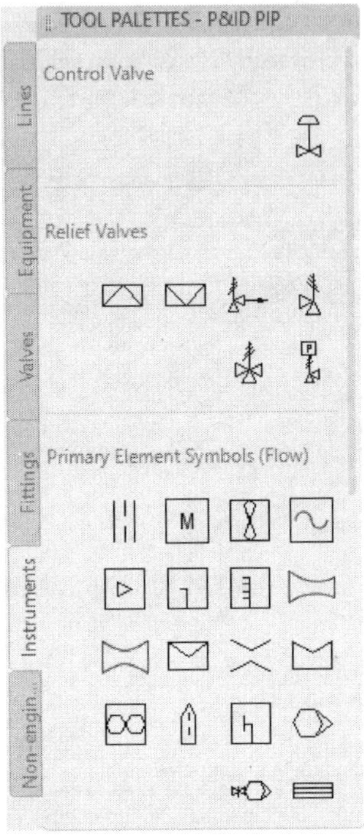

*Figure 2-20 Instrument symbols available in the **Instruments** tab*

For example, to add a control valve, choose the **Control Valve** tool from the **Instruments** tab; the **Control Valve Browser** dialog box will be displayed, if you are adding a control valve for the first time. Select the required valve body from the **Select Control Valve Body** tree view. Next, select the actuator type from the **Select Control Valve Actuator** tree view, refer to Figure 2-21 and choose the **OK** button; the dialog box will be closed and you will be prompted to select an insertion point. Select an insertion point on the line; the control valve will be added and you will be prompted to select annotation position.

Note

*If the **Control Valve Browser** dialog box is not displayed, choose the **Change body or actuator** option from the command bar to change the actuator of the valve body.*

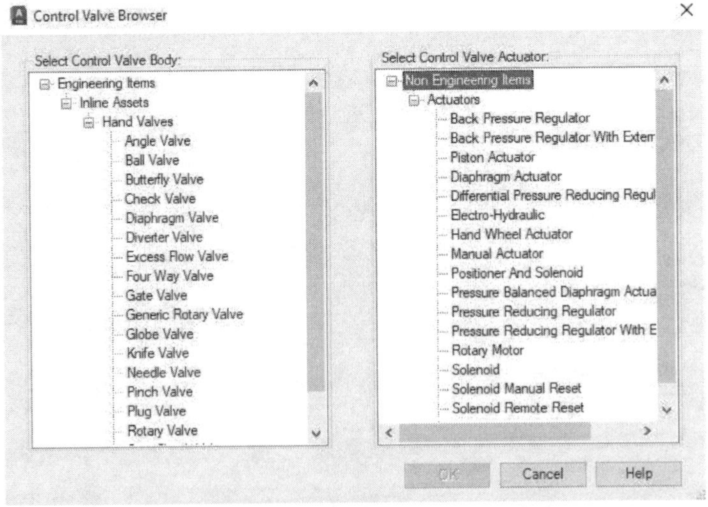

*Figure 2-21 The **Control Valve Browser** dialog box*

Move the cursor and click to position the annotation; the **Assign Tag** dialog box will be displayed. Enter values in the **Area**, **Type**, and **Loop Number** edit boxes, and clear the **Place annotation**

after assigning tag check box. Next, choose the **Assign** button from the **Assign Tag** dialog box; an annotation bubble will be placed at the specified location.

After adding the control valve, you need to add the signal lines connecting the instruments and the process line. Figure 2-22 shows various instrument lines used in a P&ID. For example, to add an electrical signal, choose the **Electric Signal** line from the **Instrument Lines** area in the **Lines** tab; you will be prompted to select the start point. Select the point on the process line; you will be prompted to select the next point. Move the cursor toward left and connect the line to the control valve.

Next, you need to place control instruments on the signal lines. For example, to add a temperature controller, choose the **Field Descrete Instrument** tool from the **General Instruments** area of the **Instruments** tab in the **TOOL PALETTES**; you will be prompted to specify the insertion point. Select a point on the signal line connecting the control valve; the instrument will be placed and the **Assign Tag** dialog box will be displayed.

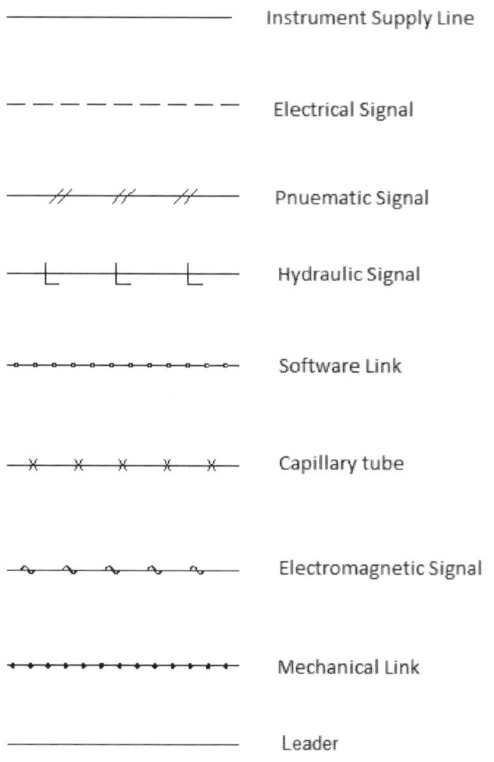

Figure 2-22 *The instrument line symbols*

In the **Assign Tag** dialog box, enter appropriate values in the **Area** and **Loop Number** edit boxes. Next, select the **TC - TEMPERATURE CONTROLLER** option from the **Type** drop-down list and choose the **Assign** button to close the dialog box. Figure 2-23 shows a typical instrumentation loop.

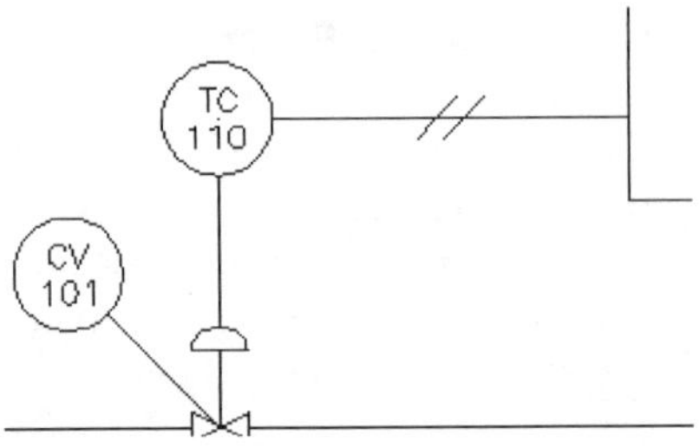

Figure 2-23 A typical instrumentation loop

Adding Fittings

Fittings are used to connect pipe lines that have different sizes or shapes and they are also used for regulating or measuring flow. You can add a fitting to a P&ID drawing from the **Fittings** tab in the **TOOL PALETTES - P&ID PIP**. This tab contains symbols of **Piping Fittings**, **Piping Specialty Items**, and **Nozzles**. For example, to add a concentric reducer, choose the **Concentric Reducer** tool from the **Piping Fittings** area in the **Fittings** tab; you will be prompted to specify the insertion point. Select a point on the line; the reducer will be placed at the specified position.

Adding the Off Page Connectors

The off page connectors are used to indicate the continuation of the process line from one drawing sheet to another. To add an off-page connector, choose the **Off Page Connector** tool from the **Off Page Connectors and Tie-In Symbol** area in the **Non-engineering** tab of **TOOL PALETTES - P&ID PIP**; you will be prompted to specify the insertion point. Select the endpoint of the line, as shown in Figure 2-24; the off page connector will be placed at the specified location.

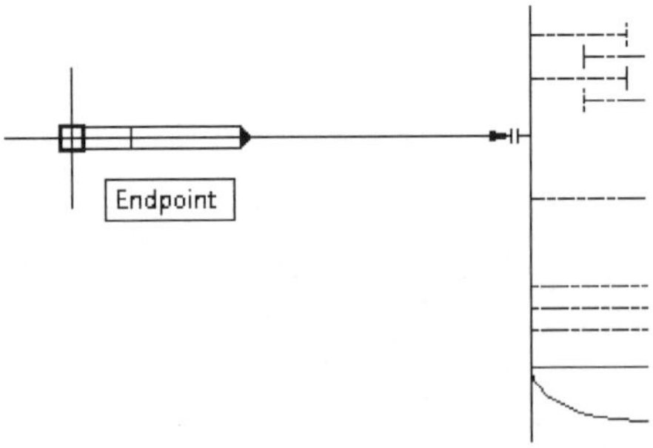

Figure 2-24 Placing the off-page connector

Connecting the Off Page Connectors

To connect two off page connectors, select any one of the off page connectors; a plus symbol will be displayed on it, refer to Figure 2-25. Select the plus symbol; the **Connect To** option will be displayed near the cursor. Choose the **Connect To** option, refer to Figure 2-26; the **Create Connection** dialog box will be displayed, as shown in Figure 2-27.

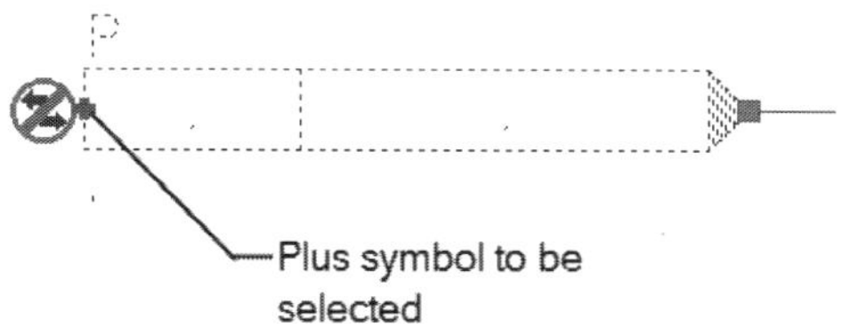

Figure 2-25 The plus symbol displayed on the off page connector

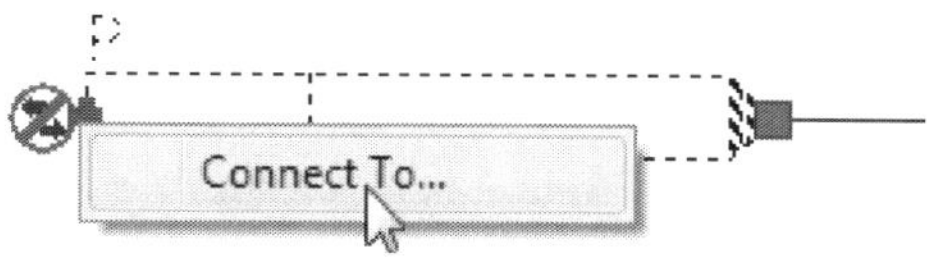

Figure 2-26 Choosing the ***Connect To*** *option*

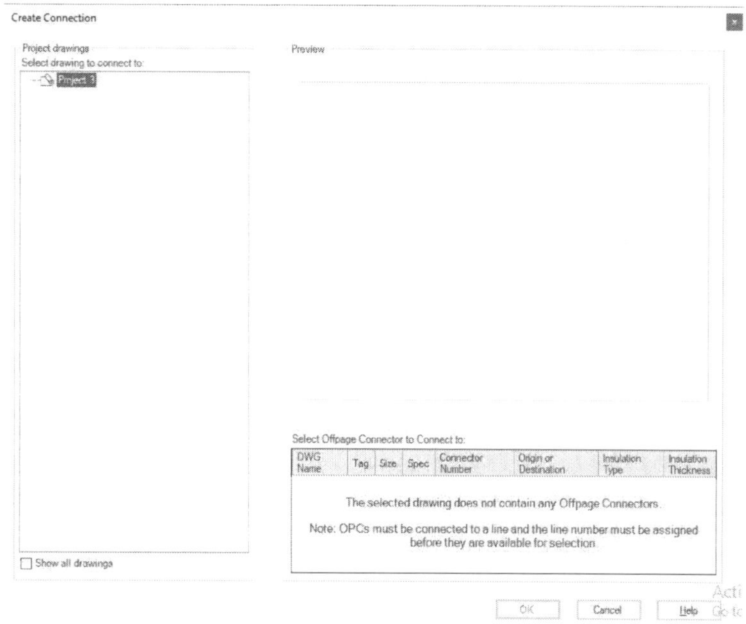

Figure 2-27 The ***Create Connection*** *dialog box*

In this dialog box, select the drawing to which the off page connector is to be connected from the **Project drawings** area. Next, select the off page connector to be connected from the **Select Offpage Connector to Connect to** table. Choose the **OK** button from the **Create Connection** dialog box; the off page connectors will be connected.

VALIDATING THE DRAWING

You need to validate the drawing to detect the errors and to correct them. To do so, first you need to select the conditions to validate a P&ID. Choose the **Validate Config** tool from the **Validate** panel in the **Home** tab; the **P&ID Validation Settings** dialog box will be displayed, as shown in Figure 2-28. Select the check boxes under the **P&ID objects** and the **Base AutoCAD objects** nodes and choose the **OK** button; the **P&ID Validation Settings** dialog box will be closed.

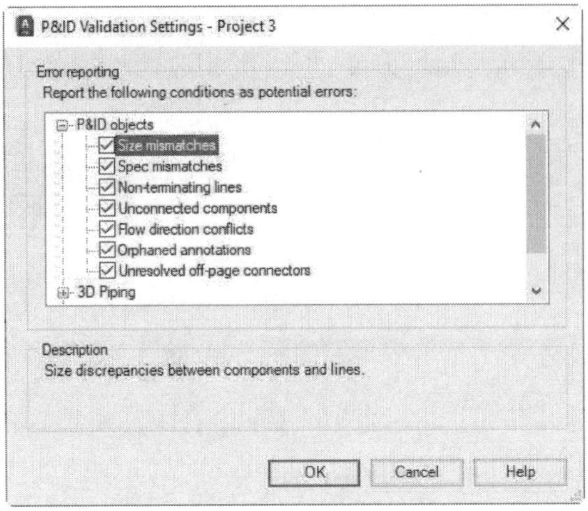

*Figure 2-28 The **P&ID Validation Settings** dialog box*

Checking for Errors

To run the checks for errors, choose the **Run Validation** tool from the **Validate** panel; the validation process will start and the errors will be checked. The list of errors detected will be displayed in the **VALIDATION SUMMARY** palette, as shown in Figure 2-29. Click on the error type in the **VALIDATION SUMMARY** palette; the drawing will zoom in where you can see the error location. Next, to correct the individual errors, select the respective drawing file node from the **VALIDATION SUMMARY** tree view. Now, choose the **Revalidate Selected Node** button in the **VALIDATION SUMMARY** palette; the **Validate Progress** message box will be displayed and errors will again be checked. You can also check for errors that are not corrected and close the **VALIDATION SUMMARY** palette. This different type of errors are discussed next.

The Base AutoCAD Object Errors

This error occurs when AutoCAD object or block is inserted in the drawing instead of AutoCAD P&ID component. You can ignore and erase a base AutoCAD object error. To do so, first you need to expand the **Base AutoCAD objects** error node. Next, click on the error; the error will zoom in the drawing.

Now, you can ignore the base AutoCAD object error. To do so, right-click on it in the **VALIDATION SUMMARY** palette; a shortcut menu will be displayed. Next, choose the **Ignore** option to ignore the error. You can also delete the object by choosing the **Erase** option.

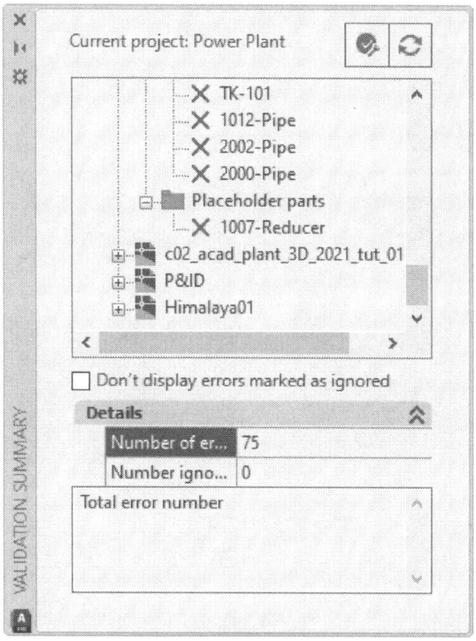

*Figure 2-29 The **VALIDATION SUMMARY** palette*

The Size mismatch Errors

This type of error occurs when size of a line and its associated component do not match. This condition is usually caused by manual changes in component properties. To correct the size mismatch error, expand the **Size mismatches** node in the **VALIDATION SUMMARY** palette and then click on the error; the error location will zoom-in in the drawing area. Next, manually change the size of the component or the size of the line connected to it.

The Spec mismatch Errors

This type of error occurs when specification properties of the connected components do not match. To correct the spec mismatch errors, zoom in the error location by clicking on the error in the **Spec mismatches** node. Next, manually change the specifications of the line by invoking the **Assign Tag** dialog box.

The Orphaned annotations Errors

This type of error occurs when the annotation tag of a component is moved away from the associated component to more than the acceptable distance. To correct the orphaned annotation error, expand the **Orphaned annotations** node in the **VALIDATION SUMMARY** palette. Next, select the error to zoom its location in the drawing area. Next, drag the annotation and place it nearer to the parent object. You can also right-click on the annotation and then choose the **Place from Parent** option from the shortcut menu displayed. Figure 2-30 shows an orphaned annotation and Figure 2-31 shows the annotation after correcting the error.

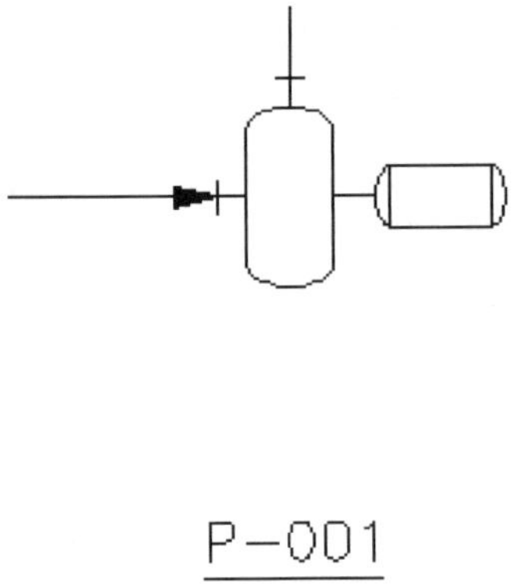

Figure 2-30 Orphaned annotation

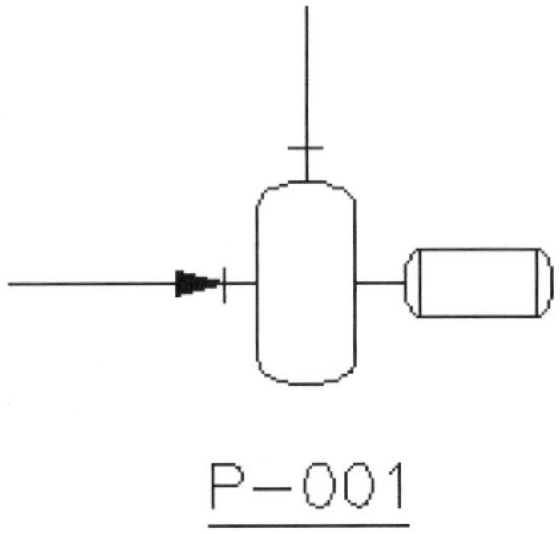

Figure 2-31 Annotation after placing it close to the component

The Unconnected Components Errors

This error occurs when the lines and components are not properly connected. To correct the unconnected component error, select the error from the **Unconnected components** node; the component in the drawing area will be zoomed in. Next, manually connect the component to the line near it.

The Unresolved Off Page Connectors Errors

This error occurs when off page connector in the drawing is not connected with the valid off page connector of the other drawing. Select the error from the **Unresolved off-page connectors** node in the **VALIDATION SUMMARY** palette; the off page connector in the drawing area will be zoomed in. Next, select and right-click on the off-page connector in the drawing and choose **Offpage Connector > Connect** from the shortcut menu displayed; the **Create Connection** dialog box will be displayed, refer to Figure 2-27. Next, select the drawing to which you want to connect the off page connector and then choose the **OK** button.

The Non-terminating lines Errors

This type of error occurs when the process line does not terminate with the valid terminator such as endline component or endline symbol. Select the error from the **Non-terminating lines** nodes to zoom it in the drawing area. Next, manually reconnect the line to the corresponding object by moving its endpoint. You can also recreate the connection.

The Flow Direction Errors

This error will occur when the direction of flow of process line is incorrect. To correct a flow direction error, click on the error under the **Flow direction conflicts** node; the error location will zoom in the drawing area. Next, click on the direction arrow; the **Flip Component** grip will be displayed. Click on this grip to flip the flow direction.

EDITING THE DRAWING

You can edit or modify a drawing while creating it or after it is created. The components of the drawing such as equipment, valves, instruments, lines, and so on can be modified as per the requirement. Various editing tasks are discussed next.

Moving an Equipment

You can move or change the location of an equipment. To do so, first you need to select the equipment by clicking on its border; the **Move Component** grip (square grip) will be displayed on it. Select it and drag the equipment and click to place it at a new location, as shown in Figure 2-32.

Moving a Valve

To move a valve, first select it to display the **Move Component** grip. Next, click on the grip, drag the valve along the horizontal or vertical line, and then click to place it at the new location. You can move the valve which is on the horizontal line to the vertical line or vice-versa, refer to Figure 2-33.

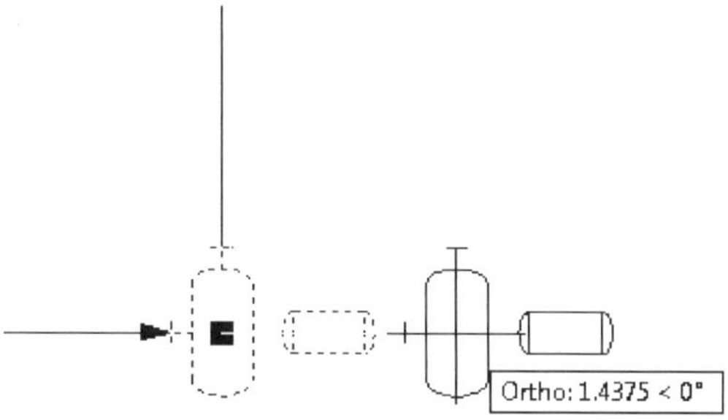

Figure 2-32 *Moving an equipment*

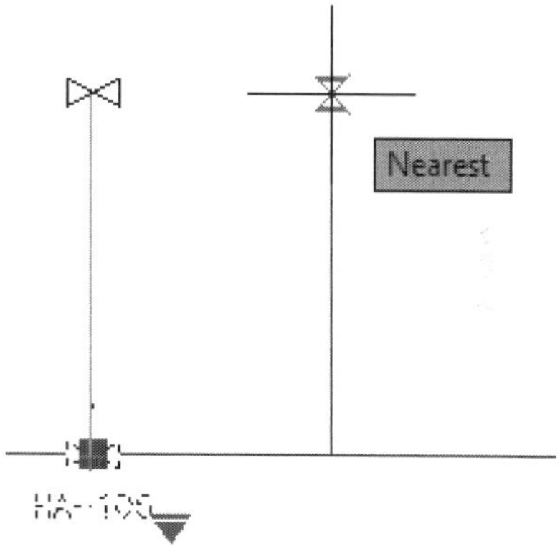

Figure 2-33 *Moving a valve from horizontal line to the vertical line*

Moving a Line

To move a line, first you need to select it; the stretch grip will be displayed on the line and the components connected to the line will be automatically selected. Select it from the middle of the line and drag it to the new location, as shown in Figure 2-34; the valves and the bubbles on the line will be moved. But the equipment connected to the line will not move.

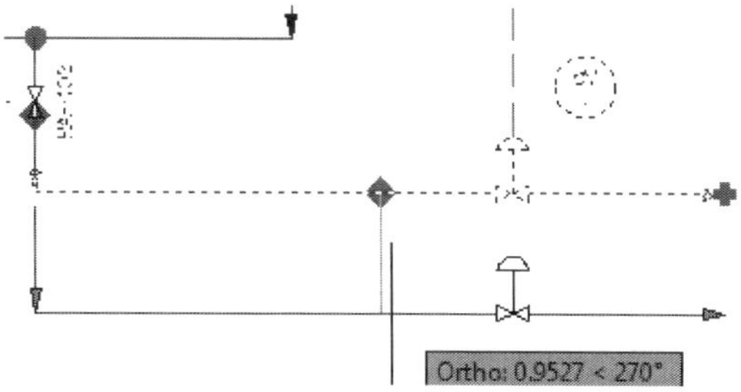

Figure 2-34 *Moving a line using the stretch grips*

Editing a Line

To edit a line, choose the **Edit** tool from the **Schematic Line** panel in the **Home** tab; you will be prompted to select a sline. Select a schematic line or a signal line from the drawing area; you will be prompted to enter an option at the command prompt. The options available at the command prompt are discussed next.

Attach

This option is used to attach the line to a component without physically connecting them. On invoking this option, you will be prompted to select a component to attach to the line. Select an equipment, valve, or any other P&ID component; you will be prompted to select an endpoint on the schematic line. Select an endpoint on the line and press ENTER; the component will be attached to the line.

Detach

This option is used to detach the attached line from the component. On invoking this option, you will be prompted to select the endpoint on the schematic line which has been attached to the source line. Select the attached endpoint on the line and press ENTER; the component will be detached from the line.

Gap

This option is used to create a gap on the line without breaking it. On invoking this option, you will be prompted to specify first gap point on the line. Specify the first gap point and move the cursor upto the required distance, and select the second gap point; a gap will be created on the line between the two specified points.

uNgap

This option removes the gap created on a line. On invoking this option, you will be prompted to select a schematic line segment with a gap or gap symbol. Select a line segment with a gap or gap symbol; the gap will be removed from the line.

Straighten

This option is used to straighten an inclined or a non-orthogonal line. On invoking this option, you will be prompted to select a line segment to be straightened. Select the line segment; you will be prompted to select an endpoint on the line to which the source line will be aligned. Select the endpoint on the line; the line will be straightened in alignment with the endpoint.

Corner

This option is used to create a corner by dividing the line into two sides. On invoking this option, you will be prompted to specify a point on the schematic line. Specify a point on the line; the line will be divided into two sides and you will be prompted to specify the second point of the corner. Move the cursor in a direction perpendicular to the line and select a point in the drawing area; you will be prompted to specify a point on the side of the line to move it. Select a point on one of the two sides; the selected side will be moved and aligned with the second point. As a result, a corner will be created.

Reverseflow

This option is used to reverse the direction of the schematic line. Invoke this option and press ENTER; the direction of flow of the line will be reversed.

Break

This option is used to break the line into two segments. On invoking this option, you will be prompted to select a break point. Select a point on the line; the line will be divided into two segments and you will be prompted to select an additional break point. Press ENTER to exit.

Join

This option is used to join two line segments. On invoking this option, you will be prompted to select one or more lines to join to the source line. Select a line segment that lies inline with the source line; the line will be joined with the source line.

Link

This option is used to link data of two line segments. On invoking this option, the following prompt will be displayed:

Select an sline segment to link to: *Select a line*

Sline segment data "?-?-?-?" will be deleted and segment will be linked to "?-?-?-?". Continue? [Yes/No]: *Enter 'Yes' to continue*

Note

*You can also choose the **Yes** option from the command prompt.*

Unlink

This option unlinks the data link created between two line segments. On invoking this option, the following prompt will be displayed:

Sline segment will be unlinked from "?-?-?-?". Continue? [Yes/No]: *Enter 'Yes' to continue*

Exit

This option is used to exit the command.

Grouping Lines

You can group schematic lines together and assign tag to the group. To do so, choose the **Make Group** tool from the **Line Group** panel in the **Home** tab; you will be prompted to select the source schematic line to group additional schematic lines to it. Select the source line; you will be prompted to select schematic lines to group. Select a single line or multiple lines and press ENTER; the lines will be grouped together.

Editing a P&ID Symbol

You can edit a P&ID symbol and also retain the original symbol by using the **Edit Block** tool. To do so, choose the **Edit Block** tool from the **P&ID** panel in the **Home** tab; you will be prompted to select a P&ID component to edit its block. Select an equipment, valve, or any other P&ID component; the **Block Editor** tab will be invoked. The drawing area of the **Block Editor** tab has a dull background and is called Authoring area. In this area, you can edit the existing entities or add new ones to the block of a P&ID component. In addition to the Authoring area, the **Block Editor** tab, **BLOCK AUTHORING PALETTES**, and **Edit P&ID Object's Block** dialog box will also be displayed. After editing the block, choose the **Save Changes and Exit Block Editor** button from the **Edit P&ID Object's Block** dialog box; the **Block Editor** will be closed and the edited P&ID component will be displayed in the drawing area.

Substituting Components

You can substitute a valve or any other component of a drawing with another similar component. For example, you can substitute a gate valve with a ball valve. To do so, first you need to select the component to display the grips on it. Next, select the substitution grip (down arrow); the **Substitute with another Component** palette containing valves will be displayed, as shown in Figure 2-35. Select the ball valve symbol from the palette; it will replace the existing gate valve.

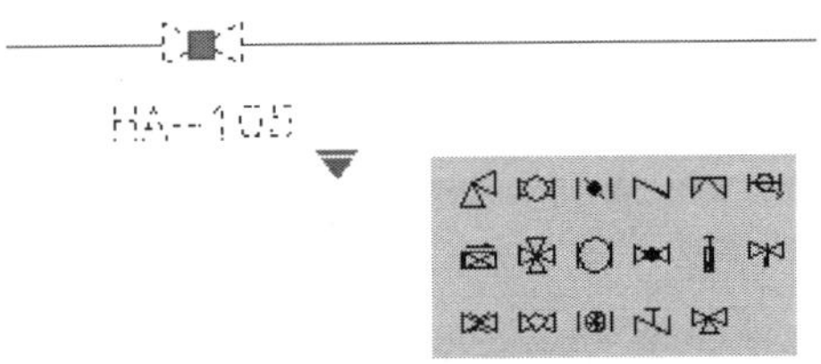

*Figure 2-35 The **Substitute with another Component** palette containing valves*

Converting AutoCAD Components into P&ID Symbols

You can add new symbols which are not available in **TOOL PALETTES - P&ID PIP** by creating a block and converting it into a P&ID symbol. To convert a block into a P&ID symbol, first you need to create a block by using the AutoCAD drawing tools. Next, select the component, right-click on it and choose **Convert to P&ID Object** from the shortcut menu displayed; the **Convert to P&ID Object** dialog box will be displayed, as shown in Figure 2-36. The tree view in the dialog box shows a list of components arranged in groups. Expand the tree view by clicking on the + sign adjacent to the required group and select the object from it. Next, choose the **OK** button; the component will be converted into the selected P&ID object.

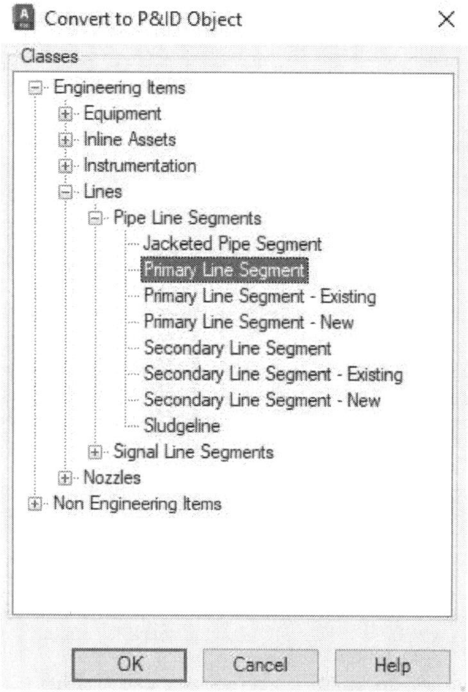

***Figure 2-36** The **Convert to P&ID Object** dialog box*

To add the converted component to the **TOOL PALETTES - P&ID PIP**, first you need to save the current file. Next, open the tab in the **TOOL PALETTES** to which you want to add the symbol. Choose the **Design Center** tool from the **Palettes** panel of the **View** tab. Choose the **Load** button at the top left corner of the **DESIGN CENTER** palette; the **Load** window will be displayed. Browse to the saved current file and choose on the **Open** button. Now, double-click on the **Blocks** icon in the **DESIGN CENTER** palette; the saved block is displayed. Now, click and drag the object and place it in the desired group in the **TOOL PALETTES**; the symbol will be displayed in the **TOOL PALETTES**.

Adding Intelligence to the Custom P&ID Symbols

You can add more intelligence to the custom P&ID symbols that you want to add to the **TOOL PALETTES**. The procedure to do so is illustrated below. Figure 2-37 shows the block of a control valve which is to be added to the TOOL PALETTES.

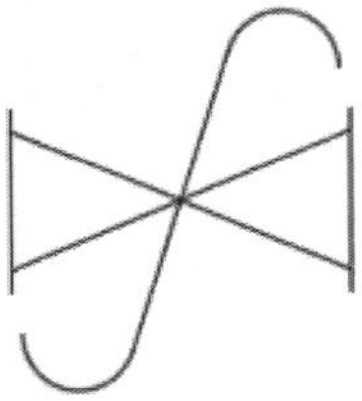

Figure 2-37 The valve symbol block to be added to the ***TOOL PALETTES***

To add the symbol shown in the Figure 2-37 to the **TOOL PALETTES**, make sure that the **P&ID PIP** workspace is activated and the tool palette is loaded with **P&ID PIP** symbols.

Create the shape of the valve symbol, as shown in Figure 2-37 and convert it into a block with the name **control valve 1**. Save the drawing by choosing the **Save** tool from the Quick Access Toolbar. Convert this shape of the valve into P&ID symbol as discussed earlier. Next, click on the **Valves** tab in the **TOOL PALETTES - P&ID PIP** to ensure that the custom symbol which is a valve here gets added to the **Valves** tab of the **TOOL PALETTES**. Choose the **Project Setup** tool from the **Project Manager** drop-down of the **Project** panel in the **Home** tab; the **Project Setup** dialog box will be displayed, refer to Figure 2-38.

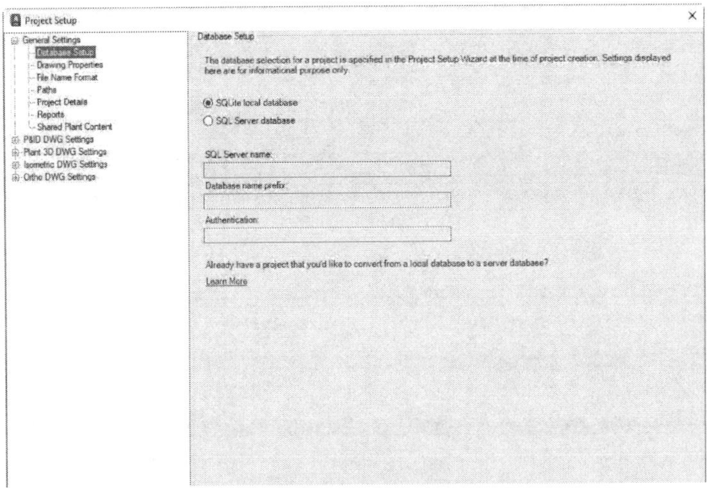

Figure 2-38 The ***Project Setup*** *dialog box*

You can also invoke this dialog box by entering **PROJECTSETUP** in the command window or by right-clicking on the project name node in the **Project** area of the **PROJECT MANAGER** and, then selecting the **Project Setup** option from the flyout displayed.

In the **Project Setup** dialog box, expand the **P&ID DWG Settings** node by clicking on the + sign and then choose, **P&ID Class Definitions > Engineering Items > Inline Assets > Hand Valves**. Click on the **Hand Valves** sub-node to highlight it. Next, right-click to display a shortcut menu and choose the **New** option; the **Create Class** dialog box will be displayed, refer to Figure 2-39.

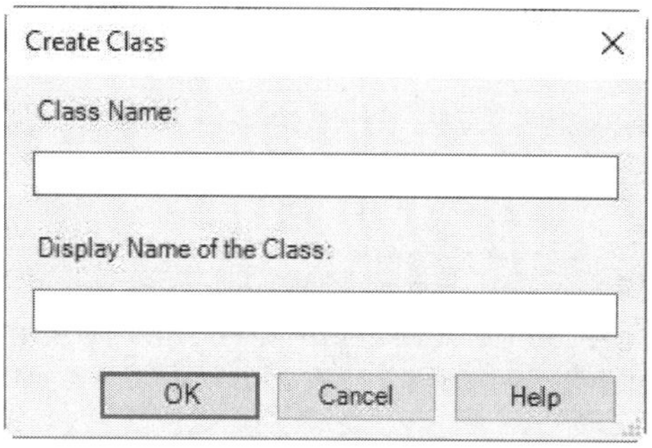

Figure 2-39 *The* ***Create Class*** *dialog box*

Enter **controlvalve1** in the **Class Name** edit box; the specified name is also displayed in the **Display Name of the Class** edit box (Note that the class name entered in the **Class Name** edit box should not have any spaces in between the alphabets). Next, choose the **OK** button to exit the dialog box; the specified class name will be listed and highlighted under the **Hand Valves** sub-node. Also, the class name (controlvalve1) will be displayed next to the **Class settings** area. Next, choose the **Add Symbols** button from the **Symbol** area; the **Add Symbols - Select Symbols** dialog box will be displayed, refer to Figure 2-40.

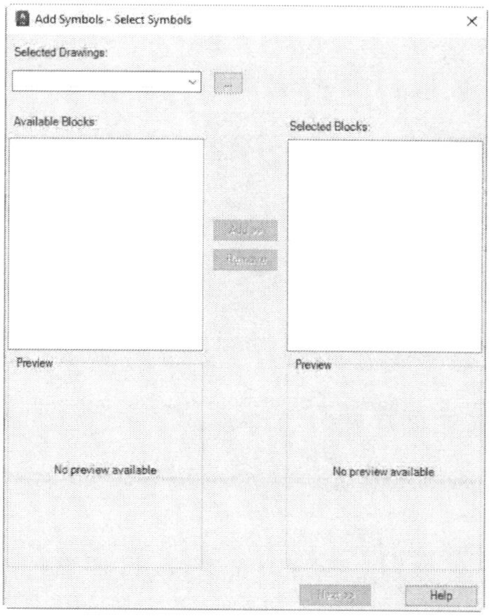

Figure 2-40 *The* ***Add Symbols - Select Symbols*** *dialog box*

In this dialog box, choose the browse button next to the drop-down in the **Selected Drawings** area; the **Select Block Drawing** dialog box will be displayed. In this dialog box, browse to the file location and open it; the **Add Symbols - Select Symbols** will be displayed again with the list of blocks contained in the drawing file in the **Available Blocks** list box. Select **control valve 1** from the **Available Blocks** list box; it will be highlighted. Also a preview of the valve block will be displayed in the **Preview** window available below the **Available Blocks** list box.

Next, choose the **Add** button; the block name will be displayed in the **Selected Blocks** list box of the dialog box. Choose the **Next** button; the **Add Symbols - Edit Symbol Settings** dialog box will be displayed. Enter **control valve 1** in the **Symbol Name** edit box of the **Symbol Properties** area. Choose the **Yes** option from the **Auto Nozzle** drop-down list and **Flanged Nozzle Style** from the **Auto Nozzle Style** drop-down list of the **Other Properties** area. Next, choose the **Finish** button; the **Add Symbols - Edit Symbol Settings** dialog box will be closed and the **Project Setup** dialog box will be displayed again. Next, choose the **Add to Tool Palette** button from the **Project Setup** dialog box; the **Create Tool** message box will be displayed with a message that the control valve 1 tool has been added to the current tool palette. Choose the **OK** button to exit the message box. Next, choose the **Apply** and then the **OK** button to exit the **Project Setup** dialog box and add the control valve 1 symbol to the **Valves** area of the **Valves** tab.

Editing the Properties of the Custom P&ID Symbol

Using the **Project Setup** dialog box, you can also edit the symbol settings, block settings, properties like tag format, annotation style and so on. Consider that the newly created **control valve 1** P&ID symbol needs some editing as per the requirement. The steps to do various editing operations on it are listed below.

Invoke the **Project Setup** dialog box and follow the path, **P&ID DWG Settings > P&ID Class Definitions > Engineering Items > Inline**

Assets > Hand Valves > controlvalve1. On doing so, the **controlvalve1** will be displayed in the **Class settings** area and its preview will be displayed in the Preview window in the **Symbol** area. Also, various properties related to the symbol will be listed in the table in the **Properties** area of the dialog box. Choose the **Edit Symbol** button in the **Symbol** area; the **Symbol Settings** dialog box will be displayed, refer to Figure 2-41.

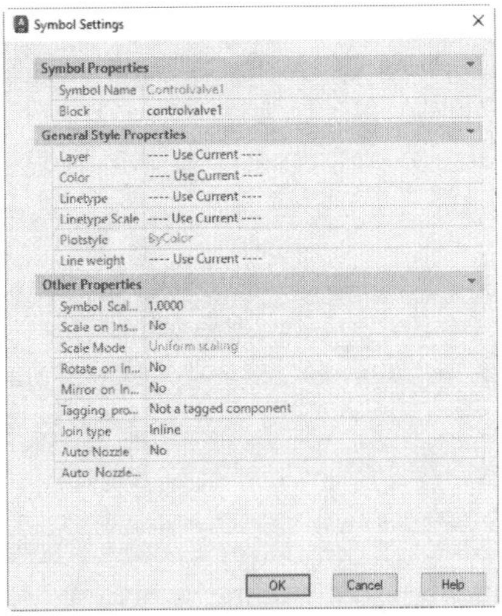

*Figure 2-41 The **Symbol Settings** dialog box*

You can change various properties such as layer, color, linetype, symbol scale, tagging prompt, join type by using the areas of this dialog box. Choose the **OK** button to exit the dialog box. Choose the **Edit Block** button in the **Symbol** area; the **Block Editor** will be invoked displaying the symbol of the custom block. You can use various tools of the **Block Editor** to make changes in the shape, size and so on of the block and then exit the **Block Editor** by choosing the **Close**

Block Editor button from the **Close** panel of the **Block Editor** tab. In the **Properties** area of the dialog box, you can set various properties related to the valve such as class name, model number, description, size, spec. The name of the property will be listed in the **Property Name** column of the table in the **Properties** area. You can also add a new property to the **Property Name** list. To do so, choose the **Add** button from the **Properties** area; the **Add Property** dialog box will be displayed, refer to Figure 2-42.

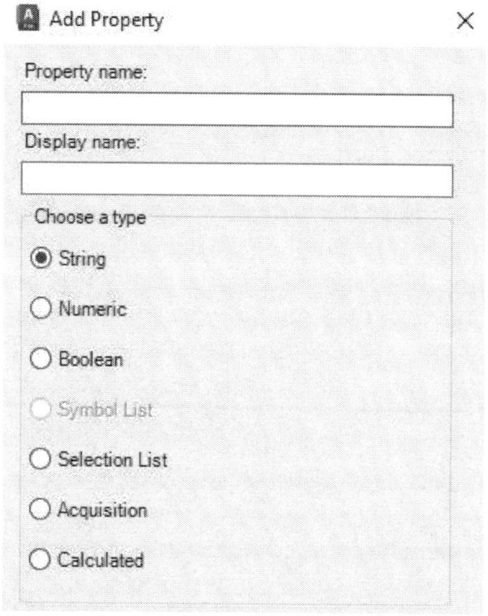

Figure 2-42 *The* ***Add Property*** *dialog box*

Enter the property name in the **Property name** edit box; the specified name will also be displayed in the **Display name** edit box. Next, choose a property type from the **Choose a type** area by selecting the respective radio button. Next, choose the **OK** button; the dialog box is closed and the property will be added to the **Property Name** column of the table. In the **Tag format** area of the dialog box, a list box will be available displaying the **Hand Valve Tag [Code-Number]** by default. You can add a new tag format to the list or modify the existing one. To add a new tag format, choose the **New** button from the **Tag format** area; the **Tag Format Setup** dialog box will be displayed, refer to Figure 2-43.

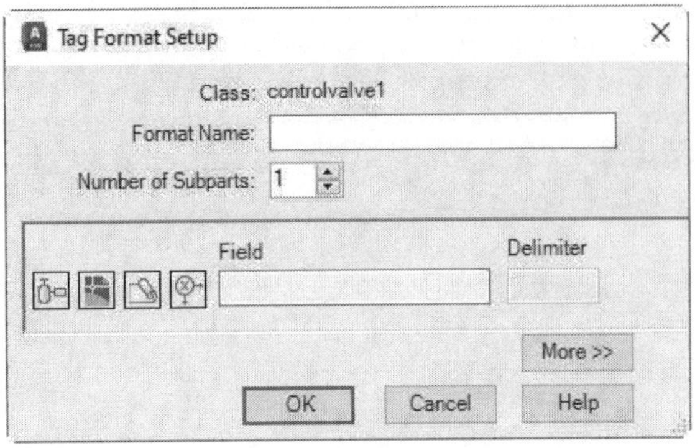

***Figure 2-43** The **Tag Format Setup** dialog box*

In this dialog box, enter the format name in the **Format Name** edit box. You can add subparts to the tag format by entering the desired value in the **Number of Subparts** edit box either manually or by using the spinner. The default value in the **Number of Subparts** edit box is **1**. The information that a tag format will be carrying can be controlled by using the buttons available in the list box of the dialog box, refer to Figure 2-43. The buttons in this list box are: **Select Class Properties**, **Select Drawing Properties**, **Select Project Properties**, and **Define Expression**. Using these buttons, you can set a new tag format for a class. After making all the necessary changes, choose the **OK** button; the dialog box will be closed and the newly created tag format will be listed and highlighted in the list box in the **Tag format** area of the **Project Setup** dialog box. You can modify or delete the newly created tag format by choosing the **Modify** or **Delete** button, respectively. You can also add, modify and change the annotation styles using the options available in the **Annotation** area of the **Project Setup** dialog box. After making all the necessary changes, exit the **Project Setup** dialog box by choosing **Apply** and then the **OK** button.

TUTORIAL

Tutorial 1

In this tutorial, you will create a P&ID, shown in Figure 2-44. You will create it by placing equipments, instruments, and inline components and then connecting them with lines.

(Expected time: 2hr)

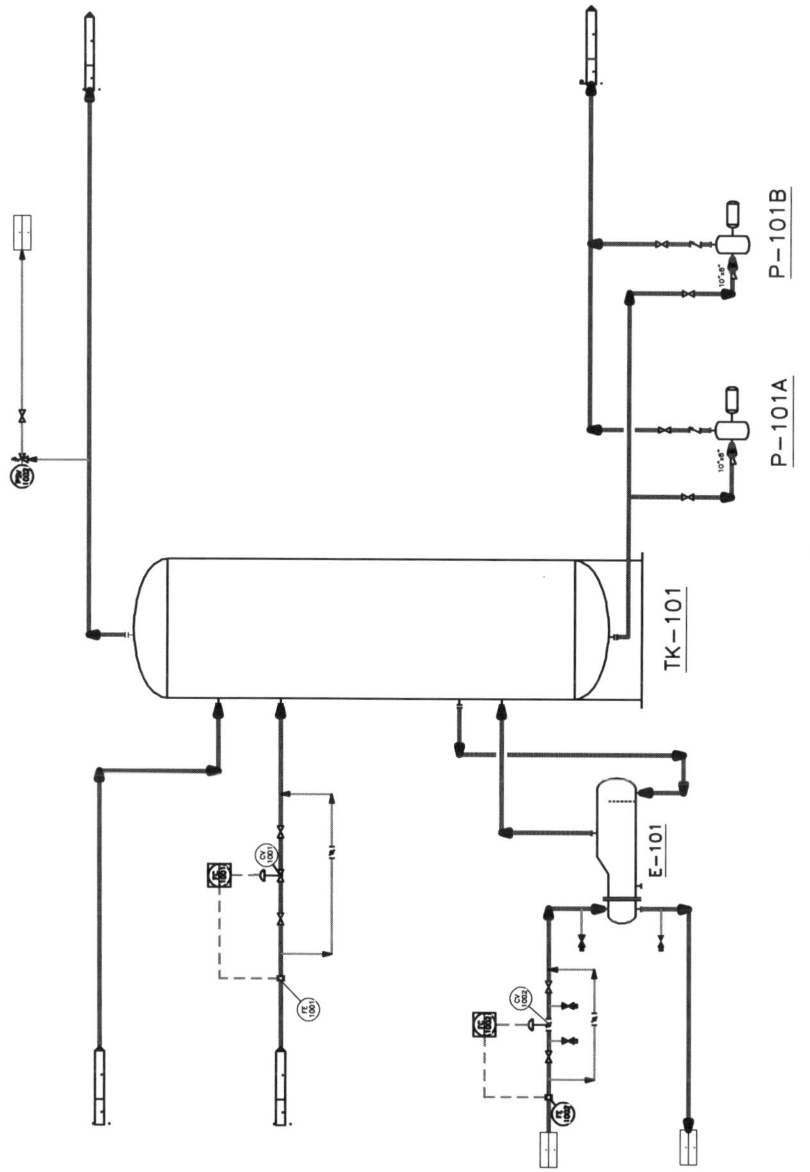

Figure 2-44 *P&ID for Tutorial 1*

The following steps are required to complete this tutorial:

a. Start AutoCAD Plant 3D and then create a new project.
b. Create a new P&ID drawing.
c. Place equipments into the drawing.
d. Connect equipments using schematic lines.
e. Place valves on the lines connecting the components.
f. Add off-page connectors to the drawing.
g. Add tags to schematic lines.
h. Validate the drawing.
i. Save and close the drawing file.

Starting AutoCAD Plant 3D and Creating a New Project

1. Double-click on the **AutoCAD Plant 3D 2023- English** icon from the desktop of your computer to start AutoCAD Plant 3D 2023.

2. In the **PROJECT MANAGER**, select the **New Project** option from the drop-down list available in the **Current Project** area; the **Project Setup Wizard** is displayed.

3. Enter **CADCIM** in the **Enter a name for this project** edit box and choose the **Next** button; the **Specify unit settings** page is displayed.

4. Select the **Imperial** radio button, if not selected by default and then choose the **Next** button; the **Specify P&ID settings** page is displayed.

5. Select the **PIP** option from the **Select the P&ID symbology standard to be used** list box, if not selected by default and choose the **Next** button; the **Specify Plant 3D directory settings** page is displayed.

6. Accept the default directory settings and choose the **Next** button; the **Specify database settings** page is displayed.

7. Make sure the **Single User - SQLite local database** radio button is selected and choose the **Next** button; the **Finish** page is displayed.

8. Choose the **Finish** button; you will notice that the CADCIM project is displayed in the drop-down list available in the **Current Project** area of the **PROJECT MANAGER**. Also, a node named CADCIM is created in the **Project** area.

Creating a New Drawing

1. In the **PROJECT MANAGER**, right-click on the **P&ID Drawings** node in the project tree and then choose the **New Drawing** option; the **New DWG** dialog box is displayed.

2. Enter **P&ID1.dwg** in the **File name** edit box and then specify the **PID ANSI D -Color Dependent Plot Styles.dwt** template in the **DWG template** edit box.

3. Choose the **OK** button from the **New DWG** dialog box; the new P&ID drawing file is created with the **TOOL PALETTES** displayed on the right side of the screen.

4. Choose the **Workspace Switching** button on the right-side of the Status Bar; a flyout is displayed. Choose the **PID PIP** option from the flyout; the **P&ID PIP** workspace is invoked and the **TOOL PALETTES** is loaded with the P&ID symbols. Note that, if **Tool Palette** is not displayed, then choose the **Tool Palettes** button from the **Palettes** panel in the **View** tab.

Placing Equipment in the Drawing

1. Choose the **Vessel** tool from the **Vessel and Miscellaneous Vessel Details** area in the **Equipment** tab of the **TOOL PALETTES - P&ID PIP**; you are prompted to specify an insertion point. Next, specify **(13,5)** as the insertion point of the vessel; you are prompted to specify the scale factor.

2. Enter **1.5** as the scale factor; the **Assign Tag** dialog box is displayed. Enter **101** in the **Number** edit box and make sure that the **Place annotation after assigning tag** check box is selected. Next, choose the **Assign** button and specify the position of the annotation tag near the vessel, as shown in Figure 2-45.

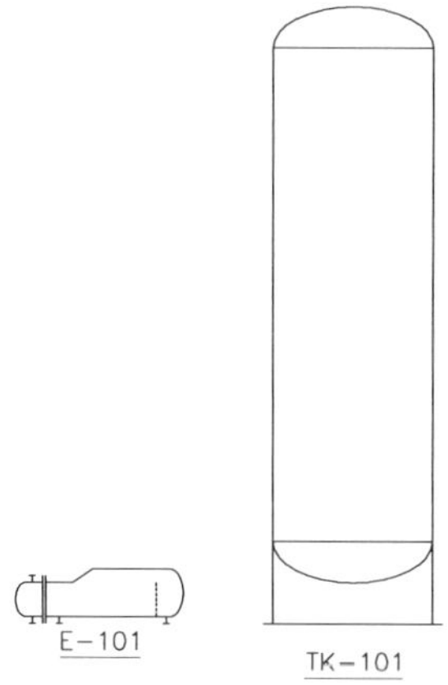

Figure 2-45 *P&ID after placing the vessel and the heat exchanger*

3. Choose the **TEMA type BKU Exchanger** from the **TEMA Type Exchangers** area in the **Equipment** tab of the **TOOL PALETTES - P&ID PIP** and place it at the point **(7,5)**. Next, assign tag **E-101** to the heat exchanger. Figure 2-45 shows the P&ID after placing the vessel and the heat exchanger.

4. Choose **Horizontal Centrifugal pump** from the **Pumps** area in the **Equipment** tab of the **TOOL PALETTES - P&ID PIP** and place at the point **(17,3)**. Next, assign tag **P-101A** to the pump.

5. Similarly, place another horizontal centrifugal pump at the point **(21,3)**. Next, assign the tag **P-101B** to the pump. Figure 2-46 shows the P&ID after placing the centrifugal pumps.

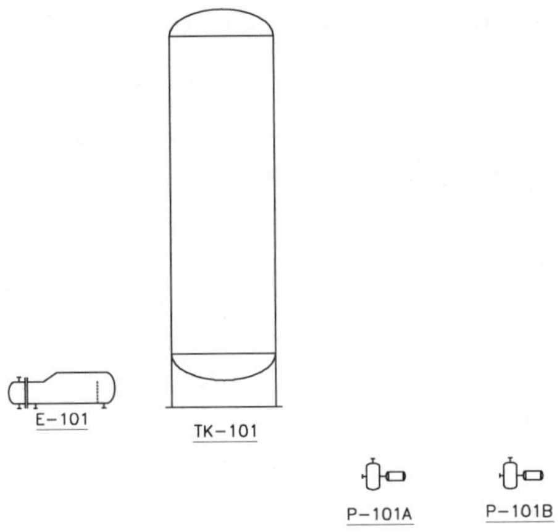

Figure 2-46 *P&ID after placing two centrifugal pumps*

Connecting the Heat Exchanger and the Vessel

Now, you need to create lines connecting the equipment. Before doing that, you need to place some nozzles at the required locations on the equipment.

1. Choose **Single Line Nozzle** from the **Nozzles** area in the **Fittings** tab of the **TOOL PALETTES - P&ID PIP**; you are prompted to select the asset to place nozzle.

2. Select the heat exchanger from the drawing sheet; the nozzle is attached to the cursor and you are prompted to specify the insertion point.

3. Specify the insertion point, refer to Figure 2-47, and then specify the rotational angle as 90-degrees.

   Next, you need to connect the heat exchanger and the vessel by drawing pipe lines. To do so, you need to first make sure that the **ORTHOMODE** is turned on to draw straight lines. Also, the **Object Snap** should be turned on for easy selection of the points.

4. Choose the **Lines** tab from the **TOOL PALETTES** and choose the **Primary Line Segment** line type from the **Pipe Lines** area; you are prompted to select the start point.

5. Specify the start point on the newly created nozzle of the heat exchanger. Next, move the cursor upward and enter **2** at the command prompt. Next, press ENTER.

6. Move the cursor horizontally toward right and specify the endpoint on the vessel; the line connecting the vessel and the heat exchanger is created, refer to Figure 2-47.

   Next, you need to assign a tag to the line.

7. Choose the **Assign Tag** tool from the **P&ID** panel in the **Home** tab and select the previously created line.

8. Press the ENTER key; the **Assign Tag** dialog box is displayed. Specify the following values in the dialog box:

Size	**8"**
Spec	**CS150**
Pipe Line Group.Service	**P-GENERAL PROCESS**
Pipe Line Group.Line Number	**1007**

9. Clear the **Place annotation after assigning tag** check box. Choose the **Assign** button; the tag is assigned to the line.

Note

*You can also annotate the line with the assigned tag. To do so, you need to select the **Place annotation after assigning tag** check box and then select the **Pipeline Tag** option from the **Annotation style** drop-down list in the **Assign Tag** dialog box.*

10. Invoke the **Primary Line Segment** tool from the **Lines** tab of the **TOOL PALETTES** and connect the vessel to the nozzle located at the bottom of the heat exchanger. The line created should be similar to the one shown in Figure 2-48.

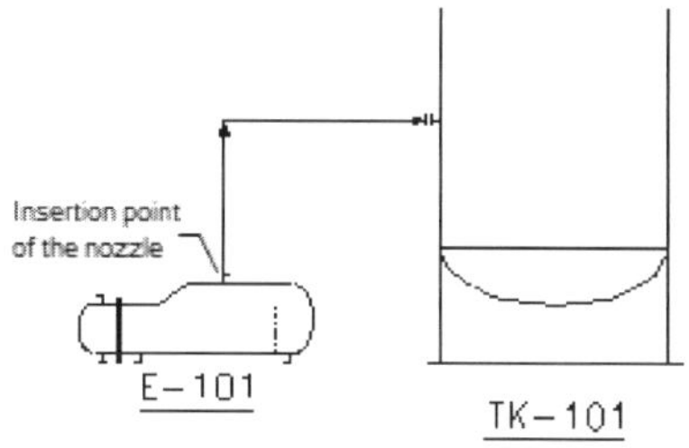

Figure 2-47 *Insertion point of the nozzle, and line connecting the vessel and the heat exchanger*

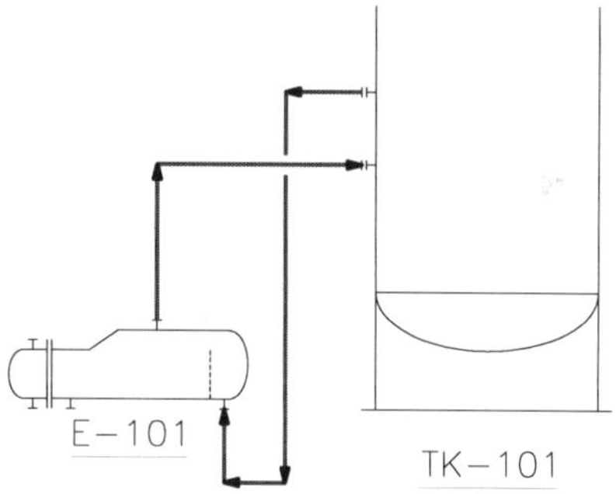

Figure 2-48 *Line connecting the vessel and the bottom nozzle of the heat exchanger*

11.Assign the tag **6"-CS150-P-1006** to the line.

12. Similarly, create lines connecting the other two nozzles (steam inlet and outlet nozzles) of the heat exchanger, as shown in Figure 2-49. Next, assign tags **6"-CS150-P-2000** and **3"-CS150-SC-2002** to steam inlet line and condensate lines, respectively.

13. Choose the **Secondary Line Segment** tool from the **Lines** tab in the **TOOL PALETTES - P&ID PIP** and create a loop on the line connecting the steam inlet nozzle of the heat exchanger, as shown in Figure 2-50.

Note that in this tutorial, the newly created line is referred as outlet line of the pump.

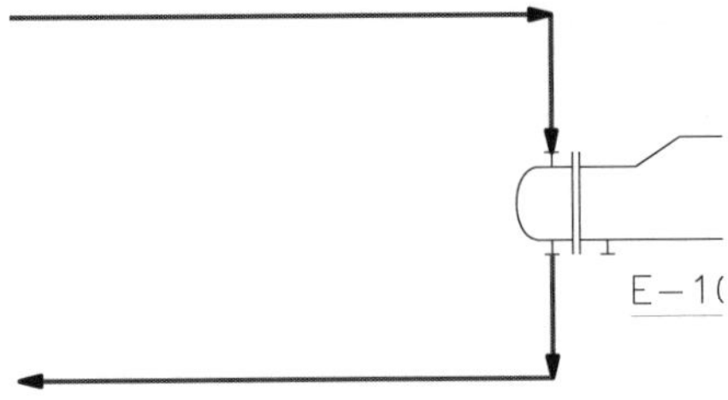

Figure 2-49 The steam inlet line and condensate line of the heat exchanger

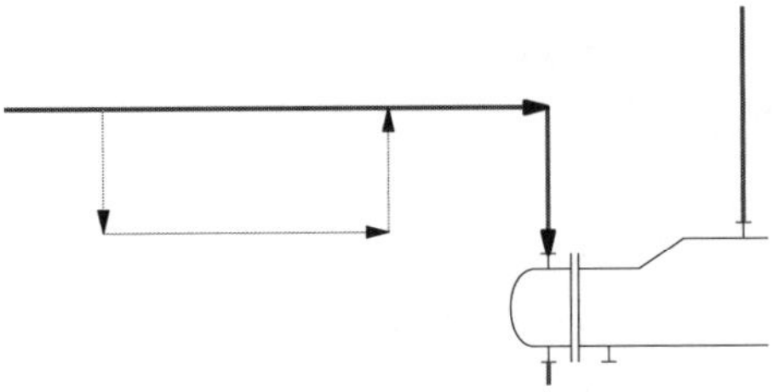

Figure 2-50 Loop created on line connecting the steam inlet nozzle

Connecting the Vessel to Pumps

Next, you need to create pipe lines connecting the vessel and pumps.

1. Invoke the **Primary Line Segment** tool from the **Lines** tab in the **TOOL PALETTES - P&ID PIP** and specify the start point at the bottom of the vessel, as shown in Figure 2-51.

2. Connect the line to the horizontal nozzle of the pump tagged **P-101B**. The line created should be similar to the one shown in Figure 2-52. In this tutorial, this line is referred as inlet line of the pump.

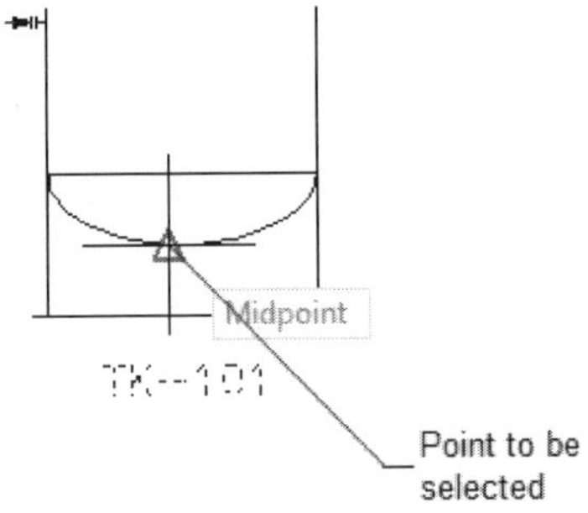

Figure 2-51 Point to be selected on the vessel

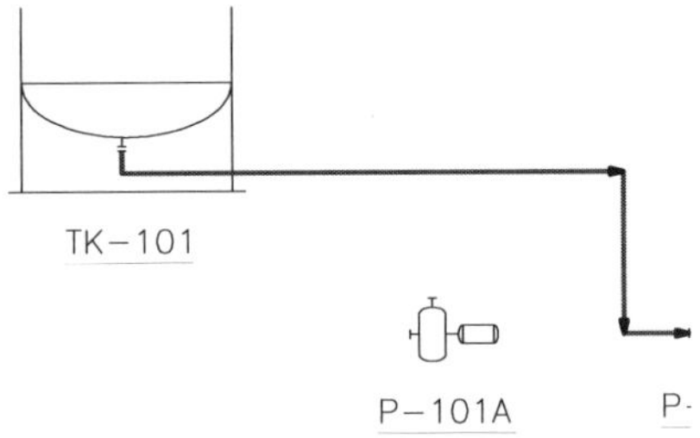

Figure 2-52 Line connecting the vessel and the pump

3. Assign tag **10"-CS150-P-1004** to the line.

4. Create a line connecting the previously created line to the horizontal nozzle of the pump tagged **P-101A**, refer to Figure 2-53. Assign the tag **10"-CS150-P-1004** to the line.

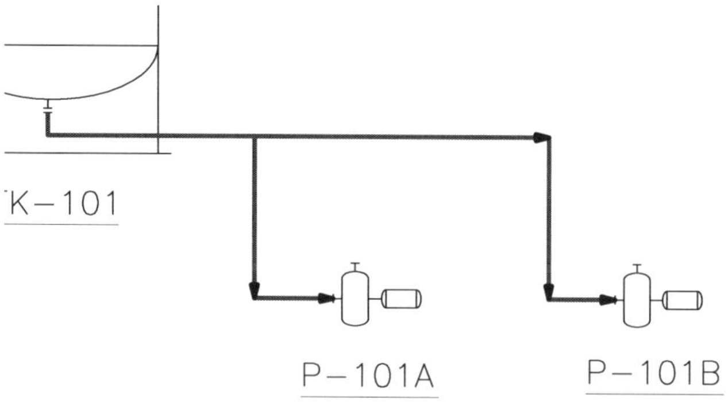

*Figure 2-53 Line connecting the horizontal nozzle of the pump tagged **P-101A***

5. Invoke the **Draw** tool from the **Schematic Line** panel in the **Home** tab; you are prompted to specify the start point of the line.

6. Specify the start point on the vertical nozzle of the pump tagged **P-101A**, and then create a line, as shown in Figure 2-54.

7. Assign tag **8"-CS150-P-1012A** to the line.

8. Next, connect the vertical nozzle of the other pump to the previously created line, refer to Figure 2-55. Assign the tag **8"-CS150-P-1012B** to the line.

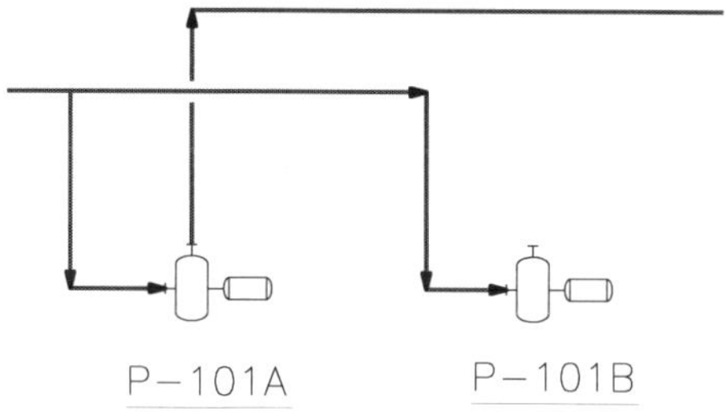

Figure 2-54 Line connecting the vertical nozzle of the pump

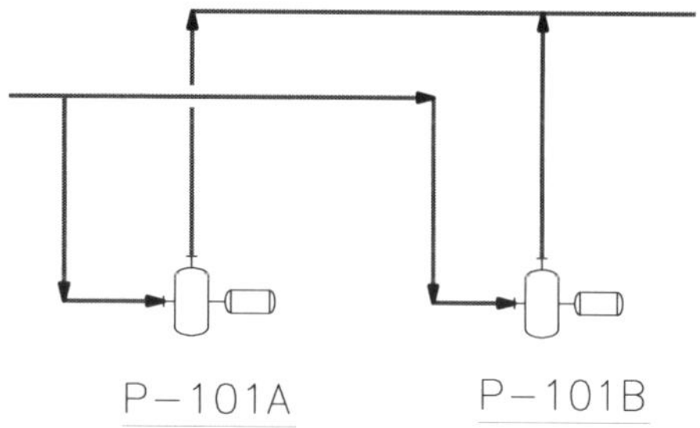

Figure 2-55 Line connecting another pump and previously created line

Note that in this tutorial, the newly created line is referred as outlet line of the pump.

Creating the Remaining Lines Connecting the Vessel

1. Invoke the **Primary Line Segment** tool and specify the start point as (4,13). Next, move the cursor horizontally toward right and specify the endpoint on the vessel; the line is created, as shown in Figure 2-56. Note that this line will be referred as feed line in this tutorial.

2. Next, assign tag **12”-CS150-P-1001** to the line.

3. Choose **Secondary Line Segment** from the **Lines** tab in the **TOOL PALETTES - P&ID PIP** and create the loop, as shown in Figure 2-57.

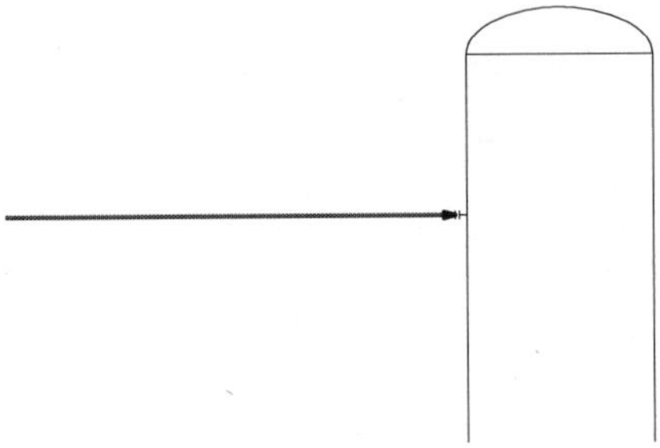

Figure 2-56 Line connected to the vessel

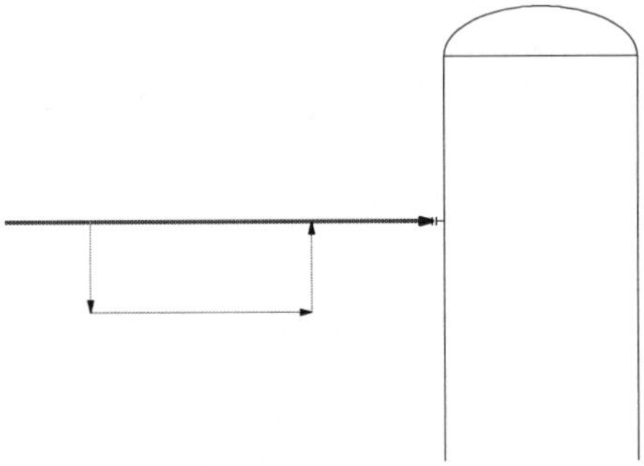

Figure 2-57 Loop created on the previous line

4. Create another primary line by specifying the start point at 4,17. It should be similar to the line shown in Figure 2-58. Assign tag **4”-CS150-P-1000** to the line.

   This line is referred as reflux line.

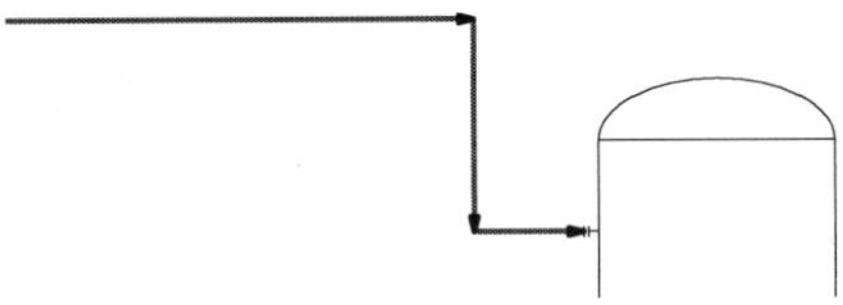

Figure 2-58 Reflux line connected to the vessel

Next, you need to create a line originating from the top of the vessel.

5. Invoke the **Primary Line Segment** tool and specify the start point on the top of the vessel.

6. Move the cursor upward and enter **1** at the command prompt. Next, press ENTER.

7. Move the cursor horizontally toward right and create a line, as shown in Figure 2-59. Assign the tag **12”-CS150-P-1006T** to the line.

8. Invoke the **Secondary Line Segment** tool and create a line, as shown in Figure 2-60.

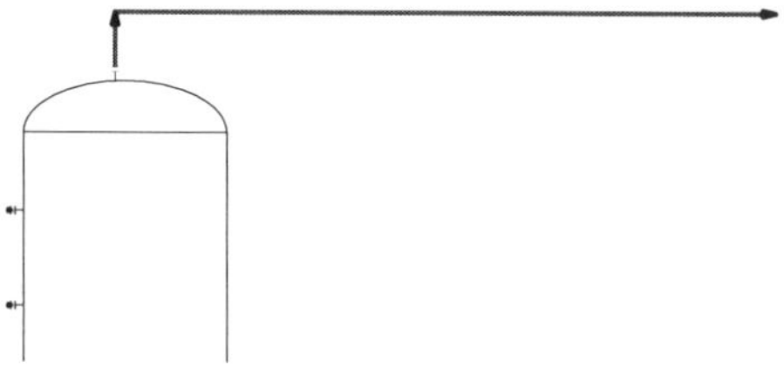

Figure 2-59 Line drawn from the top of the vessel

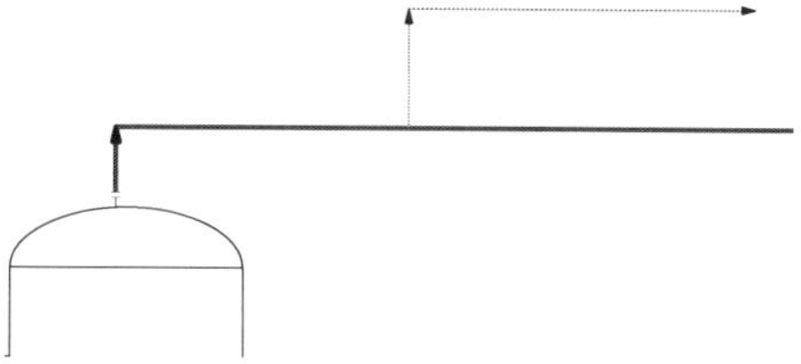

Figure 2-60 Drawing the secondary line connecting the previous line

Adding Valves to the P&ID

1. Choose the **Control Valve** tool from the **Control Valve** area of the **Valves** tab of the **TOOL PALETTES**; you are prompted to pick the insertion point.

2. Choose the **Change body or actuator** option from the command prompt to invoke the **Control Valve Browser** dialog box. Note that if you are adding a control valve for the first time, the **Control Valve Browser** dialog box will be displayed by default.

3. Select the **Globe Valve** option from the **Select Control Valve Body** tree view and **Diaphragm Actuator** from the **Select Control Valve Actuator** tree view and then choose the **OK** button.

4. Insert the control valve on the feed line, as shown in Figure 2-61.

5. Specify the position of the annotation balloon near the control valve; the **Assign Tag** dialog box is displayed.

6. Enter the values given next in this dialog box and then choose the **Assign** button.

Area	**01**
Type	**CV**
Loop Number	**1001**

7. Similarly, place a control valve on the steam inlet pipe of the heat exchanger, as shown in Figure 2-62. Note that you need to specify **Butterfly Valve** and **Diaphragm Actuator** as the valve body and actuator, respectively.

8. Enter the following values in the **Assign Tag** dialog box:

Area	**01**
Type	**CV**
Loop Number	**1002**

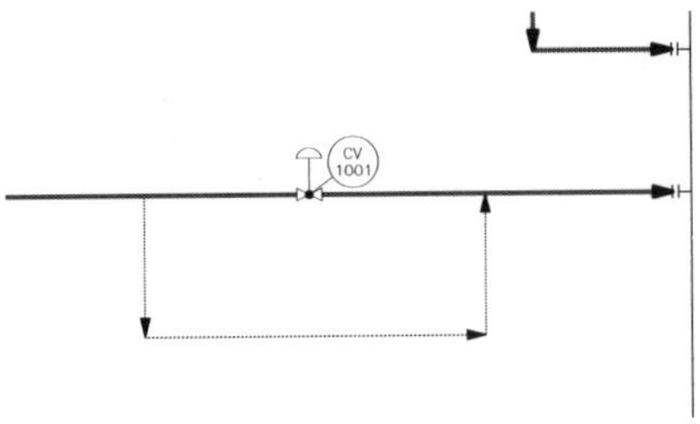

***Figure 2-61** Location of the control valve*

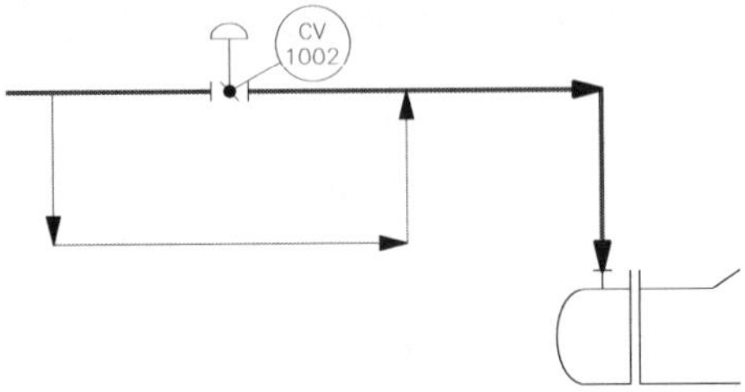

***Figure 2-62** Location of the control valve on the steam inlet line*

9. Choose **Gate Valve** from the **Valves** area in the **Valves** tab of the **TOOL PALETTES - P&ID PIP**, and place it at two locations, as shown in Figure 2-63.

Note

*On inserting a valve, an annotation is displayed by default next to it. To avoid this, you need to hide objects. To do so, right-click on the annotation and choose **Isolate > Hide Objects**.*

10. Next, choose **Butterfly Valve** and place it at the location shown in Figure 2-63.

11. Similarly, place two gate valves and a butterfly valve on the pipe which is connected to the heat exchanger, as shown in Figure 2-64.

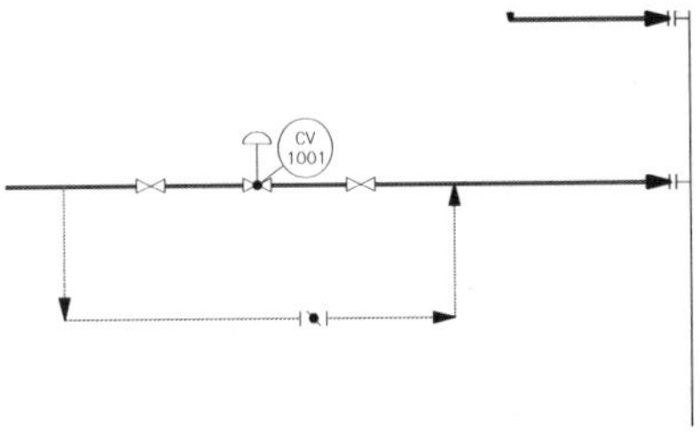

***Figure 2-63** Location of the gate valve and butterfly valve on the line connected to the vessel*

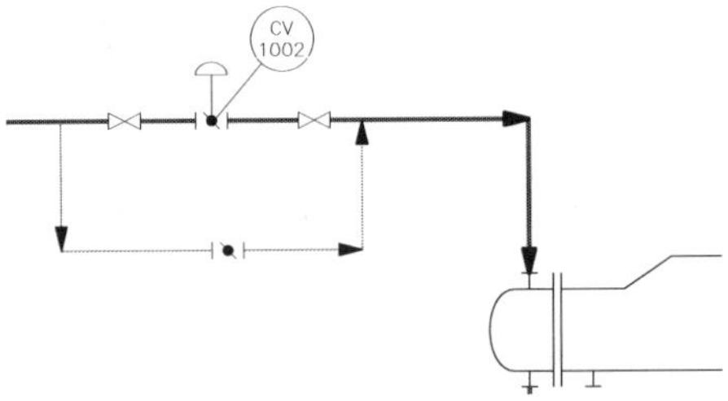

***Figure 2-64** Location of gate valves and the butterfly valve*

12. Next, create two secondary lines connecting the steam inlet pipe, as shown in Figure 2-65.

13. Choose **Plug** from the **Piping Fittings** area in the **Fittings** tab of the **TOOL PALETTES - P&ID PIP**, and then place it at the endpoints of the secondary lines, refer to Figure 2-65.

14. Next, delete the arrows on the secondary lines and place gate valves, as shown in Figure 2-66.

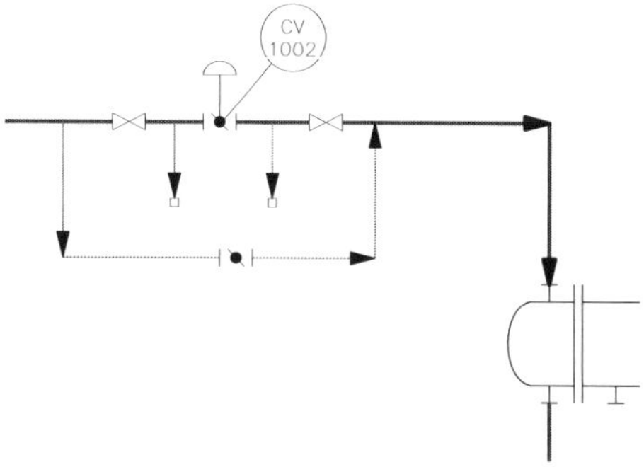

Figure 2-65 *Secondary lines with plugs placed at their endpoints*

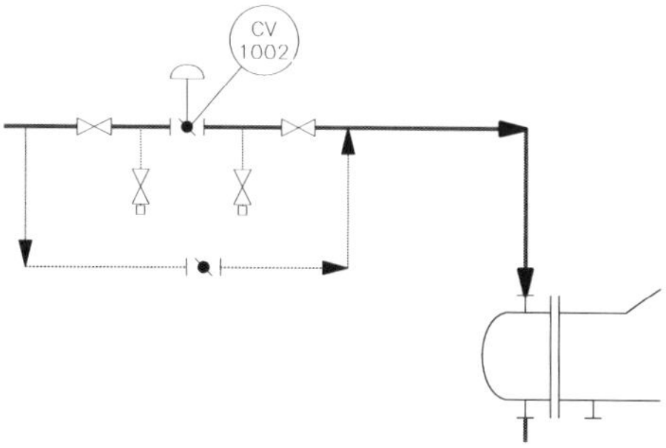

Figure 2-66 *Gate valves placed on the secondary lines*

15. Invoke the **PROPERTIES** palette of the gate valves by double-clicking on them.

16. Select the **Gate Valve Closed Style** option from the **Graphical style** drop-down list of the **Styles** area, refer to Figure 2-67; the gate valve is closed, as shown in Figure 2-68.

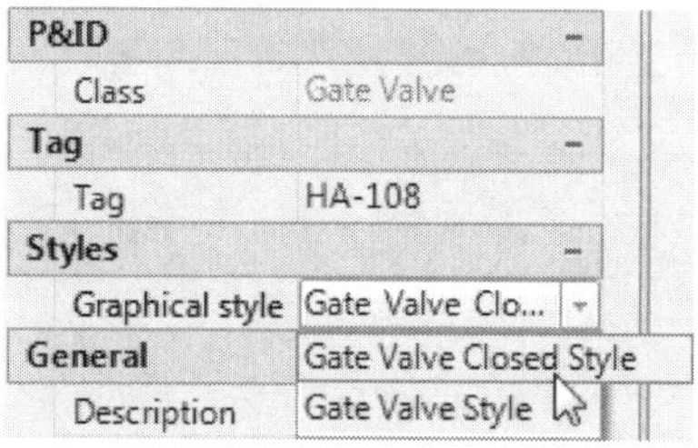

Figure 2-67 *Selecting the* ***Gate Valve Closed Style*** *option from the* ***Graphical style*** *drop-down list*

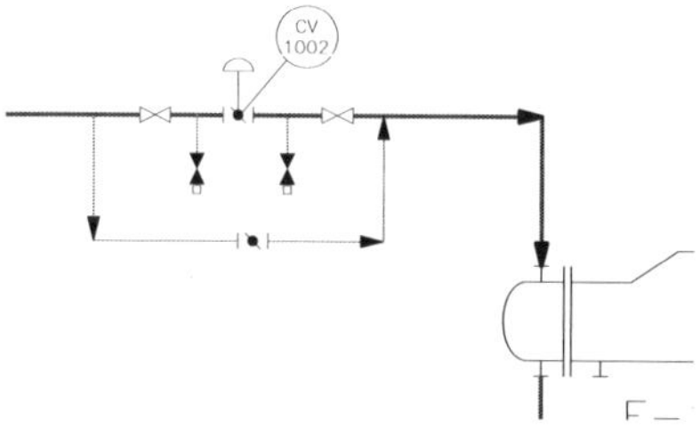

Figure 2-68 *The graphical style of gate valves changed to the closed type*

17. Similarly, create secondary lines and place the plugs and gate valves at the locations shown in Figure 2-69.

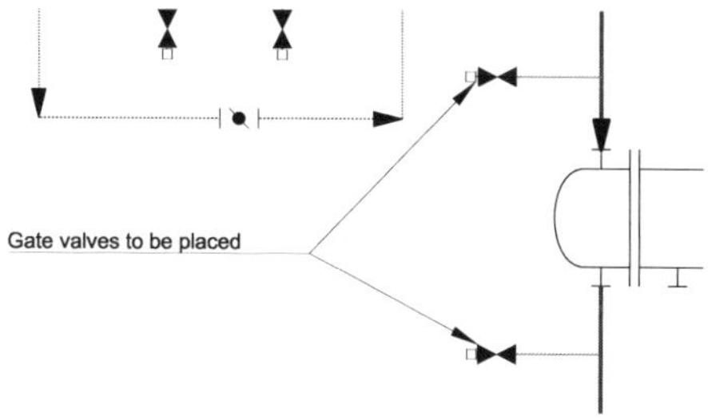

Figure 2-69 *Locations of the secondary lines and gate valves*

Note

If required, you can change the attachment point of the ***Plug*** *by choosing the* ***Point*** *option from the command prompt.*

Adding Instruments to the P&ID

Next, you need to add instruments to the P&ID.

1. Choose the **Restriction Orifice** tool from the **Primary Element Symbols (Flow)** area of the **Instruments** tab in the **TOOL PALETTES - P&ID PIP**; you are prompted to specify the insertion point.

2. Place the restriction orifice on the pipe which is connected to the vessel, as shown in Figure 2-70.

3. Enter the values given next in the **Assign Tag** dialog box and place the annotation near the orifice.

Area	**01**
Type	**FE-FLOW ELEMENT**
Loop Number	**1001**

4. Similarly, place another restriction orifice on the steam inlet pipe of the heat exchanger, refer to Figure 2-71. Next, enter the following values in the **Assign Tag** dialog box:

Area	**01**
Type	**FE-FLOW ELEMENT**
Loop Number	**1002**

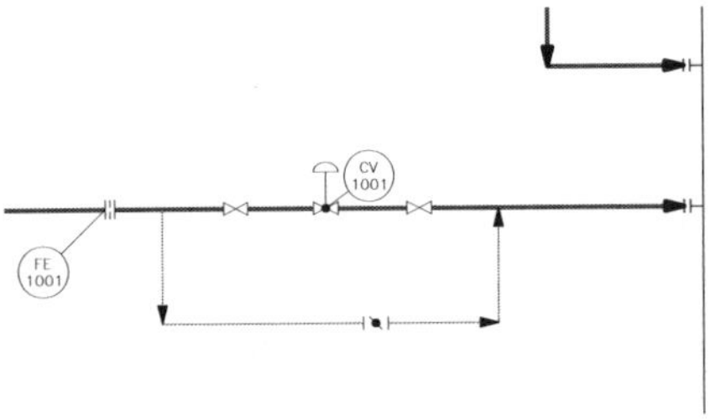

Figure 2-70 *Location of the orifice*

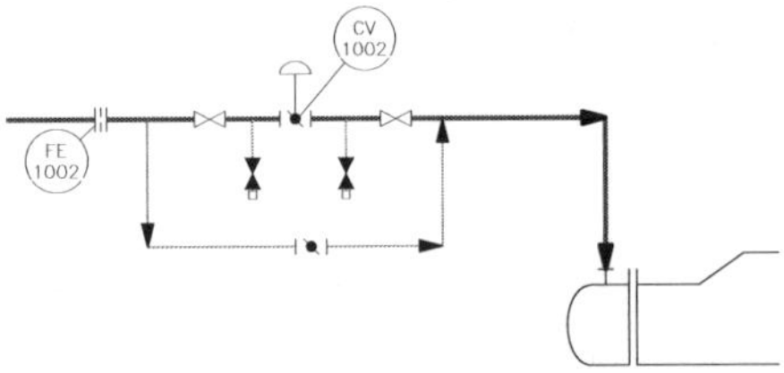

Figure 2-71 *Location of the orifice on the steam inlet line of the heat exchanger*

Next, you need to create instrumentation line connecting the orifice and the control valve.

5. Choose the **Electric Signal** tool from the **Instrument Lines** area in the **Lines** tab of the **TOOL PALETTES - P&ID PIP**.

6. Specify the start point on the orifice located on the pipe connected to the heat exchanger and move the cursor upward upto a distance of 1.5. Next, click to specify the endpoint.

You need to make sure that the **Object Snap** and **Object Snap Tracking** options are activated in the Status Bar.

7. Move the cursor toward right and click when the cursor snaps to the mid point of the control valve, as shown in Figure 2-72.

8. Next, move the cursor downward and specify the endpoint on the control valve, refer to Figure 2-73.

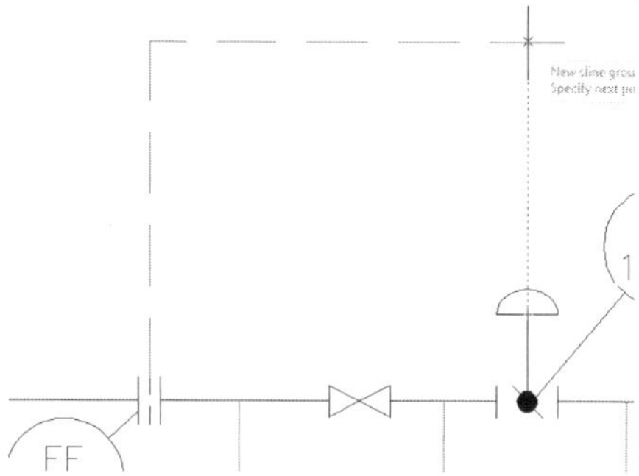

Figure 2-72 *Snapping to the midpoint of the control valve*

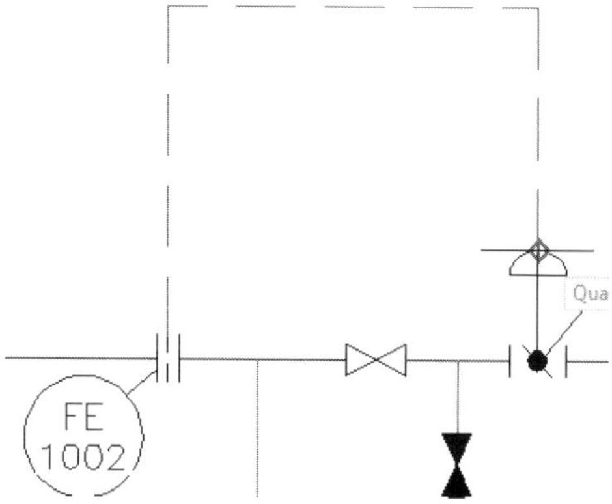

Figure 2-73 *Specifying the endpoint of the electric signal line*

9. Similarly, create an electrical signal on the pipe connecting the vessel, as shown in Figure 2-74.

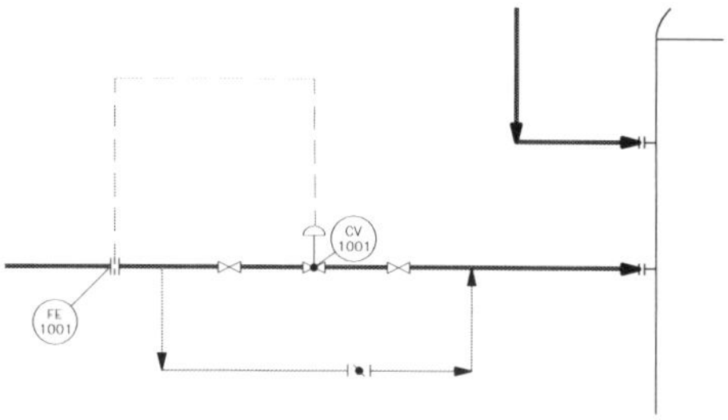

Figure 2-74 *Electric signal created on the feed line*

10. Choose **Primary Accessible DCS** from the **General Instruments** area of the **Instruments** tab in the **TOOL PALETTES - P&ID PIP**.

11. Place the instrument symbol on the electric signal line, as shown in Figure 2-75; the **Assign Tag** dialog box is displayed.

12. Enter the values given next in the **Assign Tag** dialog box and choose the **Assign** button.

Area	**01**
Type	**FC - FLOW CONTROLLER**
Loop Number	**1002**

13. Similarly, place another **Primary Accessible DCS** on the electric signal connected to the feed pipe, refer to Figure 2-76.

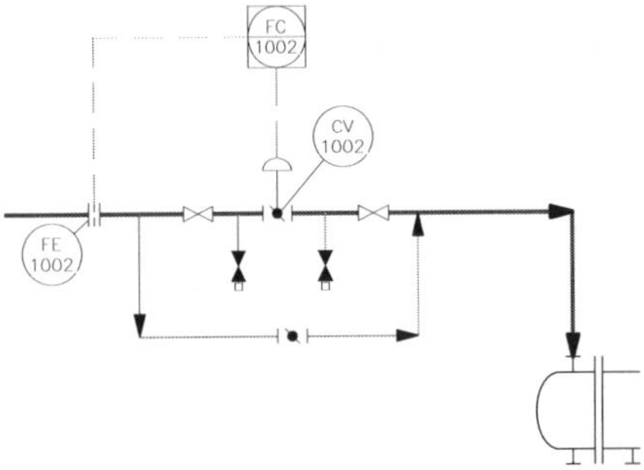

*Figure 2-75 Position of the **Primary Accessible DCS***

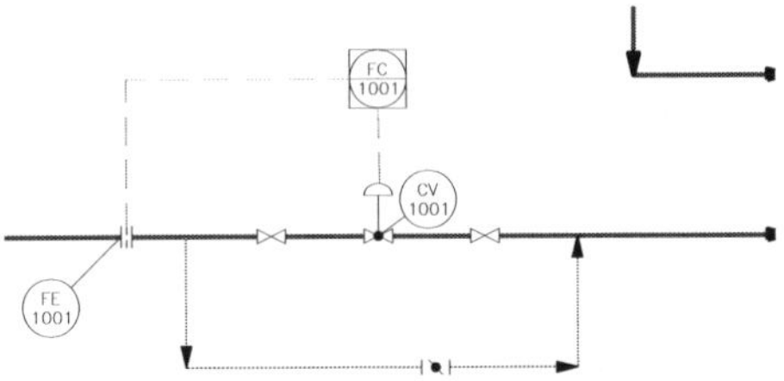

*Figure 2-76 Position of another **Primary Accessible DCS** on the electric signal line of the feed line*

14. Assign tag **01-FC-1001** to the instrument.

Placing Valves and Fittings on the Lines Connecting the Pumps

Next, you need to place valves and fittings on the lines connecting the pumps.

1. Choose **Eccentric Reducer** from the **Piping Fittings** area in the **Fittings** tab of the **TOOL PALETTES - P&ID PIP** and place it on the line connecting the horizontal nozzle of the pump, as shown in Figure 2-77.

2. Place a gate valve on the same line, refer to Figure 2-78.

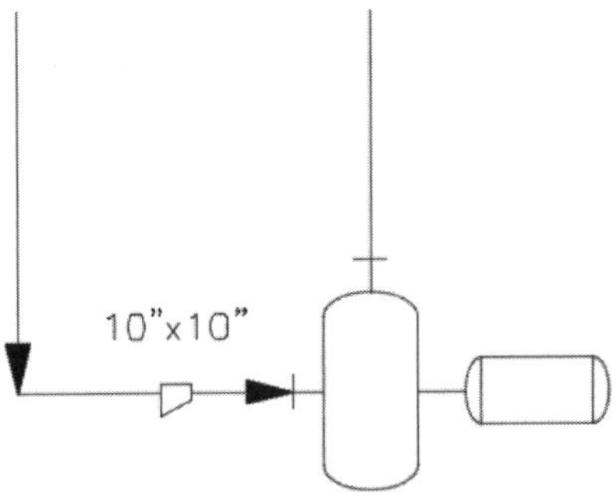

***Figure 2-77** An eccentric reducer placed on the line connecting the first pump*

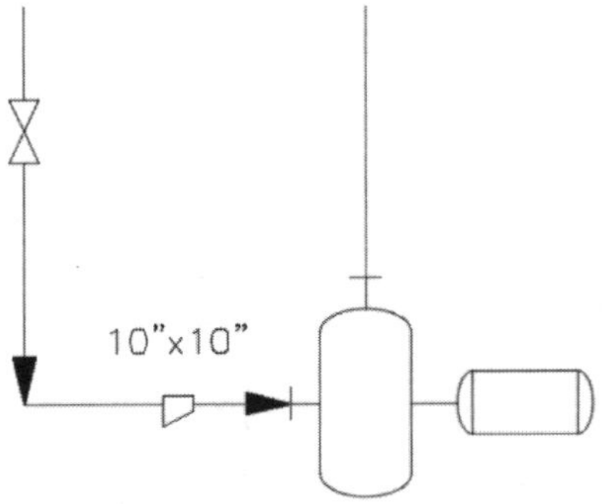

***Figure 2-78** A gate valve placed on the line connecting the first pump*

3. Choose the **Assign Tag** button from the **P&ID** panel and select the line connected to the reducing end of the eccentric reducer and press ENTER; the **Assign Tag** dialog box is invoked.

4. Modify the size of the line to **8"**, refer to Figure 2-79 and choose the **Assign** button from the dialog box.

5. Similarly, place an eccentric reducer and a gate valve on the pipe connecting the other pump, as shown in Figure 2-80.

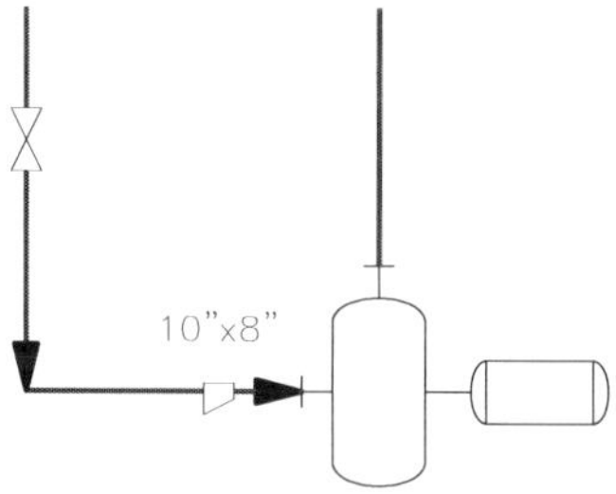

Figure 2-79 *Modified size of a line*

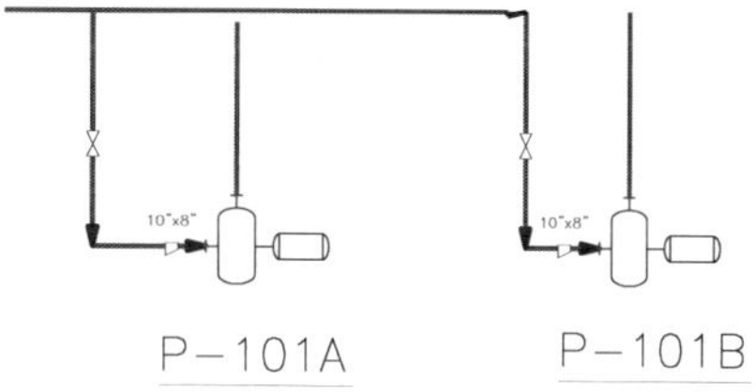

Figure 2-80 *A gate valve and an eccentric reducer placed on the line connecting the second pump*

6. Change the size of the pipe line connecting the reducing end of the eccentric reducer to 8".

7. Place Check Valves and Gate Valves on the outlet lines of the two pumps, refer to Figures 2-81 and 2-82.

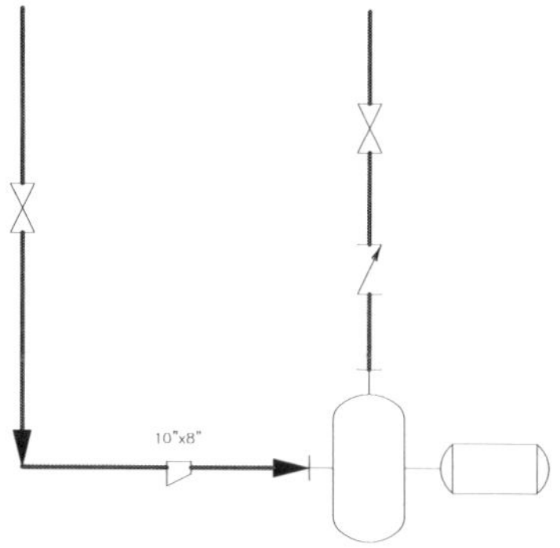

Figure 2-81 *Check valve and gate valve placed on the outlet line of the first pump*

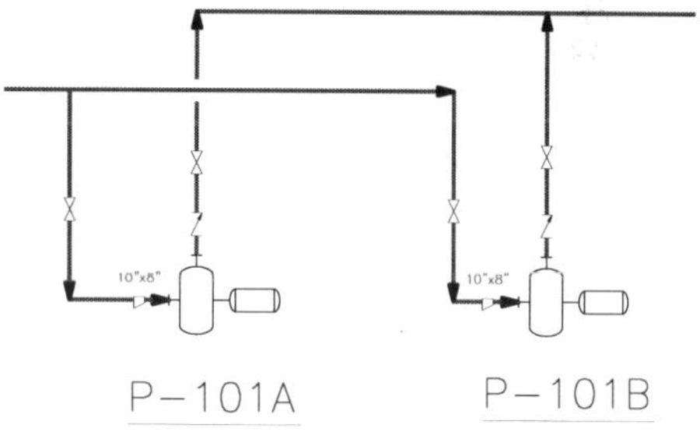

Figure 2-82 *Check valves and gate valves placed on outlet lines of both the pumps*

Next, you need to place a pressure relief valve on the pipe connecting the top nozzle of the vessel.

8. Choose **Pressure Relief Valve** from the **Relief Valves** area in the **Instruments** tab of the **TOOL PALETTES - P&ID PIP**.

9. Place the pressure relief valve on the secondary line of the line connecting the top nozzle of the vessel, as shown in Figure 2-83, and give the required parameter in the **Assign Tag** dialog box.

10. Place a gate valve on the same line, as shown in Figure 2-84.

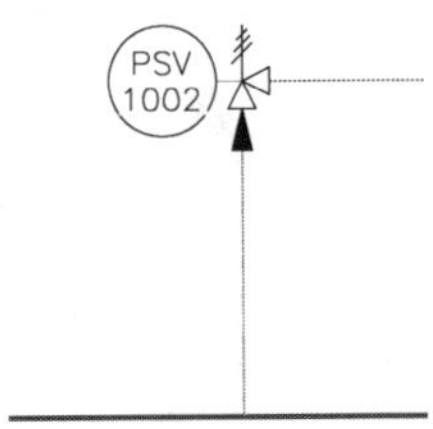

Figure 2-83 *Location of the pressure relief valve*

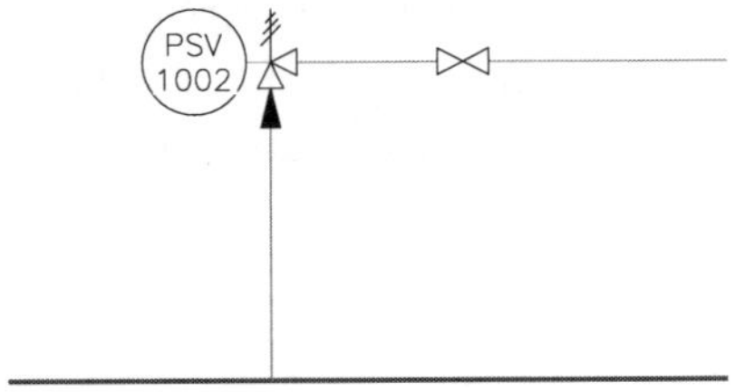

Figure 2-84 *Location of the gate valve*

Adding Off Page Connectors

1. Choose the **Off Page Connector** tool from the **Off Page Connectors and Tie-In Symbol** area of the **Non-engineering** tab of the **TOOL PALETTES - P&ID PIP**; you are prompted to specify the insertion point.

2. Place the off page connector at the endpoint of the line connecting the pumps, as shown in Figure 2-85.

3. Similarly, place off page connectors at the locations shown in Figures 2-86 and 2-87.

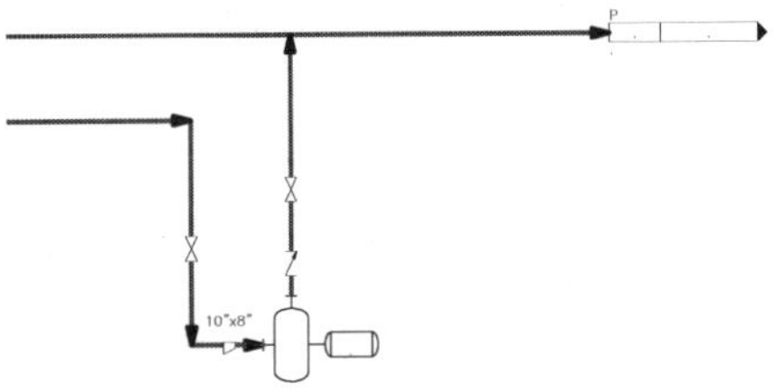

Figure 2-85 *Off page connector placed at the endpoint of the pump outlet line*

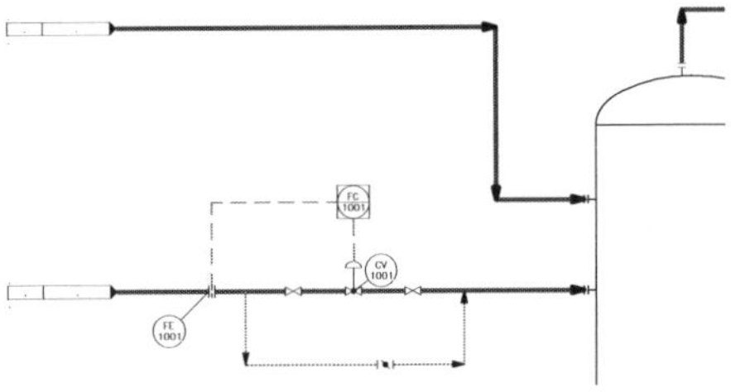

Figure 2-86 *Off page connectors placed at the endpoints of the feed line and reflux line*

4. Choose the **Utility Connector** tool from the **Off Page Connectors and Tie-In Symbol** area of the **Non-engineering** tab of the **TOOL PALETTES - P&ID PIP** and place then at the locations shown in Figures 2-87 and 2-88.

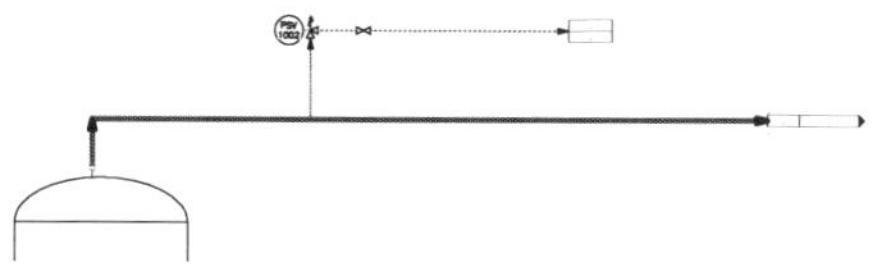

Figure 2-87 *Location of utility connector and off page connector*

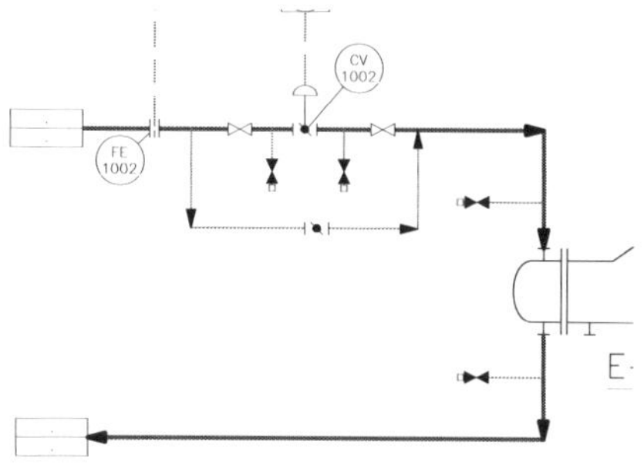

Figure 2-88 *Utility connectors placed at the endpoints of the steam inlet line and the condensate line*

Validating the Drawing

1. Choose the **Validate Config** button from the **Validate** panel in the **Home** tab; the **P&ID Validation Settings** dialog box is displayed.

2. In this dialog box, expand the **P&ID objects** node and clear the **Unresolved off-page connectors** check box. Similarly, expand the **3D Model to P&ID checks** node and clear all the check boxes under it. Next, choose the **OK** button to close the dialog box.

Note

In this tutorial, the off page connectors are not validated because they are not connected to any other P&ID.

3. Run validation process by choosing the **Run Validation** tool from the **Validate** panel; the **VALIDATION SUMMARY** palette is displayed with a tree list of errors in the drawing.

4. Right-click on any error in the **VALIDATION SUMMARY** palette and choose **Ignore** to ignore the error. Similarly, ignore all the errors and exit the **VALIDATION SUMMARY** palette.

Saving the Drawing

1. Choose the **Save** option from the **Application menu** or choose **File > Save** in the menu bar to save the drawing file *c02tut01.dwg*.

2. Choose **Close > Current Drawing** from the **Application Menu** to close the drawing file.

Self-Evaluation Test

Answer the following questions and then compare them to those given at the end of this chapter:

1. You can substitute a component by invoking the _________.

2. You need to use _________ to indicate the continuation of the process line from one drawing sheet to the another.

3. You can specify the control valve body and the actuator in _________.

4. To convert an AutoCAD component into a P&ID symbol, you need to choose the _________ option.

5. You should select the **P&ID Drawings** node in the Project tree and then choose the **New Drawing** button to create a new P&ID file. (T/F)

6. When you move a line, the equipment connected to the line also moves with the line. (T/F)

7. When you move a line, the inline components such as valves and balloons also move with the line. (T/F)

8. If the diameter of the line on either side of the reducer is changed, the reducer will be reoriented. (T/F)

9. You can add converted AutoCAD component to the **TOOL PALETTE - P&ID PIP** by clicking and dragging it into the **TOOL PALETTE**. (T/F)

10. A nozzle will be placed automatically when you connect a line to an equipment. (T/F)

Review Questions

Answer the following questions:

1. The _________ dialog box is displayed if you choose the **Control Valve** tool for the first time.

2. A project file is a _________ file.

3. You can edit a P&ID symbol using the _________ tool.

4. The _________ option in the **Edit** tool is used to attach a line to a component without physically connecting to it.

5. The __________ option in the **Edit** tool is used to divide a line into two.

6. The **Make Group** tool is used to group __________.

7. The __________ button in the **VALIDATION SUMMARY** palette is used to run validation on selected node.

8. You can place a symbol from the PID ISO Tool palette even while working in PIP standard file. (T/F)

9. The **PROJECT MANAGER** is used to manage the project files only. (T/F)

10. You can activate more than one project at a time.

EXERCISE

Exercise 1

In this exercise, you will create the P&ID shown in Figure 2-89.

(Expected time: 30 min)

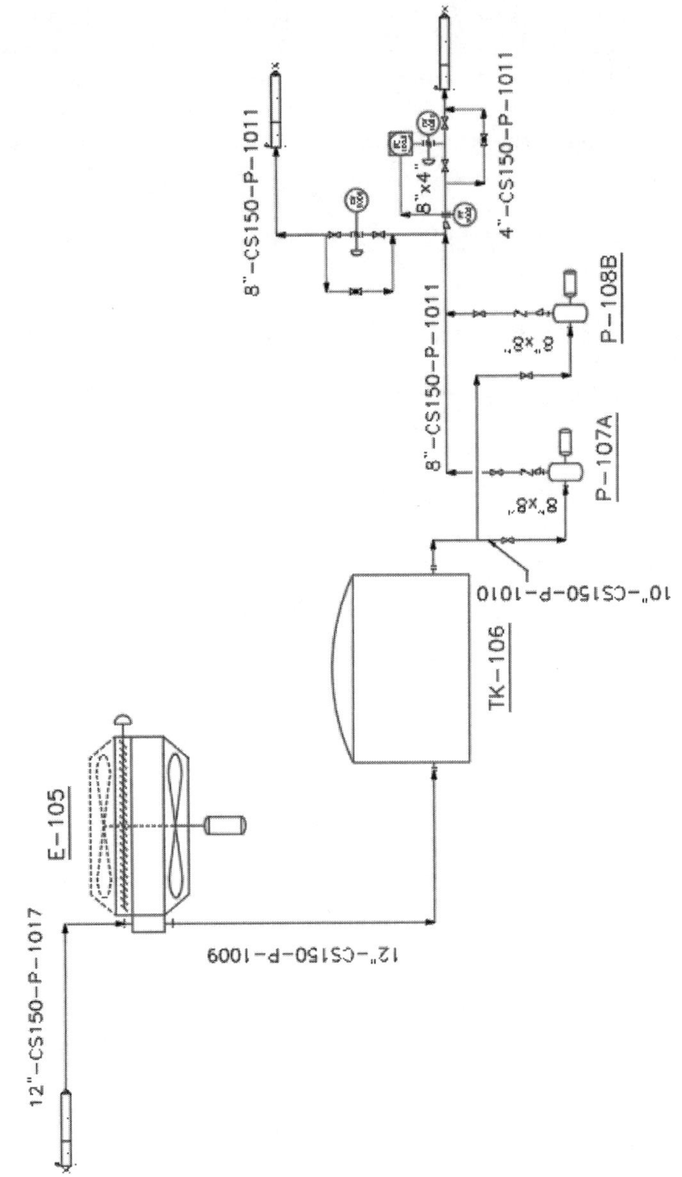

Figure 2-89 *P&ID for Exercise 1*

Answers to Self-Evaluation Test

1. Substitution palette, **2.** Off-Page Connector, **3. Control Valve Browser**, **4. Convert to P&ID object**, **5.** T, **6.** F, **7.** T, **8.** T, **9.** T, **10.** T

Chapter 3

Creating Structures

Learning Objectives

After completing this chapter, you will be able to:

- *Add members*
- *Configure member settings*
- *Add stairs*
- *Add ladders*
- *Add railings*
- *Add footings*
- *Modify the structural members*

INTRODUCTION

In this chapter, you will learn to create a structural layout that is an integral part of an industrial plant. To create a structural layout, first you need to create a new drawing file under the **Plant 3D Drawings** node using the **PROJECT MANAGER**. The procedure to create a new drawing has been discussed in Chapter 2. In this chapter, the tools used to create a structural layout will be discussed. These tools are available in the 3D Piping workspace.

Note

*If the file is opened in a workspace other than 3D Piping, you need to switch to this workspace. To do so, select the **3D Piping** option from the **Workspace** drop-down list in the **Quick Access** toolbar.*

CREATING A GRID

Ribbon:	Structure > Parts > Grid

Grid

A grid helps you to place structural components at appropriate location. To create a grid, choose the **Grid** tool from the **Parts** panel in the **Structure** tab; the **Create Grid** dialog box will be displayed, as shown in Figure 3-1.

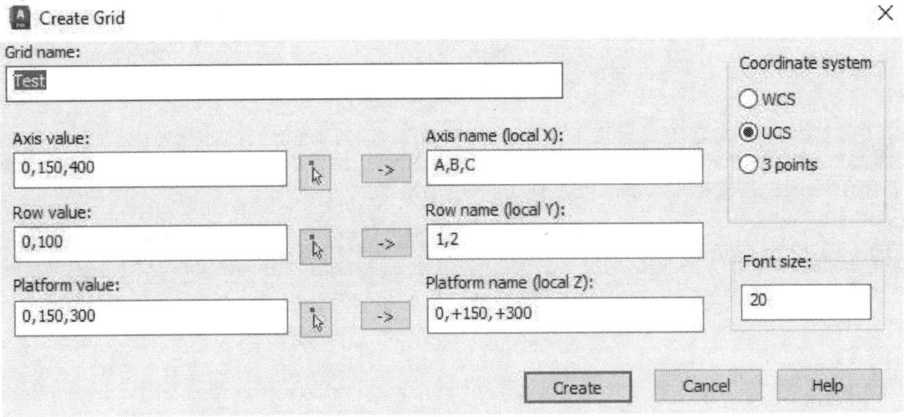

***Figure 3-1** The **Create Grid** dialog box*

In this dialog box, enter the name of the grid in the **Grid name** edit box. Next, specify the type of coordinate system to be used to set the orientation of the grid. You can specify the coordinate system by selecting any one of the radio buttons available in the **Coordinate system** area. The **WCS** radio button is used to define the coordinate system of the grid using the world coordinate system. The **UCS** radio button is used to define a user coordinate system. If you select the **3 points** radio button, you will be prompted to specify the origin point. After specifying the origin point, you need to specify one point on the X-axis, one point on XY plane, and then one point on the Z-axis. These three points specify the directions of three axes in a grid. After specifying the coordinate system, enter values for the axis points in the **Axis value** edit box. You can use @ symbol in order to enter a value relative to the preceding value. You can also use the pick point button adjacent to this edit box to directly specify the points in the workspace. Next, specify names of the axes in the **Axis name (local X)** edit box. Similarly, enter values for the row points in the **Row value** edit box. Also, enter the names of the row points in the **Row name (local Y)** edit box. Specify the platform value points (in Z-direction) and their names in the **Platform value** and **Platform name (local Z)** edit boxes, respectively. Next, enter the font size in the **Font Size** edit box for the names displayed in the grid. Choose the **Create** button; a grid will be created in the model space, refer to Figure 3-2.

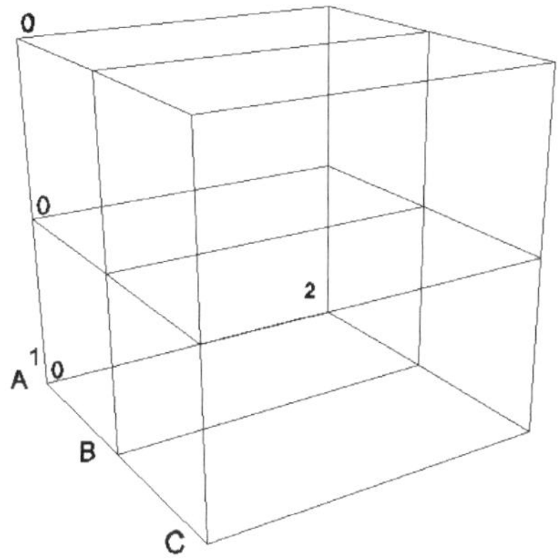

Figure 3-2 *A 3D grid created*

Editing Grids

To edit a grid, invoke the **Structure Edit** tool from the **Modify** panel of the **Structure** tab and select the grid; the **Edit Grid** dialog box will be displayed. The options in this dialog box are similar to the options in the **Create Grid** dialog box. Set the parameters in this dialog box as required and choose the **OK** button; the grid will be modified accordingly.

SETTING THE REPRESENTATION OF THE STRUCTURAL MEMBER

You can set the representation of the structural members as required by using the tools available in the **Line Model** drop-down of the **Parts** panel, refer to Figure 3-3. Different representations that you can set for the structural members are discussed next.

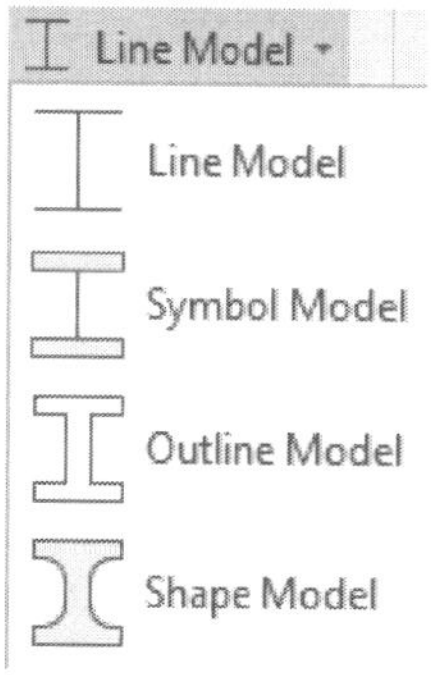

Figure 3-3 *The* ***Line Model*** *drop-down*

The **Line Model** is the default representation set for the member. In this mode, the members are represented as lines along with a symbol, as shown in Figure 3-4. This makes it easier to select the insertion point of a new member. In the **Symbol Model** mode, the members are represented as symbols, as shown in Figure 3-5. In the **Outline Model** mode, the members are represented as structural outlines, as shown in Figure 3-6. The **Shape Model** mode gives a real time look to the member with fillets added to the structural outline, as shown in Figure 3-7.

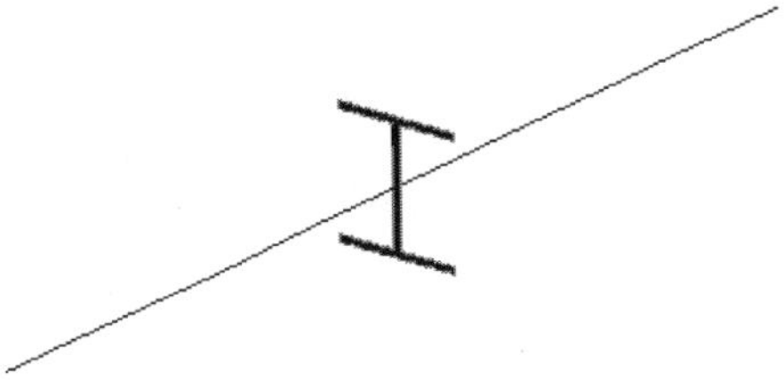

***Figure 3-4** The member in the **Line Model** mode*

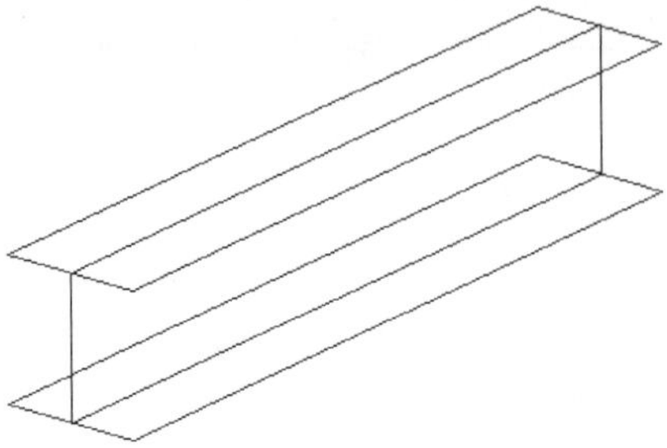

***Figure 3-5** The member in the **Symbol Model** mode*

***Figure 3-6** The member in the **Outline Model** mode*

***Figure 3-7** The member in the **Shape Model** mode*

ADDING MEMBERS

In AutoCAD Plant 3D, a column or a beam is referred to as a structural member. A structural member can be placed in the plant 3D layout by specifying its start point and endpoint. Before adding a structural member, you need to specify the member settings such as shape, size, orientation, material standard, and so on. To add a member, choose the **Member Settings** tool from the **Structural Settings** drop-down in the **Parts** panel of the **Structure** tab, refer to Figure 3-8; the **Member Settings** dialog box will be displayed, as shown in Figure 3-9.

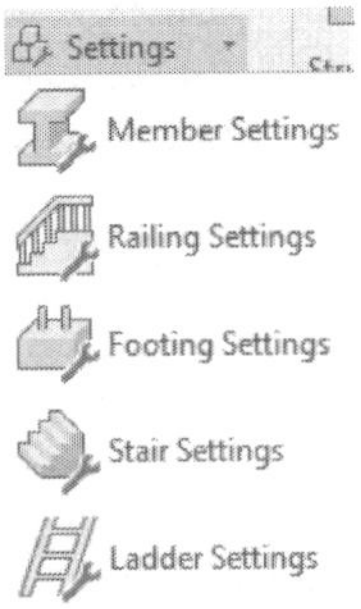

***Figure 3-8** The **Structural Settings** drop-down*

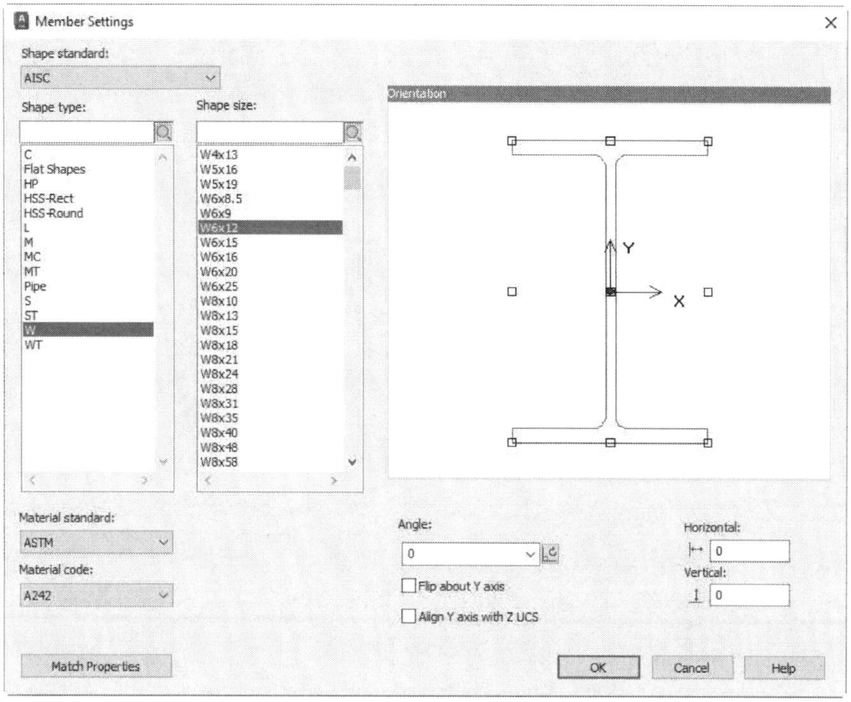

***Figure 3-9** The **Member Settings** dialog box*

In this dialog box, select the **AISC**, **CISC**, or **DIN** standard for the shape of a member from the **Shape standard** drop-down list. Next, select the shape type from the list box available in the **Shape type** area. You can search for a shape type using the search box available above the list box. After selecting the shape, you need to specify the size from the list box available in the **Shape size** area. You can search for a shape size using the search box available. Next, select the required material standard and material code from the respective drop-down lists. Specify the orientation angle using the **Angle** drop-down list. You can use the increment button available adjacent to this drop-down list to increase the angle by 90 degrees. Now, specify the justification of the member by selecting a point in the preview window. You can select the **Flip about Y axis** check box to flip the orientation about Y axis. Also, you can select the **Align Y Axis with Z UCS** check box to align the Y axis of the member with the Z UCS. After specifying all the parameters in the **Member Settings** dialog box, choose the **OK** button to set the specified member settings.

Next, choose the **Member** tool from the **Parts** panel; you will be prompted to specify the start point of the structural member. You can do so either by clicking in the drawing area or entering its coordinates at the command prompt. After specifying the start point, you will be prompted to specify the endpoint of the member. Specify the endpoint where you want to end the member. At this point, you may continue to specify the points or terminate member creation by pressing ESC.

Tip

*You can align the structural member with an already created line. To do so, enter L in the '**Specify start point of the structural member or [Line/Settings]:**' prompt in the command bar; you will be prompted to select a line to align a member with. Select a line from the drawing area and then press ENTER; the member will be aligned to the selected line.*

CREATING STAIRS

To create stairs, first you need to specify the settings. To do so, choose the **Stair Settings** tool from the **Structural Settings** drop-down; the **Stair Settings** dialog box will be displayed, as shown in Figure 3-10. In this dialog box, you can specify the settings for creating stairs.

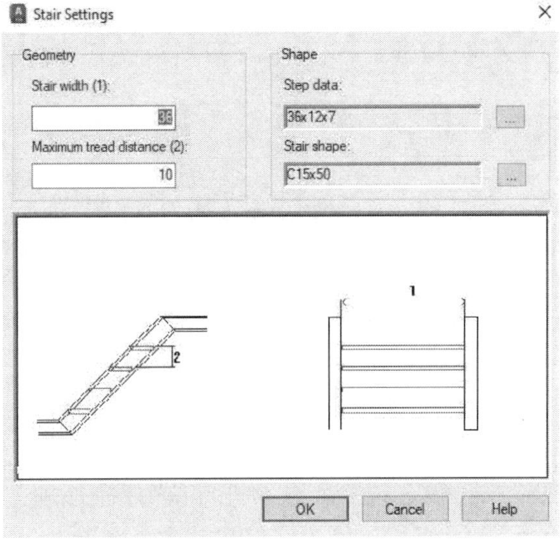

***Figure 3-10** The **Stair Settings** dialog box*

Specify the width of the stairs and the tread distance by entering desired values in the **Stair width (1)** and **Maximum tread distance (2)** edit boxes, respectively. Next, you need to specify the step shape. To do so, click on the Browse button adjacent to the **Step data** field; the **Select Step** dialog box will be displayed. In this dialog box, select the required tread standard from the **Tread standard** list box; the tread shapes under the selected standard will be displayed in the **Tread shape** list box. Select the required shape type from the **Tread shape** list box and then, choose the **OK** button; the **Select Step** dialog box will be closed. Next, you need to specify the stair shape. To do so, click on the Browse button adjacent to the **Stair shape** field; the **Select Stair Shape** dialog box will be displayed. The options in this dialog box are same as to that available in the **Member settings** dialog box. Specify the settings in this dialog box and choose the **Select** button to close this dialog box. Next, choose the **OK** button to close the **Stair Settings** dialog box.

After specifying the settings for the stairs, choose the **Stairs** tool from the **Parts** panel; you will be prompted to specify the first point. Select the bottom point for the stairs; you will be prompted to select the next point, refer to Figure 3-11. Specify the top point of the stairs and press ENTER; the stairs will be created, as shown in Figure 3-12.

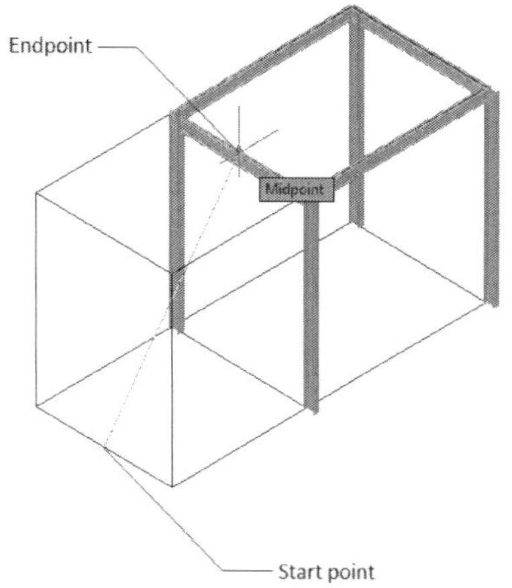

Figure 3-11 Selecting the start and end points of the stair

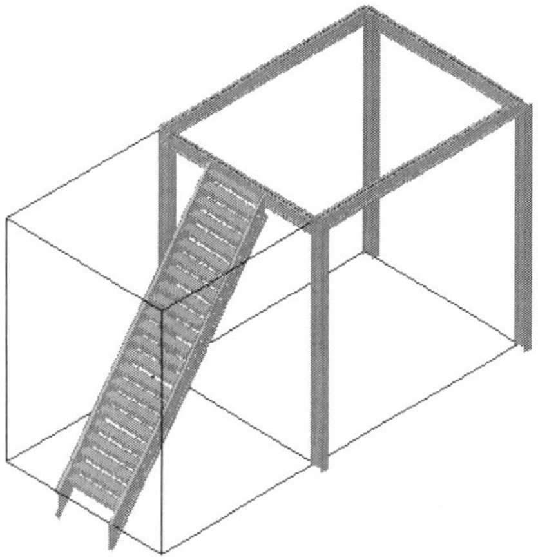

Figure 3-12 Stairs created from the selected points

Editing Stairs

To edit a stair, choose the **Structure Edit** tool from the **Modify** panel and then select the stair to be edited; the **Edit Stair** dialog box will be displayed. Modify the settings and choose the **OK** button to close the dialog box. You can also edit a stair using grips. To do so, select the stair to be edited; grips will be displayed on it, refer to Figure 3-13. You can use the center grips to change the location or length of the stairs and the side grips to increase or decrease the stair width.

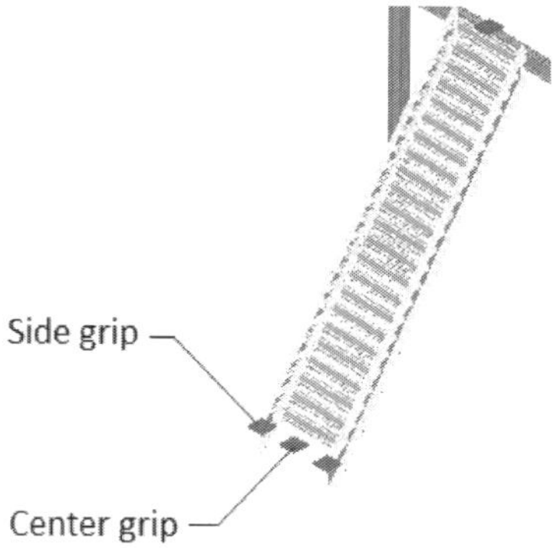

Figure 3-13 Stair with grips displayed on it

CREATING RAILINGS

To add railing to a stair, you need to specify parameters in the **Railing Settings** dialog box. To do so, choose the **Railing Settings** tool from the **Structural Settings** drop-down; the **Railing Settings** dialog box will be displayed, as shown in Figure 3-14.

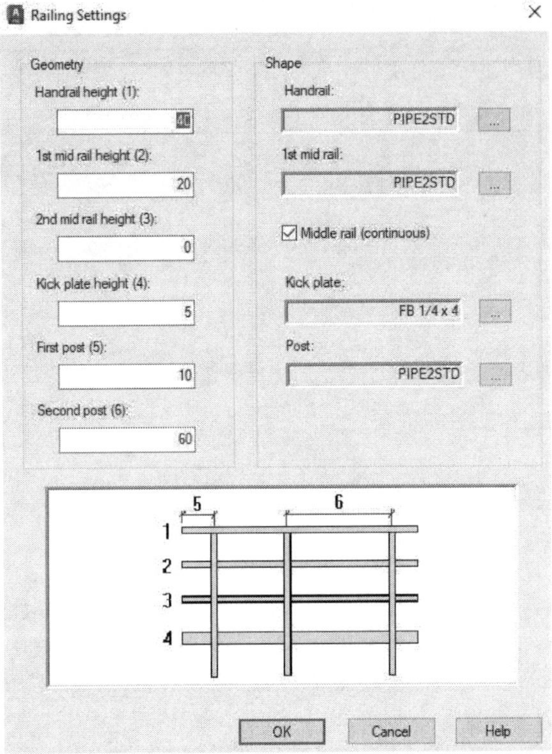

Figure 3-14 *The* ***Railing Settings*** *dialog box*

To specify different geometric values or parameters for the railing, enter desired values in the edit boxes available in the **Geometry** area in the dialog box. Refer to the preview window in Figure 3-14 for these parameters. Next, specify the shape properties by clicking on the Browse button available adjacent to the corresponding rails. Select the **Middle rail (continuous)** check box, if not selected by default, to make the mid rail continuous. Choose the **OK** button to close the dialog box.

After defining the settings for the railing, choose the **Railing** tool from the **Parts** panel; you will be prompted to specify the start point. Select the point on the stair from where you want the railing to start; you will be prompted to specify the end point of the railing. Specify the end point, you will notice that as soon as you specify the endpoint of the first railing, you will be prompted to select the next endpoint. Press ENTER to terminate the command.

You can also add a railing to the stairs or members by selecting them from the drawing area. To do so, choose the **Railing** tool and then enter **Object** at the command prompt; you will be prompted to select a stair or a structural member for applying the railing. Select the stair or a structural member from the drawing area; the railing will be added to the selected member. Figure 3-15 shows a railing added to the stairs and other structural members.

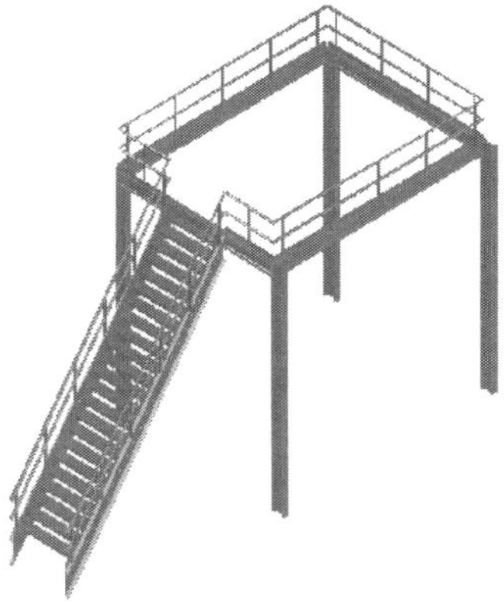

Figure 3-15 Railing added to the stairs and other structural members

CREATING LADDERS

To create a ladder, first specify the settings for the ladder. To do so, invoke the **Ladder Settings** dialog box by choosing the **Ladder Settings** tool from the **Structural Settings** drop-down. In the **Ladder Settings** dialog box, choose the **Ladder** tab, if not chosen by default, to display its options, as shown in Figure 3-16.

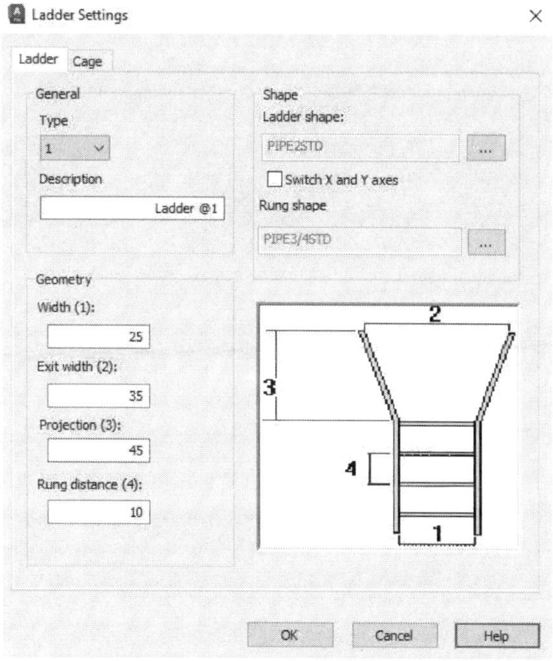

*Figure 3-16 The **Ladder Settings** dialog box*

Ladder Tab

The **Ladder** tab has three areas: **General**, **Geometry**, and **Shape**, refer to Figure 3-16. In the **General** area, you can specify the ladder type and description using the **Type** drop-down list and the **Description** edit box, respectively.

The options in the **Shape** area allow you to specify the settings for ladder shape and rung shape. To change the ladder shape settings, choose the Browse button adjacent to the **Ladder Shape** edit box; the **Select Ladder Shape** dialog box will be displayed. Modify the settings and choose the **Select** button to close the dialog box. Also, you can select the **Switch X and Y axes** check box to rotate the ladder shape by 90 degrees. Similarly, you can change the rung shape settings by invoking the **Select Rung Shape** dialog box.

The **Geometry** area contains options to define the geometry of the ladder. The preview window explains the use of the options available in this area.

Cage Tab

The **Cage** tab in the **Ladder Settings** dialog box contains the **General** and **Geometry** areas. You can see a preview of the ladder geometry in the preview area, refer to Figure 3-17.

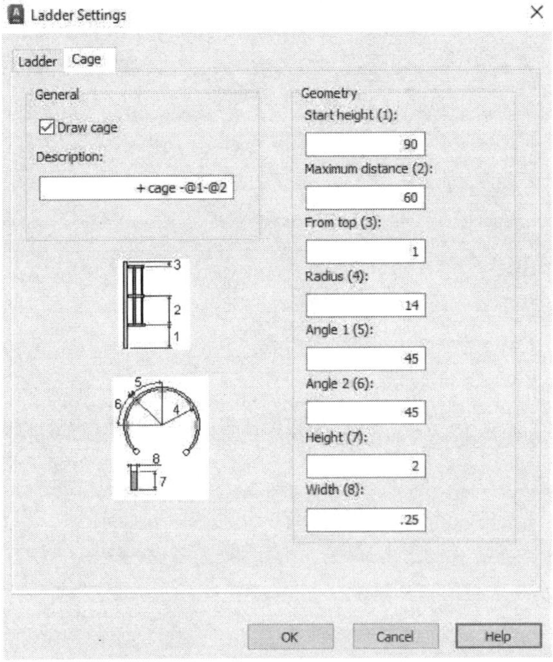

*Figure 3-17 The **Cage** tab of the **Ladder Settings** dialog box*

In the **General** area, select the **Draw cage** check box to create a ladder with a cage. You can enter the description of the cage in the **Description** edit box.

The **Geometry** area contains options to define the geometry of the cage. The preview window explains the use of the options available in this area. After specifying all the values, choose the **OK** button; the **Ladder Settings** dialog box will be closed.

After specifying the settings for the ladder, choose the **Ladder** tool from the **Parts** panel; you will be prompted to specify the start point. Select the point from where you want to start the ladder; you will be prompted to specify the end point. Select the point upto which you want the ladder to be extended; you will be prompted to specify the directional distance. Move the cursor in the perpendicular direction and enter a distance at the command prompt and press ENTER; a ladder will be created.

CREATING A PLATE/GRATE

Ribbon: Structure > Parts > Plate

To create a plate or a grate, choose the **Plate** tool from the **Parts** panel; the **Create Plate/Grate** dialog box will be displayed, as shown in Figure 3-18.

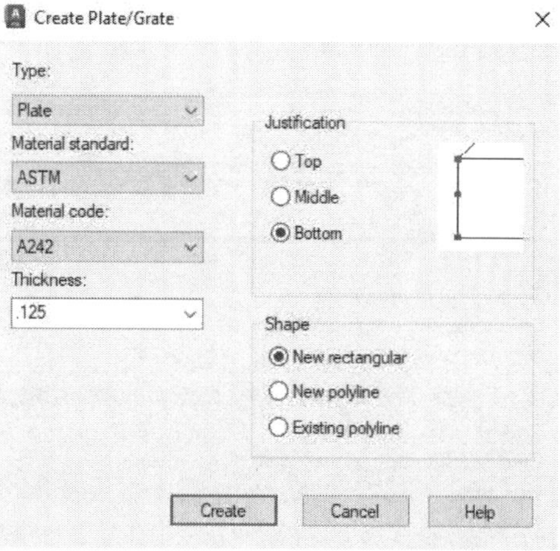

***Figure 3-18** The **Create Plate/Grate** dialog box*

Use the **Type** drop-down to specify whether you want to create a grate or a plate. Next, specify the material of the plate/grating by selecting the required material standard and material code from the respective drop-down lists. After specifying the material, select the plate/grating thickness from the **Thickness** drop-down list. If you have selected the **Grating** option from the **Type** drop-down list, you need to select the required hatch pattern and hatch scale of the grating from the **Hatch pattern** and **Hatch scale** drop-down lists, respectively.

Next, you need to specify the justification of a grating or plate using the options in the **Justification** area. In this area, three radio buttons **Top**, **Middle**, and **Bottom** are available. Select an appropriate radio button to specify the justification of the grating or the plate. Now, select an option from the **Shape** area. In this area, three radio buttons, **New rectangular**, **New polyline**, and **Existing polyline** are available. The **New rectangular** option is used to create a rectangular shaped plate/grating. The **New polyline** option is used to specify the shape by drawing a closed sketch using the **Polyline** tool. The **Existing polyline** option is used to select an already existing polyline sketch to create the plate/grating.

After specifying the options in the **Create Plate/Grate** dialog box, choose the **Create** button; a prompt will be displayed at command prompt depending upon the option selected in the

Shape area. For example, if you select the **New rectangular** option, you will be prompted to specify the first corner point of the plate. Specify the first corner and then the second corner to create a plate or a grating.

CREATING FOOTINGS

Ribbon:	Structure > Parts > Footing

Footing is one of the important components in a structural model. The function of footing in an industrial project is to provide support to the entire structural layout. To create a footing, first define the settings for it. To do so, invoke the **Footing Settings** dialog box by choosing the **Footing Settings** tool from the **Structural Settings** drop-down list. The **Footing Settings** dialog box has two areas: **Geometry** and **Material**, and a preview window, as shown in Figure 3-19.

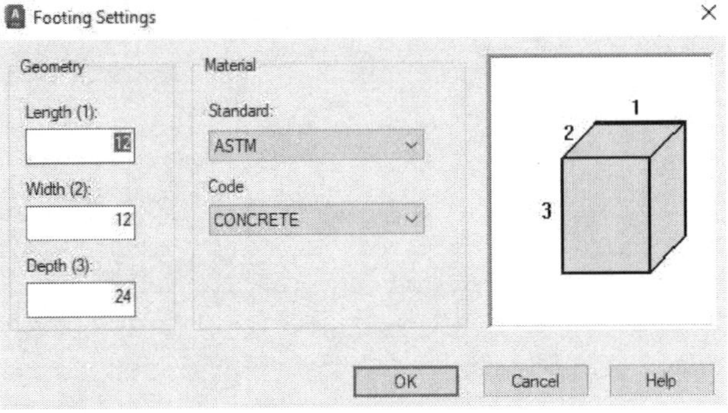

***Figure 3-19** The **Footing Settings** dialog box*

The **Geometry** area contains options to define the geometry of the footing. The preview window displays different parameters of the **Geometry** area.

The options in the **Material** area allow you to specify the material standard and code for the footing. After specifying all the values, choose the **OK** button; the **Footing Settings** dialog box will be closed.

Next, choose the **Footing** tool from the **Parts** panel and specify a point to insert the footing.

EDITING THE STRUCTURAL MEMBERS

In the design process, you may need to edit the structural members. Various editing operations are discussed next.

Changing the Length of a Member

Ribbon:	Structure > Cutting > Lengthen Member

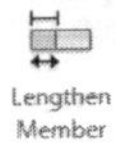

You can change the length of a structural member. To do so, choose the **Lengthen Member** tool from the **Cutting** panel; you will be prompted to select a member and the following prompt sequence will be displayed at the command prompt.

Command: PLANTSTEELLENGTHEN
Select a structural member or [Delta/Total]: *Select a member or enter an option*
The options in the command prompt are discussed next.

Delta

This option allows you to enter a value of the length that is to be added to the existing length of a member. When you choose this option from the command prompt, you will be prompted to specify a delta length. Enter a value and select a member; the length of the selected member will be changed. If you have specified a positive value, the length of the member will be increased. The length of the member will be decreased if you have entered a negative value.

Total

This option allows you to specify a value for the total length of a member.

Restoring the Member to its Original Length

Ribbon: Structure > Cutting > Restore Member

To restore the member to its original length, choose the **Restore Member** tool from the **Cutting** panel; you will be prompted to select the member. Select the member to restore it to its original length; the selected member will be restored to its original length.

Cutting Member at Intersections

Ribbon: Structure > Cutting > Cut Back Member

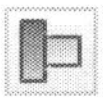

When two structural members intersect, you can remove the unwanted portion at the intersection using the **Cut Back Member** tool. To do so, invoke this tool from the **Cutting** panel; you will be prompted to select the limiting member. Also, the **cut Both** and **Gap** options will be displayed at the command prompt. The **cut Both** option is used to cut both the intersecting members. The **Gap** option is used to create a gap at the intersection point. Note that you need to specify the gap value at the command prompt. Select the limiting member. The limiting member will act as a boundary while trimming the unwanted portion of the other intersecting structural member. Next, you will be prompted to select the structural member to cut. Select the member from which you want to remove the unwanted material. You will notice that the member selected second is trimmed upto the limiting member. Figure 3-20 shows a cut created using the **cut Both** option. Figure 3-21 shows a cut created by specifying a gap distance.

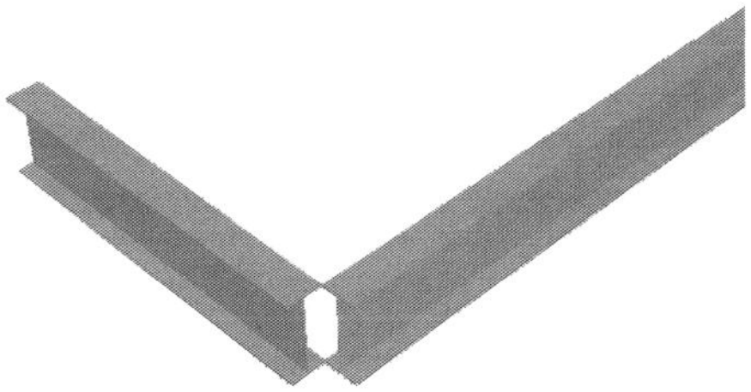

Figure 3-20 *A cut created using the* ***cut Both*** *option*

Figure 3-21 *A cut created by specifying the gap value*

Creating Miter Joints

Ribbon: Structure > Cutting > Miter Cut Member

A miter joint is created at the corner of two structural members by beveling them at an angle of 45-degrees. To create a miter joint between two structural members, choose the **Miter Cut Member** tool from the **Cutting** panel; you will be prompted to select the first structural member. Also, the **Align edges** and **Gap** options will be displayed at the command prompt. The **Align edges** option allows you to create a miter joint with the edges of both the members aligned, as shown in Figure 3-22. The **Gap** option is used to create a miter joint with a gap between the edges, as shown in Figure 3-23. After choosing the required option, select the two intersecting members; the miter joint will be created. Next, press ESC to exit the tool.

Figure 3-22 *Miter joint with aligned edges*

Figure 3-23 *Miter joint with a gap*

Trimming/Extending a Member

Ribbon: Structure > Cutting > Trim Member/Extend Member

You can trim or extend members upto a specified plane. To do so, choose the **Trim Member** tool from the **Cutting** panel; the **Trim to Plane** dialog box will be displayed, as shown in Figure 3-24. The options in this dialog box are discussed next.

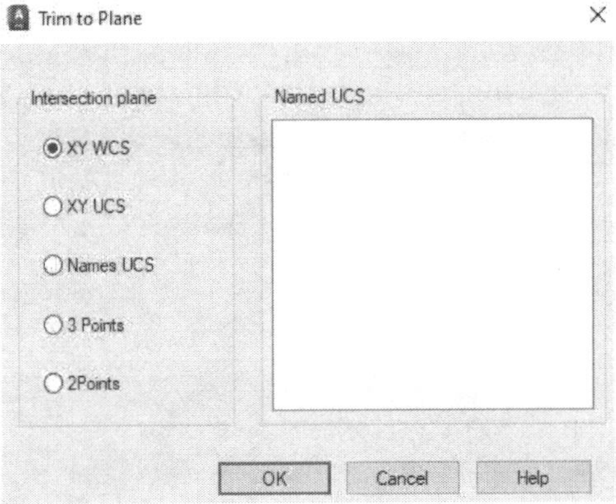

Figure 3-24 *The **Trim to Plane** dialog box*

Intersection plane Area

The options in this area are used to specify the plane about which the member is to be trimmed.

XY WCS
This radio button is selected by default. As a result, the XY plane of the World Coordinate System is selected as the cutting plane.

XY UCS
Select this radio button to specify XY plane of the User Coordinate System as the cutting plane.

Names UCS
Select this radio button to specify XY plane of the named User Coordinate System as the cutting plane.

3 Points
Select this radio button to specify XY plane of new coordinate system created by specifying 3 points.

2 Points
Select this radio button to create a cutting plane by drawing a line. The cutting plane is a vector of the line drawn.

Named UCS Display Box
This box displays the Named User Coordinate System.

Choose the **OK** button after selecting the required option and select a member that is to be trimmed; the selected member will be trimmed using the defined plane.

You can also extend a structural member. To do so, choose the **Extend Member** tool from the **Cutting** panel in the **Structure** tab; the **Extend to Plane** dialog box will be displayed. The options in this dialog box are same as in the **Trim to Plane** dialog box. Select an option from the **Intersection plane** area to define the plane upto which the member should be extended, and then choose the **OK** button; the dialog box will be closed and you will be prompted to select a member to extend. Select a structural member from the drawing area; it will extend upto the specified plane.

Exploding a Structure

Ribbon: Structure > Modify > Structure Explode

You can explode structural objects into individual elements. Note that you can only explode stairs, railings, or ladders. To explode an object, choose the **Structure Explode** tool from the **Modify** panel; you will be prompted to select a structural object. Select a stair, railing, or ladder; the selected object will be exploded into individual elements that can be modified individually.

VISIBILITY OPTIONS
When you create a plant layout, whether it is large or small, you may need to toggle the visibility of various objects in the layout. You can do so by hiding the components at any stage. The method to toggle the visibility of an object is discussed next.

Hiding and Displaying Components

To hide a component placed in a plant layout, select the component from the graphics area and choose the **Hide Selected** tool from the **Visibility** panel in the **Structure** tab; the display of the component will be turned off. To hide all other components except the selected component, choose the **Hide Others** tool from the **Visibility** panel. On doing so, the selected component will be displayed while all other components will be hidden.

To show all the hidden components, choose the **Show All** button; all the hidden components will be displayed again in the layout.

EXCHANGING DATA WITH OTHER APPLICATIONS

Ribbon: Structure > Export > Advance Steel XML Export

The **Advance Steel XML Export** tool allows the exchange of steel structure data between two applications. To create an Advance Steel XML report, choose the **Advance Steel XML Export** tool from the **Export** panel; the **Advance Steel XML Export** dialog box will be displayed, as shown in Figure 3-25. Specify the path for the output file by using the Browse button available at the right side of the **Output file** edit box in this dialog box.

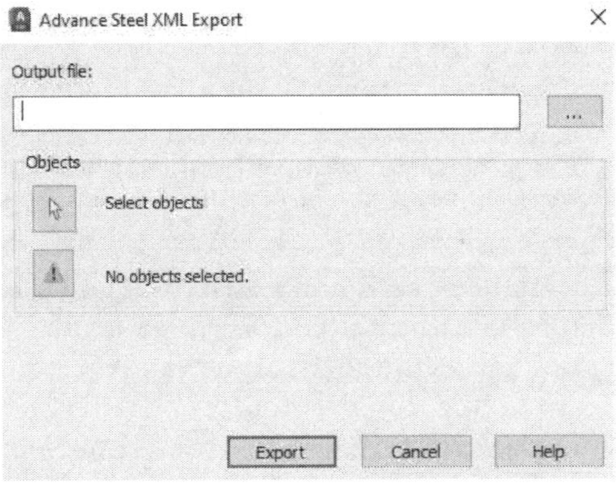

***Figure 3-25** The **Advance Steel XML Export** dialog box*

The **Select objects** button in the **Objects** area is used to select objects from the workspace. The selection status will be displayed in the **Objects** area. After selecting the objects, choose the **Export** button to export the structure to .smlx format.

TUTORIALS

Tutorial 1

In this tutorial, you will open the CADCIM project created in Chapter 2 and then create a new AutoCAD Plant 3D file. In this file, you will create a pipe rack, as shown in Figure 3-26. You need to use W10 x 12 structural members to create this pipe rack. The dimensions for the model are given in Figure 3-27. **(Expected time: 45 min)**

Note

You can also download the CADCIM project from www.cadcim.com by following the path: Textbooks > CAD/CAM > AutoCAD Plant 3D > AutoCAD Plant 3D 2023 for Designers > Input Files.

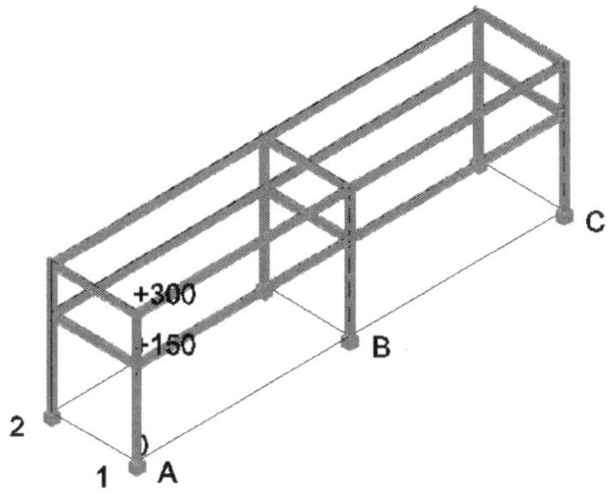

Figure 3-26 *Model for Tutorial 1*

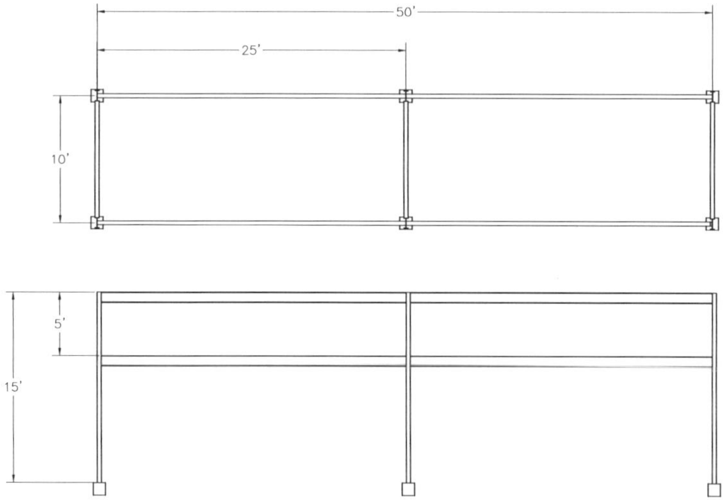

Figure 3-27 *Dimensions for the model*

Footing Dimensions:

Length= 24 inch
Width = 24 inch
Depth = 24 inch

The following steps are required to complete this tutorial:

a. Open a new Plant 3D drawing file in the current project.
b. Create a grid by specifying the axis, row, and platform values.
c. Add footings at the bottom of the grid.
d. Add columns and beams to the model.
e. Cut members at intersections.

Opening a New Plant 3D File

1. Double-click on the **AutoCAD Plant 3D 2023 - English** icon from the desktop of your computer to start AutoCAD Plant 3D 2023.

   Next, you need to create a new AutoCAD Plant 3D file.

2. Select the **CADCIM** project that you have created in Tutorial 1 of Chapter 2 from the drop-down list in the **Current Project** area of the **PROJECT MANAGER** palette.

3. Select the **Plant 3D Drawings** node in the **Project** area and choose the **New Drawing** button; the **New DWG** dialog box is displayed.

4. Enter **Pipe_rack.dwg** in the **File name** edit box and choose the **OK** button; the new file is created.

5. Select the **3D Piping** option from the **Workspace** drop-down list located in the **Quick Access Toolbar**.

Creating Grid

1. Choose the **Grid** tool from the **Parts** panel in the **Structure** tab; the **Create Grid** dialog box is displayed.

Grid

2. In this dialog box, specify the parameters as given below and retain the default settings for other parameters.

   Axis value: 0, 25', 50' **Row value: 0, 10'**
   Platform value: 0, 10', 15'

3. Next, choose the **Create** button from the dialog box; the grid is created, as shown in Figure 3-28. Note that the view orientation in the figure is set to **SW Isometric**. You can set the view orientation to **SW Isometric** by selecting the **SW Isometric** option from the **3D Navigation** drop-down list in the **View** panel of the **Structure** tab.

Creating Footings

1. Choose the **Footing Settings** tool from the **Structural Settings** drop-down in the **Parts** panel; the **Footing Settings** dialog box is displayed.

2. Enter **24** in the **Length(1)**, **Width(2)**, and **Depth(3)** edit boxes. Next, accept the default values in the **Material** area and choose the **OK** button; the settings of the footings are changed and the **Footing Settings** dialog box is closed.

3. Choose the **Footing** tool from the **Parts** panel; you are prompted to select the insertion point. Place all the footings at the bottom grid points, as shown in Figure 3-29.

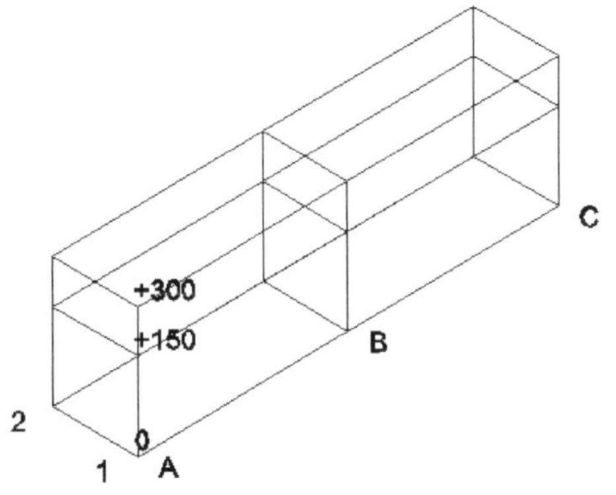

Figure 3-28 *Model after creating the grid*

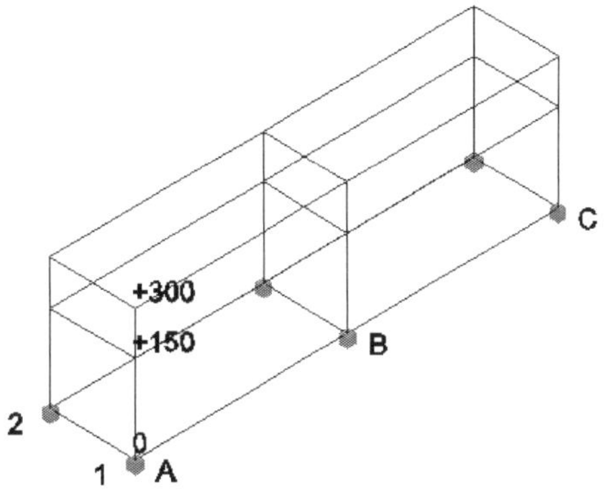

Figure 3-29 *Model after adding footings*

Note

After placing the first footing at the required point, you can copy it using the ***Copy*** *tool and place it at other desired points.*

Creating Columns and Beams

To create structural members, first you need to set the properties of the member. Also, you need to turn on the object snap and 3D object snap in order to make the selection of grid points easier.

1. Choose the **Member Settings** tool from the **Structural Settings** drop-down in the **Parts** panel; the **Member Settings** dialog box is displayed.

2. Select **W** and **W 10x12** from the **Shape type** and **Shape size** list boxes, respectively. Next, accept the default values of the other options and choose the **OK** button; the settings of the structural member are changed and the **Member Settings** dialog box is closed.

3. Choose the **Member** tool from the **Parts** panel; you are prompted to select the start point of the member.

   Make sure that the **Ortho Mode** is turned on in the Status Bar.

4. Select the grid point, as shown in Figure 3-30; you are prompted to specify the endpoint.

5. Move the cursor vertically upward and snap to the grid point, refer to Figure 3-30. Next, click to select it; the column is created between the

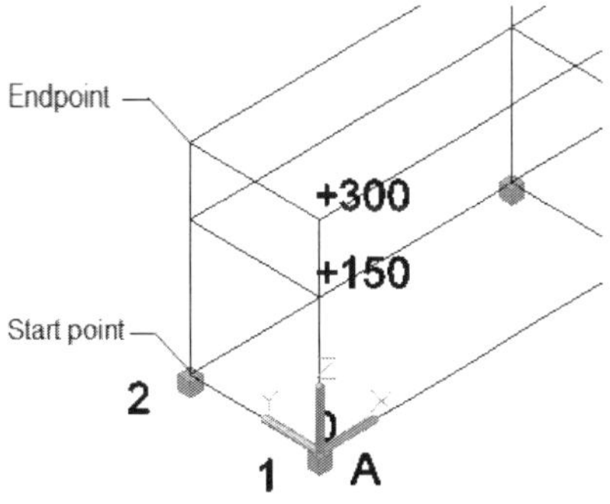

Figure 3-30 *Selecting the start point and the endpoint of the column*

two selected points. Similarly, place columns on other footings, as shown in Figure 3-31. You can also use the **Copy** command to copy the created column to the required locations.

Next, you need to add beams. But before that, you need to change the justification of the member to top.

6. Choose the **Member Settings** tool from the **Structural Settings** drop-down in the **Parts** panel; the **Member Settings** dialog box is displayed.

7. In the **Member Settings** dialog box, choose the top center justification point in the **Orientation** window. Next, choose the **OK** button to close the dialog box.

8. Invoke the **Member** tool; you are prompted to select the start point. Select the start point and the end point, as shown in Figure 3-32; a beam is created between the two selected points and a rubber-band line is displayed between the cursor and the specified point. Also, you are prompted to specify the end point of the new member. Press ENTER to exit the tool. Similarly, create rest of the beams, as shown in Figure 3-33.

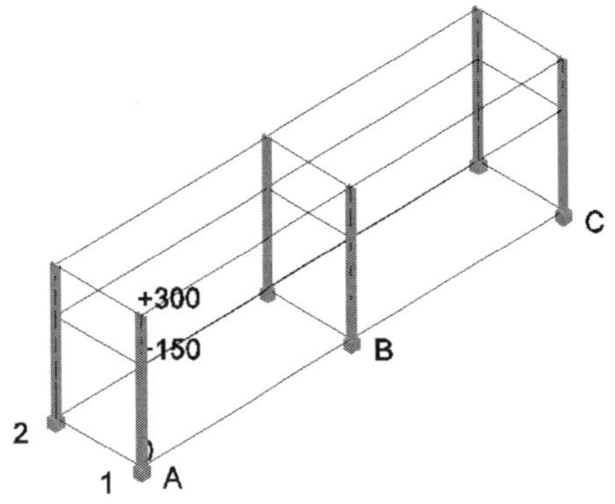

Figure 3-31 *Model after creating columns*

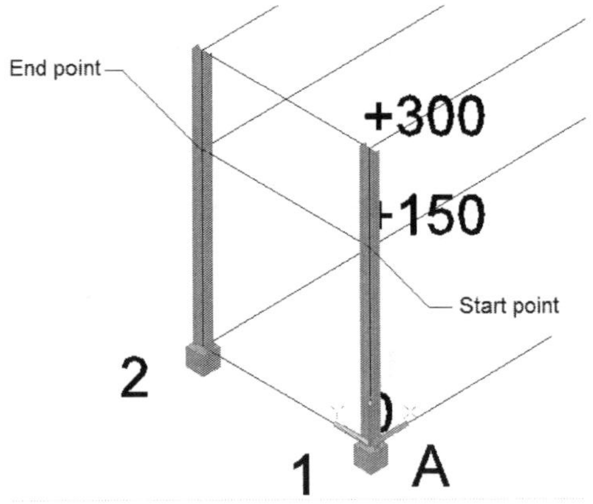

Figure 3-32 *The start point and end point of the beam*

Note

You can hide the columns such that only the grid points are visible while specifying the start and endpoints of the beam.

9. Similarly, create the second level. The model after placing all the beams is shown in Figure 3-34.

Tip

You can select all the beams at the first level and then use the ***COPY*** *command to create the second level.*

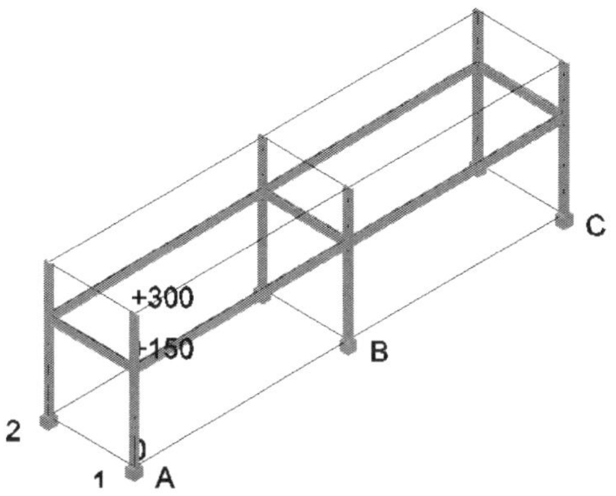

Figure 3-33 *Model after creating the lower level*

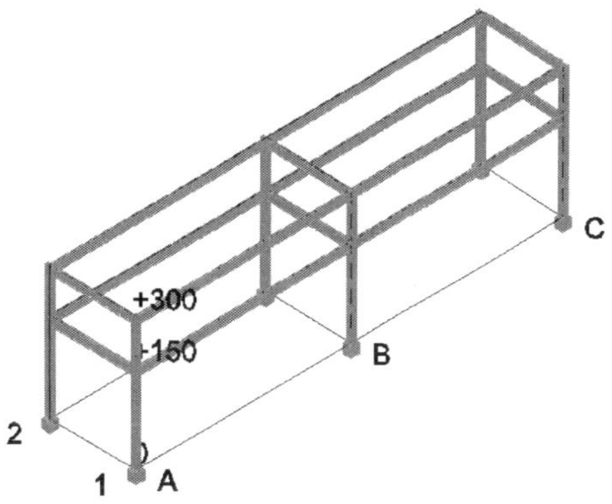

Figure 3-34 *Model after adding two levels of beams*

Next, you need to cut intersections between members.

10. Choose the **Cut Back Member** tool from the **Cutting** panel; you are prompted to select the limiting member.

11. Select the first column on the left side of the grid; you are prompted to select the structural member to be cut.

12. Select any one of the beams intersecting the column; the beam is cut at the intersection. Similarly, cut beams at the intersections by using columns as the limiting members.

Saving the Model

1. Choose the **Save** tool from the **Quick Access Toolbar**; the file is saved.

2. Choose **Application Menu > Close** to close the file.

Tutorial 2

In this tutorial, you will open the CADCIM project created in Chapter 2 and then create a new AutoCAD Plant 3D file. In this file, you will create a structural model, as shown in Figure 3-35. The layout uses W10x12 structural members. The dimensions of the model are shown in Figure 3-36.

(Expected time: 45 min)

Note

You can also download the CADCIM project file of Tutorial of Chapter 2 from www.cadcim.com by following the path: Textbooks > CAD/CAM > AutoCAD Plant 3D > AutoCAD Plant 3D 2023 for Designers > Input Files.

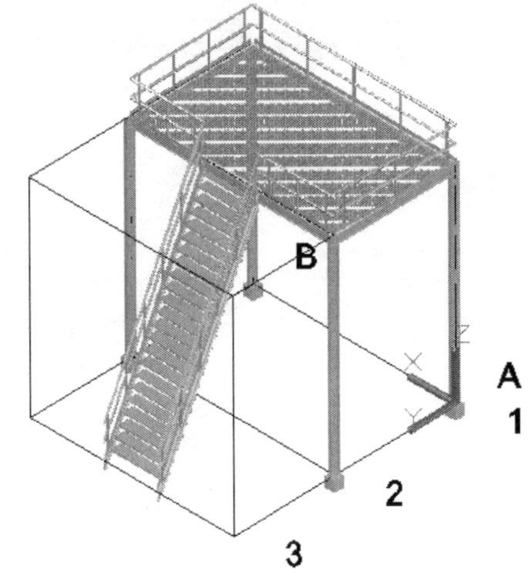

Figure 3-35 Structural model for Tutorial 2

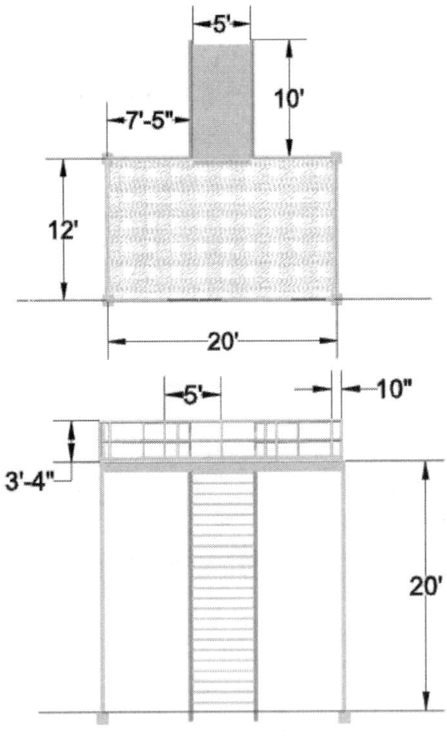

Figure 3-36 Dimensions of the model

The other specifications of the model to be created are as follows:

Grating Specifications

Material standard	ASTM
Material code	A242
Thickness	1"
Hatch pattern	ZIGZAG
Hatch scale	10"
Justification	Bottom
Shape	New rectangular

Footing Dimensions:

Length= 12 inch
Width = 12 inch
Depth = 12 inch

Railing Specifications

Handrail Height	40 inch
1st mid rail height	20 inch
2nd mid rail height	0
Kick plate height	5 inch
First post	10 inch
Second post	5 feet
Handrail Shape	PIPE2STD
Kick plate Shape	FB 1/4x4
Post Shape	PIPE2STD

Stair Specifications

Stair width	60 inch
Maximum tread distance	10 inch
Step Data	36x12x7
Stair Shape	C15x50

The following steps are required to complete this tutorial:

a. Open a new Plant 3D drawing file in the current project.
b. Create a grid by specifying the axis, row, and platform values.
c. Add footings at the bottom of the grid.
d. Add columns and beams to the model.
e. Cut members at intersections.
f. Add stair to the model.
g. Create platform by using the **Plate** tool.
h. Add railings to stair and platform.

Opening a New Plant 3D File

1. Double-click on the **AutoCAD Plant 3D 2023 - English** icon available on the desktop of your computer to start AutoCAD Plant 3D 2023.

   Next, you need to create a new AutoCAD Plant 3D file.

2. Select the **CADCIM** project that you created in Chapter 2 from the drop-down list in the **Current Project** area of the **PROJECT MANAGER**.

3. Select the **Plant 3D Drawings** node in the **Project** area and choose the **New Drawing** button available; the **New DWG** dialog box is displayed.

4. Enter **c03tut02.dwg** in the **File name** edit box and choose the **OK** button; the new file is created.

Creating the Grid

1. Choose the **Grid** tool from the **Parts** panel in the **Structure** tab; the **Create Grid** dialog box is displayed.

2. In this dialog box, specify the values, as shown in Figure 3-37.

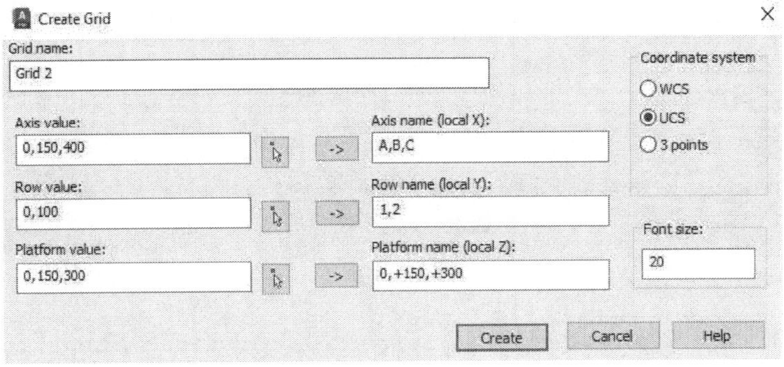

***Figure 3-37** Values specified in the **Create Grid** dialog box*

3. Next, accept the default values for other options and choose the **Create** button; the grid is created. Next, change the view orientation to **SW Isometric,** refer to Figure 3-38.

Creating Footings

1. Choose the **Footing Settings** tool from the **Structural Settings** drop-down in the **Parts** panel; the **Footing Settings** dialog box is displayed.

2. In the **Footing Settings** dialog box, specify the following values:

   Length(1): 1' **Width(2): 1'** **Depth(3): 1'**

   Next, accept the default values in the **Material** area and choose the **OK** button to close the dialog box.

3. Invoke the **Footing** tool and place the footings at the bottom grid points, as shown in Figure 3-39.

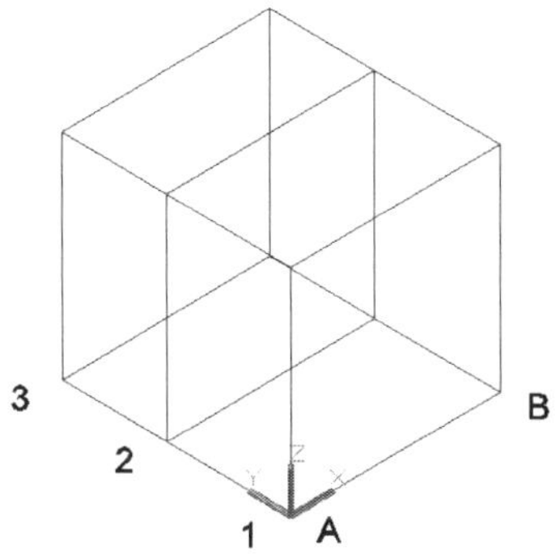

Figure 3-38 *Model after creating the grid*

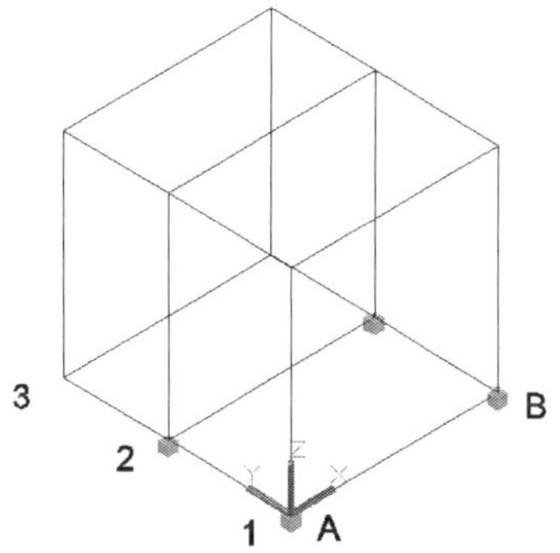

Figure 3-39 *Model after adding footings*

Creating Columns and Beams

To create structural members, first you need to set the properties of the member and then place a member. Also, you need to turn on the object snap and 3D object snap so that you can easily select the grid points.

1. Choose the **Member Settings** tool from the **Structural Settings** drop-down in the **Parts** panel; the **Member Settings** dialog box is displayed.

2. In the **Member Settings** dialog box, select **W** and **W 10x12** from the **Shape type** and **Shape size** edit boxes, respectively. Next, choose the middle center justification point in the **Orientation** window and accept default values of the other options. Choose the **OK** button; the settings of the structural member are changed and the **Member Settings** dialog box is closed.

3. Choose the **Member** tool from the **Parts** panel and then place columns on all the footings, as shown in Figure 3-40. Use grid points for the precise placement of the columns.

   Next, you need to add beams. But before that, you need to change the justification of the member to top.

4. Invoke the **Member Settings** dialog box and choose the top center justification point in the **Orientation** window. Next, choose the **OK** button to close the dialog box.

5. Choose the **Member** tool and place beams on both levels of the grid. The model after placing all the beams is shown in Figure 3-41.

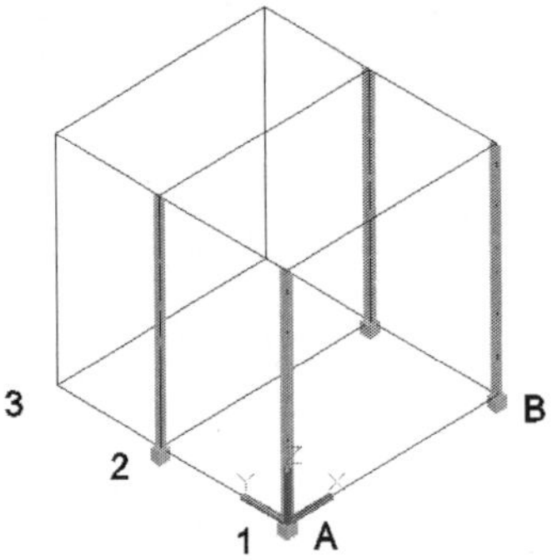

Figure 3-40 Model after adding columns

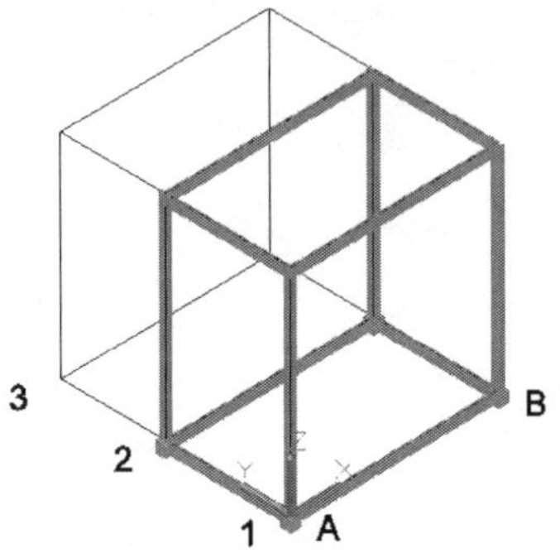

Figure 3-41 Model after adding all beams

6. Trim the intersections between the members.

Creating Stairs

Next, you need to add stairs to the model. To do so, first you need to set the stair settings and then place stairs.

1. Choose the **Stair Settings** tool from the **Structural Settings** drop-down in the **Parts** panel; the **Stair Settings** dialog box is displayed.

2. Enter **5'** in the **Stair width(1)** edit box and **10"** in the **Maximum tread distance(2)** edit box. Next, specify the **Step data** and **Stair shape** as given in the **Stair Specification**.

3. Choose the browse button located next to the **Step data** edit box; the **Select Step** dialog box is displayed. Next, choose **US Standards** and **36x12x7** from the **Tread standard** and **Tread shape** areas, respectively and then choose the **OK** button.

4. Choose the browse button given next to the **Stair shape** edit box; the **Select Stair Shape** dialog box is displayed. Now, select **C** and **C15x50** from the **Shape type** and **Shape size** list boxes, respectively and then choose the **Select** button. Next, choose the **OK** button to close the **Stair Settings** dialog box.

   Next, you need to create stairs. Make sure that the **Object Snap** and the **3D Object Snap** are turned ON.

5. Change the view orientation of the model to **NW Isometric**, and choose the **Stairs** tool from the **Parts** panel; you are prompted to specify the first point of the stair.

6. Select the midpoint on the top edge of the member, as shown in Figure 3-42; you are prompted to specify the second point.

7. Select the midpoint at the bottom grid line, as shown in Figure 3-43; a line is placed between the two specified points. Press ENTER to create the stairs on the specified line, as shown in Figure 3-44.

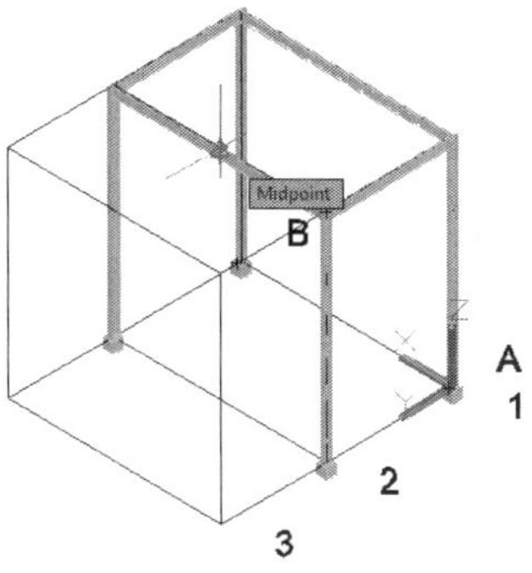

Figure 3-42 Selecting midpoint on the top edge

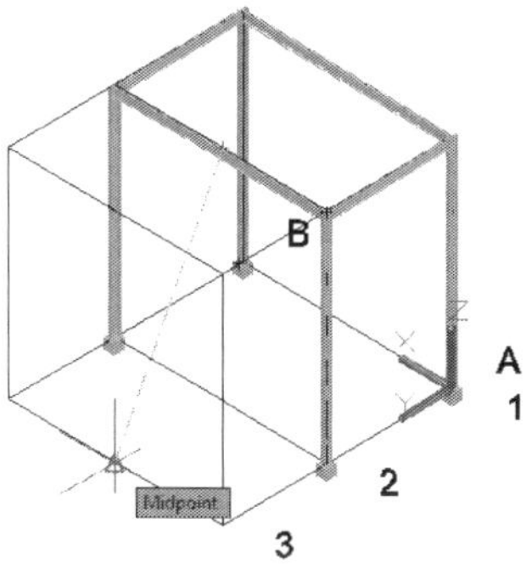

Figure 3-43 Selecting midpoint at the bottom grid line

Adding Gratings

You need to create platforms by adding gratings.

1. Change the view orientation of the model to **SW Isometric** and choose the **Plate** tool from the **Parts** panel; the **Create Plate/Grate** dialog box is displayed.

2. Specify the parameters for the grating as given below and then choose the **Create** button; you are prompted to specify the first corner of the grate.

Type	**Grating**
Material standard	**ASTM**
Material code	**A242**
Thickness	**1”**
Hatch pattern	**ZIGZAG**
Hatch scale	**10”**
Justification	**Bottom**
Shape	**New rectangular**

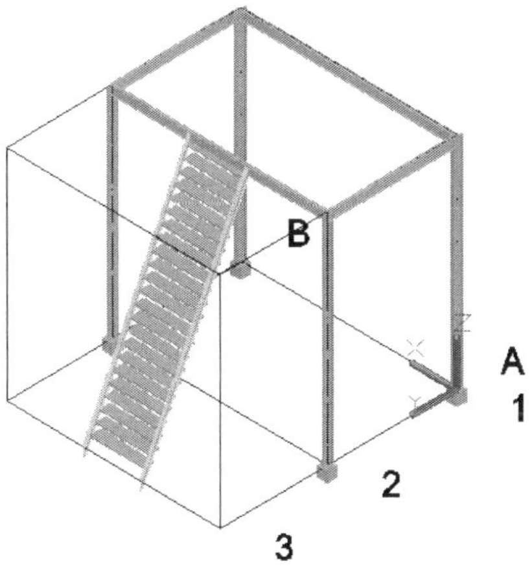

Figure 3-44 Model after creating stairs

3. Select the corner points, as shown in Figure 3-45; a rectangular grating is created on the platform, as shown in Figure 3-46.

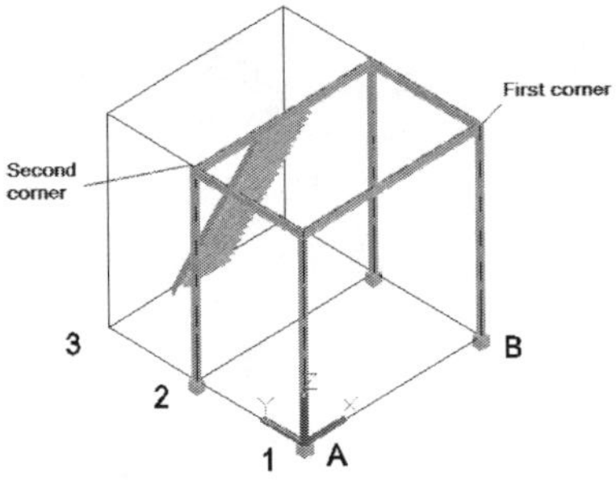

Figure 3-45 Selecting corner points to create a grating

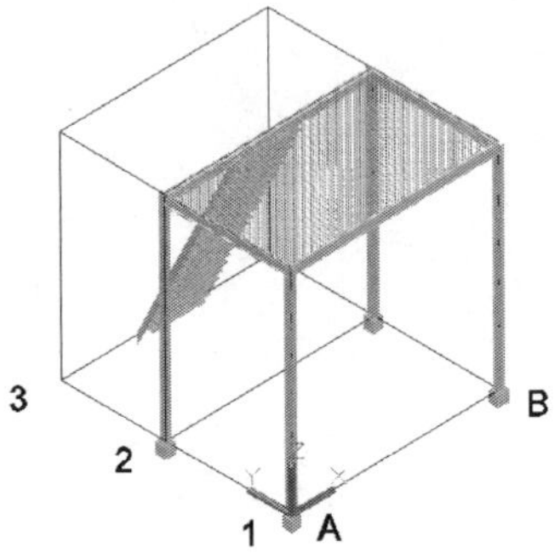

Figure 3-46 Grating created on the model

Adding Railing

Next, you need to add railing to the stairs and the platforms.

1. Choose the **Railing Settings** tool from the **Structural Settings** drop-down list; the **Railing Settings** dialog box is displayed. Next, specify settings as given in the **Railing Specifications** at the start of this tutorial and choose the **OK** button to exit the dialog box.

2. Choose the **Railing** tool from the **Parts** panel; you are prompted to select the start point of the railing. Enter **Object** at the command prompt; you are prompted to select an object to align the railing.

3. Select the stairs; a railing is added to the stairs, as shown in Figure 3-47. Note that the view orientation in the figure is set to **NW Isometric.**

   Next, you need to add railing to the beams.

4. Invoke the **Railing** tool again and select the points, as shown in Figure 3-48; a railing is created between the selected points. Next, press ENTER.

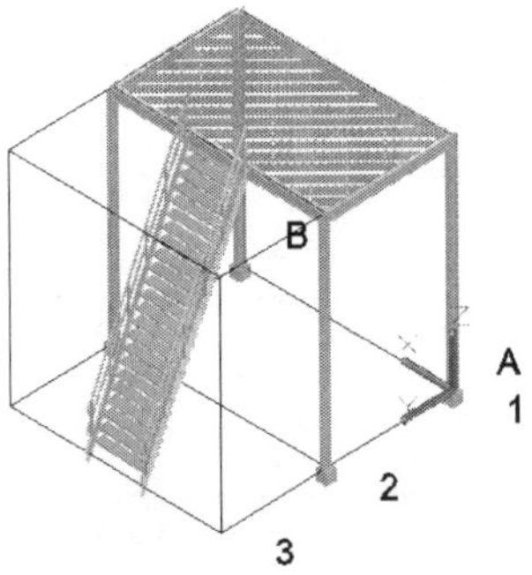

Figure 3-47 Railing added to the stairs

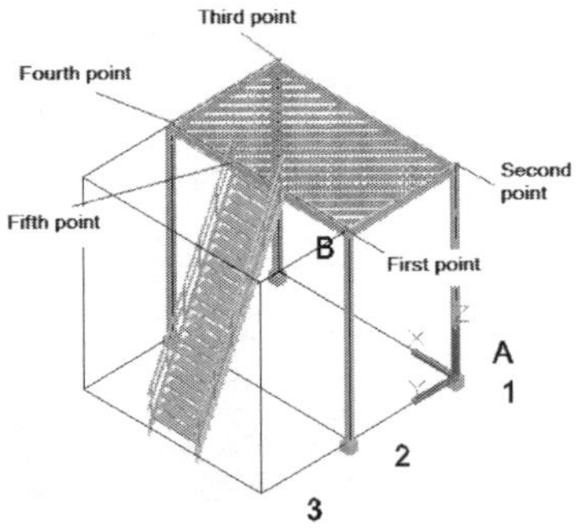

Figure 3-48 Points to be selected to create the railing

5. Invoke the **Railing** tool again and select the start and end points, as shown in Figure 3-49. The model after adding the railing is shown in the Figure 3-50.

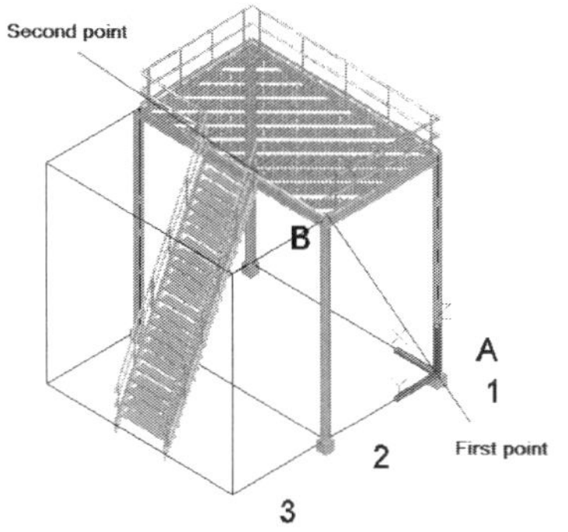

Figure 3-49 *Points to be selected to create the railing*

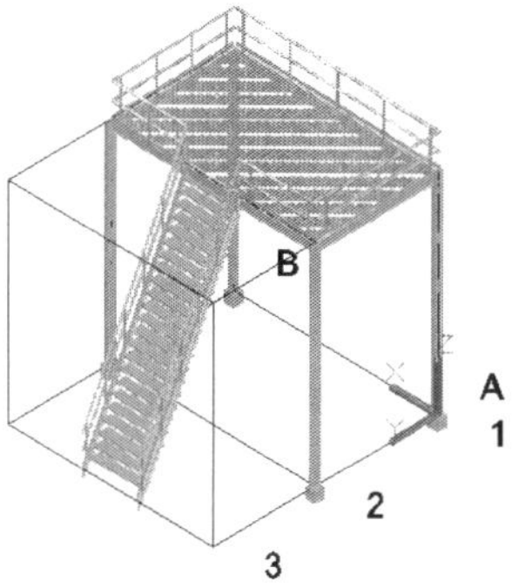

Figure 3-50 *Model after adding the railing*

Saving the Model

1. Choose the **Save** tool from the **Quick Access Toolbar**; the file is saved.

2. Choose **Application Menu > Close** to close the file.

Self-Evaluation Test

Answer the following questions and then compare them to those given at the end of this chapter:

1. In the _________ mode, the members are represented as lines with a shape symbol.

2. The _________ mode gives a 3D look to the members with fillets added to the structural outline.

3. You can add a railing to a structural member or a stair by invoking the _________ option and selecting them directly.

4. The _________ allows you to exchange steel structure data between two applications.

5. You need to specify _________ to place a ladder at some distance from the specified location.

6. The _________ option creates a miter joint with a gap between the edges.

7. You need to select the _________ check box in the **Ladder Settings** dialog box to create a ladder with a cage.

8. You can use the _________ button to match the properties of an already existing member with a new member.

9. You can modify individual elements of a stair, railing, or a ladder after exploding them. (T/F).

10. You can convert a line into a structural member by invoking the **Member** tool and then selecting the **Line** option from the command prompt. (T/F)

Review Questions

Answer the following questions:

1. The __________ check box in the **Railing Settings** dialog box is selected to make the middle rail continuous.

2. The __________ are used to increase or decrease the width of the stairs.

3. You can hide all the components except the selected one by using the __________ button.

4. You can select an existing polyline to convert it into a grating or a plate if the __________ radio button is selected in the **Create Plate/Grate** dialog box.

5. The __________ check box in the **Member Settings** dialog box is used to orient the Y axis in the opposite direction.

EXERCISE

Exercise 1 Skid

In this exercise, you will create the model shown in Figure 3-51. Its orthographic views are given in the drawing shown in Figure 3-52. The layout uses W10 x12 structural members.

(Expected time: 30 min)

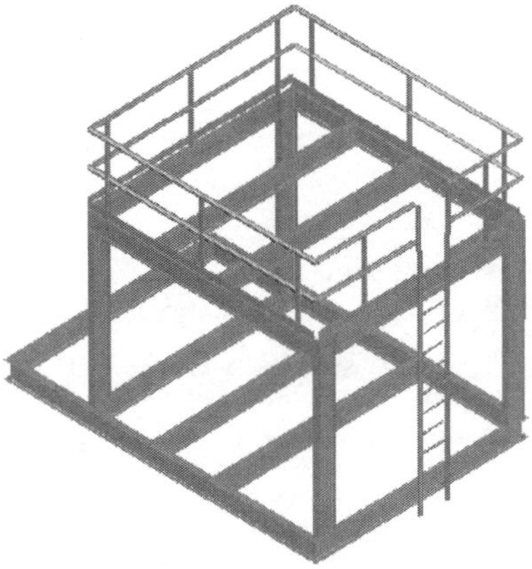

Figure 3-51 *The model for Exercise 1*

Railing Specifications

Handrail Height	40 inch
1st mid rail height	20 inch
2nd mid rail height	0
Kick plate height	5 inch
First post	10 inch
Second post	5 inch
Handrail Shape	PIPE2STD
Kick plate Shape	FB 1/4x4
Post Shape	PIPE2STD

Ladder Specifications

Width	25 inch
Exit width	35 inch
Projection	45 inch
Rung Distance	10 inch
Ladder shape	PIPE2STD
Rung Shape	PIPE3/4STD

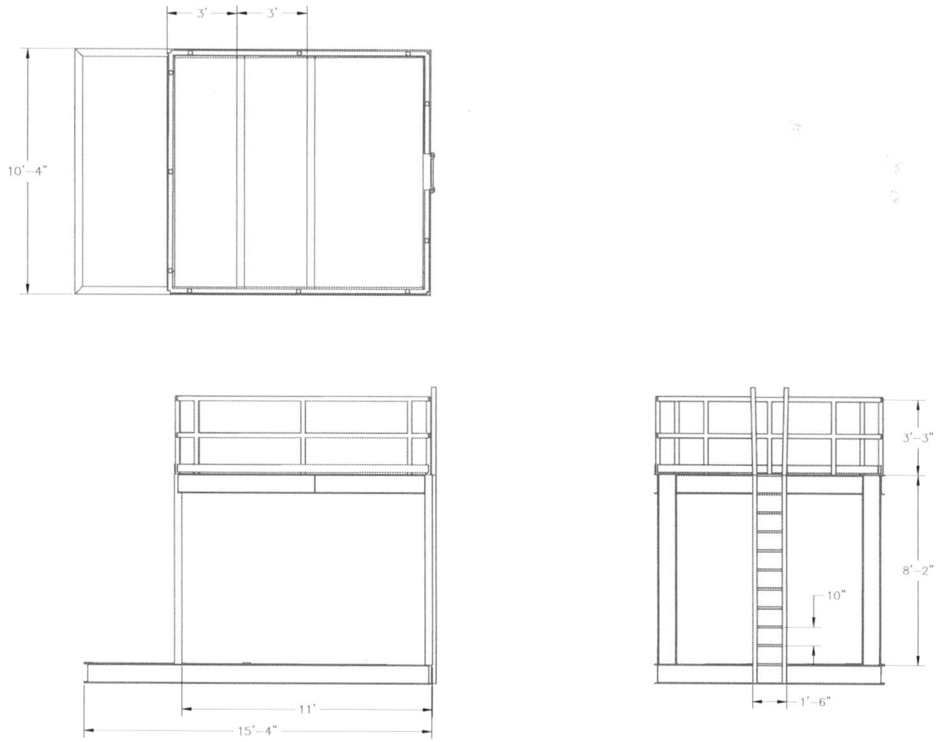

Figure 3-52 *Orthographic views for Exercise 1*

Answers to Self-Evaluation Test

1. Line Model, **2. Shape Model**, **3. Object**, **4. Advance Steel XML Export**, **5.** directional distance, **6. Gap**, **7. Draw cage**, **8. Match Properties**, **9.** T, **10.** T

Chapter 4

Creating Equipment

Learning Objectives

After completing this chapter, you will be able to:

- *Create equipment*
- *Create equipment supports*
- *Create customized equipment*
- *Modify equipment*
- *Convert solid models into equipment*

INTRODUCTION

In this chapter, you will learn how to create and place equipment in the 3D Piping workspace. In addition, you will learn how to convert solid models into equipment.

CREATING EQUIPMENT

You can create any equipment in AutoCAD Plant 3D by using the **Create Equipment** tool available in the **Equipment** panel of the **Home** tab. On choosing this tool, the **Create Equipment** dialog box will be displayed, as shown in Figure 4-1. The options in this dialog box are discussed next.

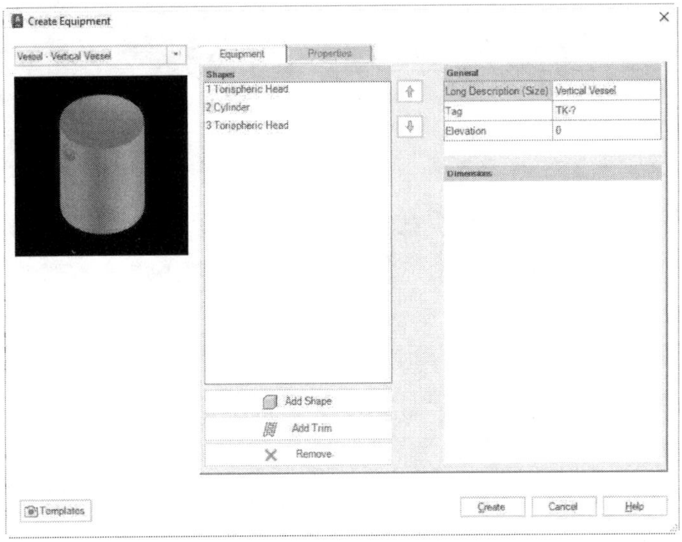

Figure 4-1 *The **Create Equipment** dialog box*

Equipment drop-down

This drop-down displays a list of equipment classes. On selecting an equipment class, a cascading menu will be displayed with a list of predefined equipment types and options to create a new custom equipment. Note that the predefined equipment models are available only for heater, heat exchanger, pump, vessel, tank, and strainer equipment.

Equipment Tab

The **Equipment** tab displays options to determine the shape and size of the equipment. In addition, you can also specify the general data. The options in this tab are discussed next.

Shapes

This **Shapes** area displays a list of shapes piled up to create a piece equipment. You can add shapes to an existing equipment and also remove shapes. Additionally, you can change the stacking order by using the up arrow and down arrow buttons. This can only be done while adding a new shape to the existing equipment.

General

There are three fields available in this area: **Long Description (Size)**, **Tag**, and **Elevation**. You can enter description and elevation of the equipment in their respective fields. To assign a tag, click in the **Tag** field; the **Assign Tag** dialog box will be displayed. Enter the desired text in the edit boxes available in this dialog box, and then choose the **Assign** button; the tag will be assigned to the equipment.

Dimensions

This area displays dimensions of the equipment. You can modify them as required. The preview window displays the annotated view of the equipment.

Properties Tab

The **Properties** tab displays properties of nozzles and fields to specify the data for the new equipment models. There are two areas in this tab, which are discussed next.

Nozzles

This area displays the properties of the nozzle such as its size, pressure class, description, and so on. Note that you cannot modify the data under this area.

Data

This area displays the data related to the equipment such as its manufacturer, material, material code, and so on. You can also specify a long description of the equipment in this area.

Templates

On choosing this button, a drop-down list is displayed. This drop-down list has the options to load the existing templates and save the new settings as a template.

PLACING EQUIPMENT IN THE DRAWING

To place an equipment in the drawing area, select it from the Equipment drop-down and then choose the **Create** button from the **Create Equipment** dialog box; the selected equipment will be attached to the cursor and you will be prompted to choose an insertion point in the drawing area. Specify the insertion point; the equipment will be placed in the drawing area and a compass will be displayed at the bottom of the equipment. Use the compass tool to specify the orientation angle of the equipment. The procedures to add various equipment are discussed next.

Note

*The compass will be displayed at the bottom of the equipment after specifying the insertion point only when the **Toggle Compass** button is activated. This button is available in the **Compass** panel of the **Home** tab. You will learn more about the **Toggle compass** button in the later chapters.*

Adding a Vessel

To add a vessel, choose **Vessel > Horizontal Vessel/Vertical Vessel** option from the Equipment drop-down of the **Create Equipment** dialog box; the options related to the vessel are displayed in the **Equipment** tab. You can increase or decrease the length or the diameter of the vessel by specifying values in the **Dimensions** area. You can also add more cylinders to increase the length of the vessel. After making all the necessary modifications in the shape and size of the vessel, click in the **Tag** edit box to assign a tag to the vessel. After assigning a tag, you can specify the elevation in the **Elevation** edit box. Next, choose the **Create** button and place the vessel in the geometry area; the compass will be displayed, refer to Figure 4-2, and you will be prompted to specify the rotation angle. Specify the rotation angle at the command prompt to orient the vessel according to the requirement.

Adding a Heat Exchanger

Heat exchangers are an important part of a process plant. They maintain the heat balance in the plant through addition or removal of heat. They exchange heat either with outside sources such as cooling towers or with streams of fluids operating at different temperatures. The heat exchanger can be classified into a cooler, exchanger, reboiler, condenser, heater, or a chiller. There are five predefined heat exchangers available in the **Create Equipment** dialog box. To create a heat exchanger, choose the **Heat Exchanger** option from the Equipment drop-down and then choose the desired predefined heat exchanger type. Next, click on the shapes in the **Shapes** area; the dimensions of the shapes will be displayed in the **Dimensions** area. You can edit their dimensions, if required. Next, choose the **Create** button and place the heat exchanger in the drawing area, refer to Figure 4-3. You can also create a new heat exchanger. The procedure to create a user-defined equipment is discussed later in this chapter.

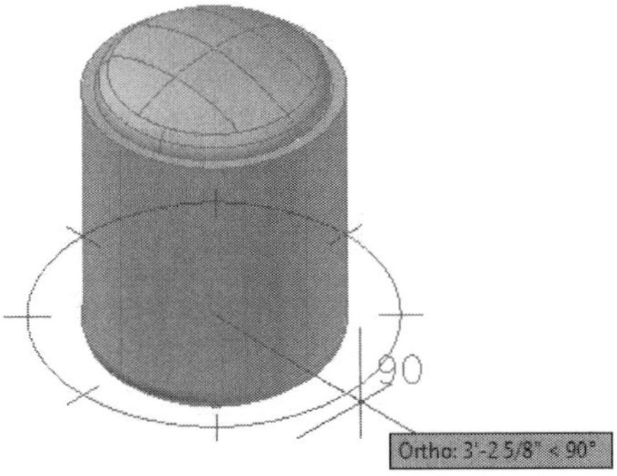

Figure 4-2 *A vessel placed in the drawing area*

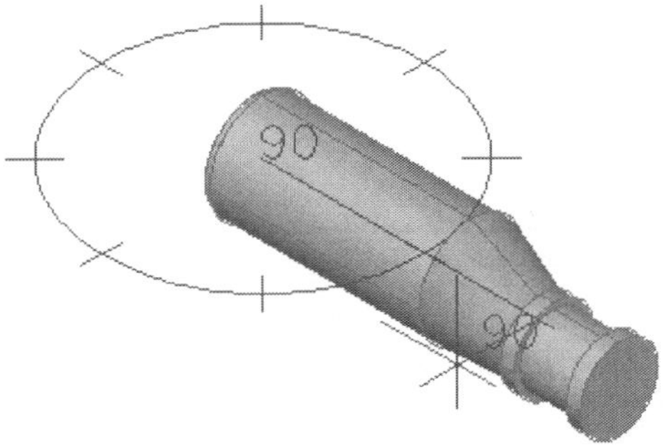

Figure 4-3 *A heat exchanger placed in the drawing area*

Adding a Pump

Pump is a mechanical device that is used to supply fluid to the desired point by using mechanical force. There are eight types of predefined pumps available in the **Create Equipment** dialog box. Choose any of the pumps; the preview of the selected pump will be displayed in the preview window, refer to Figure 4-4. You can modify the dimensions of the pump in the **Dimensions** area as per your requirement and then choose the **Create** button; the pump will be attached to the cursor and you will be prompted to specify the insertion point in the drawing area. Place the pump and orient it using the compass, refer to Figure 4-5.

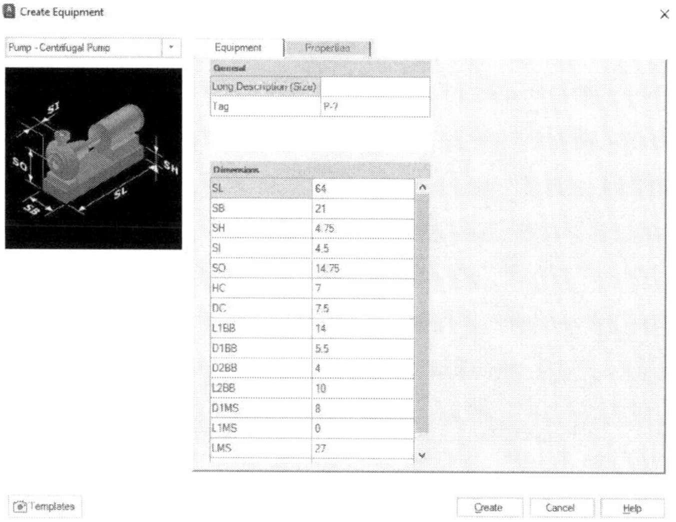

Figure 4-4 *The* ***Create Equipment*** *dialog box displaying a pump*

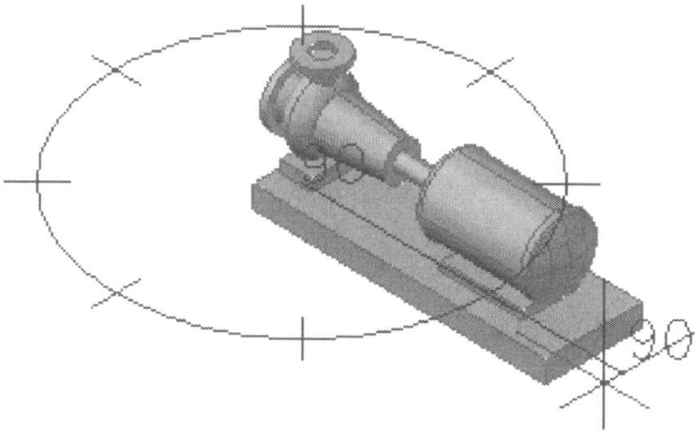

Figure 4-5 *A centrifugal pump placed*

Adding a Heater

Heaters are used to raise the temperature of a fluid to meet specific process requirements. There are two types of predefined heaters available, the **Box Type Heater** and the **Cylindric Heater**. To add a heater to a plant model, invoke the **Create Equipment** dialog box. Next, choose the **Heater** option from the **Equipment** drop-down and then choose any of the predefined heater types. Edit the dimensions, if required and then choose the **Create** button; you will be prompted to specify the insertion point. Specify the insertion point and then orient it using the compass, refer to Figure 4-6.

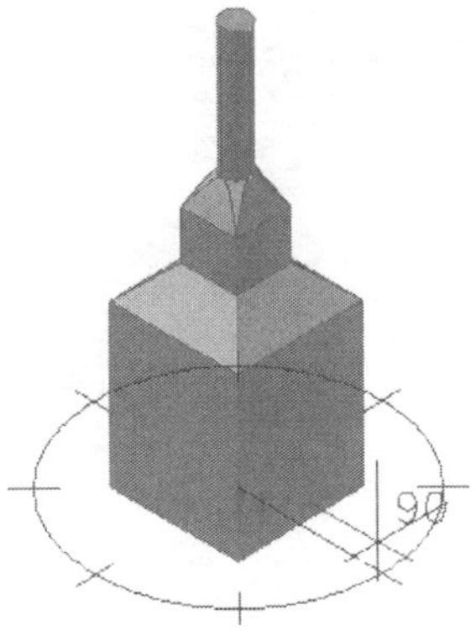

Figure 4-6 A heater placed in the drawing area

CREATING A CUSTOMIZED EQUIPMENT

To create a customized equipment, invoke the **Create Equipment** dialog box and follow the steps given below.

1. Choose the required equipment from the **Equipment** drop-down; a cascading menu will be displayed.

2. Choose the New Horizontal or New Vertical equipment type from the menu.

3. Choose the **Add Shape** button available at the bottom of the **Shapes** area; a flyout will be displayed with a list of shapes, as shown in Figure 4-7.

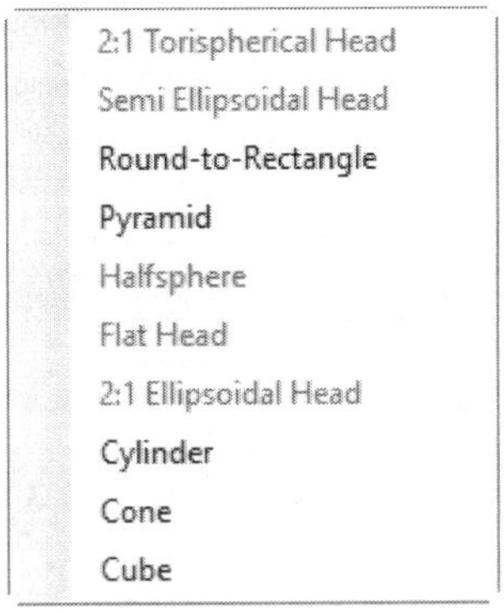

Figure 4-7 Flyout showing the list of shapes

4. Choose the desired shape from the available options in this flyout; the selected shape will be added under the **Shapes** area.

5. Similarly, add other shapes and arrange them in a systematic manner using the up arrow and down arrow buttons. Note that the shapes should be in a definite form and should be compatible with each other.

6. Modify the dimensions of the shapes by using options in the **Dimensions** area. Also, assign a tag and add description and elevation in the respective fields from the **General** area.

7. Choose the **Create** button and place the model in the drawing area.

8. Set the orientation of the model using the **Compass** tool. Figure 4-8 shows a customized equipment created.

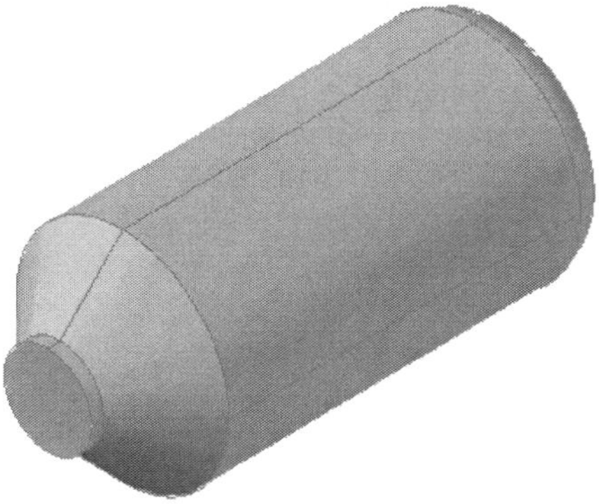

Figure 4-8 *A customized equipment*

MODIFYING EQUIPMENT

To modify an equipment, choose the **Modify Equipment** tool from the **Equipment** panel of the **Home** tab; you will be prompted to select an equipment. Select the equipment to be modified; the **Modify Equipment** dialog box will be displayed. The options in this dialog box are same as those in the **Create Equipment** dialog box. Modify the parameters of the equipment and choose the **OK** button; the dialog box will be closed and the equipment will be modified accordingly. Figures 4-9 and 4-10 show a cylindrical heater before and after modification.

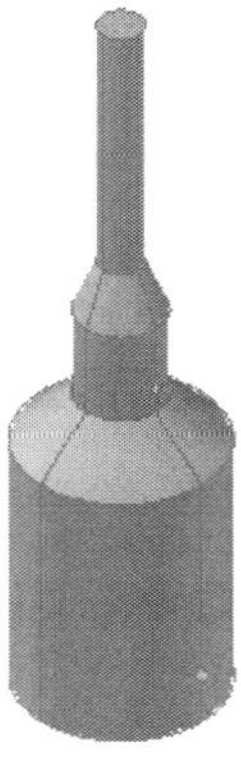

Figure 4-9 *The cylindrical heater before modifications*

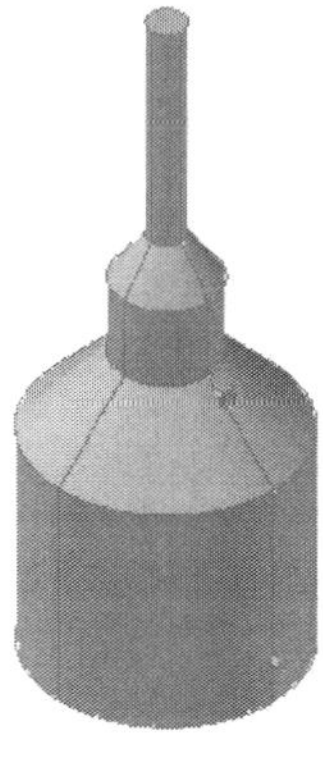

Figure 4-10 *The cylindrical heater after modifications*

CONVERTING SOLID MODELS INTO EQUIPMENT

To convert a solid model into an equipment, you need to create a solid model by using the tools that are available in the **Modeling** tab of the **Ribbon**. You can also convert solid models into equipment from other AutoCAD drawing files. To do so, right-click on the **Plant 3D Drawings** folder in the **PROJECT MANAGER** tree and choose the **Copy Drawing to Project** option from the shortcut menu displayed. Next, select the drawing file which contains the solid model; the file will be added to the **Plant 3D Drawing** folder. Now, double-click on the file in the **PROJECT MANAGER** and choose the **Convert Equipment** tool from the **Equipment** panel of the **Home** tab; you will be prompted to select a solid model. Select the solid model and press ENTER; the **Convert to Equipment** dialog box will be displayed, as shown in Figure 4-11. Select an equipment type from the **Equipment** tree and choose the **Select** button; you will be prompted to specify the insertion base point. Select a point on the model by using snaps; the **Modify Equipment** dialog box will be displayed, as shown in Figure 4-12. In the **General** area, enter the long description and assign a tag by clicking in the **Tag** field. In the **Properties** tab, enter data in the fields related to manufacturer, material, material code, size, and so on. You can also save the equipment as a template by choosing the **Templates** button. Choose the **OK** button to close the **Modify Equipment** dialog box. Figure 4-13 shows a selected AutoCAD object converted into an equipment.

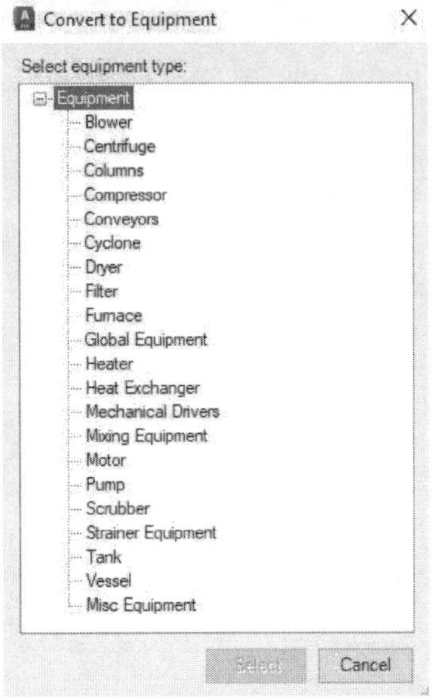

Figure 4-11 The ***Convert to Equipment*** *dialog box*

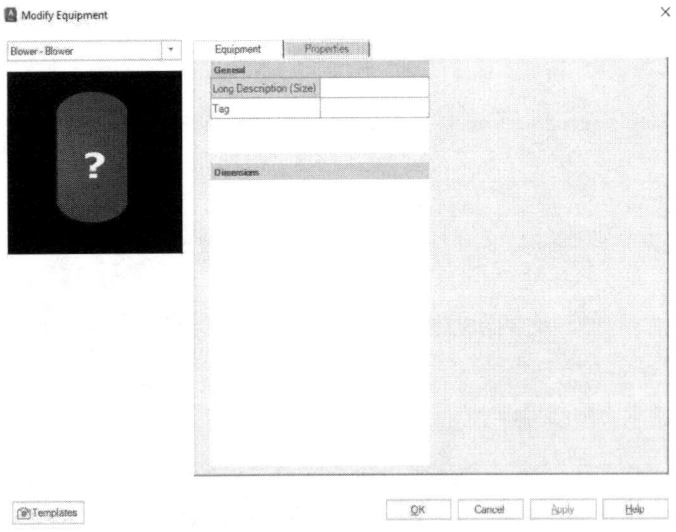

Figure 4-12 The ***Modify Equipment*** *dialog box*

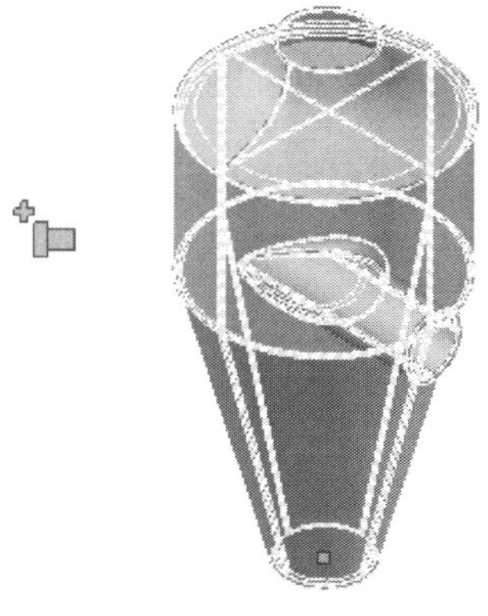

Figure 4-13 Selected AutoCAD object after converting into an equipment

CONVERTING INVENTOR MODELS INTO EQUIPMENT

To convert an Inventor model into an equipment, you need to import Inventor ADSK file into Plant 3D. To do so, choose the **Convert Inventor Equipment** tool from the **Equipment** panel of the **Home** tab; the **Import Inventor Component** dialog box will be displayed. Select the Inventor file with *.adsk* format and then choose the **Open** button from the dialog box; you will be prompted to specify the insertion point. Click in the drawing area to specify the insertion point; you will be prompted to orient the model. Orient the model by using the compass; the **Convert to Equipment** dialog box will be displayed. Select an equipment type from the **Equipment** tree and choose the **Select** button; the model will be inserted in the drawing. Now, you can modify the equipment by using the **Modify Equipment** tool.

ATTACHING OBJECTS TO AN EQUIPMENT

You can attach solid objects to an equipment. To do so, choose the **Attach Equipment** tool from the **Equipment** panel; you will be prompted to select an equipment. Select the required equipment, refer to Figure 4-14; you will be prompted to select the objects to be attached. Select the objects from the graphics window, refer to Figure 4-14, and press ENTER; the selected objects will be attached to the equipment.

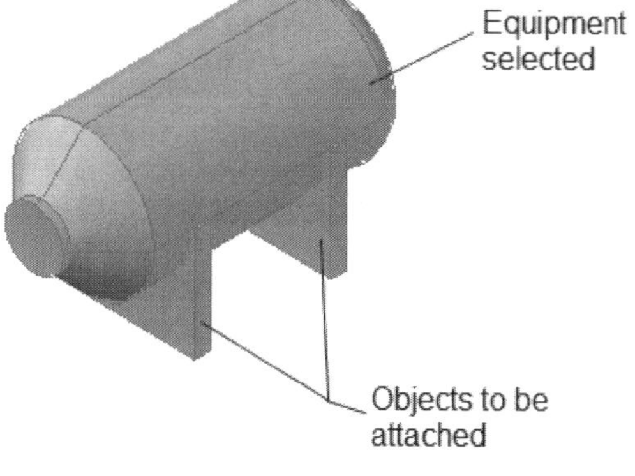

Figure 4-14 Equipment and objects to be selected

DETACHING OBJECTS FROM AN EQUIPMENT

To detach the added objects from an equipment, choose the **Detach Equipment** tool from the **Equipment** panel; you will be prompted to select the equipment to which the objects have been attached. Select the object; you will be prompted to detach all the attached objects. Choose **Yes** from the command prompt to detach all the components.

ADDING NOZZLES TO A CUSTOMIZED EQUIPMENT

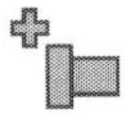

You can add a nozzle to an equipment after creating or modifying it. To do so, first select the equipment and then click on the **Add Nozzle** grip; a nozzle preview will be displayed on the equipment, refer to Figure 4-15. Also, a dialog box for adding and modifying nozzles will be displayed with the **Change Location** tab chosen, refer to Figure 4-16. The options in this dialog box are discussed next.

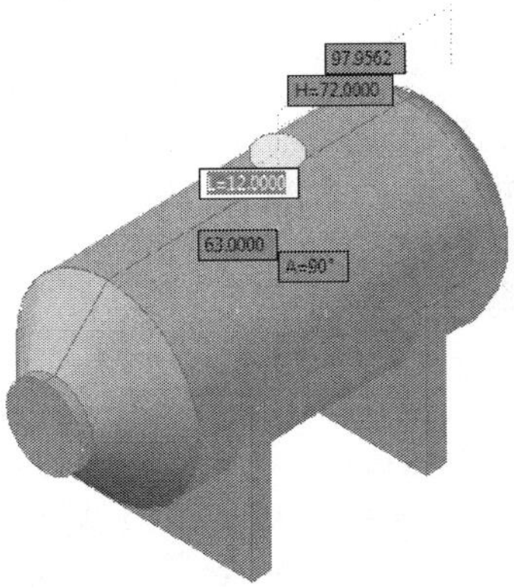

Figure 4-15 The preview of the nozzle displayed on the equipment

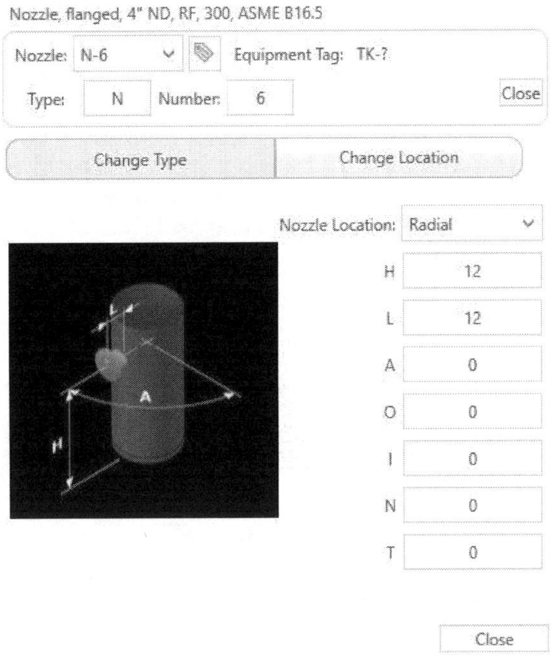

*Figure 4-16 The dialog box with the **Change Location** tab chosen*

Nozzle

This drop-down list contains the total number of nozzles in an equipment. The currently selected nozzle is displayed in it by default. You can select any nozzle from this drop-down list and modify it.

Tag Icon

On choosing this icon, the dialog box for adding or modifying nozzles will expand. You can edit the nozzle type and number in this dialog box.

Change Location Tab

The change location tab displays options to define the location of the nozzle. The options in this tab are discussed next.

Preview Window

The preview window explains the usage of options available in the **Change location** tab.

Nozzle Location

The different types of nozzle locations that can be specified on an equipment are available in this drop-down list. You can specify the location using the **Top**, **Bottom**, **Radial**, **Line**, and **Nozzle** options. On selecting the **Line** option, you will be prompted to select a line from the equipment. On doing so, the nozzle will be created in line with the selected equipment. On selecting different options from this drop-down list, various options are enabled. These options are discussed next.

Radius(R)

This edit box is available when the **Top** or **Bottom** location options are selected. You can specify the radial distance of the nozzle from the center of the vessel in this edit box.

Angle(A)

This edit box is used to specify the radial angle between the initial location and the new location of the nozzle. This edit box is available when the **Top**, **Bottom**, or **Radial** option is selected.

Length(L)

This edit box is used to specify the length of the nozzle.

Offset(O)

This edit box is used to specify the offset distance at which the nozzle will be placed. This edit box is available only when the **Radial** option is selected.

Perpendicular(P)

On selecting this check box, the nozzle will be created perpendicular to the equipment. Note that this check box is available only if you select the **Top** or **Bottom** option.

Inclination(I)

This edit box is used to specify the angle of inclination to the nozzle axis. This edit box is available when the **Top**, **Bottom**, or **Radial** option is selected.

Rotation(N)

This edit box is used to specify the rotation angle of the nozzle on the horizontal plane. This edit box is available when the **Top**, **Bottom**, or **Radial** option is selected.

Twist(T)

This edit box is used to specify the twist angle along the vertical plane. This edit box is available only when the **Top**, **Bottom**, or **Radial** option is selected.

Height(H)

This edit box is used to specify the height of the nozzle from the base of equipment. This edit box is available only when the **Radial** option is selected.

Note

The options in the ***Change Location*** *tab which have been discussed above are not available for converted equipment.*

Change Type Tab

The change type tab contains a nozzle list and the options to filter it, refer to Figure 4-17. You can choose the required nozzle by filtering parameters such as the nozzle type, size, unit, and pressure class. The options in this tab are discussed next.

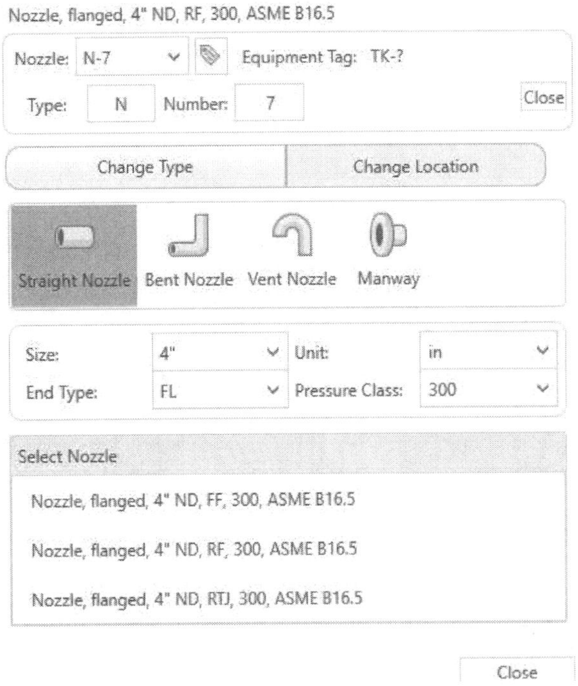

Figure 4-17 *The dialog box with the* ***Change Type*** *tab chosen*

Nozzle Type Area

There are four types of nozzles available in this area, **Straight Nozzle**, **Bent Nozzle**, **Vent Nozzle**, and **Manway**. Choose a nozzle type; different sizes of the selected nozzle type are displayed.

Size

This drop-down list is used to specify the size for the chosen nozzle type. On selecting a size, the **Select Nozzle** list is filtered to display nozzles of the selected size.

Unit

This drop-down list is used to select a unit type for the nozzle. When you select a unit type from this drop-down list, the **Select Nozzle** list displays nozzles available under the selected unit.

End Type

This drop-down list displays various end types. When you select the **FL(**Flanged) end type, the **Select Nozzle** list will be filtered to display nozzles with flanged end type.

Pressure Class

On selecting a pressure class from the **Pressure Class** drop-down list, the **Select Nozzle** list will be filtered to display the nozzles with the selected pressure class.

Select Nozzle

The nozzles available for the selected nozzle type will be displayed in the **Select Nozzle** list. On specifying the parameters such as size, end type, unit and pressure class, this list gets filtered. Select the required nozzle from the list and choose the **Close** button; the nozzle will be added at the specified location.

ADDING NOZZLES TO A CONVERTED EQUIPMENT

After converting a solid model into an equipment, you can add nozzles to it. To do so, select the equipment; a nozzle grip will be displayed on it. Click on the nozzle grip; you are prompted to specify the center of the nozzle. To specify the center of the nozzle, snap to the location where you want to add the nozzle and then select a point, refer to Figure 4-18. On doing so, a rubber band line will be attached to the cursor. Also, you will be prompted to specify the direction by selecting the second point, refer to Figure 4-19. Specify the second point; the dialog box for adding or modifying nozzles will be displayed. Specify the nozzle type using the options available in the **Change Type** tab of this dialog box and then close the dialog box.

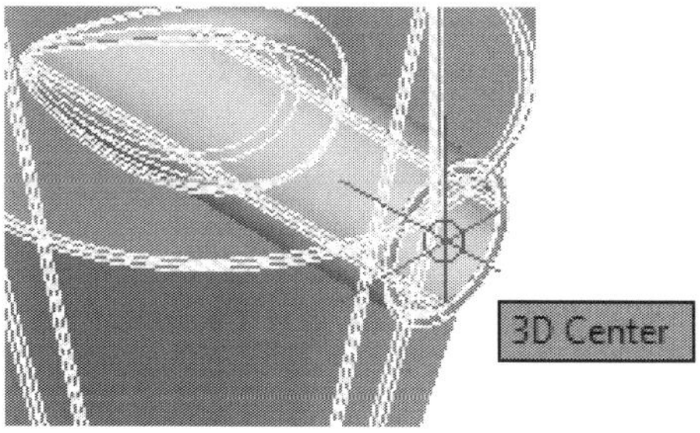

Figure 4-18 *Selecting a point to add a nozzle*

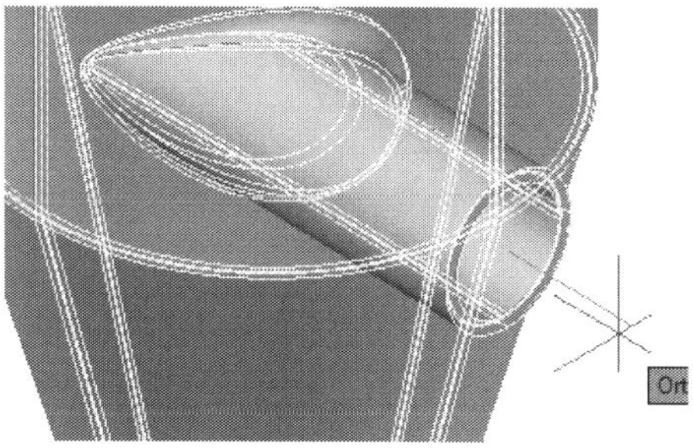

Figure 4-19 *Specifying the direction of the nozzle*

MODIFYING NOZZLES

To modify the type and location of a nozzle, you need to click on it while holding the CTRL key. On doing so, a pencil grip will be displayed on it, as shown in Figure 4-20. Release the CTRL key and click on the pencil symbol; the dialog box for adding or modifying nozzles will be displayed. But in case of converted equipment, pumps, and strainer, you need to select the equipment first. On doing so, a pencil symbol will be displayed on the nozzle. Click on the pencil symbol; the dialog box for adding or modifying nozzles will be displayed. Now, you can change the location as well as the type of nozzle, refer to Figure 4-21.

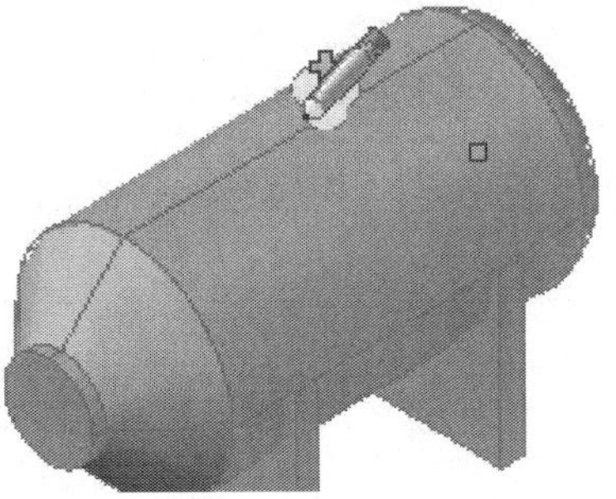

***Figure 4-20** Pencil grip displayed on the nozzle*

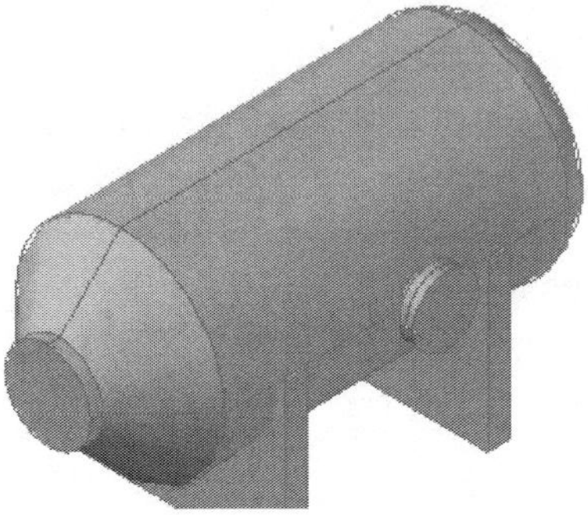

***Figure 4-21** Nozzle after modifying the location and type*

Note

1. You can remove the nozzle of a component. To do so, press the CTRL key, select the nozzle, and then press the DELETE key.

2. You can modify the nozzles of pump and strainer but cannot add or remove them.

3. In case of parametric equipment, you can provide only dimension values for nozzles. In this case, the number and position of nozzles are fixed.

TUTORIALS

Tutorial 1

In this tutorial, you will open the CADCIM project created in Chapter 2 and then start a new AutoCAD Plant 3D file. In this file, you will create equipments, as shown in Figure 4-22. The dimensions and location of equipments are shown in Figures 4-23 through 4-26.

(Expected time: 1 hr)

Figure 4-22 The model for Tutorial 1

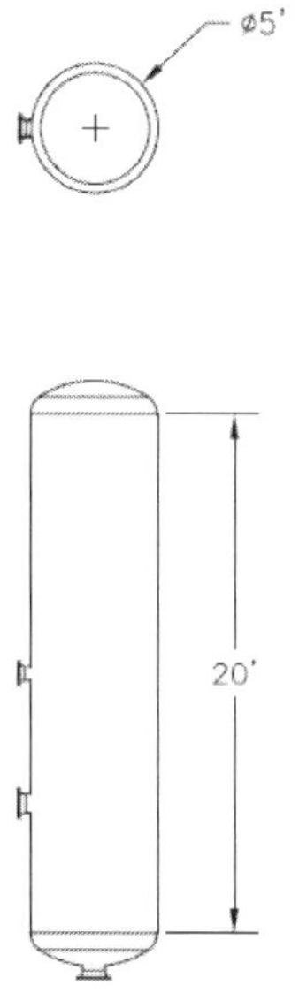

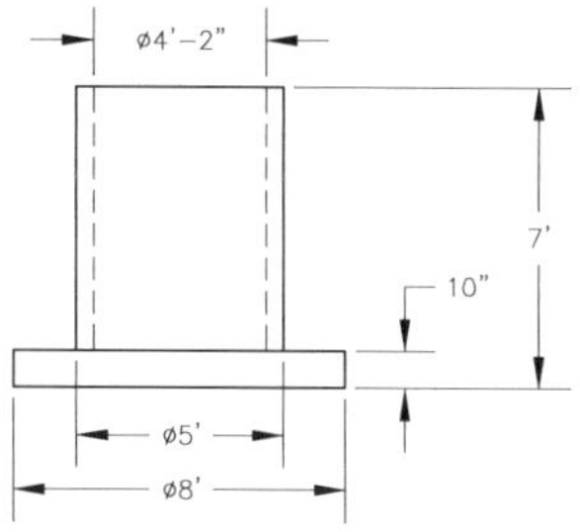

Figure 4-23 Dimensions of the Vertical vessel

Figure 4-24 Dimensions of the support

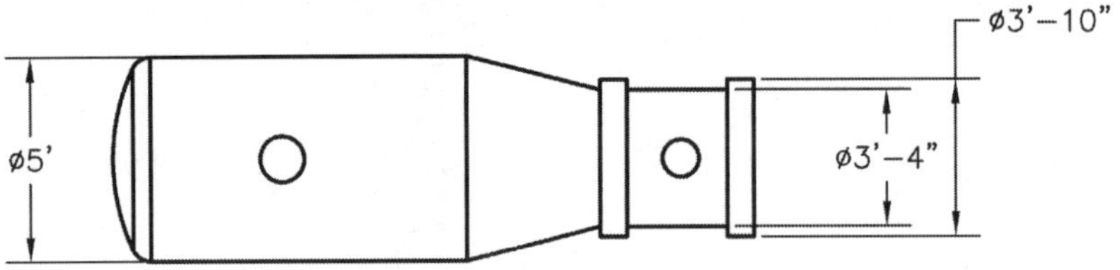

Figure 4-25 Dimensions of the Reboiler

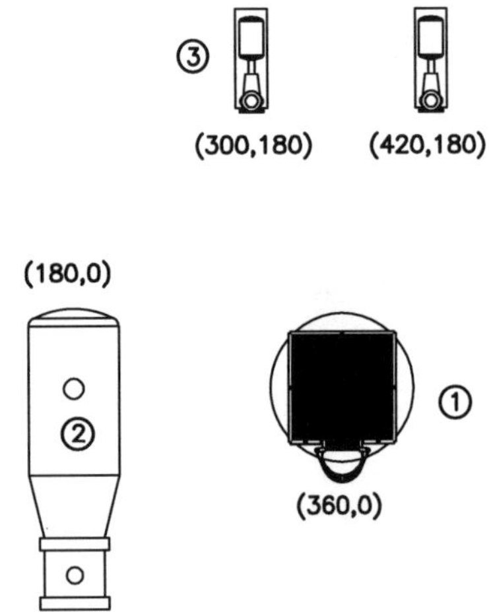

S.NO.	EQUIPMENT NAME	QTY
1	Vertical Equipment	1
2	Reboiler	1
3	Centrifugal Pump	2

Figure 4-26 Coordinate dimensions of the model

Grating Specifications

Material standard:	ASTM
Material code:	A242
Thickness:	1"
Hatch pattern:	ZIGZAG
Hatch scale:	10"
Justification:	Top
Shape:	New rectangular
Size:	70"x70"

Railing Specifications

Handrail Height(1):	40 inch
1st mid rail height(2):	20 inch
2nd mid rail height(3):	0 inch
Kick plate height(4):	5 inch

First post(5):	10 inch
Second post(6):	5'
Handrail Shape:	PIPE2STD
1st mid rail Shape:	PIPE2STD
Kick plate Shape:	FB 1/4x4
Post Shape:	PIPE2STD

Ladder Specifications

Width(1):	25 inch
Exit width(2):	35 inch
Projection (3):	45 inch
Rung Distance (4):	10 inch
Ladder shape:	PIPE2STD
Rung Shape:	PIPE3/4STD

Cage Specifications

Start Height:	7'-6"
Maximum distance:	5'
Distance From top:	1 inch
Radius:	14 inch

The following steps are required to complete this tutorial:

a. Open a new Plant 3D drawing file in the current project.
b. Create a vertical vessel and place it in the model space.
c. Create a grating at the top of the vertical vessel. Also, add a ladder and a railing.
d. Place the reboiler.
e. Create supports and attach them to the reboiler.
f. Place centrifugal pumps.
g. Save the model.

Creating a New AutoCAD Plant 3D File

1. Double-click on the **AutoCAD Plant 3D 2023 - English** icon; AutoCAD Plant 3D starts.

   Next, you need to start a new Plant 3D file.

2. Select **CADCIM** from the drop-down list in the **Current Project** area of the **PROJECT MANAGER** palette.

3. Select the **Plant 3D Drawings** node in the **Project** area and choose the **New Drawing** button available; the **New DWG** dialog box is displayed.

4. Enter **Piping Model.dwg** in the **File name** edit box and choose the **OK** button; a new file is created.

Creating a Support for the Vessel

In this section, you will create a support before adding the vessels.

1. Create support for the vessel using the modeling tools, as shown in Figure 4-27. The dimensions of the support are given in Figure 4-24. You can also download this model from *www.cadcim.com* by visiting the following link: *Textbooks > CAD/CAM > AutoCAD Plant 3D > AutoCAD Plant 3D 2023 for Designers > Input Files.*

Figure 4-27 *Support for the vertical vessel*

2. Place the support at **360, 0** considering the base point at the center of bottom face.

3. Change the view orientation to SW Isometric.

Creating a Vertical Vessel

1. Choose the **Create Equipment** tool from the **Equipment** panel in the **Home** tab; the **Create Equipment** dialog box is displayed.

2. Choose **Vessel > Vertical Vessel** option from the drop-down located at the top left of the dialog box; a list of shapes stacked together to form the vessel is displayed in the **Shapes** area.

3. Arrange the shapes in the **Shapes** area according to the given sequence.

   Torispheric Head
   Cylinder
   Torispheric Head

4. Select the **Torispheric Head** from the **Shapes** area; the diameter of the torispheric head is displayed in the **Dimensions** area. Change the diameter to **60 inch**.

5. Similarly, change the diameter of the other torispheric head to **60 inch**.

6. Select **Cylinder** from the **Shapes** area and change its dimensions as given next.

   D 60"
   H 240"

   Next, you need to assign a tag to the vessel.

7. Click in the field next to the **Tag** box in the **General** area; the **Assign Tag** dialog box is displayed.

8. Enter **TK** in the **Type** edit box and **101** in the **Number** edit box and choose the **Assign** button; a tag is assigned to the vessel.

9. Enter **84** as the value of elevation in the field next to the **Elevation** box and choose the **Create** button; the vessel is attached to the cursor and you are prompted to choose an insertion point.

10. Specify the insertion point of the vessel as **360,0** at the command prompt; the compass tool is displayed at the bottom of the vessel. Set the orientation of the vessel to 0 degree; the vessel is placed at the specified point.

Adding a Platform and a Ladder to the Vessel

Next, you need to add a platform and a ladder to the vessel.

1. Choose the **Plate** tool from the **Parts** panel in the **Structure** tab of the **Ribbon**; the **Create Plate/Grate** dialog box is displayed.

2. In this dialog box, select the **Grating** option from the **Type** drop-down list. Next, select the **New rectangular** radio button from the **Shape** area.

3. Select the **Top** radio button from the **Justification** area.

4. Specify the settings given at the start of the tutorial and choose the **Create** button; you are prompted to specify the first corner of the rectangular grating.

5. Enter **325,-35,340** at the command prompt to specify the first corner; you are prompted to specify the second corner.

6. Enter **395,35,340** at the command prompt; the rectangular grating is created, as shown in Figure 4-28.

Figure 4-28 Rectangular grating created

7. Choose the **Ladder Settings** tool from the **Structural Settings** drop-down in the **Parts** panel of the **Structure** tab and specify the settings in the **Ladder Settings** dialog box as given in the Ladder specification at the beginning of the tutorial. Next, choose the **OK** button to exit the **Ladder Settings** dialog box. Now, invoke the **Ladder** tool from the **Parts** panel in the **Structure** tab and then select the first point, refer to Figure 4-29.

8. Now, snap the second point, as shown in Figure 4-30. Make sure the **Object Snap** and **Object Snap Tracking** buttons are chosen in the Status Bar and move the cursor in the -Y direction. Enter **5** at the command prompt and then press ENTER; the ladder is placed, as shown in Figure 4-31.

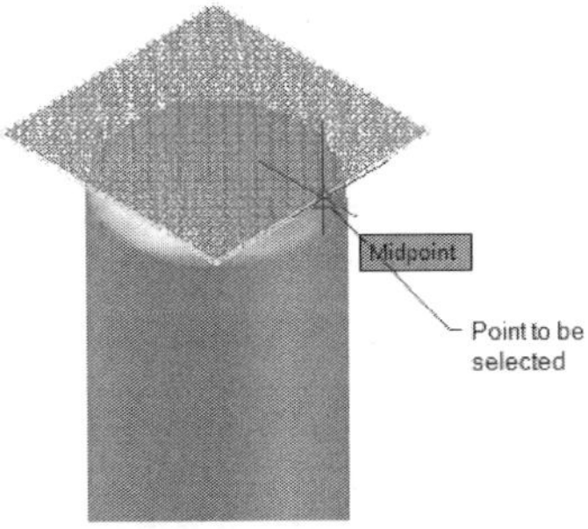

Figure 4-29 Selecting the first point of the ladder

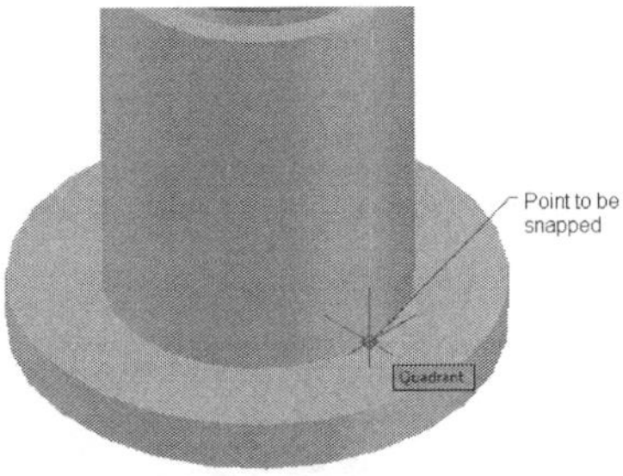

Figure 4-30 Selecting the second point of the ladder

Adding Railing to the Grating

1. Invoke the **Railing** tool from the **Parts** panel and create the railing on the edges of the grating, as shown in Figure 4-32. For dimensions, refer to the specifications given in the tutorial description.

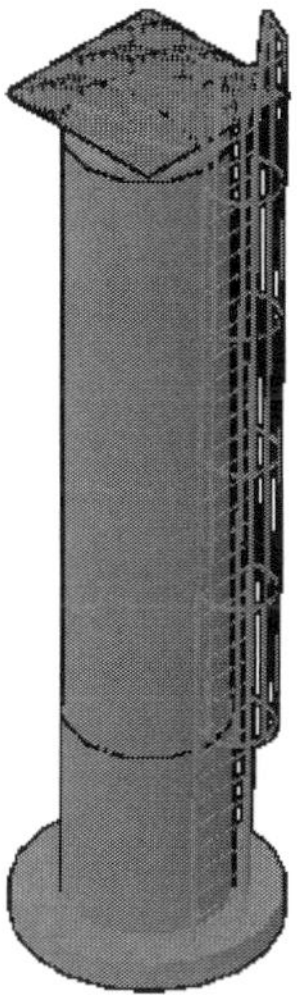

Figure 4-31 Vertical vessel after placing the ladder

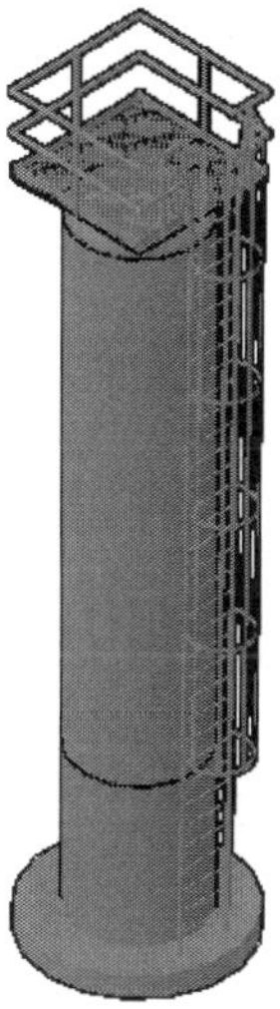

Figure 4-32 Vertical vessel after adding railing to the grating

Placing the Reboiler

1. Invoke the **Create Equipment** dialog box and choose **Heat Exchanger > Reboiler** option from the drop-down located at the top left of the dialog box; a list of shapes is displayed in the **Shapes** area.

2. Select the **Cylinder** located at the top in the **Shapes** area; the **Dimensions** area is displayed in the **Create Equipment** dialog box.

3. Enter the following values in the **Dimensions** area:

   Cylinder1
   D: 46" **H: 8"**

4. Similarly, specify the values for the other shapes, as given next.

   Cylinder2
   D: 40" **H: 30"**

   Cylinder3
   D: 46" **H: 8"**

   Cone
   Orientation: Upwards **D1: 60"** **D2: 40"**
   H: 40" **E: 10"** **A: 0**

Cylinder4
D: 60" **H: 96"**

Torispheric Head
D: 60"

Note
While specifying the diameter values in the ***Dimensions*** *area of the* ***Create Equipment*** *dialog box, you need to fix these values. To do so, click in the diameter field; a swatch will be displayed at the right of the field box. Next, click on the swatch; a shortcut menu will be displayed. Choose the* ***Override mode*** *option from the shortcut menu. The respective value in the field will be fixed.*

Next, you need to assign a tag to the reboiler.

5. Click in the **Tag** field in the **General** area of the **Create Equipment** dialog box; the **Assign Tag** dialog box is invoked. Enter **101** in the **Number** edit box and choose the **Assign** button; the **Assign Tag** dialog box is closed. Next, enter **48** in the **Elevation** field in the **Create Equipment** dialog box.

6. Choose the **Create** button and insert the reboiler at the point **180, 40**. Next, set the orientation to **270** degrees by entering 270 at the command prompt. Figure 4-33 shows the model after placing the reboiler.

   Next, you need to create supports and attach them to the reboiler.

7. Invoke the **Box** tool from the **Modeling** panel in the **Modeling** tab of the **Ribbon**; you are prompted to specify the first corner. You need to make sure that the **Dynamic Input** is turned off.

8. Enter **165,36** at the command prompt; you are prompted to specify the other corner of the box. Enter **195,40** at the command prompt; you are prompted to specify the height of the box.

9. Enter **50** at the command prompt; a rectangular box is created, as shown in Figure 4-34.

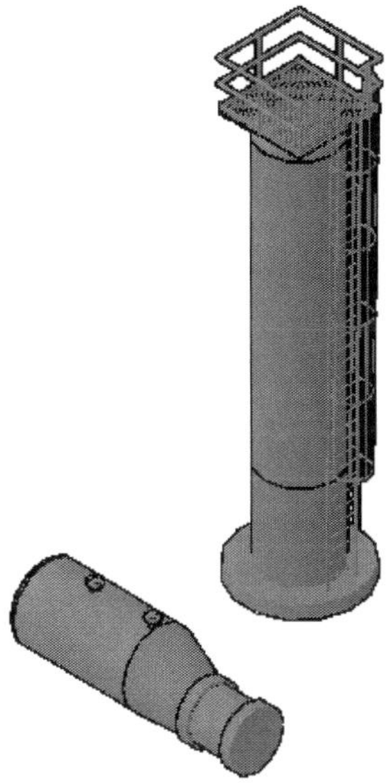

Figure 4-33 *The model after placing the heat exchanger*

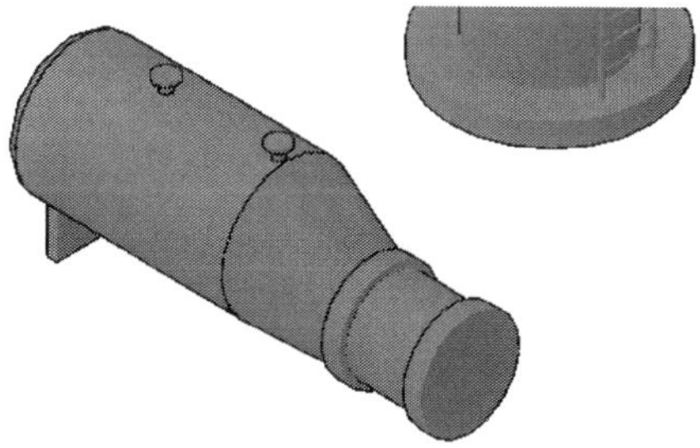

Figure 4-34 *Support created for the reboiler*

10. Copy the box and select a base point, refer to Figure 4-35. Move the cursor along the Y-direction and enter **-10'** at the command prompt; the box is copied at the specified location. Note that the **Ortho Mode** should be turned on.

11. Invoke the **Attach Equipment** tool from the **Equipment** panel in the **Home** tab; you are prompted to select a single equipment item.

12. Select the reboiler from the model space; you are prompted to select objects to be attached to the selected equipment.

13. Select the boxes below the reboiler. Next, press ENTER; the boxes are attached to the reboiler. Figure 4-36 shows the reboiler after attaching supports.

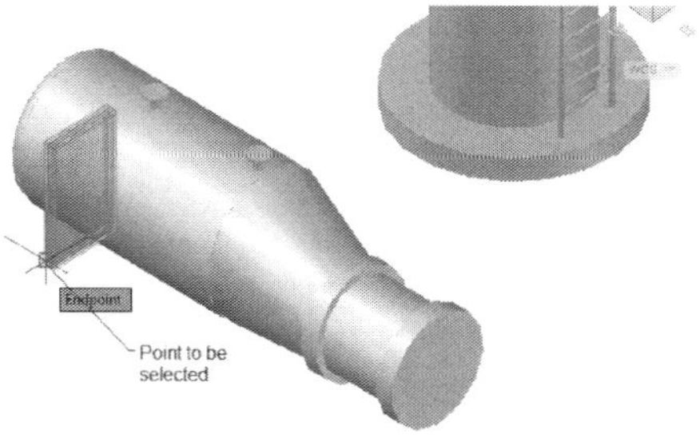

Figure 4-35 *Selecting a base point*

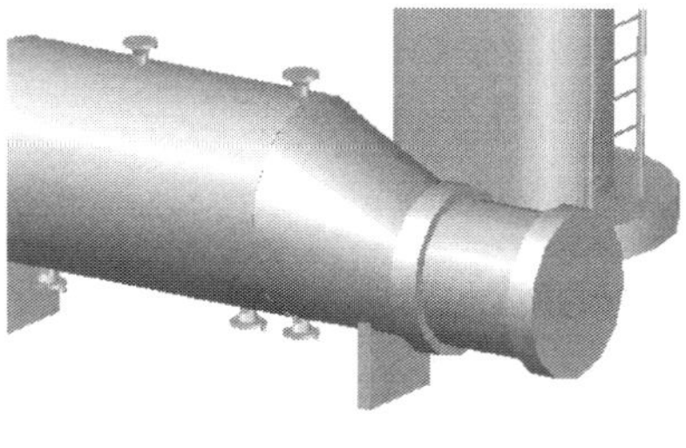

Figure 4-36 *Reboiler after attaching supports*

Placing Centrifugal Pumps

Next, you need to add centrifugal pumps to the model.

1. Invoke the **Create Equipment** dialog box and choose **Pump > Centrifugal Pump** option from the drop-down located at the top left of the dialog box; the **Equipment** tab along with a preview image of the centrifugal pump is displayed in the **Create Equipment** dialog box.

2. Click in the **Tag** field and enter **101A** in the **Number** edit box in the **Assign Tag** dialog box. Choose the **Assign** button to close the dialog box. Next, enter long description as **Centrifugal Pump**.

3. Choose the **Create** button; the **Create Equipment** dialog box is closed and you are prompted to select an insertion point. Enter the insertion point as **25',15'** and orient the pump to **90** degrees, as shown in Figure 4-37.

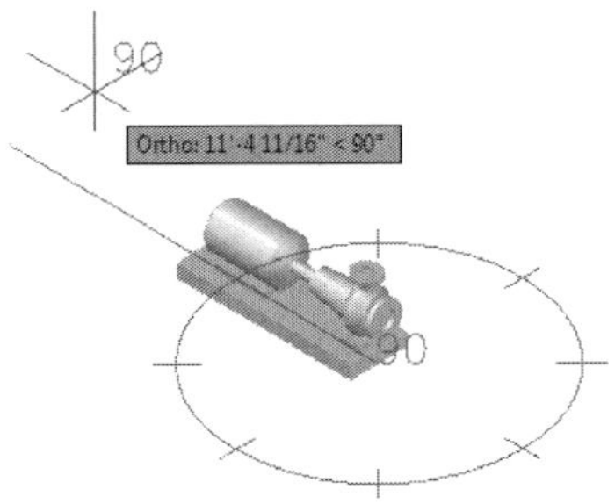

Figure 4-37 *Setting the orientation of the centrifugal pump*

4. Similarly, place another pump with the same orientation at insertion point **35',15'** and assign it the tag P-101B.

The model after placing pumps is shown in Figure 4-38.

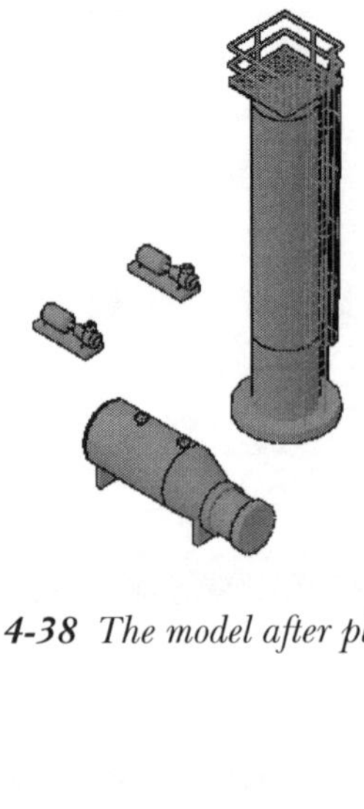

Figure 4-38 *The model after placing pumps*

Saving the Model

1. Choose the **Save** button from the **Quick Access Toolbar**; the file is saved.

2. Choose **Close > Current Drawing** from the **Application Menu**; the file is closed.

Tutorial 2

In this tutorial, you will add nozzles to the equipment created in Tutorial 1. The list of nozzles to be created is given next. **(Expected time: 45 min)**

Nozzle List of the Vertical Vessel

S.No	Nozzle Location	Long Description	Height (H)	Angle (A)	Length (L)
1	Radial	Nozzle, flanged 6" ND, RF, 150, ASME B16.5	10'	180	6"
2	Radial	Nozzle, flanged 8" ND, RF, 150, ASME B16.5	5'	180	6"
3	Bottom	Nozzle, flanged 10" ND, RF, 150, ASME B16.5		180	6"

Nozzle List of the Reboiler

S.No	Nozzle Location	Long Description	Height (H)	Angle (A)	Length (L)
1	Radial	Nozzle, flanged 8" ND, RF, 150, ASME B16.5	3'-4"	90	6"
2	Radial	Nozzle, flanged 6" ND, RF, 150, ASME B16.5	1'-3"	270	6"
3	Radial	Nozzle, flanged 6" ND, RF, 150, ASME B16.5	13'-4"	90	6"
4	Radial	Nozzle, flanged 3" ND, RF, 150, ASME B16.5	13'-4"	270	6"

Nozzle List of the Centrifugal Pump

S.No	Nozzle Type	Long Description
1	Inlet Nozzle	Nozzle, flanged 10" ND, RF, 150, ASME B16.5
2	Outlet Nozzle	Nozzle, flanged 8" ND, RF, 150, ASME B16.5

The following steps are required to complete this tutorial:

a. Open the **Piping Model.dwg** file created in the previous tutorial.
b. Add nozzles to all vertical vessels.
c. Modify nozzles of the reboiler.
d. Modify nozzles of the pumps.
e. Save the model.

Opening the File

1. Double-click on the **AutoCAD Plant 3D 2023 - English** icon; AutoCAD Plant 3D starts.

2. Choose *CADCIM> Plant 3D Drawings > Piping Model* from the **PROJECT MANAGER**; the Plant 3D model is opened.

Adding Nozzles to the Vertical Vessel

1. Click on the vertical vessel; the **Add Nozzle** grip is displayed on it.

2. Click on the **Add Nozzle** grip; a preview of the nozzle and the dialog box for adding or modifying nozzles is displayed. 

3. Choose the **Change Location** tab from the dialog box displayed, if not already chosen; the options to change the location of the nozzle are displayed.

4. Select the **Radial** option from the **Nozzle Location** drop-down list. Next, enter the following values in the edit boxes displayed in this tab:

   H: 120” **A: 180** **L: 6”**

5. Choose the **Change Type** tab and filter the **Select Nozzle** list by setting the following options:

   Size: 6” **Unit: in** **End Type: FL**
   Pressure Class: 150

6. Select **Nozzle, flanged 6” ND, RF, 150, ASME B16.5** from the **Select Nozzle** list. Next, choose the **Close** button to close the dialog box; the nozzle is added.

7. Invoke the dialog box for adding or modifying nozzles again and choose the **Radial** option from the **Nozzle Location** drop-down list in the **Change Location** tab. Next, enter the following values in the respective edit boxes:

   H: 60” **A: 180** **L: 6”**

8. Choose the **Change Type** tab and filter the **Select Nozzle** list by setting the following options:

   Size: 8” **Unit: in** **End Type: FL**
   Pressure Class:150

9. Select **Nozzle, flanged 8" ND, RF, 150, ASME B16.5** from the **Select Nozzle** list. Next, choose the **Close** button to close the dialog box; the nozzle is added.

   Next, you need to modify the nozzle at the bottom of the vessel. To do so, first you need to hide the support at the bottom.

10. Choose the **Hide Selected** tool from the **Visibility** panel in the **Home** tab of the **Ribbon**; you are prompted to select the object to be hidden.

11. Select the support of the vertical vessel and press ENTER; it gets hidden.

12. Press the CTRL key and click on the nozzle located at the bottom; a pencil symbol is displayed on it.

13. Release the CTRL key and click on the pencil symbol; the dialog box for adding or modifying nozzles is displayed. Choose the **Change Location** tab from this dialog box.

14. Select the **Bottom** option from the **Nozzle Location** drop-down list and enter **180** in the **T** edit box and **6"** in the **L** edit box.

15. Choose the **Change Type** tab and then filter the **Select Nozzle** list by setting the following options:

Size	**10"**
Unit	**in**
End Type	**FL**
Pressure Class	**150**

16. Select **Nozzle, flanged, 10" ND, RF, 150, ASME B16.5** from the **Select Nozzle** list. Next, choose the **Close** button from the dialog box; the bottom nozzle is modified.

17. Next, choose the **Show All** tool from the **Visibility** panel in the **Home** tab to show the support at the bottom.

Deleting the Nozzle on the Reboiler

1. Press CTRL and click on the nozzle located at the bottom of the reboiler, refer to Figure 4-39.

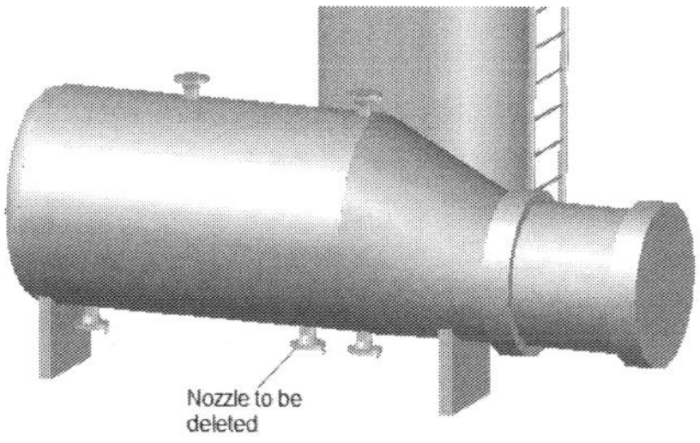

Figure 4-39 *Nozzles to be deleted*

2. Press DELETE to delete the selected nozzle.

Modifying the Nozzles of the Reboiler

1. Press CTRL and click on **Nozzle 1** located at the top of the reboiler, refer to Figure 4-40; a pencil grip is displayed on it, as shown in Figure 4-41. Note that the nozzles are numbered in the figure for reference only.

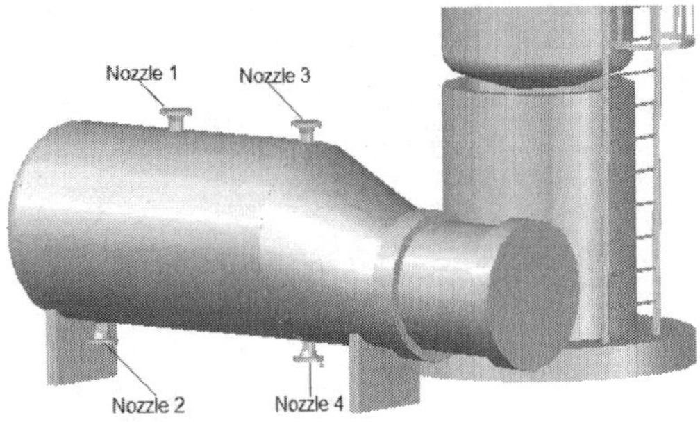

Figure 4-40 Nozzles displayed on the reboiler

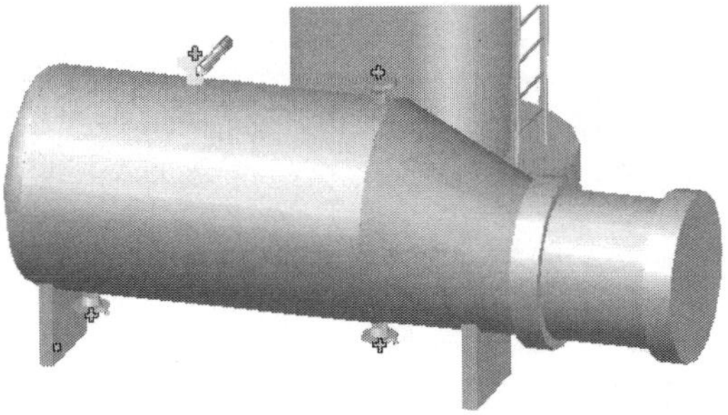

Figure 4-41 Pencil symbol displayed on the nozzle

2. Click on the pencil symbol displayed on the nozzle; the dialog box for adding or modifying nozzles is displayed.

3. Choose the **Change Type** tab from the dialog box and filter the **Select Nozzle** list by specifying the following parameters:

Size	**8”**
Unit	**in**
End Type	**FL**
Pressure Class	**150**

4. Select **Nozzle, flanged 8” ND, RF, 150, ASME B16.5** from the **Select Nozzle** list and close the dialog box.

5. Similarly, modify the other nozzles of the reboiler as given next. For nozzle sequence, refer to Figure 4-40.

Nozzle 2
Type: **Nozzle, flanged 6” ND, RF, 150, ASME B16.5**

Location

H: 15”	**A: 270**	**L: 6”**

Nozzle 3
Type: **Nozzle, flanged 6” ND, RF, 150, ASME B16.5**

Location

H: 160”	**A: 90**	**L: 6”**

Nozzle 4
Type: **Nozzle, flanged 3" ND, RF, 150, ASME B16.5**
Location
H: 160" **A: 270** **L: 6"**

Modifying the Pump Nozzles

Next, you need to modify the pump nozzles.

1. Select a centrifugal pump from the drawing area; pencil symbols are displayed on the inlet and outlet nozzles of the pump.

2. Click on the pencil symbol displayed on the inlet nozzle; the dialog box for adding or modifying nozzles is displayed.

3. Choose the **Change Type** tab from the dialog box and filter the **Select Nozzle** list by specifying the following parameters:

   Size: 10" **Unit: in** **End Type: FL**
   Pressure Class: 150

4. Select **Nozzle, flanged 10" ND, RF, 150, ASME B16.5** from the **Select Nozzle** list and close the dialog box.

   Next, you need to modify the outlet nozzle.

5. Select the pump and then click on the pencil symbol displayed on the outlet; the dialog box for adding or modifying nozzles is displayed.

6. Choose the **Change Type** tab from the dialog box and filter the **Select Nozzle** list by specifying the following parameters.

   Size: 8" **Unit: in** **End Type: FL**
   Pressure Class: 150

7. Select **Nozzle, flanged 8" ND, RF, 150, ASME B16.5** from the **Select Nozzle** list and close the dialog box.

   Similarly, modify the inlet and outlet nozzles of the other centrifugal pump.

Saving the Model

1. Choose the **Save** button from the **Quick Access Toolbar**; the file is saved.

2. Choose **Close > Current Drawing** from the **Application Menu**; the file is closed.

Self-Evaluation Test

Answer the following questions and then compare them to those given at the end of this chapter:

1. On choosing the __________ button in the **Create Equipment** dialog box, a flyout will be displayed with a list of shapes.

2. The __________ area in the **Create Equipment** dialog box displays a list of shapes piled up to create a piece equipment.

3. The __________ edit box in the dialog box that is displayed for adding or modifying nozzles is used to specify the value to create a nozzle at an inclination to the nozzle axis.

4. The __________ edit box in the dialog box that is displayed for adding or modifying nozzles is used to apply a twist angle to the nozzle.

5. On selecting the __________ check box in the dialog box that is displayed for adding or modifying nozzles, the nozzle will be created perpendicular to the equipment.

Review Questions

Answer the following questions:

1. You can attach objects to an equipment using the __________ tool.

2. The __________ grip is displayed on selecting an equipment from the model space.

3. You can create a new equipment and save it as a template. (T/F)

4. You can add shapes to a pump. (T/F)

5. The **Convert Equipment** tool is used to convert a solid model into an equipment. (T/F)

Answers to Self-Evaluation Test

1. Add Shape, 2. Shapes, 3. I (Inclination), **4. T** (Twist), **5. P** (Perpendicular)

Chapter 5

Editing Specifications and Catalogs

Learning Objectives

After completing this chapter, you will be able to:

- *Create a new spec file*
- *Add parts to a spec*
- *Edit parts added to a spec*
- *Modify spec*
- *Create a new catalog*
- *Add a new part to a catalog*
- *Modify and assign a branch table*

INTRODUCTION

In this chapter, you will learn to create and edit the specifications and the catalog files. A catalog is a list of parts that can be used while creating a piping model. The specification file contains specifications of the parts present in the catalog. When you are creating a 3D piping model, the specification file provides the required specifications for pipes, fittings, fasteners, and so on.

GETTING STARTED WITH AutoCAD Plant 3D Spec Editor

You can start Autodesk AutoCAD Plant 3D Spec Editor 2023 by double-clicking on its shortcut icon on the desktop of your computer. On doing so, the **Autodesk AutoCAD Plant 3D Spec Editor 2023** will be displayed along with the Welcome screen, as shown in Figure 5-1. You can close the Welcome screen by using the **Close Welcome Screen** button at its top right corner.

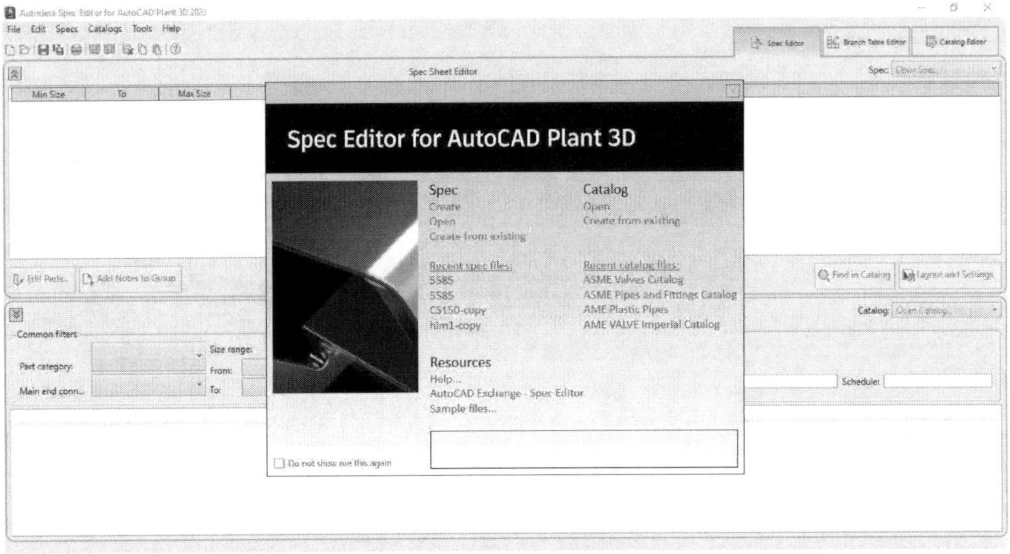

Figure 5-1 *Welcome screen of* ***Autodesk AutoCAD Plant 3D Spec Editor 2023***

The Autodesk AutoCAD Plant 3D Spec Editor contains three tabs: **Spec Editor**, **Branch Table Editor**, and **Catalog Editor**. You can use the **Spec Editor** tab to create and edit the spec sheet of the parts. The **Branch Table Editor** is used to assign a fitting for a branch size. The assigned fitting will be used while creating a branch. The **Catalog Editor** is used to add or remove parts from a catalog. Also, you can edit the general properties and size parameters of a part.

WORKING WITH SPEC FILES

In AutoCAD Plant 3D, you can start a new spec file. To do so, choose **New > Create Spec** from the **File** menu; the **Create Spec** dialog box will be displayed, as shown in Figure 5-2. Next, enter the spec name in the **New Spec name** edit box, and set the path of the file by using the Browse button next to the edit box. Enter a description of the spec in the **Spec description** text box. Next, specify the part catalog that you want to use for creating the spec from the **Load catalog** drop-down list. Then, choose the **Create** button; a new spec file will be created. The newly created spec file will be displayed in the **Spec Sheet**. The parts of the selected catalog are loaded in the **Catalog**. You can also load additional catalogs in the **Catalog** by using the **Open Catalog** option from the **Catalog** drop-down list in the **Catalog**.

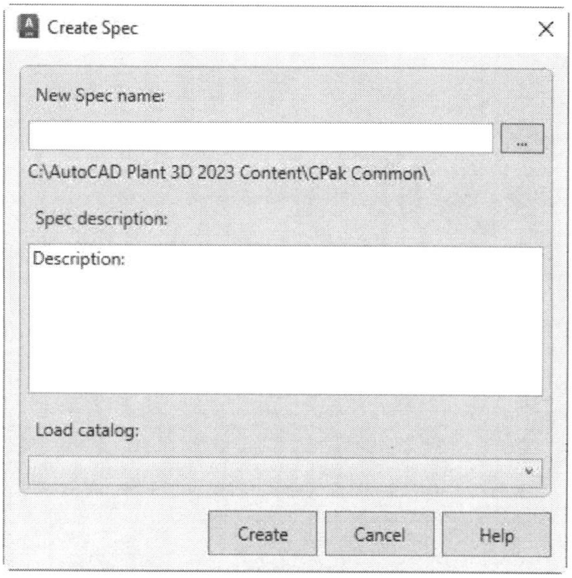

***Figure 5-2** The **Create Spec** dialog box*

> **Tip**
> *You can also start a new spec file in AutoCAD Plant 3D using the **PROJECT MANAGER**. To do so, right-click on the **Pipe Specs** sub-node of the project name node in the **Project** area of the **PROJECT MANAGER**; a flyout will be displayed. Next, choose the **New Pipe Spec** option from the flyout, the **New Spec** dialog box will be displayed. Enter the name of the spec in the **Spec name** edit box and the spec description in the **Spec description** text box and then, choose the **OK** button; the **Autodesk AutoCAD Plant 3D Spec Editor 2023** window will be displayed with the name of the spec file in the **Spec Sheet**.*

You can also add an existing spec file to the current project. To do so, right-click on the **Pipe Specs** sub-node of the **PROJECT MANAGER**; a flyout will be displayed. Choose the **Copy Specs to Project** option from the flyout, the **Select Files to Copy to Project** dialog box will be displayed. Next, browse to the location of the spec file that you want to add to the current project and then choose the **Open** button. The spec file will be added to the **Pipe Specs** sub-node of the **PROJECT MANAGER**.

Creating a New Spec File from an Existing Spec

You can create a new spec file from an existing one. On doing so, the parts of the existing spec file will be loaded to the new spec file and you can add more parts to it. To create a new spec file from the existing file, choose **New > Create Spec From Existing** from the **File** menu; the **Create Spec From Existing Spec** dialog box will be displayed, as shown in Figure 5-3. Choose the Browse button next to the **Source Spec name** edit box; the **Open** dialog box will be displayed. Browse to the location of the existing spec file and choose the **Open** button; the file path will be displayed in the **Source Spec name** and the **New Spec name** edit boxes.

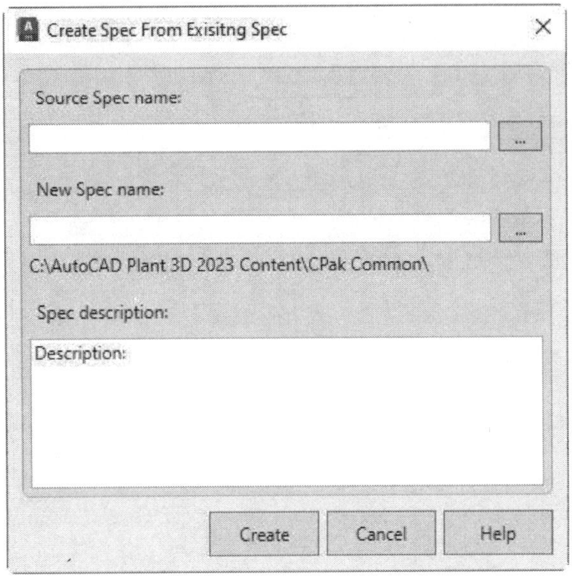

*Figure 5-3 The **Create Spec From Existing Spec** dialog box*

You can modify the name and path of the new spec in the **New Spec name** edit box. It is optional to enter the description in the **Spec description** text box. Next, choose the **Create** button; the **Create Spec From Existing Spec** dialog box will be closed and the new spec file will be created and displayed in the **Spec Editor** tab. The path of the new spec file is displayed in the **Spec Sheet**, refer to Figure 5-4. The options in the **Spec Editor** tab are discussed next.

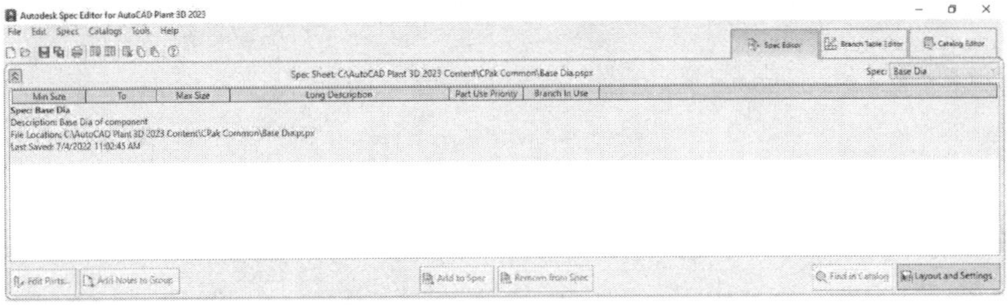

*Figure 5-4 The **Spec Sheet***

Spec Sheet

The **Spec Sheet** displays properties of the parts which are added to a spec file, refer to Figure 5-4. It displays the properties such as **Min size**, **Max size**, **Long Description**, **Part Use Priority**, and **Branch In Use**.

Catalog

The **Catalog** consists of two areas namely **Common filters** and **Property overrides**, and a table which displays the data related to the catalog, refer to Figure 5-5. The **Common filters** area contains options to filter the display of the catalog data and the options in the **Property overrides** area are used to override the property values of the part copied from the catalog to the spec sheet. In AutoCAD Plant 3D, you can only override **Material**, **Material Code**, and **Schedule**.

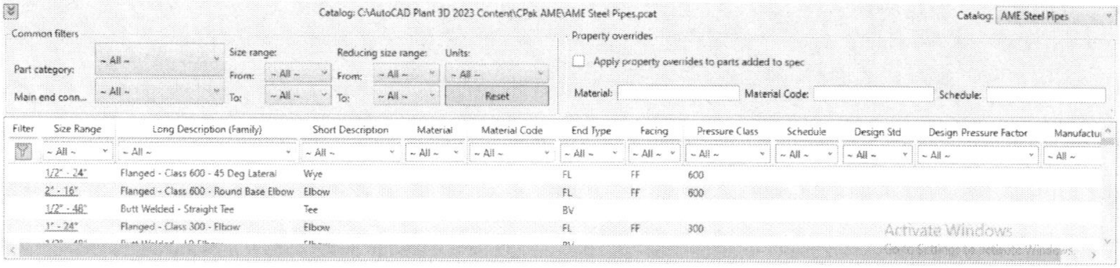

*Figure 5-5 The **Catalog***

Adding Parts to the Spec Sheet

You can add parts from a catalog to a newly created spec or an existing spec. For example, to add a pipe of size ranging from 3" to 10", select **Pipe** from the **Part category** drop-down list of the **Common filters** area in the **Catalog**, refer to Figure 5-6. Set the size range using the **From** and **To** drop-down lists in the **Size range** group; the pipes under the specified range will be displayed in the table. Select a pipe from the table and enter the property values (**Material**, **Material Code**, and **Schedule**) in the **Property overrides** area, refer to Figure 5-7. Next, select the **Apply property overrides to parts added to spec** check box and then choose the **Add to Spec** button; the pipe group will be added to the spec sheet. Similarly, you can add fittings that you want to use while routing a pipe. Also, you can add multiple parts by loading multiple catalogs using the **Catalog** drop-down list.

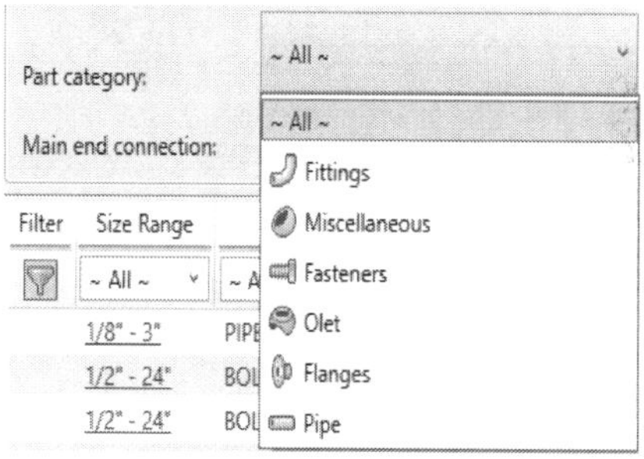

*Figure 5-6 Selecting an option from the **Part category** drop-down list*

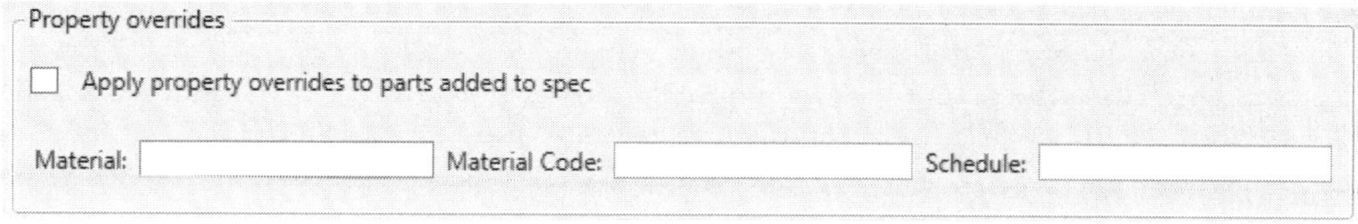

*Figure 5-7 The **Property overrides** area*

To remove a part from a spec, select it from the **Spec Sheet** and choose the **Remove from Spec** button; the part will be removed from the spec sheet.

Editing the Parts Added to a Spec

To edit a part added to a spec, select it from the spec sheet and choose the **Edit Parts** button; the **Edit Parts** dialog box will be displayed, as shown in Figure 5-8. This dialog box contains the **Parts List** tab and the **Edit Properties** tab. The options in these tabs are discussed next.

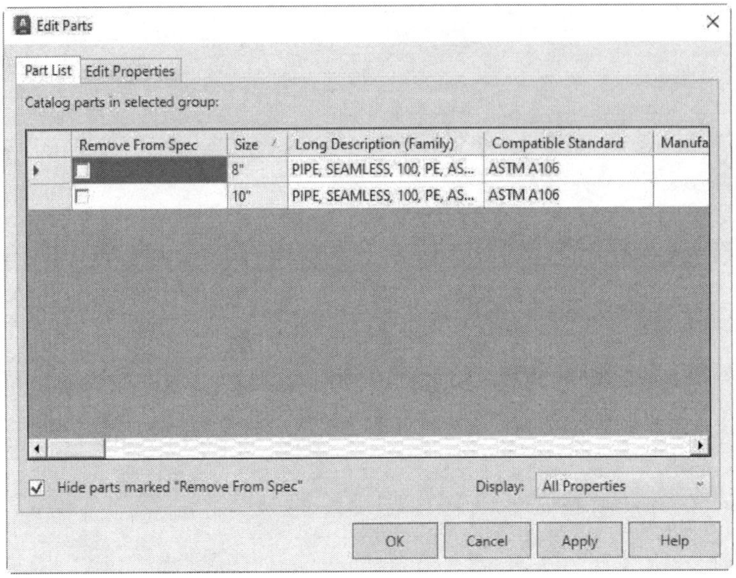

*Figure 5-8 The **Edit Parts** dialog box*

Part List Tab

This tab consists of a table that displays the properties of the parts. You can modify the display of part properties by using the options in the **Display** drop-down list. The following options are available in this drop down list: **Catalog Properties Only**, **Added Properties Only**, and **All Properties**.

The **Remove From Spec** column in the table consists of check boxes which are used to remove the parts from the spec.

The **Hide parts marked "Remove From Spec"** check box is selected by default. When you clear this check box, the parts which are not included in the selected group will be displayed. The properties of these parts are read-only. You can clear the check boxes to include them into the spec file.

Edit Properties Tab

You can add properties to the selected part using the options in this tab. Figure 5-9 shows the options in the **Edit Properties** tab. The **Property definition** area contains the options which are used to define a new property. The **Add Property to** drop-down list in this area is used to specify whether to add the property to the current part or to all parts.

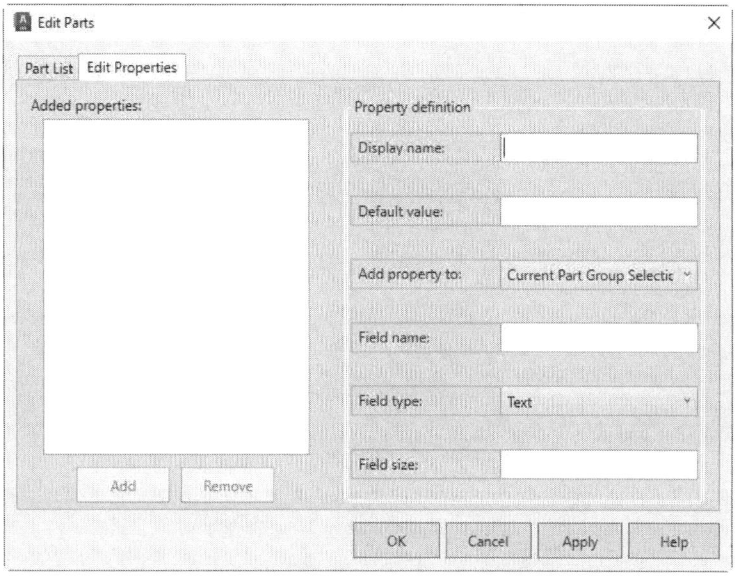

***Figure 5-9** The **Edit Properties** tab in the **Edit Parts** dialog box*

The **Added properties** list box displays the newly added properties. The **Add** button below this list box is used to add the properties defined in the **Property definition** area. Note that the **Add** button will become active only after entering values in all the edit boxes of the **Property definition** area. You can remove the newly added properties by selecting them from the **Added properties** list box and then choosing the **Remove** button. Choose the **Apply** button to accept the changes and then, choose the **OK** button to exit the **Edit Parts** dialog box.

Setting the Part Use Priority

After assigning sizes to the parts in the spec sheet, there may be a case in which there are more than one part of the same size in a spec file. In such a case, the system is unable to decide upon the part to be used first and it displays an error symbol in the spec sheet. Therefore, you need to manually set the priority of the parts. To do so, click on the error symbol in the spec sheet; the **Part Use Priority** dialog box will be displayed, as shown in Figure 5-10. Follow the steps given next to set the priority in this dialog box.

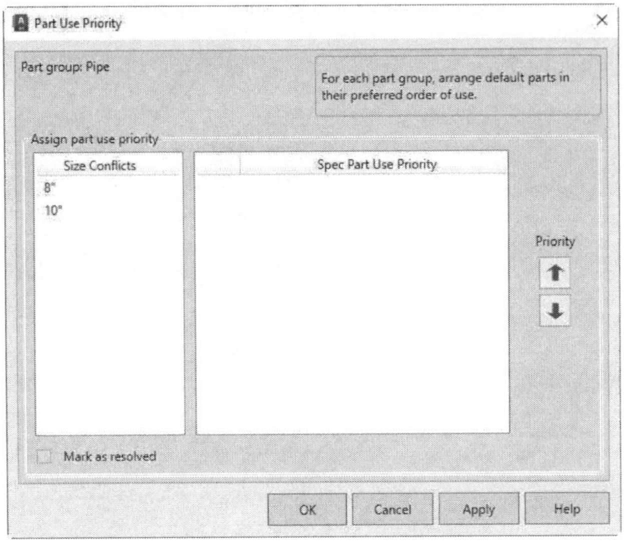

Figure 5-10 *The **Part Use Priority** dialog box*

1. Select a size from the **Size Conflicts** list in the **Assign part use priority** area; the parts under the selected size will be displayed in the **Spec Part Use Priority** list box.

2. Select a part from the **Spec Part Use Priority** list and move it up and down in the list using the up and down arrows, respectively.

3. Select the **Mark as resolved** check box and choose the **OK** button to close the dialog box.

Adding Notes to a Group

You can add notes to a part group that you have added to a spec sheet. To do so, select a part from the spec sheet and choose the **Add Notes to Group** button from the **Spec Sheet**; the **Add Notes To Group** dialog box will be displayed, as shown in Figure 5-11. This dialog box consists of a text box in which you can enter notes. The entered note will be displayed when the spec sheet is printed. You can edit the notes by again choosing the **Add Notes to Group** button from the **Spec Sheet** and can also check the spellings by right-clicking in the text box.

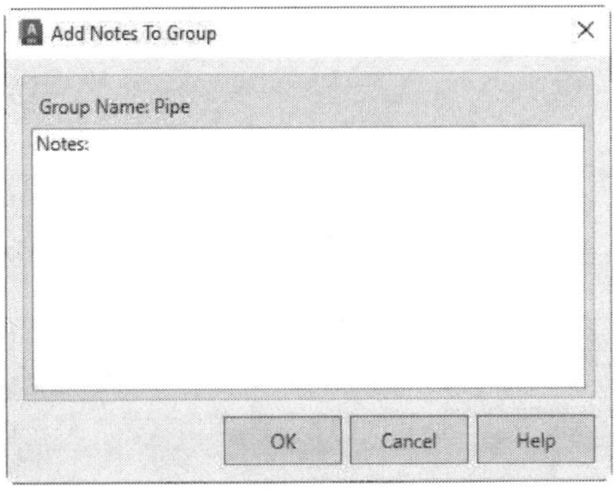

Figure 5-11 *The **Add Notes To Group** dialog box*

Editing the Long Description Styles

In **AutoCAD Plant 3D Spec Editor 2023**, you can customize the long description styles of parts as per the requirement of the project. For example, the default long description style of a Y-Type Strainer is shown in Figure 5-12. You can modify the long description, as shown in Figure 5-13.

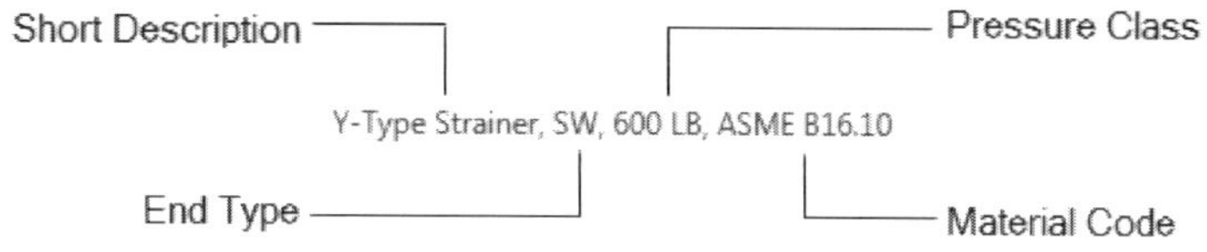

Figure 5-12 *A default long description style of a Y-Type Strainer*

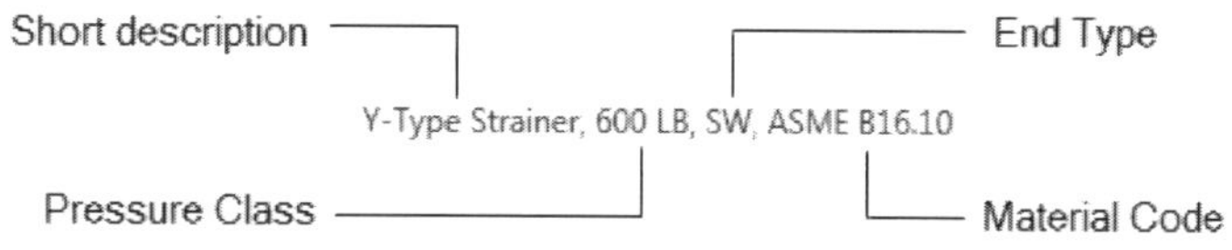

Figure 5-13 *Modified long description style*

To modify the long description style of a piping component, choose the **Layout and Settings** button in the **Spec Sheet** of the **Spec Editor** tab; the **Spec Editor Layout and Settings** dialog box will be displayed, as shown in Figure 5-14. In this dialog box, choose the **Edit long description styles** button; the **Edit Long Description Style** dialog box will be displayed, as shown in Figure 5-15.

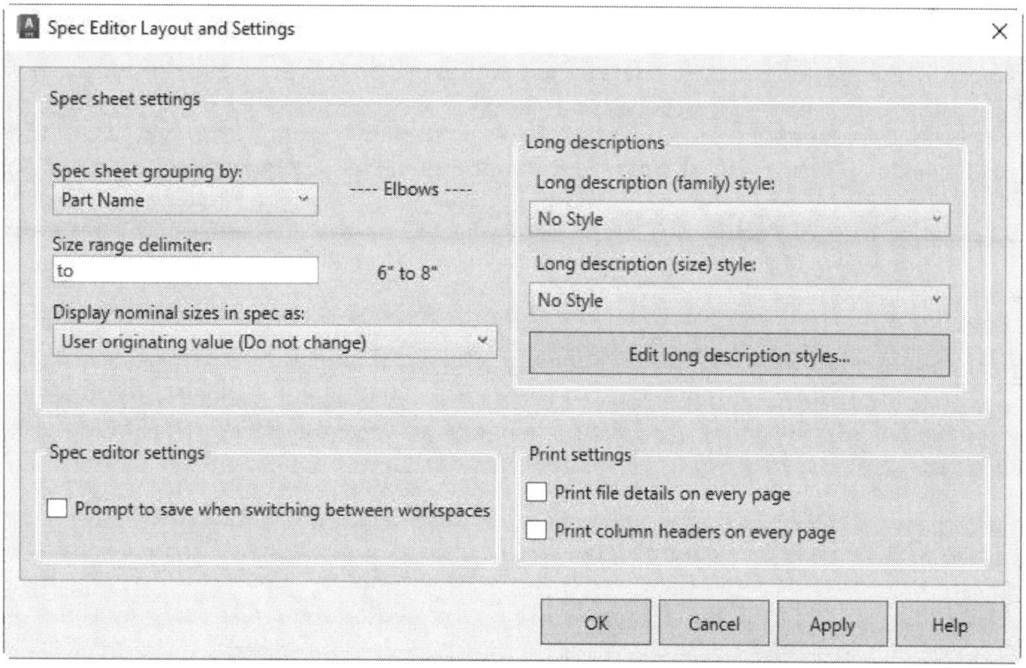

Figure 5-14 *The **Spec Editor Layout and Settings** dialog box*

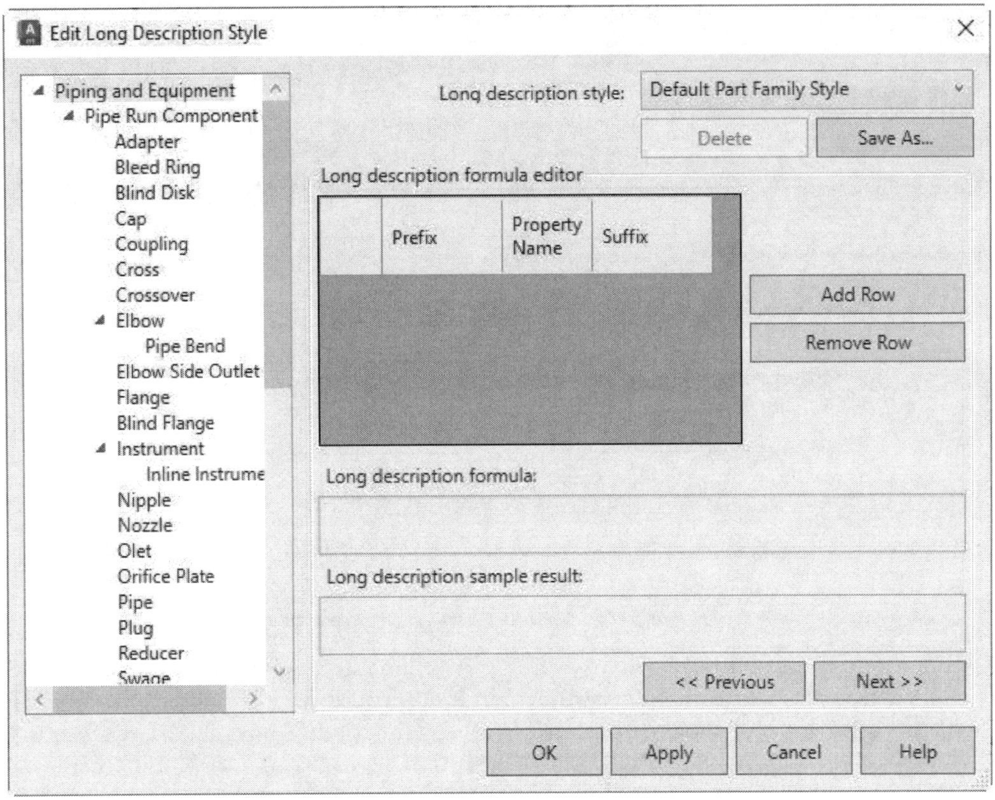

*Figure 5-15 The **Edit Long Description Style** dialog box*

Follow the steps given next to edit the long description.

1. Select a component from the Part list available on the left side of the **Edit Long Description Style** dialog box; the default long description settings will be displayed.

2. Select the **Default Part Family Style** option from the **Long description style** drop-down list, if not selected by default; the editor table in the **Long description formula editor** area displays the property names arranged.

3. You can edit the formula editor table by modifying the names of the properties and by adding new prefixes and suffixes to the table. Also, you can add a new row to this table by choosing the **Add Row** button. The resultant long description style will be displayed in the **Long description formula** text box.

4. Choose the **Save As** button located below the **Long description style** drop-down list; the **Save As New Style** dialog box will be displayed.

5. Enter the name in the **Style name** edit box and choose the **Create** button; a new long description style will be created.

6. Choose the **OK** button; the **Edit Long Description Style** dialog box will be closed and the **Spec Editor Layout and Settings** dialog box will be activated.

7. Select the newly created long description style from the **Long description (family) style** or the **Long description (size) style** drop-down list and choose the **OK** button; the selected long description style will be assigned to the part.

Assigning a Long Description Style to Multiple Specs

You can assign a long description style to multiple specs at a time. To do so, choose **Specs > Batch Assign Long Description Styles** from the menu bar; the **Batch Assign Long Description Styles** dialog box will be displayed, refer to Figure 5-16. Select a long description style from the **Long description (family) style** or **Long description (size) style** drop-down list, or from both. Next, choose the **Add** button from the **Select spec files** area; the **Open** dialog box will be displayed. Browse to the spec file location and double-click on it; the spec file will be added to the **Apply long description styles to these pipe specifications** table.

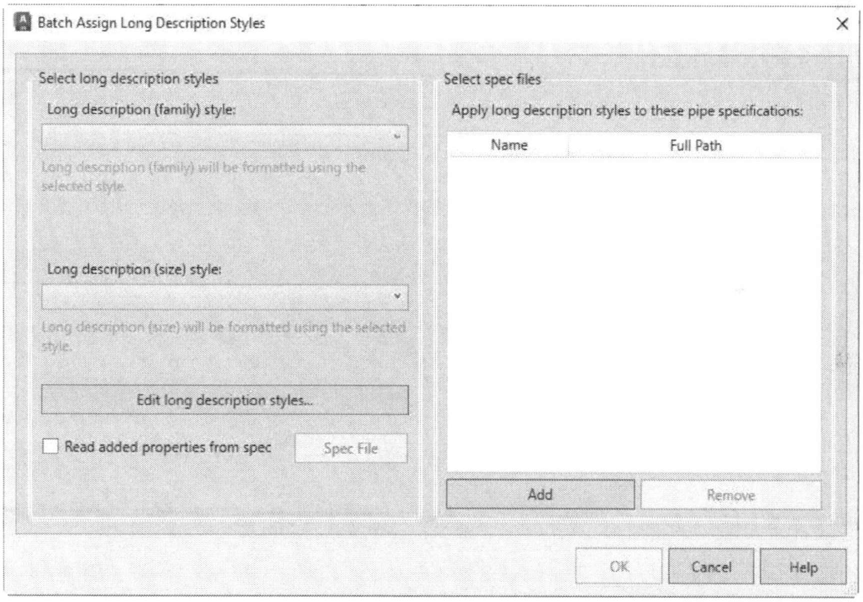

*Figure 5-16 The **Batch Assign Long Description Styles** dialog box*

Similarly, you can add more spec files to the table. You can remove any spec file from the table by selecting it and choosing the **Remove** button. Choose the **OK** button after adding all the spec files to which you want to assign a long description style; the long description style will be assigned and a message box will be displayed with the message **Batch assignment of Long Description Styles has successfully completed**. Next, choose the **Close** button to exit the message box.

Assigning Operators (Actuators) to Valves

Actuators are assigned to the valves by default but they are not displayed in the **Spec Sheet**. To assign an actuator to a valve, right-click on the valve in the **Spec Sheet** and choose the **Edit**

Valve Operator option from the shortcut menu displayed, refer to Figure 5-17; the **Override Valve Operators** dialog box will be displayed, as shown in Figure 5-18.

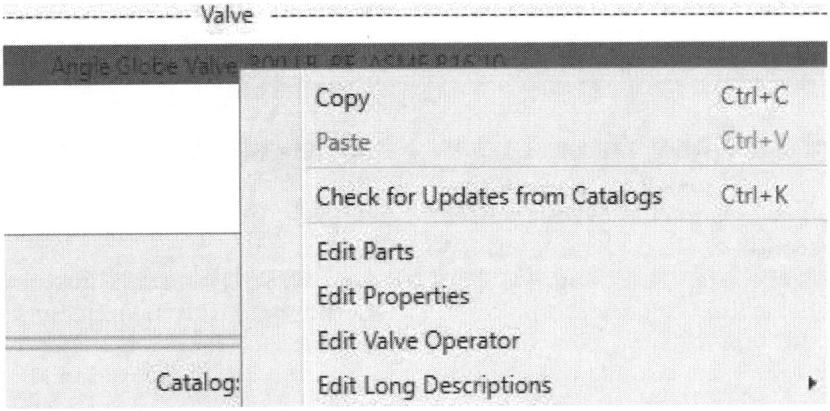

*Figure 5-17 Choosing the **Edit Valve Operator** option*

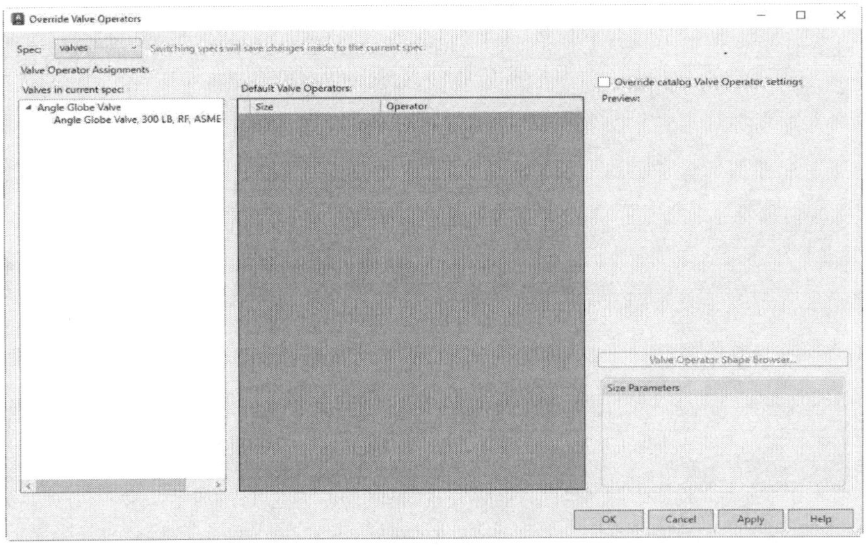

*Figure 5-18 The **Override Valve Operators** dialog box*

In this dialog box, select required valve from the **Valves in current spec** tree; the default operators of the selected valve will be displayed in the **Default Valve Operators** table. Select required size from the **Default Valve Operators** table. Next, select the **Override catalog Valve Operator settings** check box located above the **Preview** area; the **Valve Operator Shape Browser** button will be activated. Now, choose this button; the **Valve Operator Shape Browser** dialog box will be displayed, as shown in Figure 5-19. Select the required valve operator from the **Select Operator Shape** area of this dialog box and choose the **OK** button; the **Valve Operator Shape Browser** dialog box will be closed. Next, choose the **OK** button from the **Override Valve Operators** dialog box; the selected operator will be assigned to the selected valve.

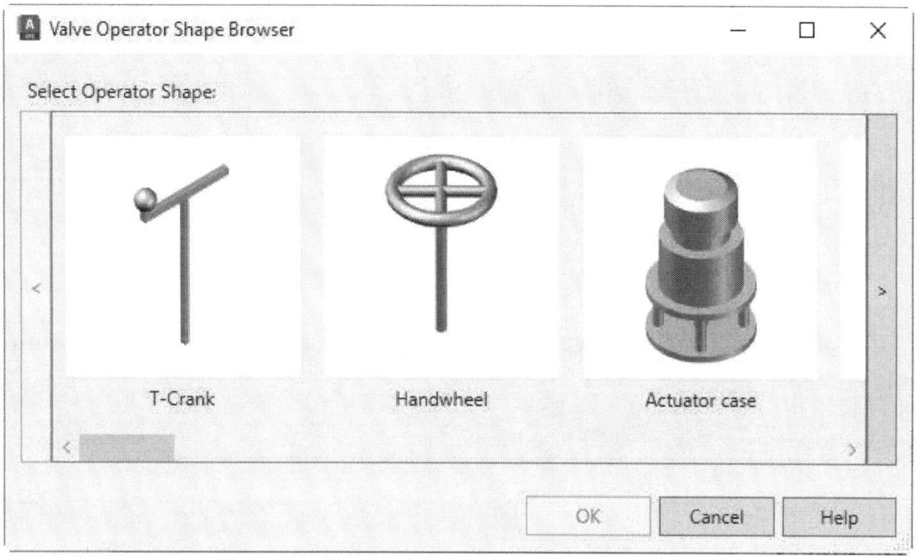

*Figure 5-19 The **Valve Operator Shape Browser** dialog box*

WORKING WITH THE CATALOG EDITOR

The **Catalog Editor** is used to modify a catalog and customize it according to the requirement. You can add or remove parts as well as change the size and properties of the existing parts in the catalog. The **Catalog Editor** tab has two windows - **Piping Component Editor** and **Catalog Browser**. The options in the **Catalog Editor** tab are discussed next.

Piping Component Editor

The **Piping Component Editor** has options to edit properties of the given component. It contains two tabs: **General Properties** and **Sizes**.

General Properties Tab

The **General Properties** tab, refer to Figure 5-20, consists of two areas - **Connection Port Properties** and **Piping Component Properties**. A preview of the selected component is also available on the left in this tab. The two areas in this tab are discussed next.

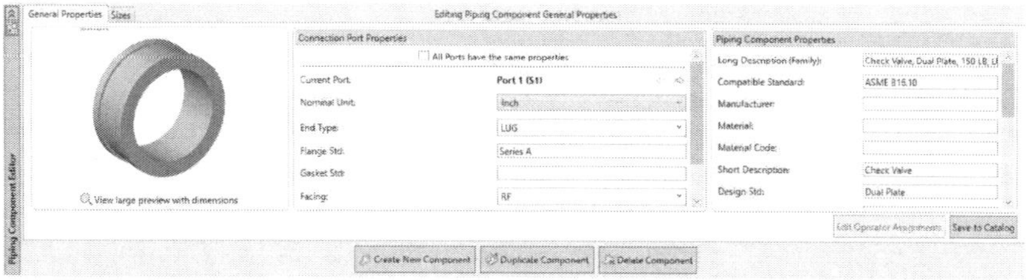

*Figure 5-20 The **General Properties** tab in the **Piping Component Editor***

Connection Port Properties Area

This area contains options to specify port properties. These port properties include nominal unit, end type, flange standard, and so on. You can apply these properties to multiple ports by selecting the **All Ports have the same properties** check box.

Piping Component Properties Area

This area is used to set the property values of the components such as material, material code, design standards, and so on.

Sizes Tab

This tab displays the size data of the component, as shown in Figure 5-21. The options in this tab are discussed next.

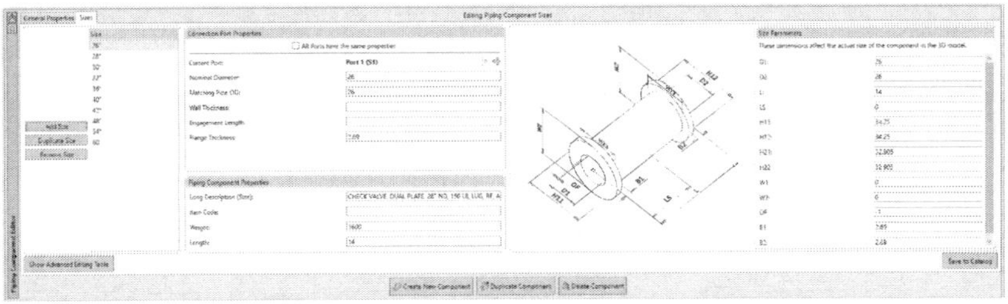

Figure 5-21** The **Sizes** tab in the **Piping Component Editor

Size Area

This area consists of a list box which displays the range of sizes available for the selected component. You can select the required size for the selected component from the list box. Also, you can add or remove a size from the list box using the **Add Size** and **Remove Size** buttons. Also, you can duplicate a size using the **Duplicate Size** button.

Size Parameters Area

This area displays dimensional parameters of the selected component. The preview area shows the parameters in the **Size Parameters** area.

The **Show Advanced Editing Table** button is used to display the component size properties in the form of a table.

Catalog Browser

This is similar to the **Catalog** of the **Spec Editor** tab. But in this browser, there are some additional options to filter the display of the parts in this browser. You can display an individual part size or all part sizes using the **Show All Part Sizes** and the **Show All Part Families** buttons respectively. These buttons are available below the **Catalog** drop-down list in the **Catalog Browser**.

CREATING A NEW CATALOG FROM AN EXISTING CATALOG

You can create a new catalog from an existing one and modify the catalog content as per your requirement. To do so, choose **New > Create Catalog from Existing** from the **File** menu; the **Create Catalog From Existing Catalog** dialog box will be displayed, as shown in Figure 5-22. Select an existing catalog file by using the Browse button next to the **Source catalog name** edit box and then choose the **Create** button; the copy of the selected catalog will be created.

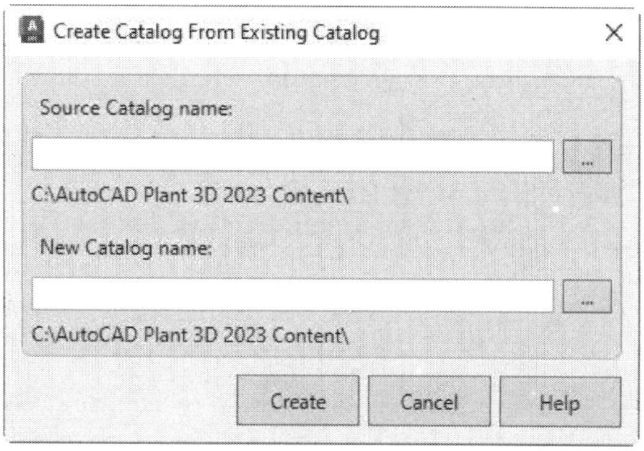

***Figure 5-22** The **Create Catalog From Existing Catalog** dialog box*

ADDING A NEW PART TO A CATALOG

To add a new part to a catalog, choose the **Create New Component** button from the **Piping Component Editor**; the **Create New Component** dialog box will be displayed, as shown in Figure 5-23. Using the options in this dialog box, you can create a new component using two different methods which are discussed next.

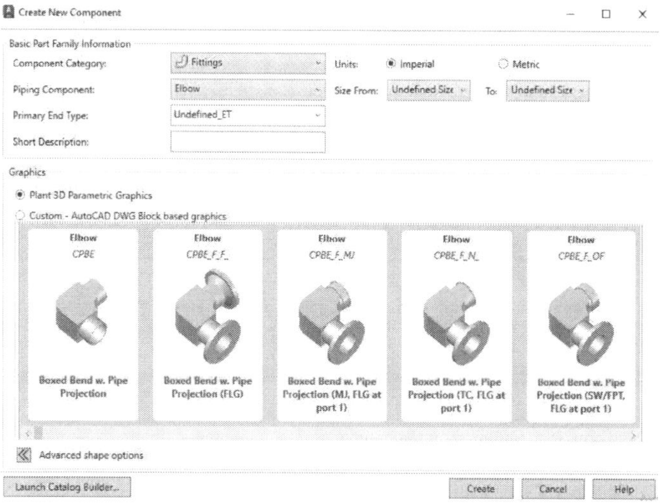

***Figure 5-23** The **Create New Component** dialog box*

Creating a New Component Using Parametric Graphics

To create a component using parametric graphics, follow the steps given next.

1. Select the **Plant 3D Parametric Graphics** radio button from the **Graphics** area of the **Create New Component** dialog box if not selected by default.

2. Select a component category (For example: **Fittings**) from the **Component Category** drop-down list; the **Piping Component** drop-down list displays a list of component types available for the selected component type.

3. Select the desired piping component type from the **Piping Component** drop-down list.

4. Select the primary end type (For example: **FL** for flanged end) from the **Primary End Type** drop-down list.

5. Enter a short description about the component in the **Short Description** edit box.

6. Select the required component from the **Graphics** area.

7. Specify units (For example: **Imperial**) by selecting the respective radio buttons from the **Units** area.

8. Select the size range by using the **Size From** and **To** drop-down lists available below the **Units** area.

9. Choose the **Create** button; the component will be added to the **Catalog Editor**.

10. Choose the **Sizes** tab from **Piping Component Editor**. Now, choose the **Show Advanced Editing Table** button.

11. Enter the outer diameter value in the **Matching Pipe OD** edit box in the **Connection Port Properties** area. Next, select the **All Ports have the same properties** check box from the **Connection Port Properties** area if the connection port properties of all ports are same. If not, select the right arrow and enter values of the matching pipe OD for all the other ports.

12. Enter the long description of the component in the **Long Description (Size)** edit box from the **Piping Component Properties** area.

13. Choose the **Save to Catalog** button; the component will be added to the catalog.

Creating a New Component Using Block Based Graphics

To create a new component using blocks, first you need to create a block and convert it into an AutoCAD Plant 3D component. Create a valve, elbow, flange, or any other piping component by using AutoCAD modeling tools and convert it into a block. Next, invoke the **PLANTPARTCONVERT** command and select the block; the prompt **Select a port operation [Add/Delete/Move/eXit] <Add>:** will be displayed at the command prompt. Enter the **Add** option at the command prompt; you will be prompted to specify the port location. Specify a point and create a port, as shown in Figure 5-24, and choose the **Accept** option and then the **eXit** option after adding the port. Similarly, add ports to the other side of the block and choose the **Accept** option and then the **eXit** option. Note that whenever you use the **eXit** option, the file in which the block is created is to be saved.

Figure 5-24 *Creating a port on an AutoCAD block*

Switch to **Catalog Editor** and choose **Open Catalog** from the **File** menu; the **Open** dialog box will be displayed. Choose a Catalog file and then select **Open** from this dialog box. After that, choose the **Create New Component** button; the **Create New Component** dialog box will be displayed. In this dialog box, choose the **Custom - AutoCAD DWG Block based graphics** radio button and follow the steps given next.

1. Select a component category (For example: **Valves**) from the **Component Category** drop-down list.
2. Select the desired piping component type from the **Component** drop-down list.
3. Enter a short description about the component in the **Short Description** edit box.
4. Select the primary end type from the **Primary End Type** drop-down list.
5. Specify the number of ports available on the block in the **Number of Connection Ports** drop-down list in the **Graphics** area, refer to Figure 5-25.
6. Specify units (For example: **Imperial**) by selecting the respective radio button.
7. Select the size range by using the **Size From** and **To** drop-down lists available below the **Units** radio buttons.
8. Choose the **Create** button; the dialog box will be closed.
9. Choose the **Sizes** tab in the **Piping Component Editor**.

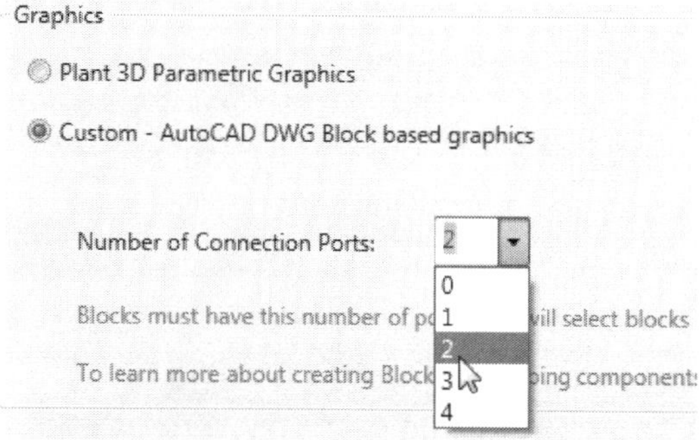

Figure 5-25 *Specifying the number of ports*

10. In the **Size** area, select sizes that are not needed and choose the **Remove size** button. You can also add sizes to the list by using the **Add size** button.

11. Choose the **Select Model** button from the **Size** area; the **Open** dialog box will be displayed. Open the file containing the block; the **Select Block Definition** dialog box will be displayed, as shown in Figure 5-26.

12. Select the block that has ports specified using the **PLANTPARTCONVERT** command and choose the **OK** button; the selected block will be displayed in the **Piping Component Editor**, as shown in Figure 5-27.

13. Enter the outer diameter value in the **Matching Pipe OD** edit box in the **Connection Port Properties** area. Next, select the **All Ports have the same properties** check box from the **Connection Port Properties** area if the connection port properties of all the ports are same. If not, select the right arrow and enter the values of the matching pipe OD for all the other ports.

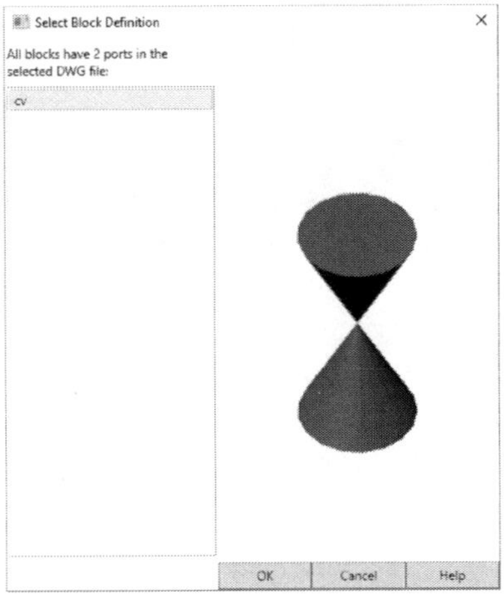

Figure 5-26 The ***Select Block Definition*** *dialog box*

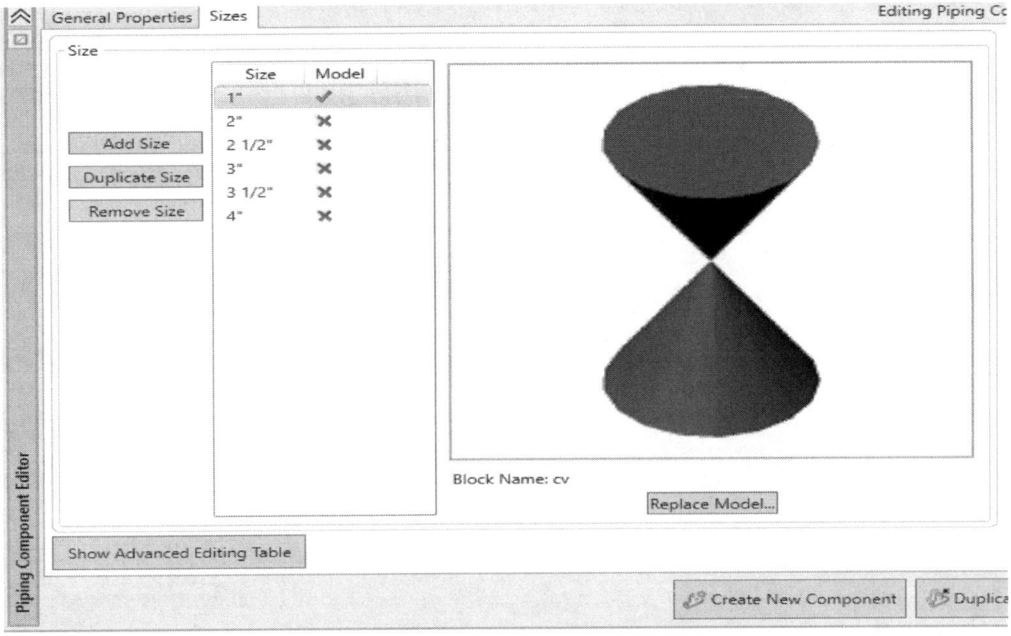

Figure 5-27 The selected block displayed in the ***Piping Component Editor***

Note

The values entered in the ***Nominal Diameter*** *edit box and* ***Matching Pipe OD*** *edit box in the* ***Connection Port properties*** *area follow a particular pattern. For example, if the value of nominal diameter entered in the* ***Nominal Diameter*** *edit box is 1 then the standard outside diameter value will be 1.315 for the* ***Matching Pipe OD*** *edit box.*

14. Enter the long description of the component in the **Long Description (Size)** edit box.

15. Choose the **Save to Catalog** button; the component will be added to the catalog.

MODIFYING THE BRANCH TABLE

The branch table determines the type of fitting to be used while joining a header pipe with a branch pipe. The branch table is prepared by arranging the header sizes in the columns and the branch sizes in the rows, as shown in Figure 5-28. While routing a pipe in AutoCAD Plant 3D, you can use the branch table information to match the header size with the branch size and insert an appropriate fitting connecting them. In the branch table, the cell where the header size column and the branch size row meet, a legend symbol is displayed that represents the type of fitting, refer to the highlighted cell shown in Figure 5-28, where the header size 24" and the branch size 2" meet. It displays two legend symbols representing the fittings (S001,O001) to be used while connecting the header pipe of 24" with the branch pipe of 2". You can also assign more symbols to the branch cell.

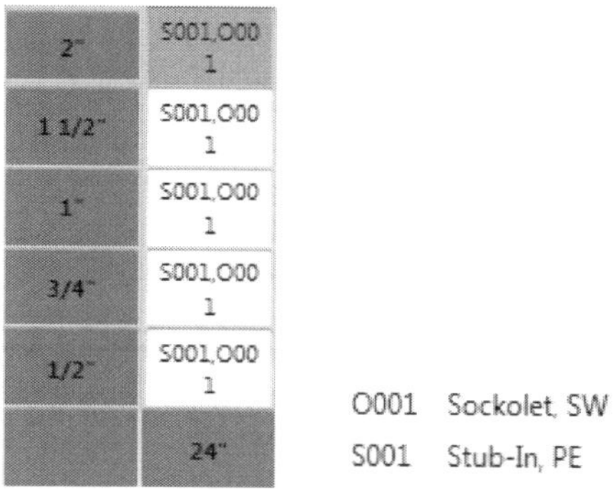

Figure 5-28 *Cells displaying fittings to be used for different branches*

Creating Branch Table Legends

To create a legend (symbol) for a branch table, first you need to identify all the branch fittings available in a spec. Next, you need to decide the type of fittings to be used for each type of branch and then create their legends. Follow the steps given next to create and add a new legend to a branch table.

1. Choose the **Edit Legend** button from the **Legend Notes** pane in the **Branch Table Editor** tab; the **Branch Table Setup** dialog box will be displayed, refer to Figure 5-29.

2. Choose the **Add Branch** button from the **Branch connection part setup** area; a new branch row will be added to the part setup table.

3. Select the desired fitting from the **Part Type** drop-down list in the newly created row.

4. Select a part from the **Spec Part** drop-down list. If you select the **Use preferred part from pipe spec** option, the part will be selected based on the priority set in the **Spec Sheet**.

5. Double-click in the **Legend Symbol** field and enter a symbol (for example: T001). Next, choose the **OK** button.

You can also add a reducer along with the fitting to connect the branch pipe. To do so, select the check box in the corresponding field in the **Add Reducer** column of the **Branch connection part setup** table; a new row will be added. You first need to select a reducer type and then the reducer part.

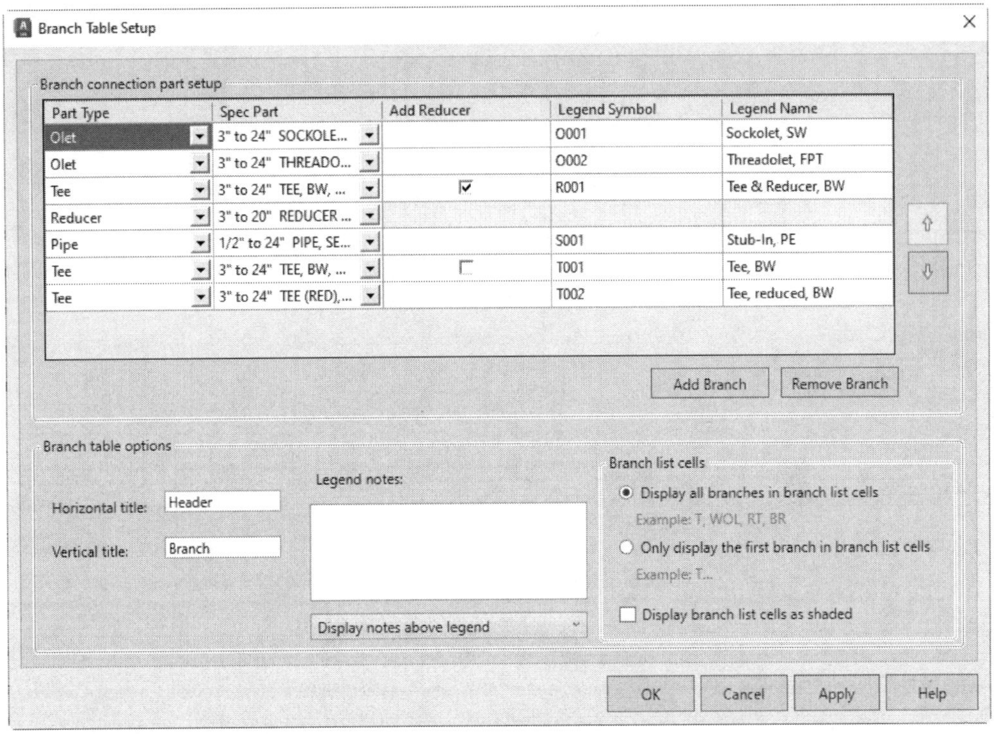

***Figure 5-29** The **Branch Table Setup** dialog box*

Assigning Legends to a Branch Table

You can assign more than one branch fitting to a branch. Also, you can change the default branch fitting and assign a new one. To do so, open the **Branch Table Editor** and follow the steps given next.

1. Select a cell that is intersecting with the header size and branch size.

2. Right-click on the selected cell and choose the **Multi Branch Selection** option from the shortcut menu displayed; the **Select Branch List** dialog box will be displayed, refer to Figure 5-30.

3. The table in this dialog box displays the available branch symbols. In the table, select the check box adjacent to the **Branch Symbol** under the **Use Branch** column; the fitting will be added to the selected cell in the **Branch Table Editor** and will be used while routing.

4. Use the up and down arrows to set the priority of the fitting. Next, choose the **OK** button to close the dialog box.

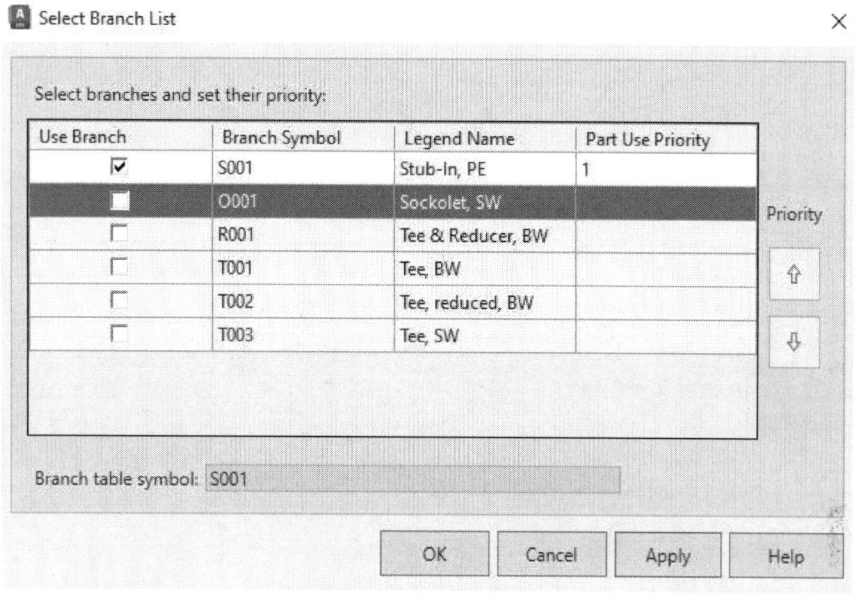

***Figure 5-30** The **Select Branch List** dialog box*

Note

1. Fittings can be assigned to multiple branches simultaneously. To do so, you need to select multiple cells. To select multiple cells which are in a sequential order, click on the first cell and then press and hold the SHIFT key and keep clicking on the cells to be selected.

2. To select multiple cells that are not in a sequence, press and hold the CTRL key and then click on multiple cells.

TUTORIALS

Tutorial 1

In this tutorial, you will create a spec file and add parts to it, as shown in Figure 5-31. You will also modify the part properties. **(Expected time: 30 min)**

The following steps are required to complete this tutorial:

a. Open the AutoCAD Plant 3D Spec Editor and create a new spec file.
b. Add parts to the **Spec Sheet** and edit the part properties.
c. Set part use priorities.
d. Edit long description styles.
e. Save the spec file.

			Cross
2"	to	8"	CROSS, 125 LB, FF, ASME B16.1
3/4"	to	1 1/2"	CROSS, BW, ASME B16.9
			Olet
2"	to	8"	SOCKOLET, 3000 LB, BWXSW, 9/16" LG, ASME B16.11
			Pipe
3/4"	to	1 1/2"	PIPE, SEAMLESS, 10S, PE, ASTM A312
2"	to	8"	PIPE, SEAMLESS, 5S, PE, ASTM A312
2"	to	8"	PIPE, SEAMLESS, 80, PE, ASTM A106
			Reducer
2"	to	8"	REDUCER (ECC), 125 LB, FF, ASME B16.1
			Tee
2"	to	8"	TEE, 125 LB, FF, ASME B16.1
3/4"	to	1 1/2"	TEE, BW, ASME B16.9

Figure 5-31 *The **Spec Sheet** after adding parts to it*

Opening the AutoCAD Plant 3D Spec Editor and Creating a new Spec File

1. Double-click on the shortcut icon of **AutoCAD Plant 3D Spec Editor 2023 - English**; the Autodesk AutoCAD Plant 3D Spec Editor 2023 starts and the welcome screen is displayed.

2. Choose the **Create** option under the **Spec** category from the welcome screen; the **Create Spec** dialog box is displayed.

3. Enter **CADCIM** in the **New Spec name** edit box and **Sample Spec** in the **Spec description** editor.

4. Select **ASME Pipes and Fittings Catalog** from the **Load catalog** drop-down list and then choose the **Create** button; the new spec file is created.

Adding Parts to the Spec Sheet and Editing their Part Properties

In this section, you will add parts to the **Spec Sheet**. To do so, you need to filter the parts in the **Catalog** under the **Spec Editor** tab by using the filter options available in it.

1. Select the **Pipe** option from the **Part category** drop-down list in the **Common filters** area and apply the following filters:
   Size range: **0.75 to 1.5**
   Units: **in**

2. Select the **PIPE, SEAMLESS** option from the **Short Description** drop-down list in the **Catalog** to show seamless pipes in it.

3. Select **PIPE, SEAMLESS, 10S, PE, ASTM A312** from the **Catalog** and choose the **Add to Spec** button from the **Spec Sheet**; the part is added to the **Spec Sheet**.

   Next, you need to add a larger pipe to the **Spec Sheet**.

4. Select the **Pipe** option if not already selected, and apply the following filters:

   Size range: **2 to 8**
   Unit: **in**
   Short Description: **PIPE, SEAMLESS**

5. Select the **Apply property overrides to parts added to spec** check box from the **Property overrides** area and enter the following values:

   Material: **CS**
   Material Code: **A106**
   Schedule: **100**

6. Press the CTRL key on the keyboard and select **PIPE, SEAMLESS, 5S, PE, ASTM A312** and **PIPE, SEAMLESS, 80, PE, ASTM A106** from the **Catalog**. Next, choose the **Add to Spec** button; the parts get added to the **Spec Sheet**.

7. Select the **Fittings** option from the **Part category** drop-down list and apply the following filters:

   Size range: **2 to 8**
   Main end connection: **FL**
   Short Description: **REDUCER (ECC)**

8. Select **REDUCER (ECC), 125 LB, FF, ASME B16.1** from the **Catalog** and choose the **Add to Spec** button from the **Spec Sheet**; the reducer gets added to the **Spec Sheet**.

   Similarly, add other parts to the **Spec Sheet** by filtering required parameters in the **Catalog** using the following filter options:

   Part Category: **Fittings**
   Main end connection: **FL**
   Size range: **2 to 8**

   Select the following parts from the **Catalog**:

   TEE, 125 LB, FF, ASME B16.1
   CROSS, 125 LB, FF, ASME B16.1

   Next, you need to add an olet to the **Spec Sheet**.

9. Select the **Olet** option from the **Part category** drop-down list and apply the following filters:

Main end connection:	**All**
Size range:	**2 to 8**
Short Description:	**SOCKOLET**

10. Select **SOCKOLET, 3000 LB, BWXSW, 3/8" LG, ASME B16.11** from the **Catalog** and choose the **Add to Spec** button; the part is added to the **Spec Sheet**.

11. Add another **Tee** and **Cross** to the **Spec Sheet**. Use the following filter options to filter the **Catalog**:

Part Category:	**Fittings**
Main end connection:	**BV**
Size range:	**0.75 to 1.5**

Select the following parts from the **Catalog**:

TEE, BW, ASME B16.9
CROSS, BW, ASME B16.9

Figure 5-32 shows the **Spec Sheet** after adding parts to it.

		Cross	
2"	to	8"	CROSS, 125 LB, FF, ASME B16.1
3/4"	to	1 1/2"	CROSS, BW, ASME B16.9
		Olet	
2"	to	8"	SOCKOLET, 3000 LB, BWXSW, 9/16" LG, ASME B16.11
		Pipe	
3/4"	to	1 1/2"	PIPE, SEAMLESS, 10S, PE, ASTM A312
2"	to	8"	PIPE, SEAMLESS, 5S, PE, ASTM A312
2"	to	8"	PIPE, SEAMLESS, 80, PE, ASTM A106
		Reducer	
2"	to	8"	REDUCER (ECC), 125 LB, FF, ASME B16.1
		Tee	
2"	to	8"	TEE, 125 LB, FF, ASME B16.1
3/4"	to	1 1/2"	TEE, BW, ASME B16.9

Figure 5-32 *The* ***Spec Sheet*** *after adding parts*

Setting the Part Use Priority

In this section, you will set the part priority for parts with conflicts. You will notice that an error symbol is displayed next to the parts with conflicts, refer to Figure 5-32. It indicates that part usage priority is not assigned to parts having same size.

1. Click on any of the error symbols displayed in the **Part Use Priority** column in the **Spec Sheet**; the **Part Use Priority** dialog box is displayed.

2. Select the **2"** size from the **Size Conflicts** list in the dialog box; the parts for the selected size are displayed in the **Spec Part Use Priority** list.

3. Move **PIPE, SEAMLESS, 80, PE, ASTM A106 Pipe** to top in the **Spec Part Use Priority** list and select the **Mark as resolved** check box. Similarly, move **PIPE, SEAMLESS, 5S, PE, ASTM A312 Pipe** to the top, select all the sizes given in the **Size Conflicts** area, and then, select the **Mark as resolved** check box again. Next, choose the **OK** button; the dialog box is closed and a green dot is displayed in the **Part Use Priority** column. The green dot indicates that the conflict is resolved.

Editing Long Description Styles

In this section, you will edit the long description style of the pipe.

1. Select **PIPE, SEAMLESS, 80, PE, ASTM A106** from the **Spec Sheet** and right-click on it; a shortcut menu is displayed. Choose **Edit Long Descriptions > Assign Long Description Styles to Spec** option from the shortcut menu; the **Spec Editor Layout and Settings** dialog box is displayed.

2. Choose the **Edit long description styles** button from the dialog box; the **Edit Long Description Style** dialog box is displayed.

3. Select **Pipe** from the component list available on the left side in the **Edit Long Description Style** dialog box; the default long description settings is displayed.

4. Select the **Default Part Size Style** option from the **Long description style** drop-down list; the editor table in the **Long description formula editor** area displays the property names arranged by default, refer to Figure 5-33.

5. Modify the formula editor table by deleting some rows and modifying the values in the **Property Name** column. You can also add new prefixes and suffixes to this table. Figure 5-34 shows the formula editor table after modification.

6. Choose the **Save As** button located below the **Long description style** drop-down list; the **Save As New Style** dialog box is displayed.

7. Enter **Long description1** in the **Style name** edit box and choose the **Create** button; a new long description style is created.

	Prefix	Property Name	Suffix
▶		Short Description	
	,	Size	ND
	,	End Type	
	,	Facing	
	,	Pressure Class	LB
	,	Compatible Standard	
	,	Material Code	
	, SCH	Schedule	

Figure 5-33 The formula editor table

	Prefix	Property Name	Suffix
▶		Short Descri...	
	,	Size	ND
	, SCH	Schedule	
	,	End Type	
	,	Compatible ...	

Figure 5-34 The formula editor table after modification

8. Choose the **OK** button; the **Edit Long Description Style** dialog box is closed and the **Spec Editor Layout and Settings** dialog box is displayed again.

9. Select the newly created long description style from the **Long description (size) style** drop-down list and choose the **OK** button; the selected long description style is assigned to the pipe component.

Saving the Spec File

1. Choose the **Save As** tool from the menu bar and specify the location as *C:\Users\User_name\Documents\CADCIM\Spec Sheets* to save the file.

2. Choose **File > Exit** from the menu bar to close the spec file.

Tutorial 2

In this tutorial, you will create a new catalog from an existing one, add a part to it, and modify the part properties. Figure 5-35 shows the part to be added to the catalog. Figures 5-36 and 5-37 show the parameters of the part. **(Expected time: 30 min)**

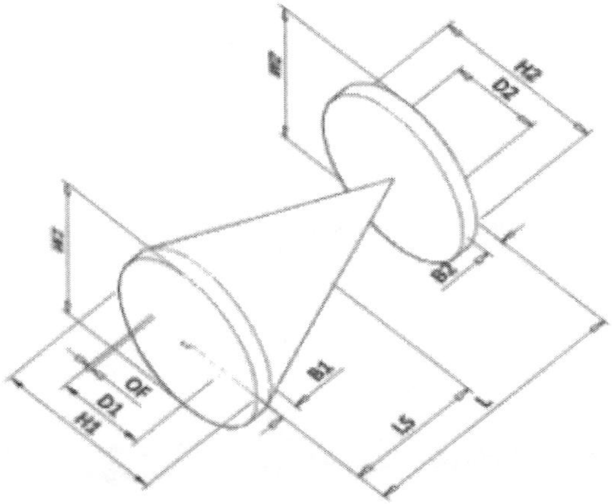

Figure 5-35 Part to be added to the catalog

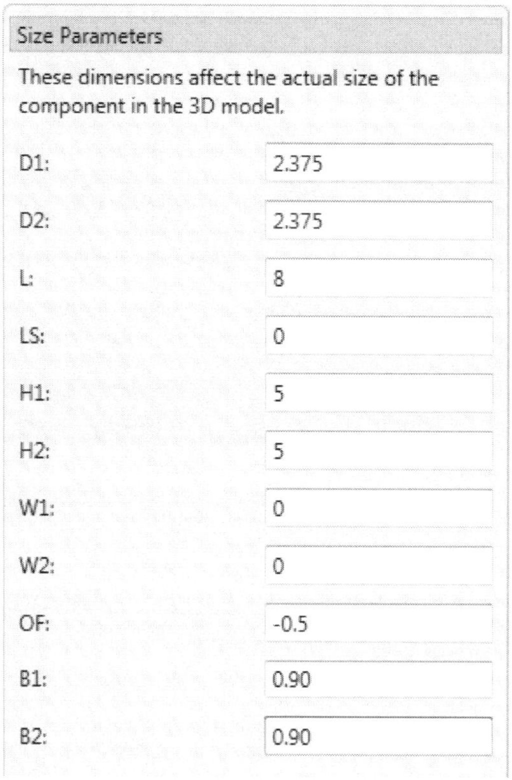

Figure 5-36 *Size parameters for 2" nominal diameter*

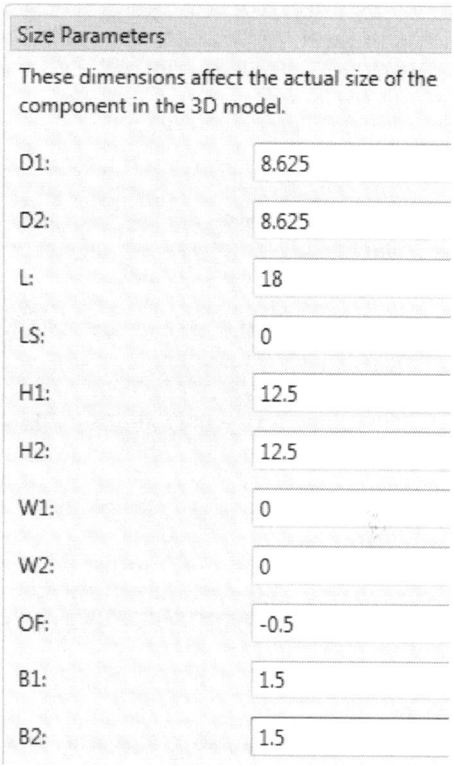

Figure 5-37 *Size parameters for 8" nominal diameter*

The following steps are required to complete this tutorial:

a. Create a new catalog file from the existing catalog.
b. Add parts to the catalog.
c. Save the catalog file.

Creating a New Catalog from the Existing Catalog

1. Double-click on the shortcut icon of **AutoCAD Plant 3D Spec Editor 2023 - English**; the AutoCAD Plant 3D Spec Editor 2023 starts and the welcome screen is displayed.

2. Close the welcome screen and choose **File > New > Create Catalog From Existing** in the menu bar; the **Create Catalog From Existing Catalog** dialog box is displayed.

3. Choose the Browse button next to the **Source Catalog name** edit box; the **Open** dialog box is displayed. Browse to: *C:\AutoCAD Plant 3D 2023 Content\CPak ASME* and double-click on *ASME Valves Catalog.pcat*; the selected catalog file is displayed in the **Source Catalog name** edit box.

4. Choose the **Browse** button next to the **New Catalog name** edit box, browse to: *C:\Users\User_name\Documents\CADCIM*. Now, enter **c05tut02** in the **File name** edit box and choose the **Save** button from the dialog box.

5. Now, choose the **Create** button from the **Create Catalog From Existing Catalog** dialog box; the dialog box is closed and a new catalog file is created.

Adding a New Part to the Catalog

In this section, you will create a new part and add it to the catalog. The part will be created using the parametric graphics.

1. Choose the **Catalog Editor** tab from the upper right corner of the **Piping Component Editor**.

2. Choose the **Create New Component** button from the **Piping Component Editor**; the **Create New Component** dialog box is displayed.

3. Choose the **Plant 3D Parametric Graphics** radio button from the **Graphics** area, if not selected.

4. Choose the **Valves** option from the **Component Category** drop-down list; the **Piping Component** drop-down list displays the **Valve** and **ValveBody** options.

5. Choose the **Valve** from the **Piping Component** drop-down list.

6. Choose the **FL** from the **Primary End Type** drop-down list.

7. Browse to the valve images in the **Graphics** area and select **Inline Valve**, **Check Valve Style (FLG/BW/PE)** from it.

8. Enter **Check Valve** in the **Short Description** edit box.

9. Select the **Imperial** radio button and specify the size range as **2”** to **8”**.

10. Choose the **Create** button; the component is displayed in the **Piping Component Editor**.

11. Choose the **General Properties** tab if not chosen.

12. Specify the connection port properties in the **Connection Port Properties** area, as shown in Figure 5-38.

    Next, you need to specify the piping component properties.

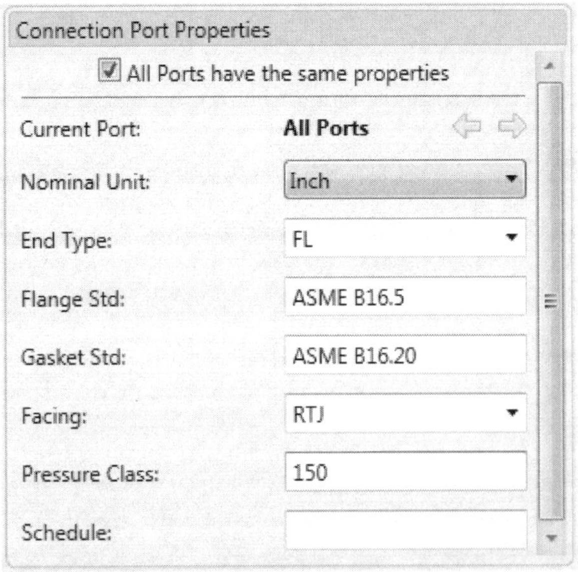

Figure 5-38 *The* ***Connection Port Properties*** *area*

13. Enter **Check Valve, Lift, 150 LB, RTJ, ASME B16.10** and **ASME B16.10** in the **Long description (Family)** and **Compatible Standard** edit boxes, respectively in the **Piping Component Properties** area. Also, specify other property values in this area, refer to Figure 5-39.

Figure 5-39 *Partial view of the* ***Piping Component Properties*** *area*

14. Choose the **Sizes** tab and remove sizes between 2" and 8" from the **Size** list. Next, choose the **Show Advanced Editing Table** button; the **Piping Component Editor** gets expanded and the **Advanced Editing Table** is displayed.

15. Select **2"** from the **Size** list and enter **2.375** in the **Matching Pipe OD** edit box in the **Connection Port Properties** area. Make sure that the **All Ports have the same properties** check box is selected.

16. Enter **CHECK VALVE, LIFT, 2" ND, 150 LB, RTJ, ASME B16.10, 8" LG** in the **Long Description (Size)** edit box and **8** in the **Length** edit box in the **Piping Component Properties** area.

Next you need to specify the size parameters for the check valve for nominal diameter 2". Figure 5-40 displays the size parameters of the check valve.

17. Specify values for nominal diameter 2" in the **Size Parameters** area, as shown in Figure 5-41.

18. Select **8"** from the **Size** list and enter **8.625** in the **Matching Pipe OD** edit box in the **Connection Port Properties** area.

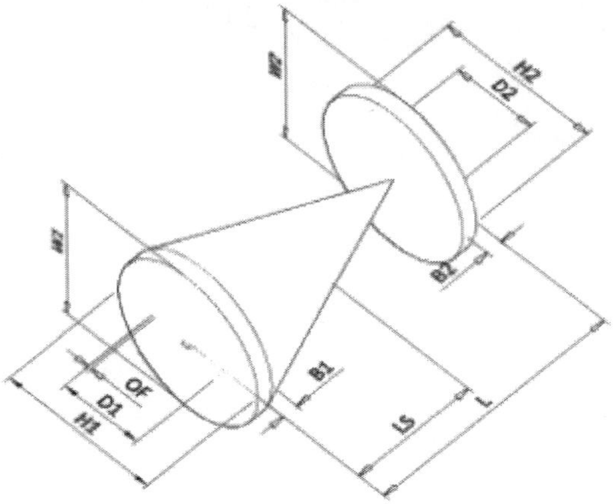

Figure 5-40 *The size parameters of a check valve*

19. Enter **CHECK VALVE, LIFT, 8" ND, 150 LB, RTJ, ASME B16.10, 18" LG** in the **Long Description (Size)** edit box and **18** in the **Length** edit box in the **Piping Component Properties** area.

20. Now, specify values in the **Size Parameters** area, as shown in Figure 5-42.

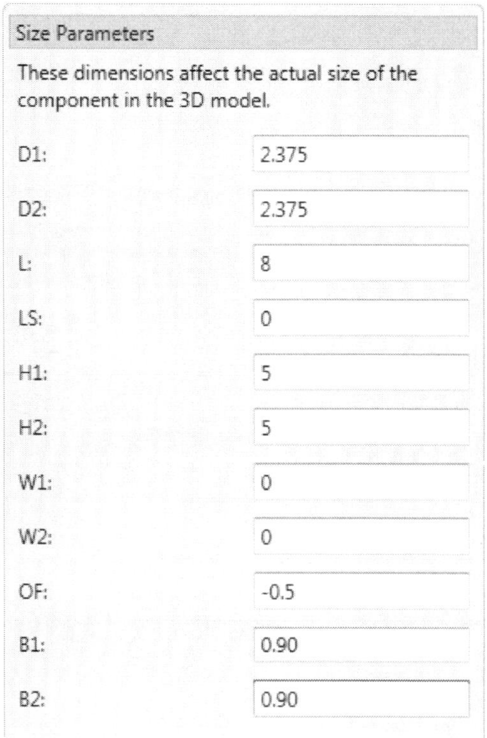

Figure 5-41 *Size parameters for 2" nominal diameter*

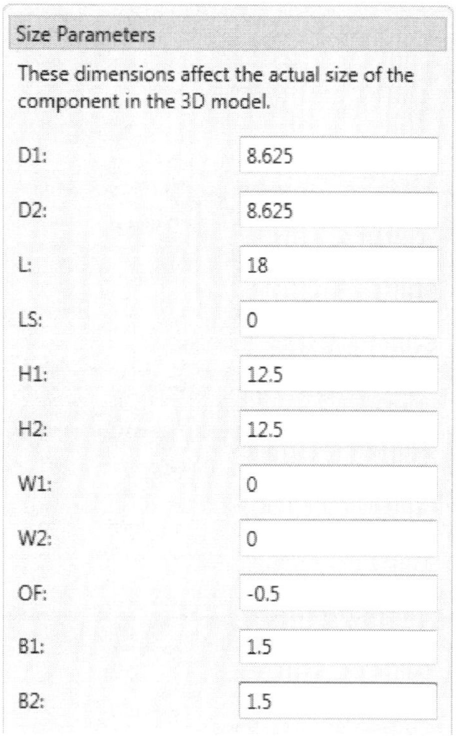

Figure 5-42 *Size parameters for 8" nominal diameter*

21. Choose the **Save to Catalog** button located below the **Size Parameters** area; the part is saved to the catalog.

Saving the Catalog File

1. Choose **File > Save** from the menu bar to save the catalog.

Tutorial 3

In this tutorial, you need to download the *c05tut03.dwg* file from *https://www.cadcim.com*. The path of the file is as follows: *Textbooks > CAD/CAM > Plant 3D > AutoCAD Plant 3D 2023 for Designers > Input Files*. The file contains the 3D model of a valve body and actuators, as shown in Figure 5-43. You will add these blocks to the catalog. The parameters of the parts are given next. **(Expected time: 45 min)**

Hand Wheel Actuator Parameters

Long Description (Family):	Hand Wheel
Long Description:	CUSTOM HAND WHEEL, H=36", W=24"
Short Description:	Custom Hand Wheel
Operator Type:	Manual
Actuator Type:	Wheel
Actuator Height:	36
Actuator Width:	24

Diaphragm Actuator Parameters

Long Description (Family):	Diaphragm
Long Description (Size):	CUSTOM DIAPHRAGM, H=30", W=16"
Short Description:	Custom Diaphragm
Operator Type:	Pneumatic
Actuator Type:	Diaphragm
Actuator Height:	30
Actuator Width:	16

Valve Parameters

Long Description (Family):	Globe Valve, 150 LB, ASME B16.10
Operator Size:	Diaphragm, H=30", W=16"
Matching Pipe Outer Diameter:	4.5

Valve Port Properties

Port Diameter:	4"
Nominal Unit:	Inch
End Type:	BV
Flange Standard:	ASME B16.10
Gasket Standard:	ASME B16.10
Facing:	RF
Pressure Class:	150

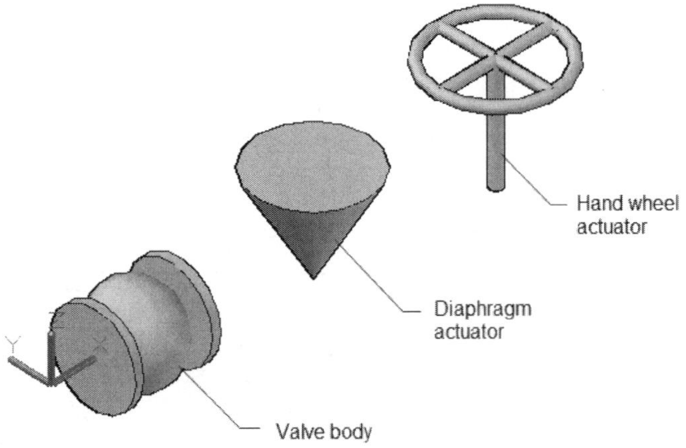

Figure 5-43 *AutoCAD blocks of valve body, diaphragm actuator, and hand wheel actuator*

The following steps are required to complete this tutorial:

a. Convert the AutoCAD blocks into a piping component.
b. Open the catalog created in Tutorial 2 and add actuator.
c. Add a valve to the catalog.
d. Assign operators to the valve.
e. Save the file.

Converting AutoCAD Blocks to Plant 3D Components

In this section, you will convert blocks of a valve body and actuators into AutoCAD Plant 3D components, refer to Figure 5-43.

1. Download the *c05tut03.dwg* file from *https://www.cadcim.com.*

2. Open the *c05tut03.dwg* file in AutoCAD Plant 3D 2023.

3. Choose the **Create Block** tool from the **Block Definition** panel in the **Insert** tab and create three separate blocks and name them as **Valve Body**, **Diaphragm**, and **Hand Wheel**. The insertion points of these blocks are shown in Figure 5-44.

4. Invoke the **PLANTPARTCONVERT** command and select the valve body block; the prompt **Select a port operation [Add/Delete/Move/eXit]** is displayed at the command prompt. Make sure that the **Ortho Mode** and **Dynamic Input** buttons are chosen in the Status Bar.

5. Choose the **Add** option from the contextual menu; you are prompted to specify the port location.

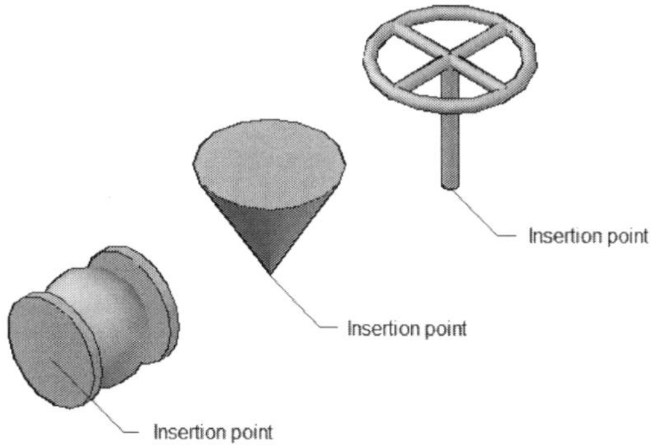

Figure 5-44 *Insertion points of the blocks*

6. Specify the port location on the block, as shown in Figure 5-45; you are prompted to specify the port direction. Move the cursor in the -x-direction and left-click; a contextual menu is displayed. Choose the **Accept** option from the contextual menu; the port is created, refer to Figure 5-45. Also, the **Select a port operation [Add/Delete/Move/eXit] <Add>:** prompt is displayed.

7. Choose the **Add** option from the contextual menu and create another port, refer to Figure 5-46. Next, choose the **Accept** option and then the **eXit** option to get out of the command.

Note

*Choose the **Save** tool whenever you use the **eXit** option.*

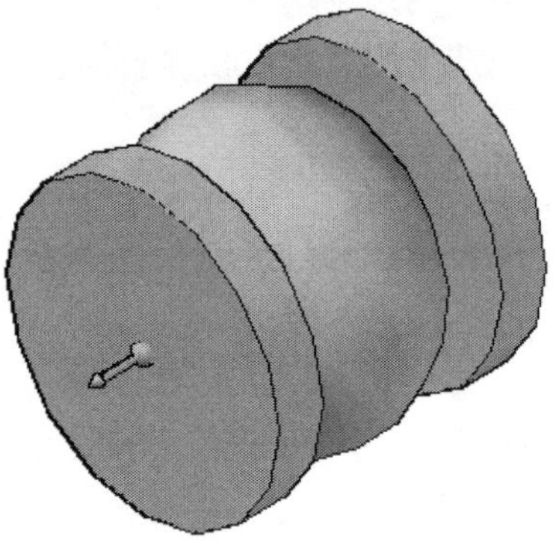

Figure 5-45 Specifying the first port location

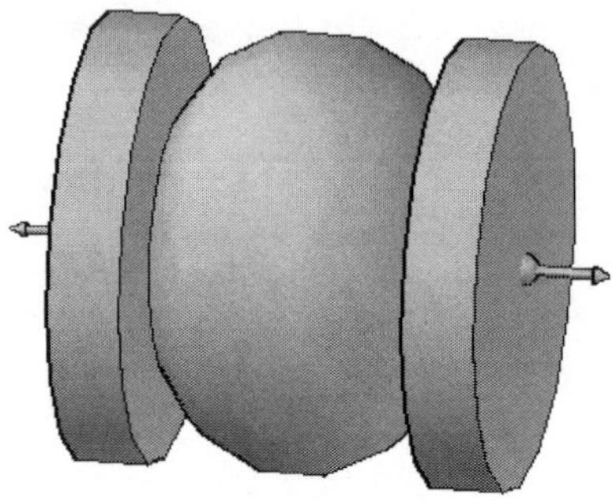

Figure 5-46 Specifying the second port location

Now convert actuators into a piping component.

8. Invoke the **PLANTPARTCONVERT** command. Select the hand wheel actuator block and then choose the **eXit** command from the contextual menu. Now, choose the **Save** tool. Similarly, convert diaphragm actuator block into piping component. Note that you should not add ports to the actuators.

9. Save this file with the name *c05tut03* at the location *C:\Users\User_name\Documents\CADCIM*.

Opening the Catalog and Adding Actuators to It

1. Start **Autodesk AutoCAD Plant 3D Spec Editor** and open the *c05tut02.pcat* file from the Welcome screen. Alternatively, choose **File > Open Catalog** from the menu bar; the **Open** dialog box is displayed. Browse to the location *C:\Users\User_name\Documents\CADCIM* and open the *c05tut02.pcat* file.

2. Choose the **Catalog Editor** tab and then choose the **Create New Component** button; the **Create New Component** dialog box is displayed.

3. Select the **Custom - AutoCAD DWG Block based graphics** radio button from the **Graphics** area.

4. Select **Actuators** from the **Component Category** drop-down list and specify other options, as shown in Figure 5-47.

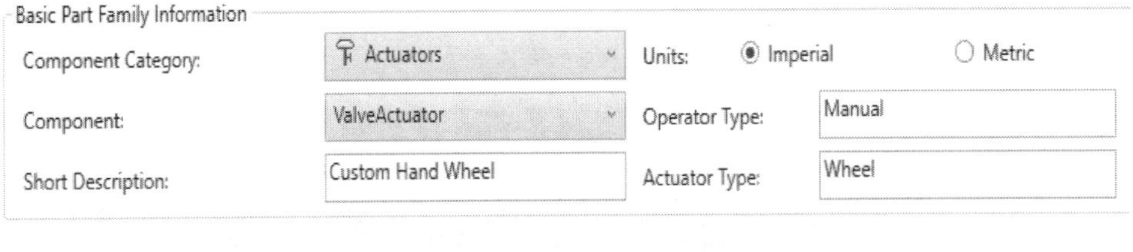

Figure 5-47 The options to be specified in the ***Basic Part Family Information*** *area*

5. Choose the **Create** button from the dialog box; the new entry is created in the catalog. Now, you need to enter the long description (family) and size description.

6. Enter **Hand Wheel** in the **Long Description (Family)** edit box and then choose the **Sizes** tab; various options to specify the size parameters are displayed.

7. Choose the **Select Model** button from the preview window located beside the size list; the **Open** dialog box is displayed.

8. Browse to the location *C:\Users\User_name\Documents\CADCIM* and open the file *c05tut03.dwg*; the **Select Block Definition** dialog box is displayed.

9. Select **Hand Wheel** from the **Select Block Definition** dialog box and choose the **OK** button; the block is displayed in the preview window.

10. Enter the values in the **Piping Component Properties** area, as shown in Figure 5-48. Next, choose the **Save to Catalog** button; the actuator is saved to the catalog.

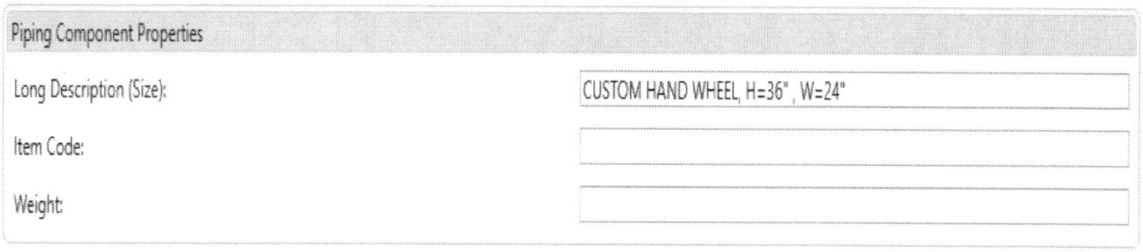

Figure 5-48 *Values to be entered in the* ***Piping Component Properties*** *area*

Next, you need to add the diaphragm actuator to the catalog.

11. Invoke the **Create New Component** dialog box and select the **Custom - AutoCAD DWG Block based graphics** radio button in the **Graphics** area.

12. Select **Actuators** from the **Component Category** drop-down list and then specify other options in the **Basic Part Family Information** area, as shown in Figure 5-49.

Figure 5-49 *The options to be specified in the* ***Basic Part Family Information*** *area*

13. Choose the **Create** button from the dialog box; the new entry is created in the catalog.

Now, you need to enter the long description (family) and size description.

14. Enter **Diaphragm** in the **Long Description (Family)** edit box of the **General Properties** tab and choose the **Sizes** tab; various options to specify the size parameters are displayed.

15. Choose the **Select Model** button from the preview window located beside the size list; the **Open** dialog box is displayed.

16. Browse to the location *C:\Users\User_name\Documents\CADCIM* and open the file *c05tut03.dwg*; the **Select Block Definition** dialog box is displayed.

17. Select **Diaphragm** from the list in the **Select Block Definition** dialog box and then choose the **OK** button; the block is displayed in the preview window.

18. Enter values in the **Piping Component Properties** area, as shown in Figure 5-50. Next, choose the **Save to Catalog** button; the actuator is saved to the catalog.

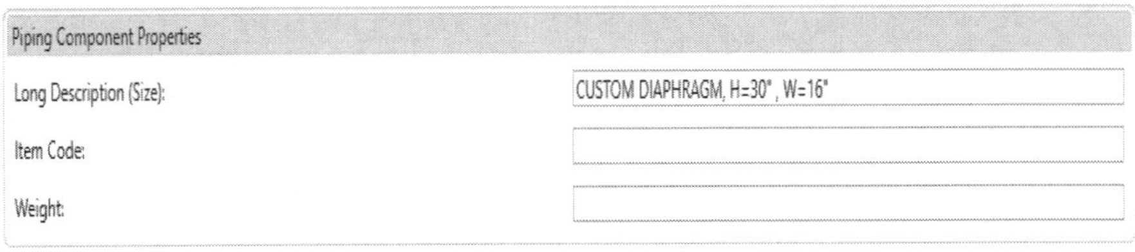

Figure 5-50 *Values to be entered in the* ***Piping Component Properties*** *area*

Adding a Valve to the Catalog

1. Choose the **Catalog Editor** tab if not chosen and then choose the **Create New Component** button; the **Create New Component** dialog box is displayed.

2. Choose the **Custom - AutoCAD DWG Block based graphics** radio button from the **Graphics** area.

3. Choose the **Valves** from the **Component Category** drop-down list.

4. Choose the **ValveBody** from the **Component** drop-down list.

5. Enter **Globe Valve** in the **Short Description** edit box and select **BV** from the **Primary End Type** drop-down list.

6. Select **2** in the **Number of Connection Ports** drop-down list and select the **Imperial** radio button from the **Units** area.

7. Select **4"** in the **Size From** and **To** drop-down lists.

8. Choose the **Create** button; the **Create Component** dialog box is closed and the **Piping Component Editor** is displayed.

9. Specify the port properties in the **Connection Port Properties** area of the **General Properties** tab, as shown in Figure 5-51.

10. Enter **Globe Valve, 150 LB, ASME B16.10** and **ASME B16.10** in the **Long description (Family)** and **Compatible Standard** edit boxes, respectively, in the **Piping Component Properties** area.

11. Choose the **Sizes** tab and remove all sizes except size **4"** if present from the **Size** list.

12. Choose the **Select Model** button from the preview window located beside the size list; the **Open** dialog box is displayed.

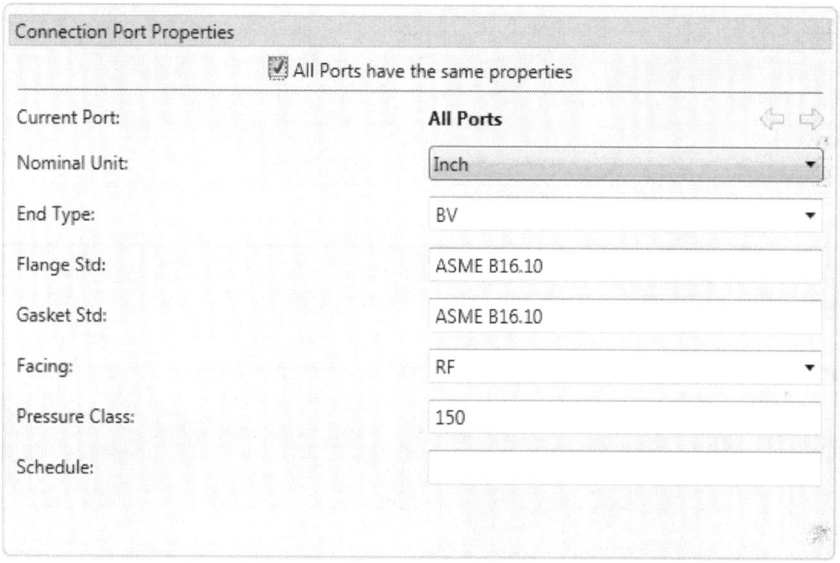

Figure 5-51 *The **Connection Port Properties** area*

13. Browse to the location *C:\Users\User_name\Documents\CADCIM* and open the file *c05tut03.dwg*; the **Select Block Definition** dialog box is displayed.

14. Select **Valve body** from the list in the **Select Block Definition** dialog box and choose the **OK** button; the block is displayed in the preview window.

15. Enter **4.5** in the **Matching Pipe OD** edit box and select the **All Ports have the same properties** check box from the **Connection Port Properties** area; the **Port Properties Differences** message box is displayed. Choose **Yes** from the message box.

16. Enter **Globe Valve, 150 LB, ASME B16.10** in the **Long Description (Size)** edit box and **24** in the **Length** edit box in the **Piping Component Properties** area.

17. Choose the **Save to Catalog** button; the component is saved to the catalog.

Assigning Operators to the Valve

1. Choose the **General Properties** tab and then choose the **Edit Operator Assignments** button which is located below the **Piping Component Properties** area; the **Valve Operator Mapping** dialog box is displayed.

2. In this dialog box, choose the **Add** button located below the **Operator Assignments** table; a new row is added to the table.

3. Select **CUSTOM DIAPHRAGM, H=30", W=16"** from the drop-down list in the **Long Description (Size)** column, refer to Figure 5-52.

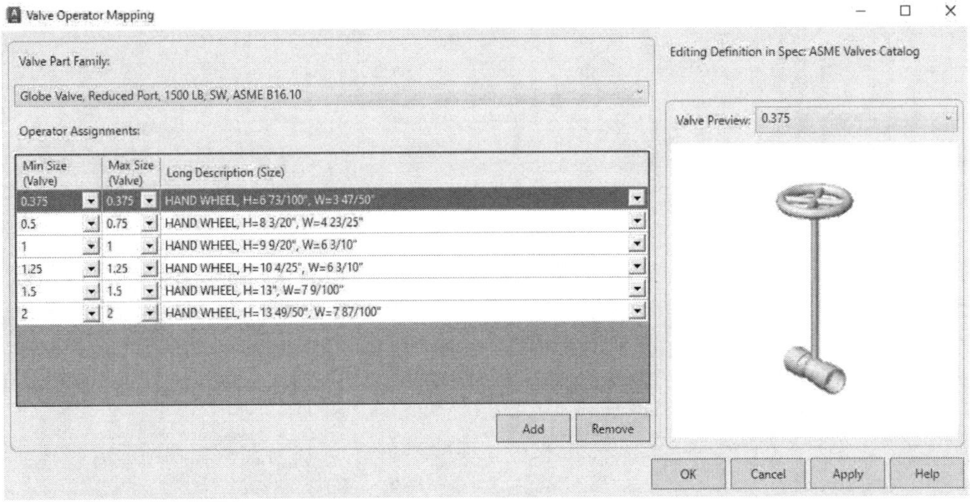

*Figure 5-52 Selecting an operator from the drop-down list in the **Long Description (Size)** column*

4. Choose the **Add** button to add a new row to the table. Next, select **CUSTOM HAND WHEEL, H=36", W=24"** from the **Long Description (Size)** column; the diaphragm actuator is assigned to the globe valve.

5. Choose the **OK** button from the **Valve Operator Mapping** dialog box; the dialog box is closed.

6. Choose the **Save to Catalog** button from the **Piping Component Editor**; the valve is saved in the catalog.

Saving the Spec File

1. Choose **File > Save** from the menu bar; the catalog is saved.

2. Choose **File > Exit** from the menu bar to close the application.

Self-Evaluation Test

Answer the following questions and then compare them to those given at the end of this chapter:

1. The _________ displays the parts which are added to a spec file.
2. The _________ is used to modify the catalog content and customize it according to the requirement.
3. You can apply the specified properties to multiple ports by selecting the _________ check box.
4. The _________ button displays the component size properties in the form of a table.
5. The _________ determines the type of fitting to be used while joining a header pipe with a branch pipe.

Review Questions

Answer the following questions:

1. To create a new component using blocks, first you need to create a block and convert it into an AutoCAD Plant 3D component using the _________ command.
2. To create a _________ for a branch table, first you need to identify all the branch fittings available in a spec.
3. You need to set the part use priority if there are more than one parts available in the same group. (T/F)
4. Actuators are displayed in the spec sheet. (T/F)
5. You can assign more than one branch fitting to a branch. (T/F)

Answers to Self-Evaluation Test

1. **Spec Sheet**, 2. **Catalog Editor**, 3. **All Ports have the same properties**, 4. **Show Advanced Editing Table**, 5. Branch table

Chapter 6

Routing Pipes

Learning Objectives

After completing this chapter, you will be able to:

- *Select a spec*
- *Route a pipe line*
- *Route a pipe using a P&ID*
- *Work with the compass*
- *Connect two pipes*
- *Route a pipe at an offset*
- *Route a pipe at a slope*
- *Create branches*
- *Create Autodesk connection points*

INTRODUCTION

In the previous chapter, you have learned to create equipment for a plant 3D model. In this chapter, you will learn various methods to route pipes.

SELECTING A SPEC

Before you start routing a pipe, you need to select a piping spec. It defines specifications of the pipes, valves, fittings, and other piping components to be used in the 3D piping model. You can select pipes with specific material standard, pressure class, pipe schedule, and so on using the spec file. In addition to pipe specifications, the piping spec also provides you the information about the type of joints, fittings, and fasteners to be used. You can also select valves and valve actuators from a spec.

The **Spec Selector** drop-down list available in the **Part Insertion** panel of the **Home** tab contains various specs, refer to Figure 6-1. Select a spec from the **Spec Selector** drop-down list; various components available in the selected spec are displayed in the **TOOL PALETTES** located on the right in the drawing window. You can select a pipe or piping component from this **TOOL PALETTES**.

After selecting the spec, you need to select the pipe size from the **Pipe Size Selector** drop-down list available in the **Part Insertion** panel, refer to Figure 6-2.

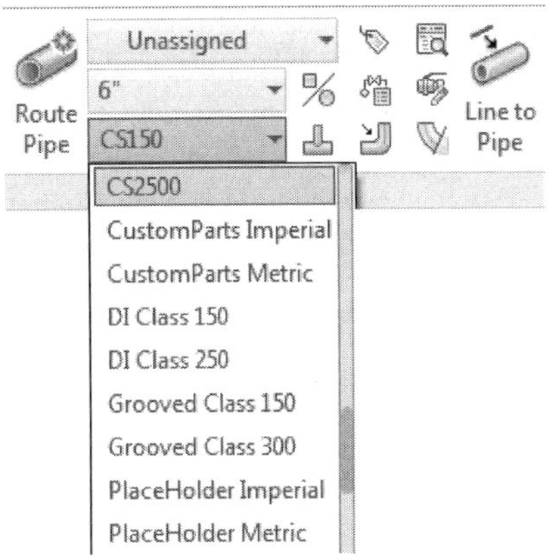

***Figure 6-1** Selecting a spec from the **Spec Selector** drop-down list*

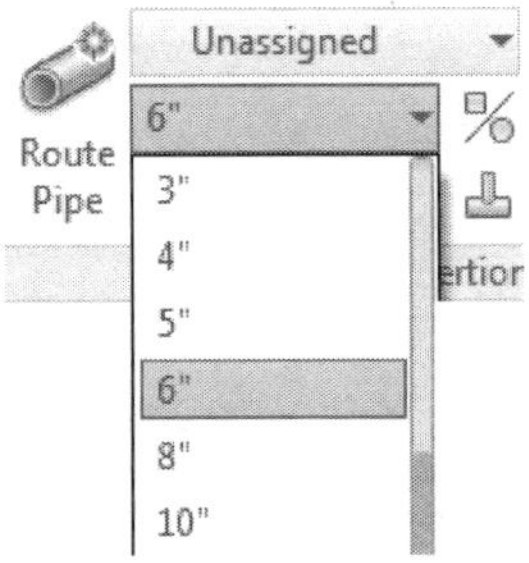

***Figure 6-2** Selecting pipe size from the **Pipe Size Selector** drop-down list*

WORKING WITH THE SPEC VIEWER

In AutoCAD Plant 3D, you can use the **Spec Viewer** palette to view the components present in the selected spec. To do so, choose the **Spec Viewer** button available in the **Part Insertion** panel; the **PIPE SPEC VIEWER** palette will be displayed, as shown in Figure 6-3. It contains the **Spec Sheet** rollout and the **Part Sizes** rollout. When you select a part from the **Spec Sheet**, the **Part Sizes** rollout displays various sizes available for the selected part. Also, the details of the component such as **Part Type**, **End Connection**, **Material Grade** and **Rating** are displayed in the **Part details** area. The **PIPE SPEC VIEWER** palette has many other options. The usage of these options is discussed next.

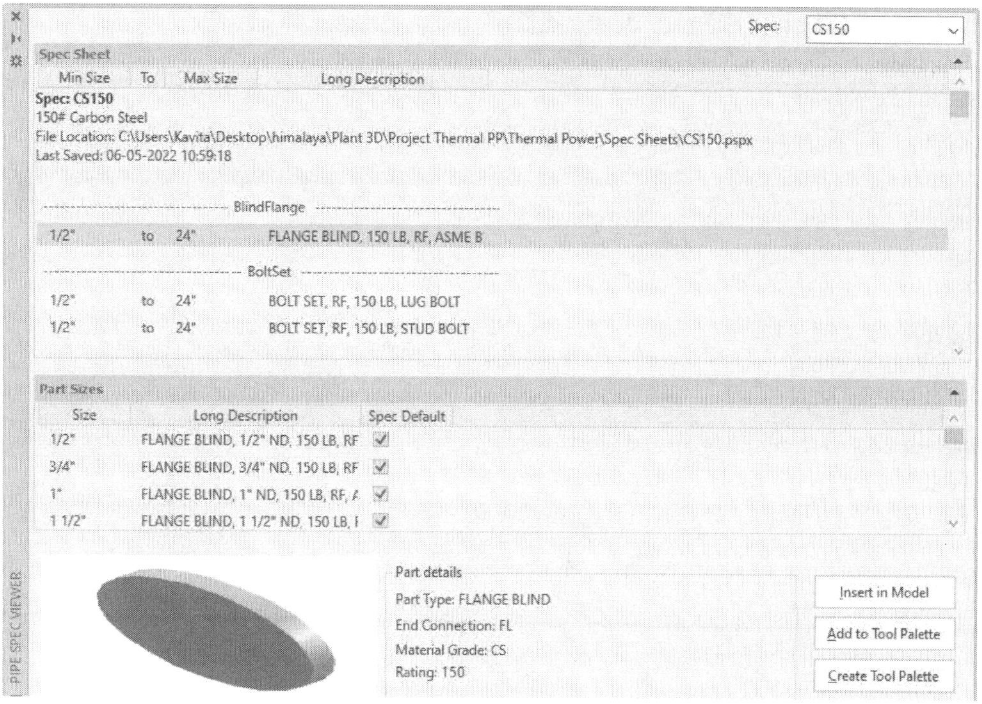

*Figure 6-3 The **PIPE SPEC VIEWER** palette*

Adding a Part to the TOOL PALETTE

To add a part to the **TOOL PALETTES**, select a part from the **Spec Sheet** rollout in the **PIPE SPEC VIEWER** palette; the **Part Sizes** rollout displays various sizes available for the selected part. Select a part size and choose the **Add to Tool Palette** button; the selected part will be added to the **TOOL PALETTES**.

Creating a New TOOL PALETTE

You can also create a new **TOOL PALETTE**. To do so, select a spec from the **Spec** drop-down list and then choose a part from the **Spec Sheet** rollout in the **PIPE SPEC VIEWER** palette. Next, choose the **Create Tool Palette** button; a new tool palette is created and it will be loaded with the parts available in the selected spec.

Inserting a Part from the Spec Viewer

You can insert a part from the **PIPE SPEC VIEWER** palette into a model by using the **Insert in Model** button from the **PIPE SPEC VIEWER** palette. However, this option will be useful to you while placing components after routing a pipe.

ROUTING A PIPE

Piping is an important part of a plant 3D model. In AutoCAD Plant 3D, you can route pipes between various equipment and add inline components to them. The methods to route a pipe are discussed next.

Routing a Pipe with a New Line Number

You can route a pipe and assign it a new line number. To do so, select the **Route New Line** option from the **Line Number Selector** drop-down list in the **Part Insertion** panel of the **Home** tab; the **Assign Tag** dialog box will be displayed, as shown in Figure 6-4. Next, follow the steps given below:

1. Specify a line number in the **Number** edit box.
2. Select a line size from the **Size** drop-down list.
3. Select a new spec from the **Spec** drop-down list, if you want to change the already selected spec.
4. Choose the **Assign** button; the **Assign Tag** dialog box will be closed and you will be prompted to specify the start point of the pipe.
5. Select a point in the drawing area; the pipe will be attached to the cursor and a compass will be displayed, refer to Figure 6-5. You can rotate the pipe at incremental angles using the compass.

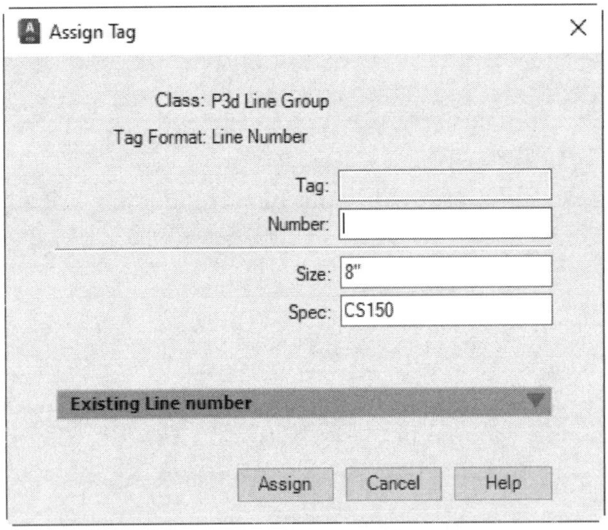

Figure 6-4 *The* ***Assign Tag*** *dialog box*

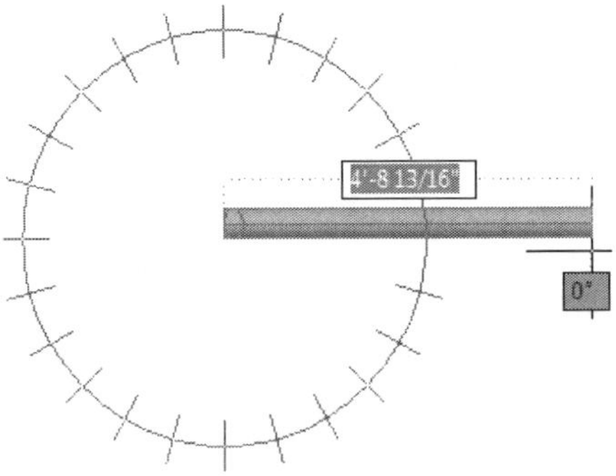

Figure 6-5 *Creating a pipe*

6. Move the cursor and select the end point of the pipe; the pipe will be created and you will be prompted to specify the next point. Press ENTER to finish the line.

Note

If the ***Dynamic Input*** *mode is on, dynamic edit boxes are displayed along with the pipe. You can enter values for the pipe length and the orientation angle in these edit boxes.*

Setting the Route Line

While routing a pipe, you are actually drawing a line. The pipe is created with reference to this line. This reference line is known as the route line. By default, the drawn route line is the center line of the pipe. However, you can change the position of the route line. To do so, select an option from the **Set Routing Line** drop-down list in the **Elevation & Routing** panel in the **Home** tab; the position of the route line changes with respect to the selected option. The options in the **Set Routing Line** drop-down list are discussed next.

Top of Pipe (TOP)

On selecting this option, the route line will be set at the top of center point of the pipe.

Center of Pipe (COP)

On selecting this option, the route line will be set at the center point of the pipe.

Bottom of Pipe (BOP)

On selecting this option, the route line will be set at the bottom of the center point of the pipe.

Top Left

On selecting this option, the route line will be set at the top left point of the pipe.

Top Right

On selecting this option, the route line will be set at the top right point of the pipe.

Center Left

On selecting this option, the route line will be set at the left of center point of the pipe.

Center Right

On selecting this option, the route line will be set at the right of center point of the pipe.

Bottom Left

On selecting this option, the route line will be set at the bottom left point of the pipe.

Bottom Right

On selecting this option, the route line will set at the bottom right point of the pipe.

Routing a Pipe from a Line

In AutoCAD Plant 3D, you can also route a pipe by converting a line into a pipe. To do so, first you need to draw a line, polyline, arc, or 3D polyline. Next, choose the **Line to Pipe** tool from the **Part Insertion** panel of the **Home** tab; you are prompted to select a line, polyline, or arc. Select a line entity and press ENTER; it will be converted into a pipe. Figure 6-6 shows a 3D polyline and Figure 6-7 shows a piping created from it. Notice that the elbows and tees are automatically placed while creating a pipe. You can also create a pipe from an arc. Figure 6-8 shows an arc and Figure 6-9 shows a piping created from that arc.

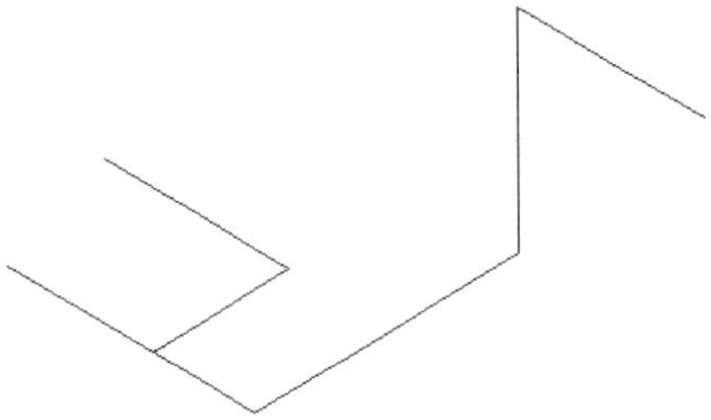

***Figure 6-6** A 3D polyline*

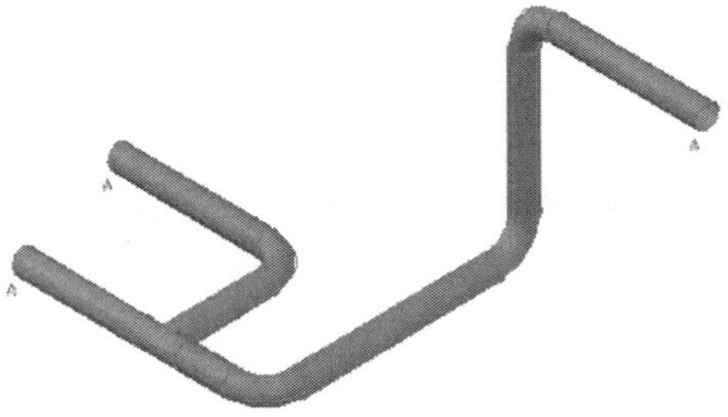

***Figure 6-7** Piping created from 3D polyline*

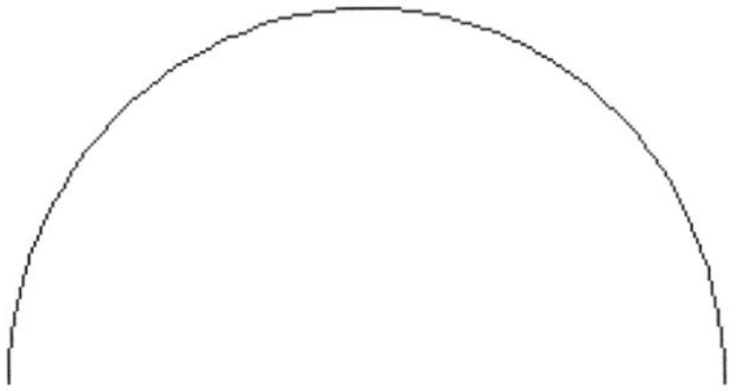

***Figure 6-8** An arc created using the **Arc** tool*

***Figure 6-9** Piping created from an arc*

The line entity gets deleted after being converted into a pipe. If you want to retain the line entity, press E and ENTER after selecting the entity; a prompt will be displayed prompting you to erase the selected entity or not. Choose the **No** option to retain it, and then press ENTER.

Routing a Pipe using a P&ID

You can route a pipe in a Plant 3D model using a P&ID. Note that the P&ID and the Plant 3D files should be located in the same project. To route a pipe using a P&ID, choose the **P&ID Line List** button from the **Part Insertion** panel, refer to Figure 6-10; the **P&ID LINE LIST** palette will be displayed, as shown in Figure 6-11. Next, select the P&ID from the drop-down list available in this window; the Pipe Line Groups of the selected P&ID will be displayed in the tree view in the **P&ID LINE LIST** palette. Expand the Pipe Line Group from the tree and then select a line segment, refer to Figure 6-11. Next, choose the **Place** button; you will be prompted to specify the start point of the line. Select a point in the drawing area; you are prompted to specify the next point of the pipe. Move the cursor upto a desired length and specify the next point. Next, press ENTER to exit the command.

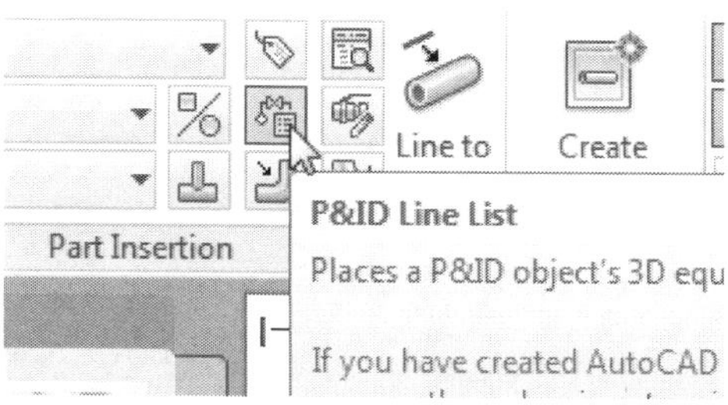

*Figure 6-10 Choosing the **P&ID Line List** button*

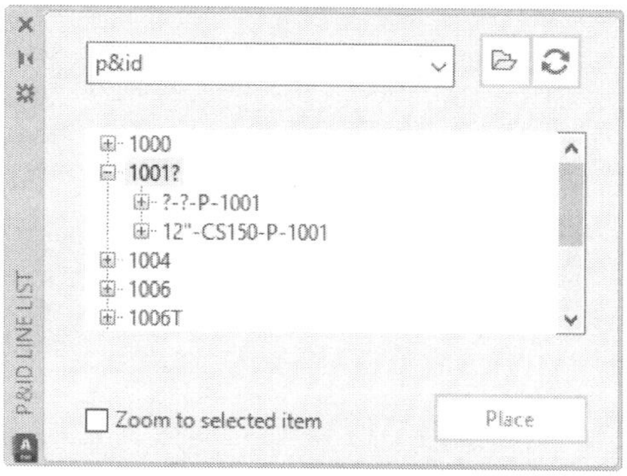

*Figure 6-11 The **P&ID LINE LIST** palette*

Routing a Pipe from an Equipment

In earlier chapters, you learned how to create equipment and add nozzles to them. Now, you will learn how to route a pipe directly from an equipment. To do so, click on an equipment in the model space; the equipment will be highlighted and a continuation grip (+ symbol) will be displayed on the nozzles located on it, refer to Figure 6-12. Click on the continuation grip and move the cursor away from the nozzle; a pipe is attached to the cursor and a compass is displayed on it, refer to Figure 6-13. Also, the size of the pipe is automatically set to match the nozzle size. Next, specify the end point of the pipe by clicking in the drawing area. Press ENTER to end the routing of the pipe.

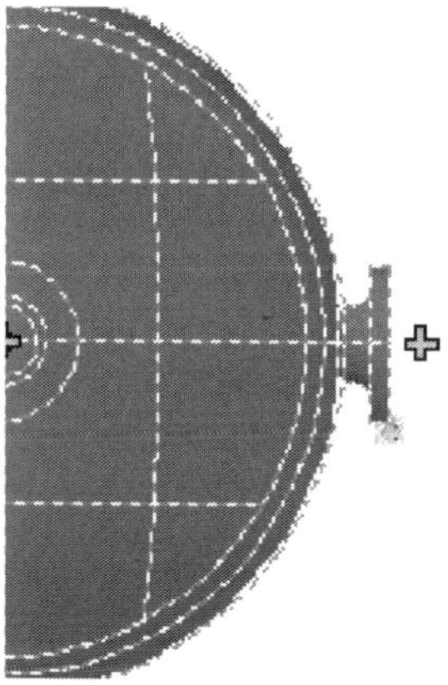

Figure 6-12 The continuation grip displayed on the nozzle

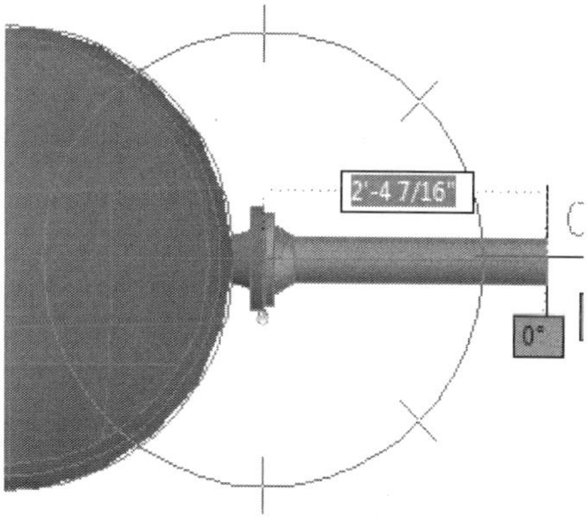

Figure 6-13 Routing a pipe from an equipment

Working with the Compass

Compass is an on-screen tool which makes it easy to route a pipe. Compass which is displayed automatically while routing a pipe helps you to rotate the pipe at a precise angle. You can change the default compass settings such as size, color, angle increments, and so on by using the options in the **Compass** panel of the **Ribbon** in the **Home** tab, refer to Figure 6-14.

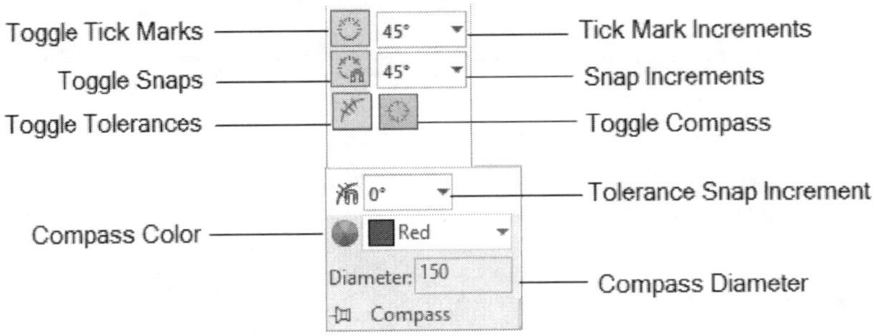

***Figure 6-14** The **Compass** panel*

The options in the **Compass** panel are discussed next.

Toggle Compass
This toggle button is used to turn the display of compass on or off while routing a pipe.

Toggle Tick Marks
This toggle button controls the display of tick marks on the compass. By default, this toggle button is chosen in the **Compass** panel. As a result, tick marks are displayed on the compass. Figure 6-15 shows a compass with tick marks and Figure 6-16 shows a compass without tick marks. This button will be activated only when the **Toggle Compass** button is chosen.

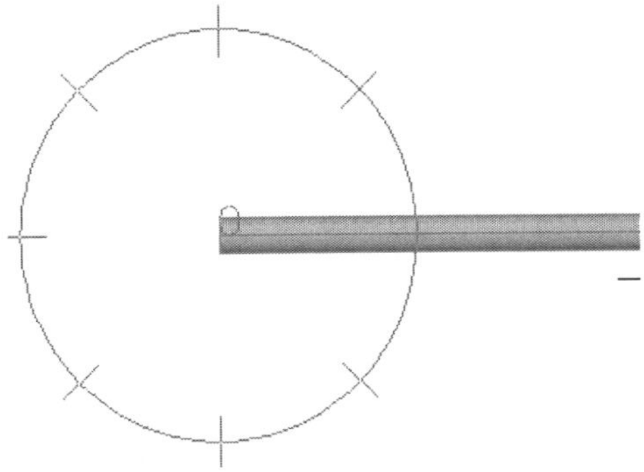

***Figure 6-15** Compass with tick marks*

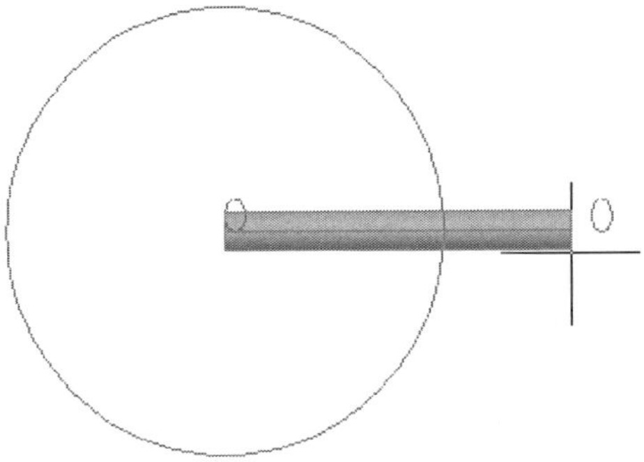

***Figure 6-16** Compass without tick marks*

Tick Mark Increments
This edit box is used to specify the angle between tick marks displayed on the Compass. This edit box will be activated when the **Toggle Tick Mark** button is chosen.

Toggle Snaps
This toggle button controls whether the angle snap is turned on or off. On choosing this button, the cursor will snap through the angle specified in the **Snap Increments** edit box. Figure 6-17 shows the compass when the angle snap is turned on and Figure 6-18 shows the compass when the angle snap is turned off. This button will be activated only when the **Toggle Compass** button is chosen.

Snap Increments
This edit box is used to specify the angular increments through which cursor will snap. For example, when you specify the snap increment as 60-degree, the cursor snaps through 0, 60, 120, and 180 degrees. Figure 6-17 shows a compass with 45-degree snap increments. This edit box will be activated when the **Toggle Snap** button is chosen.

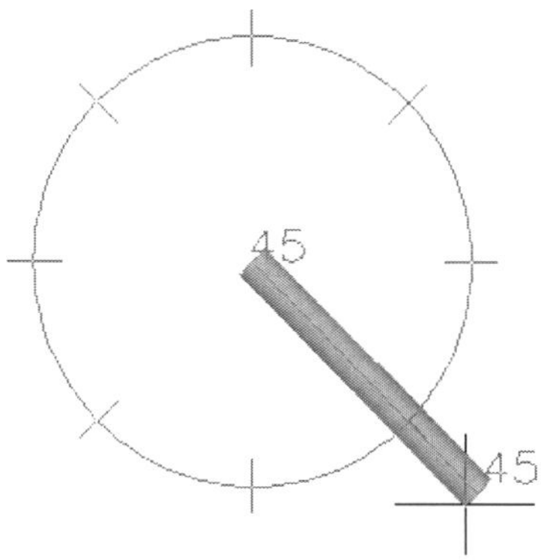

Figure 6-17 *Compass when the angle snap is turned on*

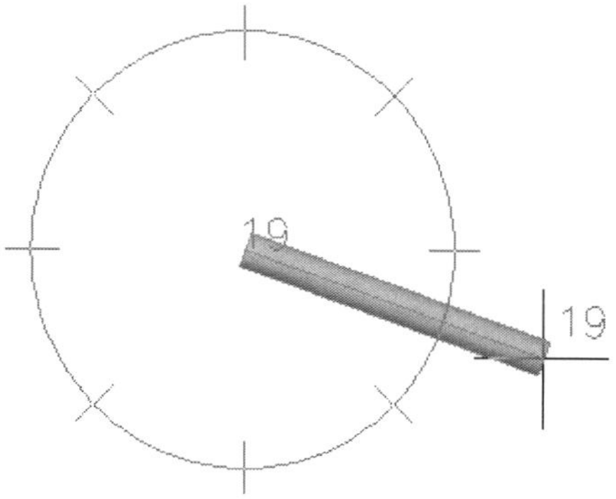

Figure 6-18 *Compass when the angle snap is turned off*

Toggle Tolerances
This toggle button is used to turn the fitting tolerance on or off. This button will be activated only when the **Toggle Compass** button is chosen.

Tolerance Snap Increment
This edit box is used to set the tolerance angle when a pipe is connected to another pipe or fitting.

Note

When a new pipe is created, the ***Toggle Tick Marks*** *and* ***Toggle Snaps*** *buttons in the* ***Compass*** *panel works as described above. To make them work in the same manner after the second point is specified for the pipe, you need to choose the* ***Toggle Pipe Bends*** *button from the* ***Part Insertion*** *panel of the* ***Home*** *tab.*

Compass Color
This drop-down list is used to specify the color of the compass.

Compass Diameter
This edit box is used to specify the diameter of the compass.

Connecting Two Open Ports of Pipes
You can connect ports of two different pipes. Before connecting these ports, you need to specify the **Object Snap** settings. To do so, right-click on the **Object Snap** button in the Status Bar and choose the **Object Snap Settings** option from the shortcut menu displayed; the **Drafting Settings** dialog box will be displayed. Select the **Node** check box from this dialog box and choose the **OK** button; the cursor will snap to the center point of the pipe.

After specifying the **Object Snap** settings, choose the **Route Pipe** tool from the **Part Insertion** panel; you will be prompted to select the start point of the pipe. Snap to the center point of the first pipe and click to select the first point, as shown in Figure 6-19; you will be prompted to select the next point. Press and hold the SHIFT key and right-click to display the shortcut menu. Choose the **Node** option from the menu and then select the center point of the second pipe, as shown in Figure 6-20; the system generates multiple solutions for connecting the two open ports. Choose the **Next** option from the command prompt until the desired solution is displayed. To accept the desired solution, choose the **Accept** option; the two ports are connected by the selected solution. Figure 6-21 shows two open ports connected.

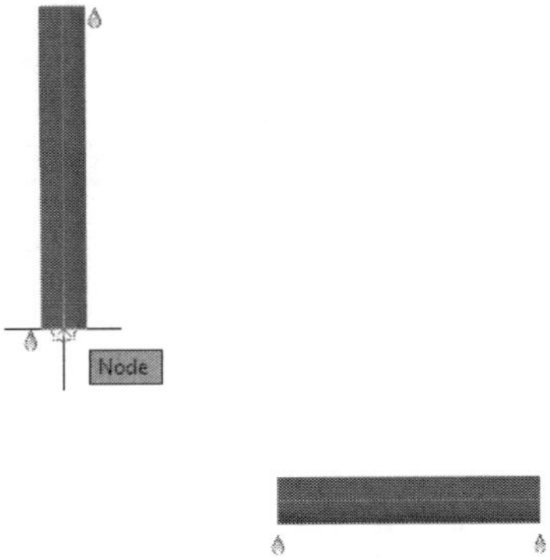

Figure 6-19 *Selecting the first port*

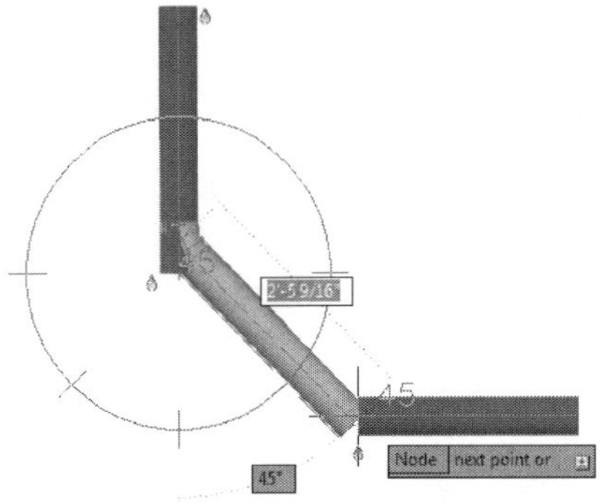

Figure 6-20 *Selecting the second pipe*

You can also connect two equipment by using the same procedure. Figure 6-22 through 6-24 show examples of two equipment connected using the automatically generated solutions.

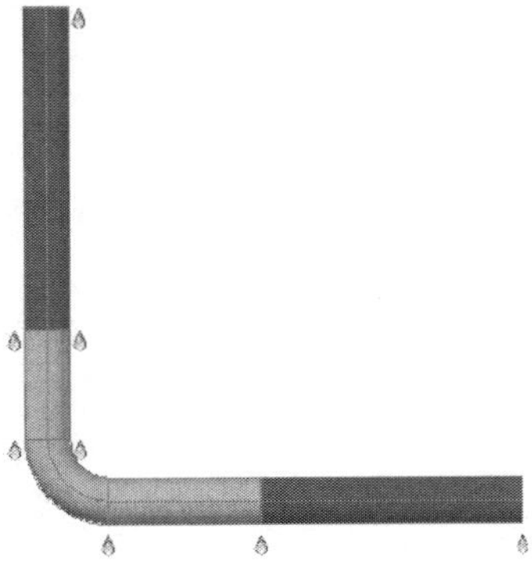

Figure 6-21 *Two pipes connected by using system generated solutions*

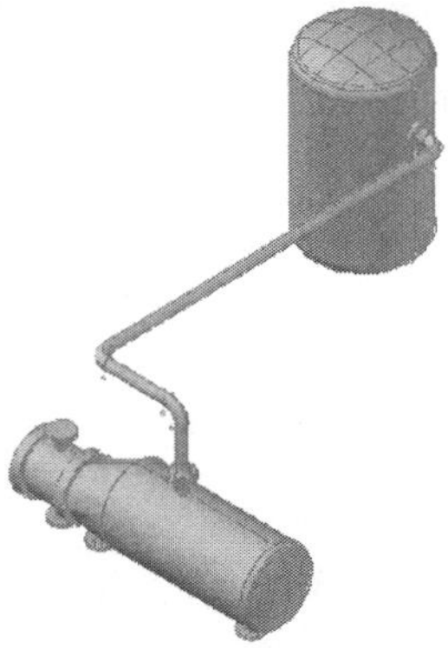

Figure 6-22 *Solution 1 for connecting two equipments*

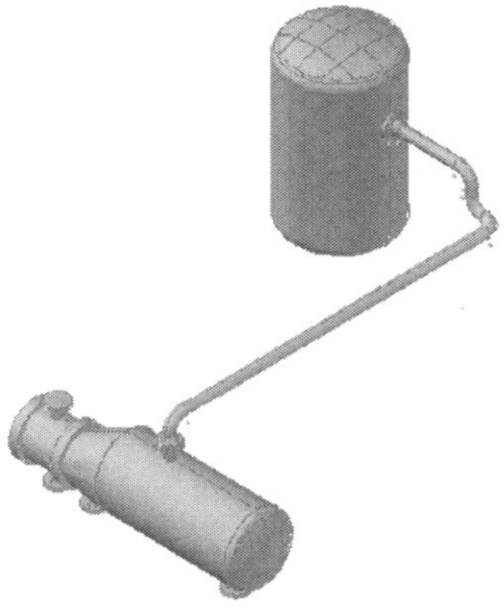

Figure 6-23 Solution 2 for connecting two equipments

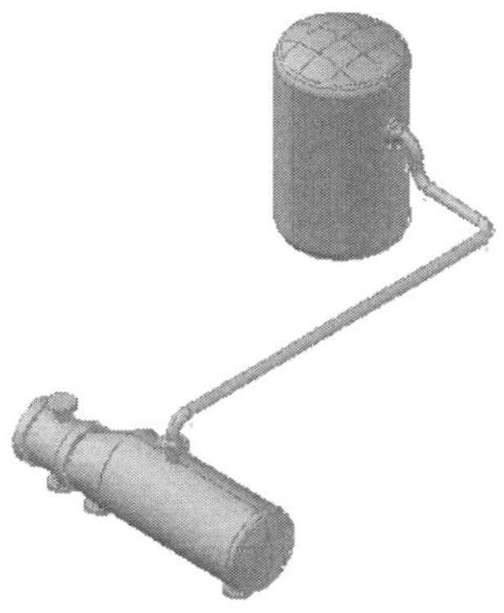

Figure 6-24 Solution 3 for connecting two equipments

Changing the Pipe Size while Routing

You can change the pipe size while routing. To do so, invoke the **Route Pipe** tool and then choose the **Size** option from the command prompt; you will be prompted to specify the nominal size of the pipe. Also, the prompt displays a symbol "?" at the command prompt. If you enter "?", the sizes available in the selected spec will be listed in the command prompt. Next, enter the desired size from the list displayed; the pipe size changes to the size specified. Figure 6-25 shows a reducer automatically added between two pipes on entering a new pipe size at the prompt. Similarly, if you set a pipe size different from the nozzle size, a reducer is placed between the pipe and the nozzle, as shown in Figure 6-26.

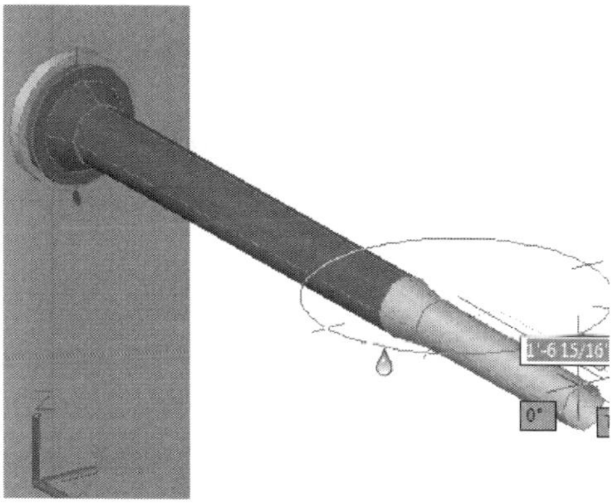

Figure 6-25 Pipes of different sizes connected by a reducer

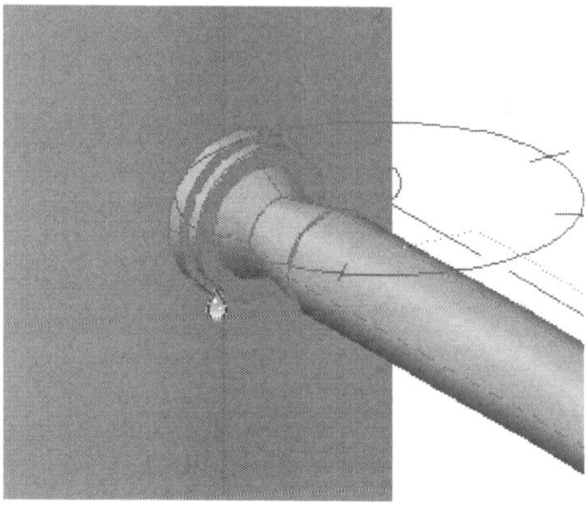

Figure 6-26 A reducer placed between the nozzle and the pipe

Changing the Orientation Plane while Routing a Pipe

While routing a pipe, you can change the plane orientation by choosing the **Plane** option from the command prompt. To do so, invoke the **Route Pipe** tool and then choose any pipe node of equipment. Then enter **P** at the Command prompt; the plane in which the pipe has been routed changes. Figures 6-27 through 6-29 show a pipe routed on different planes.

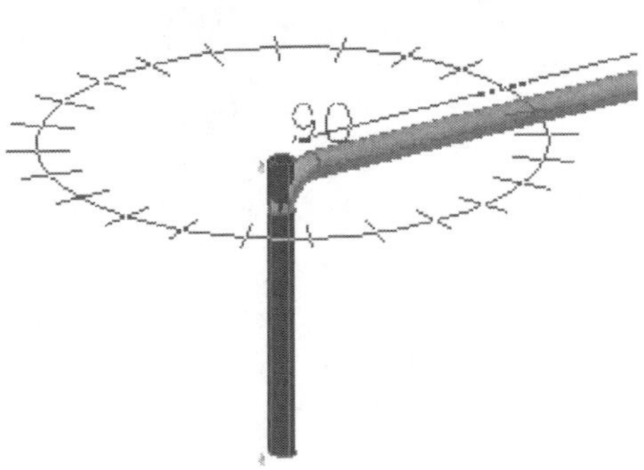

Figure 6-27 Routing a pipe on XY plane

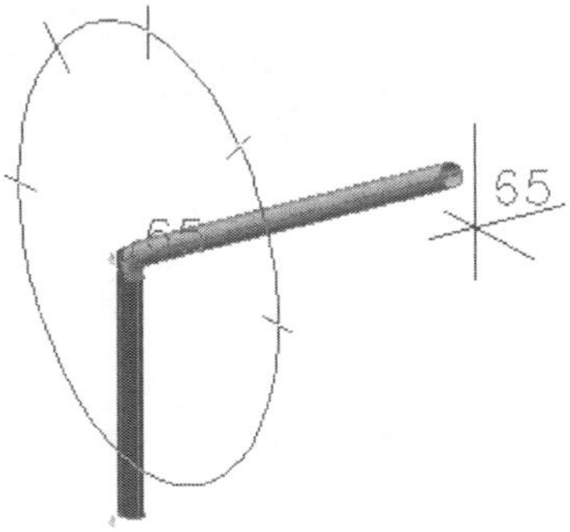

Figure 6-28 Routing a pipe on XZ plane

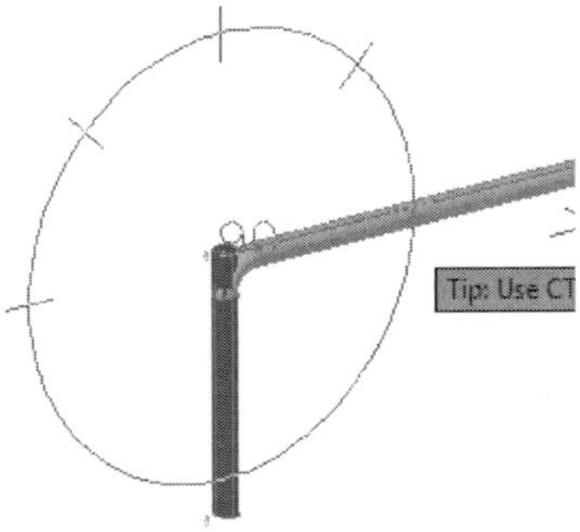

Figure 6-29 Routing a pipe on YZ plane

Creating a Cutback Elbow

When you bend a pipe, you will notice that an elbow is automatically placed. You will also notice that the elbow is created only at 45 degree and 90 degree. This is because, by default, the cutback mode is turned off. As a result, you can only place elbows available in the spec (45 or 90 degree elbows). To turn the cutback mode on, choose the **Toggle Cutback Elbows** button from the **Part Insertion** panel, refer to Figure 6-30; the cutback mode gets turned on and you can now create an elbow whose angle is different from the angle of elbow available in the spec. Figures 6-31 and 6-32 show a pipe being rotated when the cutback mode is off and on, respectively.

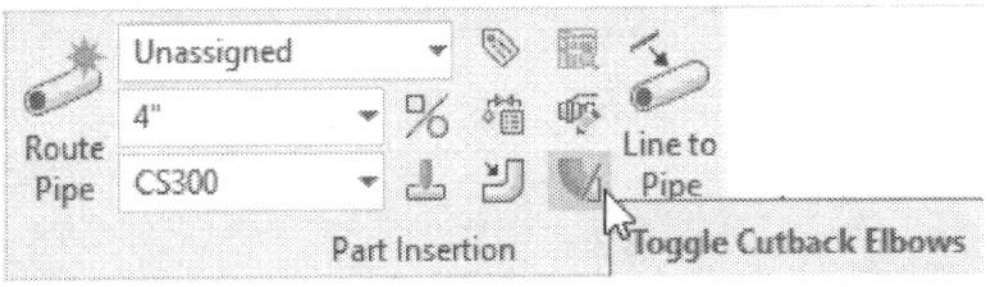

Figure 6-30 *Choosing the* ***Toggle Cutback Elbows*** *button*

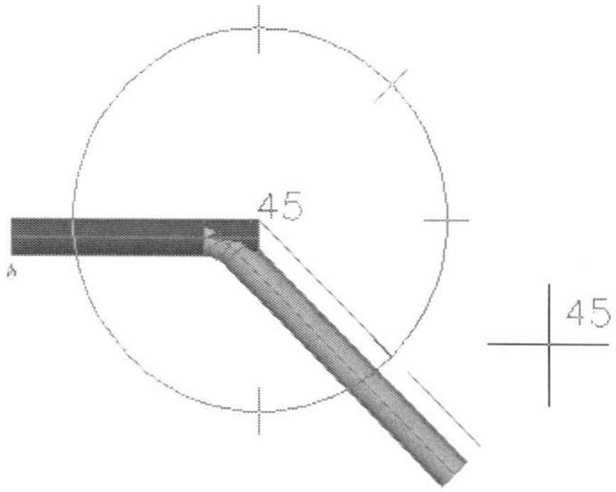

Figure 6-31 *Creating an elbow with the cutback mode turned off*

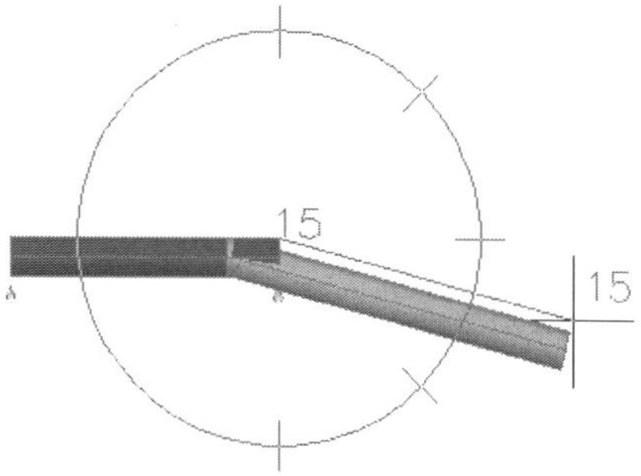

Figure 6-32 *Creating an elbow with the cutback mode turned on*

Creating a Roll Elbow

You can create a rolled elbow (twisted elbow) by specifying rotation angles in two planes which are perpendicular to each other. To do so, invoke the **Route Pipe** tool and specify the point at the end of the existing pipe. Next, choose the **Rollelbow** option from the command prompt; you will be prompted to specify the next point for the elbow angle. Bend the pipe to the required angle and click to specify the angle on the first plane, refer to Figure 6-33; you will be prompted to specify the next point for the roll angle which is perpendicular to the first plane. Rotate the pipe and click to specify the angle on the second plane, refer to Figure 6-34. Press ENTER to exit the command prompt; a rolled elbow is created after specifying angles on both the planes, refer to Figure 6-35.

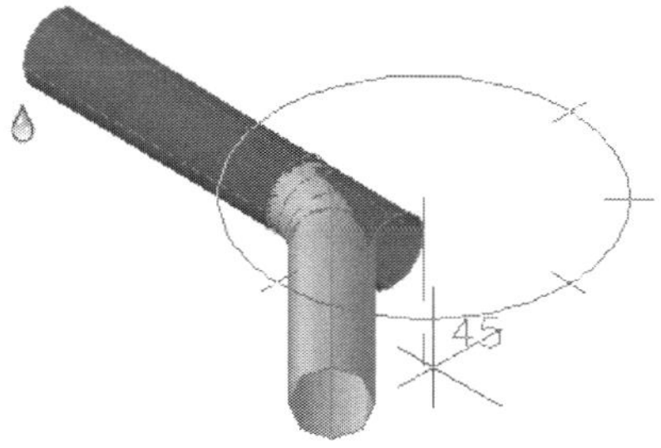

Figure 6-33 *Specifying an angle in the first plane*

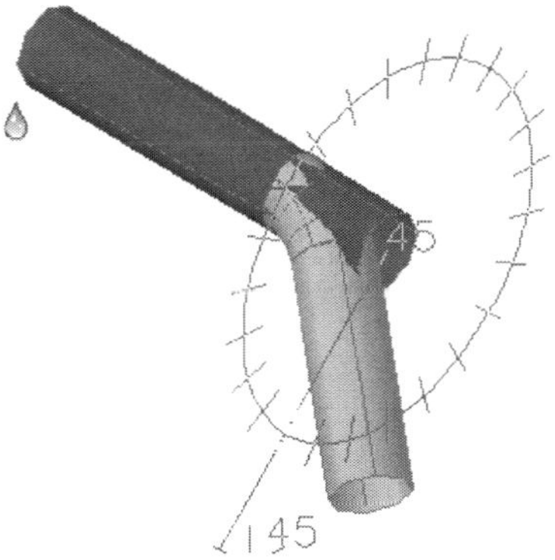

Figure 6-34 *Specifying an angle in the second plane (perpendicular plane)*

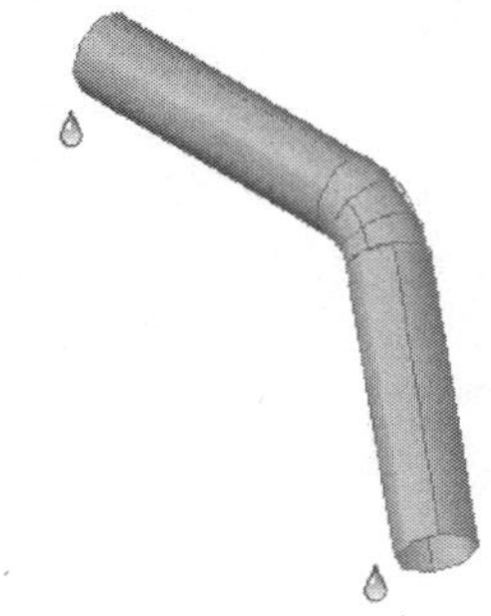

Figure 6-35 Roll elbow created

Creating Bends

You can create a pipe bend upto an angle of 180 degrees. To do so, choose the **Route Pipe** tool from the **Part Insertion** panel of the **Home** tab; you will be prompted to specify the start point of the pipe. Specify the start point and then choose the **pipeBEnd** option from the command prompt. Next, enter the angle value. Make sure that the **Toggle Pipe Bends** button is chosen in the **Part Insertion** panel, refer to Figure 6-36. Figure 6-37 shows the pipe bends created at different angles.

*Figure 6-36 Choosing the **Toggle Pipe Bends** button*

Figure 6-37 Pipe bends created at different angles

Changing the Elevation while Routing a Pipe

To change the elevation while routing a pipe, enter the desired value in the **Set Elevation** edit box located in the **Elevation & Routing** panel; the cursor elevates to the specified height. Next, you can route the pipe at the specified elevation. You can also enter an elevation value relative to the current elevation. To do so, invoke the **Elevation** option from the command prompt and then enter **R** at the command prompt; the relative option will be invoked. Next, enter an elevation value at the command prompt. Figure 6-38 shows a pipe before specifying the elevation and Figure 6-39 shows a pipe after specifying the elevation.

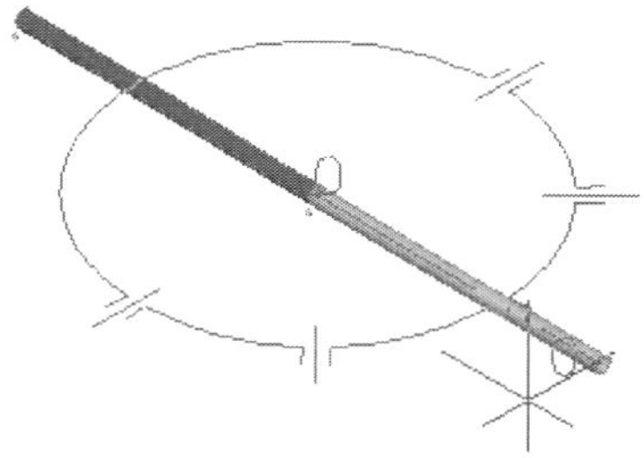

Figure 6-38 *Routing a pipe before specifying the elevation*

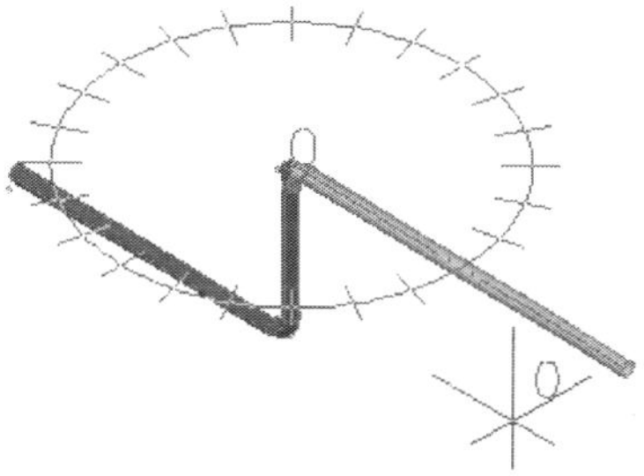

Figure 6-39 *Routing a pipe after specifying the elevation*

Routing a Pipe at an Offset

You can route a pipe at an offset distance to the routing line. To do so, first select the justification point of the pipe. By default, the justification point is at the center of the pipe. To change it, select the required justification point option from the **Set Routing Line** drop-down list in the **Elevation & Routing** panel. Next, click on the down arrow in the **Elevation & Routing** panel; it gets expanded and displays the **Horizontal Offset** and **Vertical Offset** edit boxes, as shown in Figure 6-40. Enter a required value in the **Horizontal Offset** edit box and then choose the **Route Pipe** tool. Next, specify the start point and the end point of the pipe; you will notice that the pipe is being created at an offset distance from the specified line, as shown in Figure 6-41. Continue routing the pipe. Figure 6-42 shows a pipe created at an offset. Similarly, you can also route a pipe at an elevation by entering a required value in the **Vertical Offset** edit box.

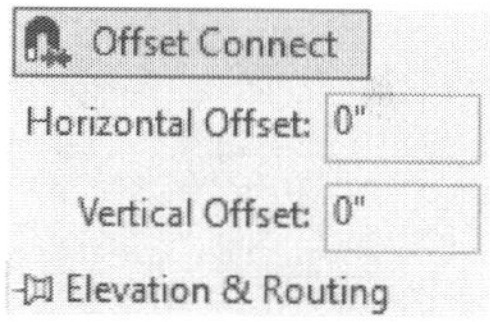

Figure 6-40 *Expanded portion of the* ***Elevation*** *&* ***Routing*** *panel*

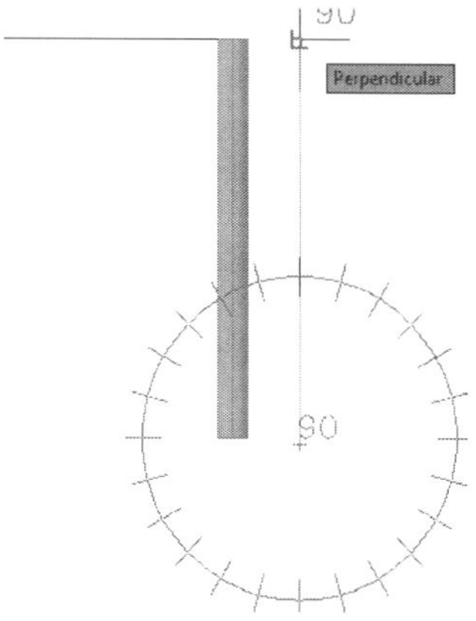

Figure 6-41 *Routing a pipe at an offset*

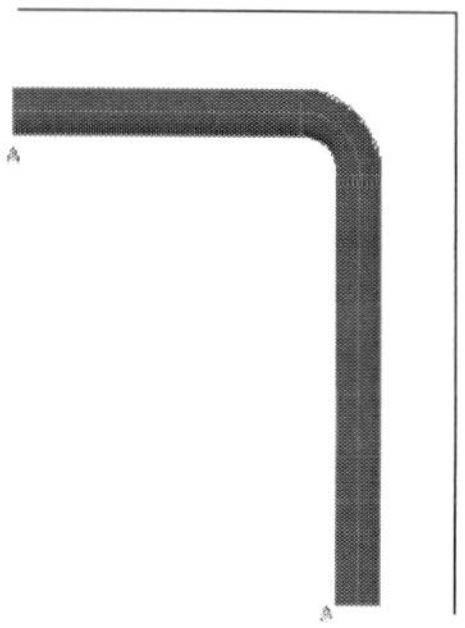

Figure 6-42 *Pipe created at an offset*

Offset Connect

This option is very useful while routing an offset pipe. The **Offset Connect** button is chosen by default in the expanded **Elevation & Routing** panel. As a result, you can connect an offset pipe to a piping component. Figure 6-43 shows an offset pipe snapped to a piping component and Figure 6-44 shows the offset pipe connected to it. Figure 6-45 shows a case when the offset connect mode is turned off.

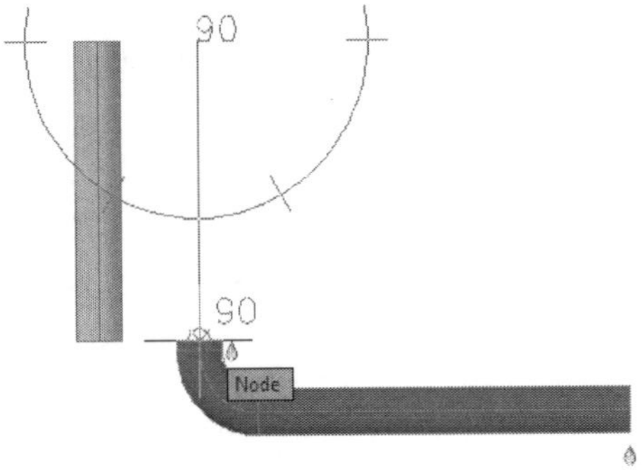

Figure 6-43 *An offset pipe snapped to a piping component*

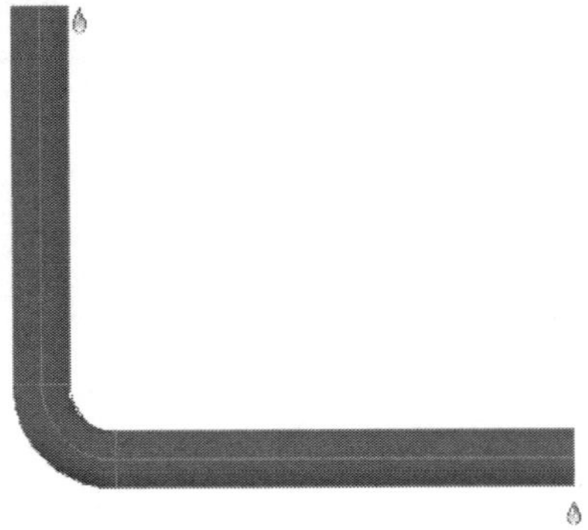

Figure 6-44 *Offset pipe connected to the snapped component*

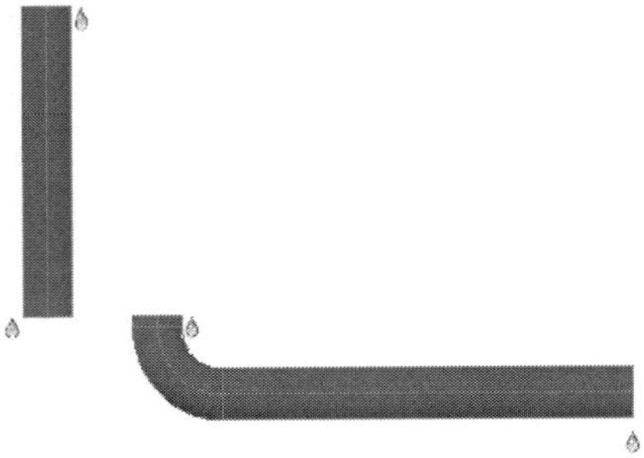

Figure 6-45 *Offset pipe not connected to the snapped component*

Routing a Pipe at a Slope

In a piping layout, there are many cases where you need to create a sloped pipe. For example, you need a sloped pipe for lifting a fluid to a higher gradient. To create such a pipe, first you need to enter a value in the **Slope Rise** edit box and then in the **Slope Run** edit box. These two edit boxes are located in the **Slope** panel of the **Home** tab. On doing so, the system automatically calculates the slope angle and displays it in the **Slope** edit box. The slope angle is calculated by the specified vertical distance (rise) and the horizontal distance (run). Next, choose the **Toggle Slope** button, if not chosen already. The **Toggle Slope** button controls whether the pipe is to be created at a slope or not. Next, choose the **Route pipe** tool and start routing the pipe; you will notice that the pipe has been created at a slope, as shown in Figure 6-46. An example of a sloped pipe is shown in Figure 6-47.

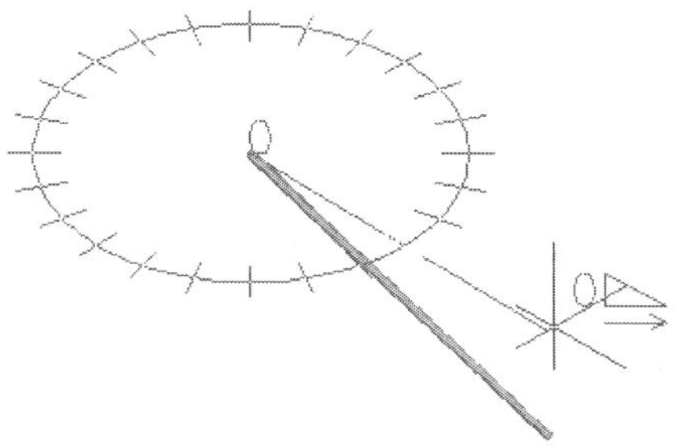

Figure 6-46 Pipe created at a slope

Figure 6-47 An example of sloped pipe

Converting a Straight Pipe to a Sloped Pipe

You can convert a straight pipe to a sloped pipe. To do so, select a horizontal straight pipe from the drawing area and right-click on it; a shortcut menu will be displayed. Choose the **Pipe Slope Editing** option from the shortcut menu; the **Edit Slope** dialog box will be displayed, as shown in Figure 6-48. Choose the Start point button located on the left of the dialog box; you will be prompted to select the start point on the highlighted component.

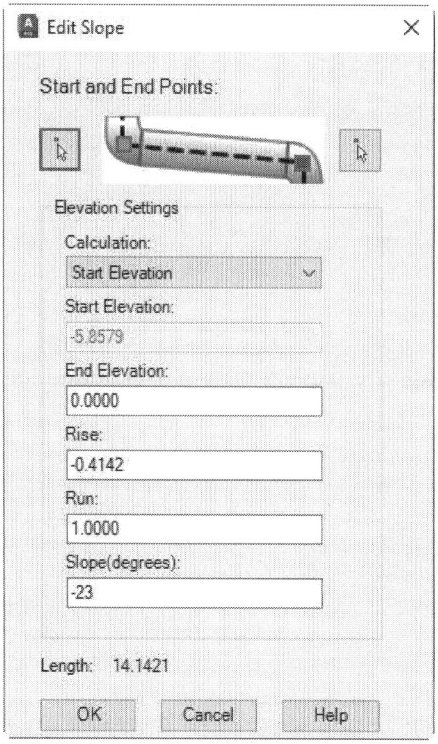

*Figure 6-48 The **Edit Slope** dialog box*

Click on the pipe to specify the start point of the slope. Next, choose the End point button located on the right in the dialog box and then, specify the end point on the pipe, refer to Figure 6-49. Next, you need to specify the slope calculation method. There are three options available in the **Calculation** drop-down list to calculate the slope, **Start Elevation**, **End Elevation**, and **Slope**.

Select the **Slope** option from the **Calculation** drop-down list and specify values in the **Start Elevation** and **End Elevation** edit boxes; the slope angle is automatically calculated. Next, choose the **OK** button; the pipe will be sloped, as shown in Figure 6-50.

Figure 6-49 *Specifying the start point and end point*

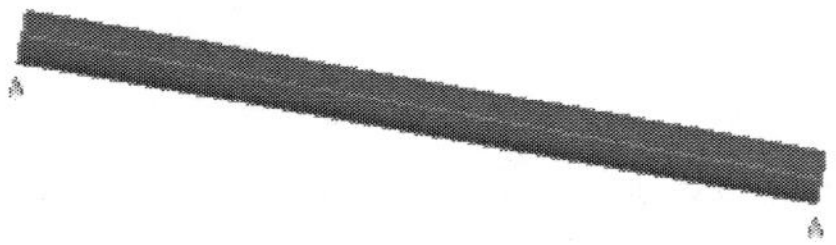

Figure 6-50 *Slope applied to a straight pipe*

Note

It is recommended that you create a sloped pipe instead of converting a straight pipe into a sloped pipe.

CREATING BRANCHES

You can create and connect branches to a header pipe. The methods to create branches are discussed next.

Creating a Tee Branch

You can create a tee branch using the continuation grip displayed at the center of the pipe. To do so, first select the pipe from the drawing area; a continuation grip is displayed at the midpoint of the pipe, as shown in Figure 6-51. Click on the continuation grip located at the midpoint of the pipe. You will notice that a tee fitting is placed at the midpoint and also a branch pipe is connected to it. You can rotate this branch pipe to a required angle using the compass. Next, drag the branch pipe upto some required distance and press ENTER; the branch will be created, as shown in Figure 6-52.

Figure 6-51 *Continuation grip displayed on selecting a pipe*

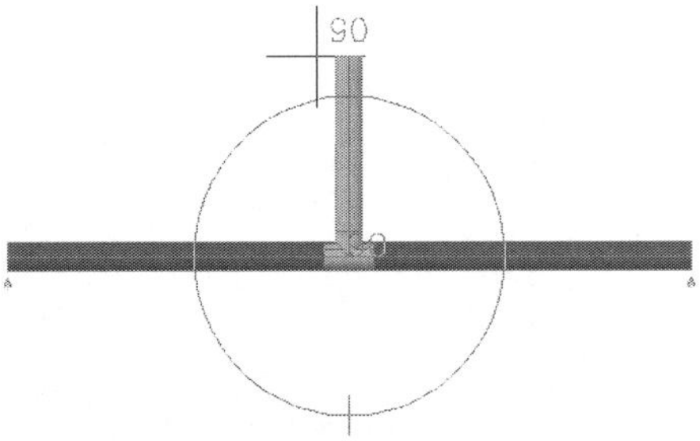

Figure 6-52 *Creating a tee branch*

You can also create a tee branch by manually placing a tee from the tool palette. To do so, scroll to the **Tee** area in the **Dynamic Pipe Spec** tool palette and select a tee fitting; a tee will be attached to the cursor. By default, the left port of the tee is attached to the cursor. You can change the cursor port attachment by pressing the CTRL key, refer to Figure 6-53. Next, select a point on the pipe in the drawing area; the tee will be placed at that point, as shown in Figure 6-54. You can rotate the tee using the compass. After placing the tee, press ESC and click on the tee; a continuation grip is displayed on it. Click on the continuation grip; a branch pipe is created, as shown in Figure 6-55.

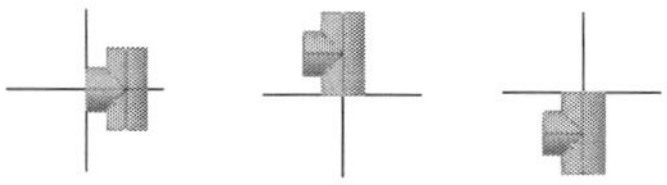

Figure 6-53 *Changing the cursor port attachment*

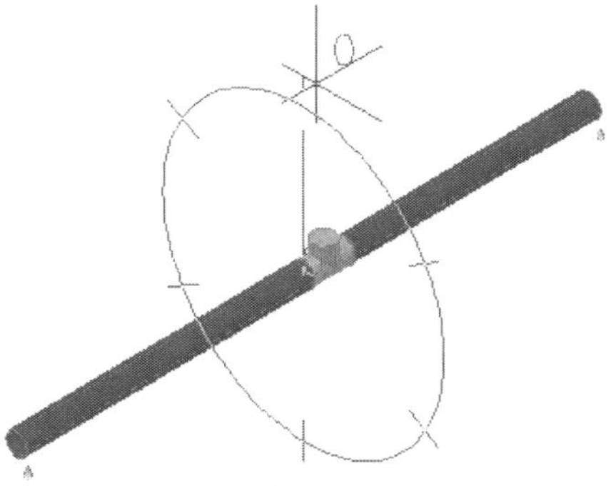

Figure 6-54 *Tee placed on the pipe*

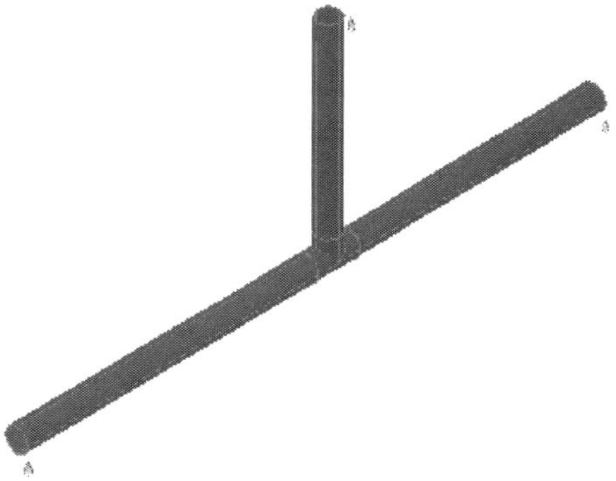

Figure 6-55 *Branch pipe created from the tee*

Creating an O-let Branch

To create an O-let branch, first choose the **Spec Viewer** button from the **Part Insertion** panel; the **PIPE SPEC VIEWER** palette will be displayed. Next, select a part from the **Olet** category in the **Spec Sheet** rollout; the sizes available for the selected olet are displayed in the **Part Sizes** rollout. Next, select the size from the **Part Sizes** rollout and choose the **Insert In Model** button; the olet will be attached to the cursor. Select a point on the pipe and place the olet, as shown in Figure 6-56. Next, press ESC and click on the olet; a continuation grip is displayed. Select it and create a branch, as shown in Figure 6-57.

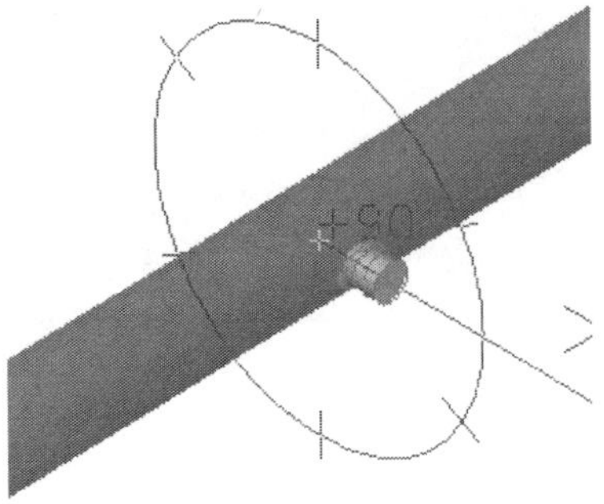

Figure 6-56 *Tee placed on the pipe*

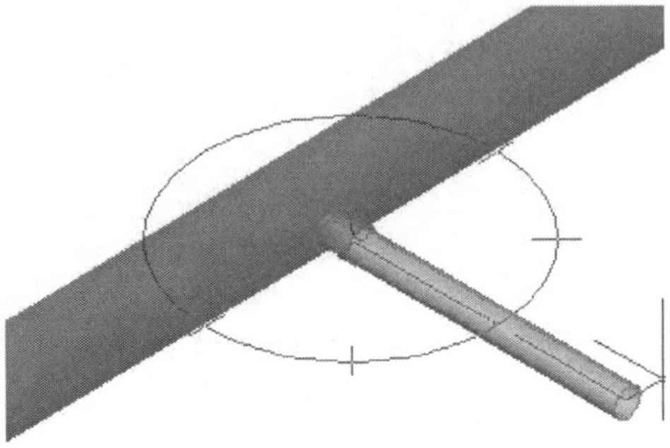

Figure 6-57 *Creating an olet branch*

Creating a Stub-In Branch

A Stub-In is a type of branch directly welded into the side of a header pipe. In this type of branch, you do not use any fitting. This is the most common and least expensive method of welding a full size pipe or reducing branch to a header pipe. To create a branch on an existing pipe using stub-in, first you need to invoke the **Route Pipe** tool. Next, choose the **Toggle Stub-In** button from the **Part Insertion** panel and specify the start point of the stub-in branch on the pipe, refer to Figure 6-58. Next, specify the end point of the branch pipe; the stub-in branch will be created, as shown in Figure 6-59.

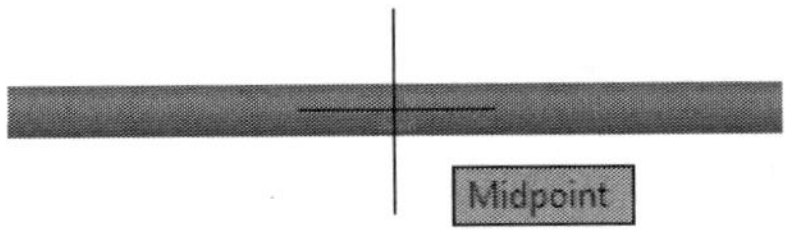

Figure 6-58 *Specifying the start point of the Stub-in branch pipe*

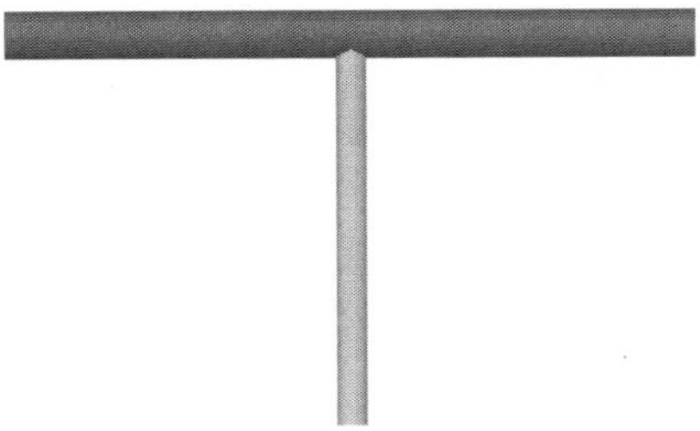

Figure 6-59 *Specify the end point of the Stub-in branch pipe*

Creating a Stub-In Branch at an Offset from the Center of the Header Pipe

You can create a stub-in branch at an offset from the center of the header pipe. To do so, first invoke the **Route Pipe** tool and then choose the **Toggle Stub-In** button from the **Part Insertion** panel. Next, select the pipe size from the **Pipe Size Selector** drop-down list and specify the start point on the header pipe. Next, choose the **connectionoffseT** option from the command prompt; the **Offset connection to [Toptangent/Bottomtangent/Distance] <Distance>:** prompt will be displayed. Choose the appropriate option from the command prompt. For example, choose the **Toptangent** option if you want to create a stub-in branch at the top of the pipe; the stub-in branch will be created at the top of the header pipe, as shown in Figure 6-60.

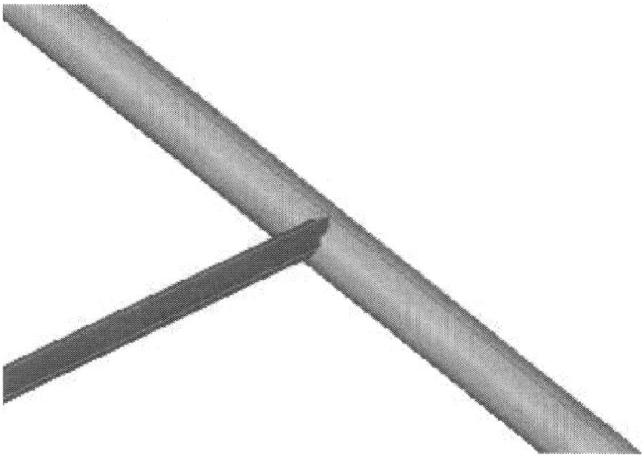

Figure 6-60 A stub-in branch created on the top of the header pipe

To create a stub-in branch at a specific offset distance from the center of the pipe, choose the **Distance** option and specify the offset distance value at the command prompt; the stub-in branch will be created at the specified offset distance from the center of the pipe.

Creating a Branch from an Elbow

You can create a branch from an elbow. To do so, click on an elbow connected to a pipe in the drawing area. On doing so, two continuation grips will be displayed on the elbow, as shown in Figure 6-61. Click on one of the grips and create a pipe. Press ENTER after routing the pipe; a branch will be created, as shown in Figure 6-62.

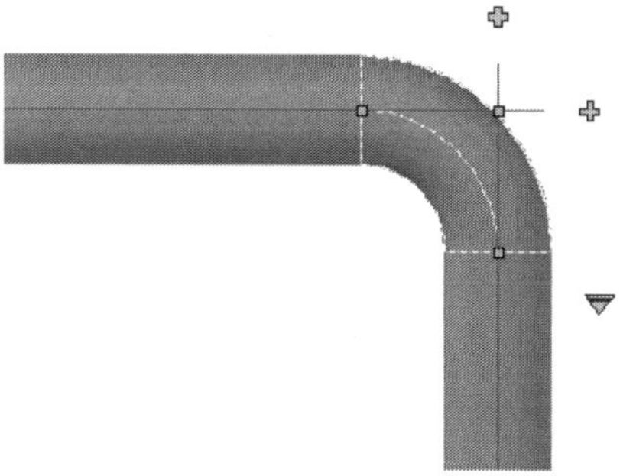

Figure 6-61 Continuation grips displayed on an elbow

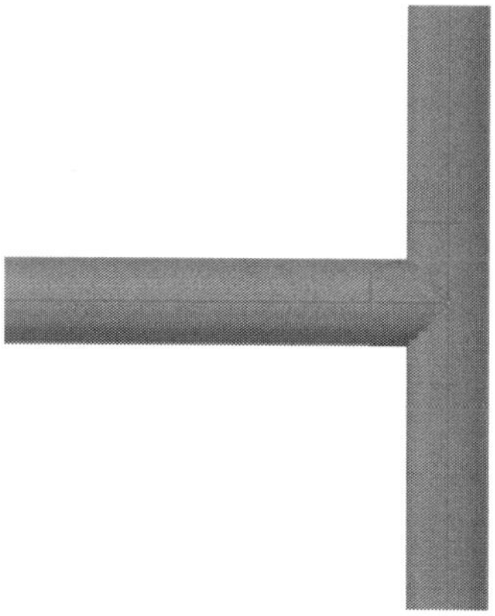

Figure 6-62 Branch created from an elbow

Creating a Stub-in Branch at a Precise Location

You can create a branch at a precise location on the header pipe. To do so, first invoke the **Route pipe** tool from the **Part Insertion** panel and make sure that the **Toggle Stub-In** button is chosen. Next, press and hold the SHIFT key and right-click to invoke the shortcut menu. Choose the **From** option from the shortcut menu displayed; you are prompted to specify a point from which the distance will be calculated. Next, move the cursor over the header pipe; the Osnap glyph will be displayed on the pipe. Next, click to select the point from which the distance will be calculated, as shown in Figure 6-63. Next, enter the value of distance from the base point of the header pipe at the command prompt; the start point of the pipe will be specified and a pipe will be connected to the cursor. Move the cursor and specify the end point of the branch pipe, and then press ENTER; a branch will be created at a precise location, as shown in Figure 6-64.

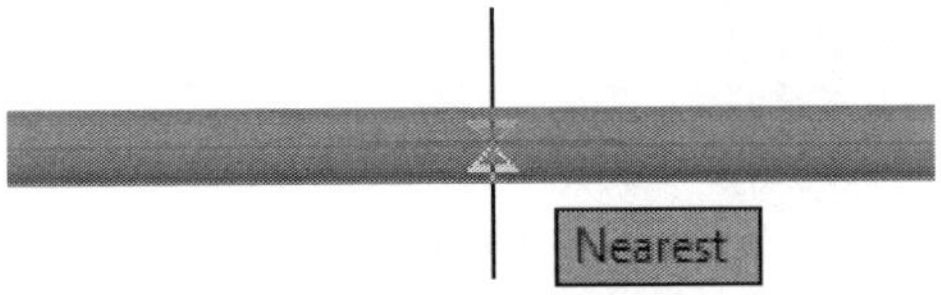

Figure 6-63 *Osnap glyph displayed on the pipe*

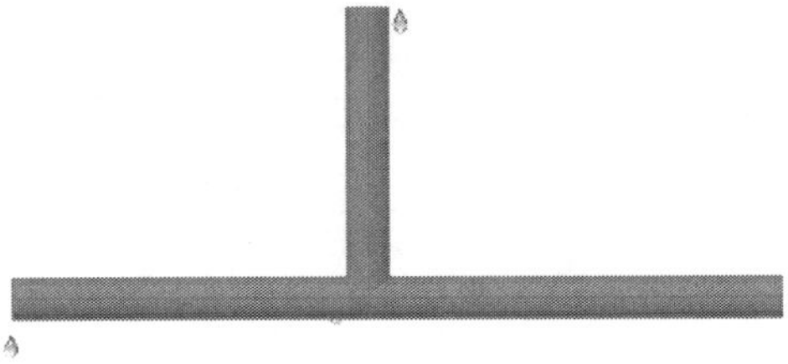

Figure 6-64 *Branch created at a precise location*

Adding a Reinforced Pad to a Stub-In Branch

You can add a reinforced pad to a stub-in branch. To do so, right-click on the stub-in branch and choose the **Add Reinforcing Pad** option from the shortcut menu displayed; a reinforcing pad will be added to the stub-in branch, as shown in Figure 6-65.

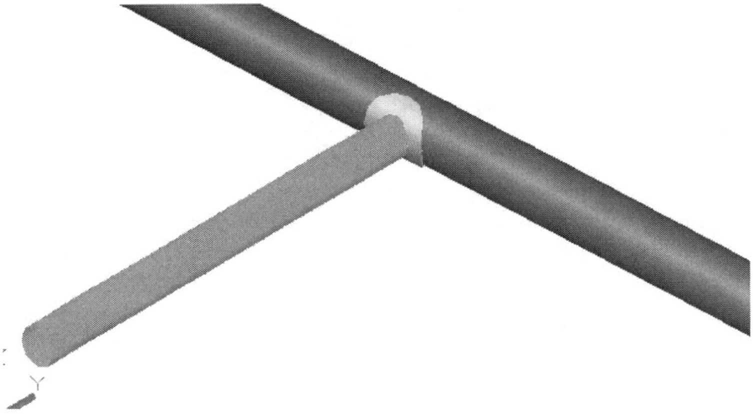

Figure 6-65 *A reinforcing pad added to a stub-in branch*

CREATING A WELD CONNECTION

Weld connections are automatically created when two piping components are connected to each other. You can also create weld connections manually at precise locations. You can break a long pipe into small pipes of a specific length by creating a weld connection. To create a welding connection, right-click on the pipe to display a shortcut menu. Next, choose the **Add Weld to Pipe** option from the shortcut menu displayed; you are prompted to specify the point location on the pipe. Specify a point on the pipe and press ENTER; a weld connection will be created, as shown in Figure 6-66.

You can also create a weld connection at a precise distance from the end point of the pipe. First, make sure that the **Dynamic Input** mode is turned on. Next, enter a distance value in the dynamic edit box, as shown in Figure 6-67; a weld will be created at the specified distance from the end point.

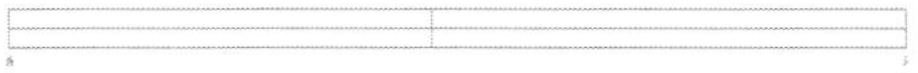

Figure 6-66 *Weld connection created on a pipe*

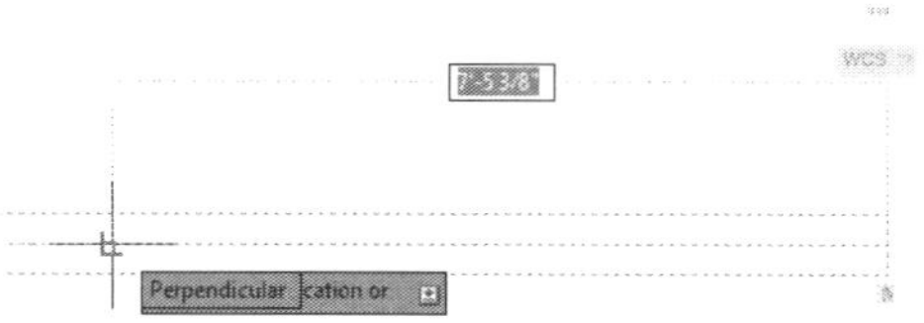

Figure 6-67 *Specifying distance in the dynamic edit box*

CREATING Autodesk CONNECTION POINT

The Autodesk Connection point contains the port information of the piping components. Using the Autodesk connection point, you can connect an AutoCAD Plant 3D pipe to piping components created in other Autodesk applications such as AutoCAD MEP, Civil 3D, and so on. You can also connect a pipe to Xref objects using the Autodesk connect points. To create an Autodesk Connection point, choose the **Insert** tool from the **Autodesk Connection Point** panel in the **Insert** tab; you will be prompted to pick an object. Choose the piping component; you will be prompted to pick an open point. Select an insertion point on the pipe and place the Autodesk connection point on it, as shown in Figure 6-68.

Figure 6-68 *The Autodesk Connection point created on an open port*

Editing an Autodesk Connection Point

You can edit the properties of an Autodesk connection point. To do so, choose the **Edit** button from the **Autodesk Connection Point** panel; you will be prompted to select an Autodesk connection point. Select a point from the drawing area; the **Autodesk Connection Point Editor** dialog box will be displayed, as shown in Figure 6-69. Edit the property values as per your requirement in the dialog box and choose the **OK** button; the dialog box will be closed and the properties of the Autodesk Connection point will be changed.

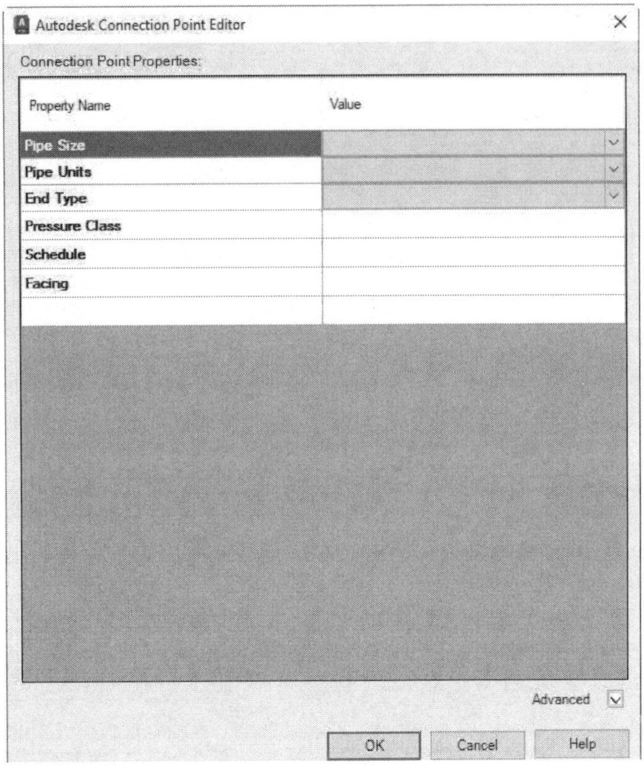

*Figure 6-69 The **Autodesk Connection Point Editor** dialog box*

Routing a Pipe from an Autodesk Connection Point

You can also route a pipe from the Autodesk connection point. To do so, choose the **Route Pipe From Point** button from the **Autodesk Connection Point** panel. Next, select an Autodesk connection point and start routing a pipe.

TUTORIAL

Tutorial 1

In this tutorial, you will open the file created in Tutorial 2 of Chapter 4. Next, you will connect the equipment by routing pipes, as shown in Figure 6-70. Figure 6-71 shows the list of pipes used in the model. The dimensions of the model are given in Figures 6-72 through 6-74.

(Expected time: 45 min)

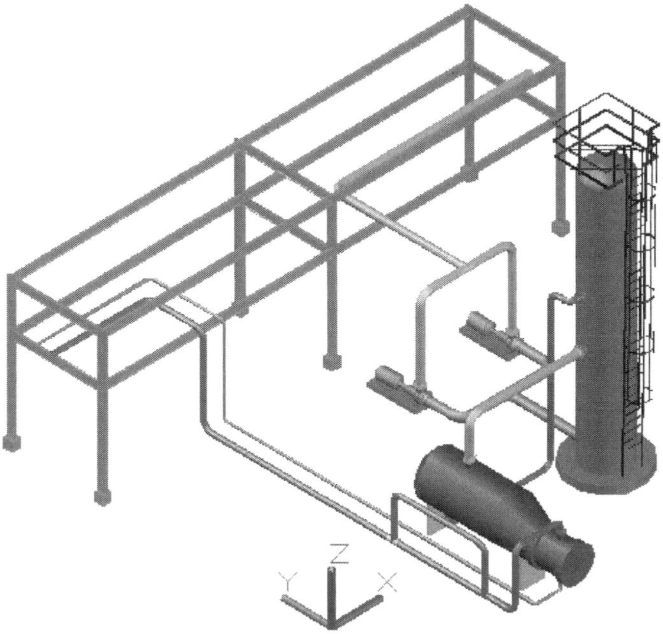

Figure 6-70 *Model for Tutorial 1*

The following steps are required to complete this tutorial:

a. Open the **Piping Model** file from the **PROJECT MANAGER**.
b. Select the required spec and select the pipe size.
c. Insert the **Pipe_rack** file into the model space.
d. Connect centrifugal pumps and vertical vessel.
e. Route pipes from outlet nozzles of the pump.
f. Connect the reboiler to the vertical vessel.
g. Route pipes from the other nozzles of the reboiler.

Pipe List			
Pipe Number	Spec	Size	Equipment connected
1004	CS150	10"	Centrifugal Pumps and Vertical Vessel
1012	CS150	8"	Centrifugal Pumps
1007	CS150	8"	Vertical Vessel and Reboiler
2002	CS150	3"	Reboiler
2000	CS150	6"	Reboiler

Figure 6-71 *The Pipe list*

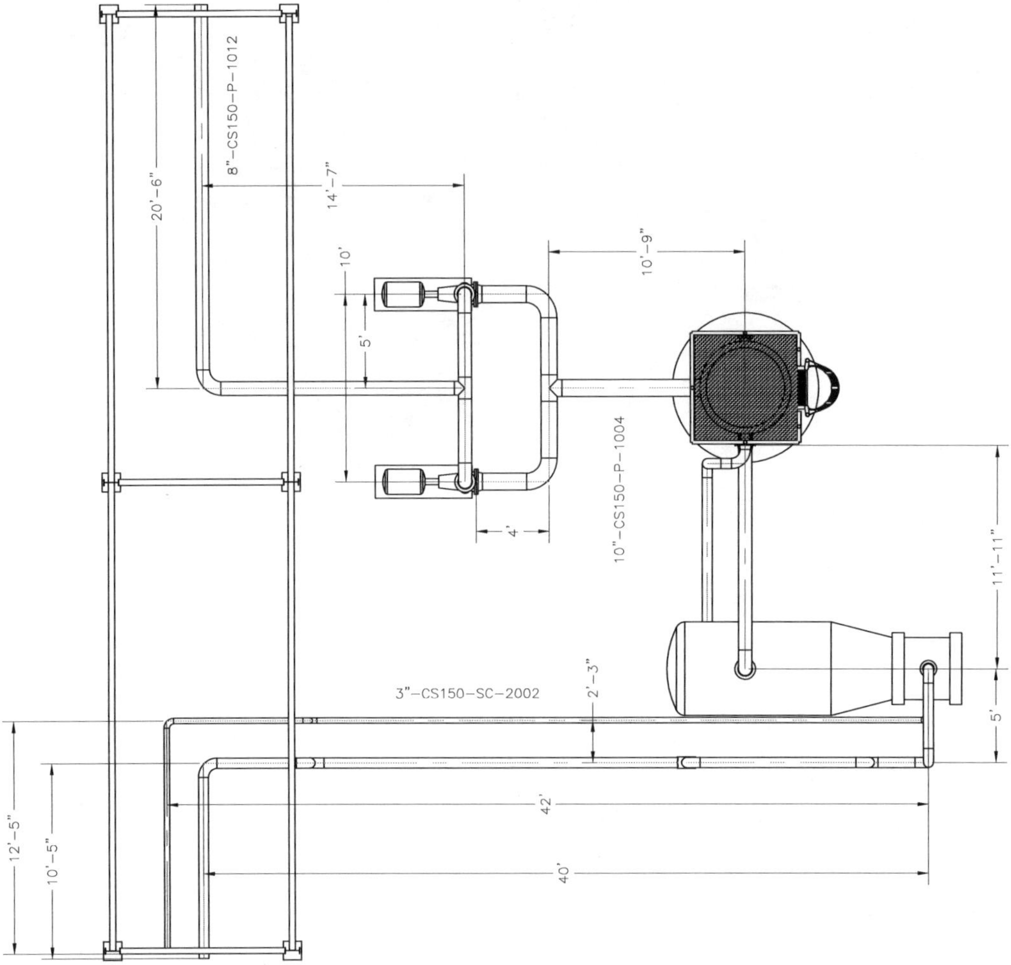

***Figure 6-72** Top view of the model*

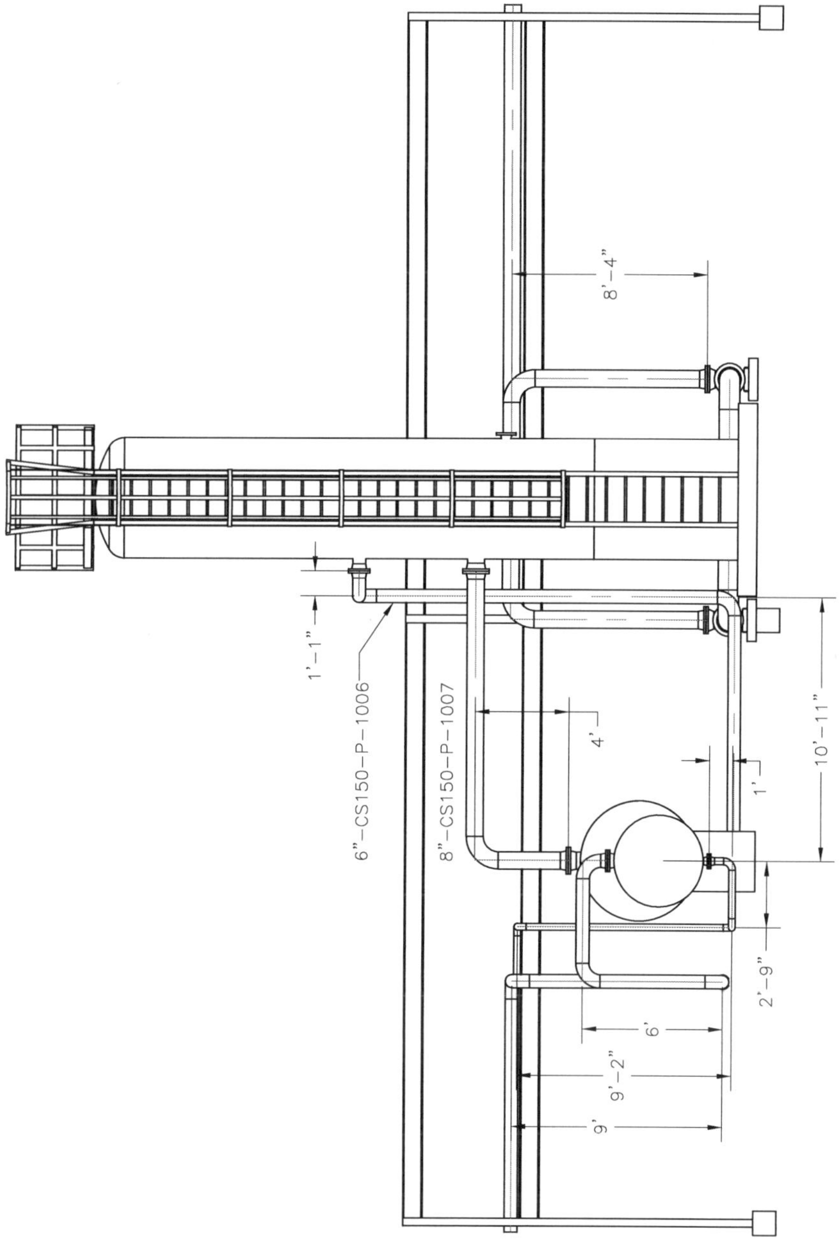

Figure 6-73 *Front view of the model*

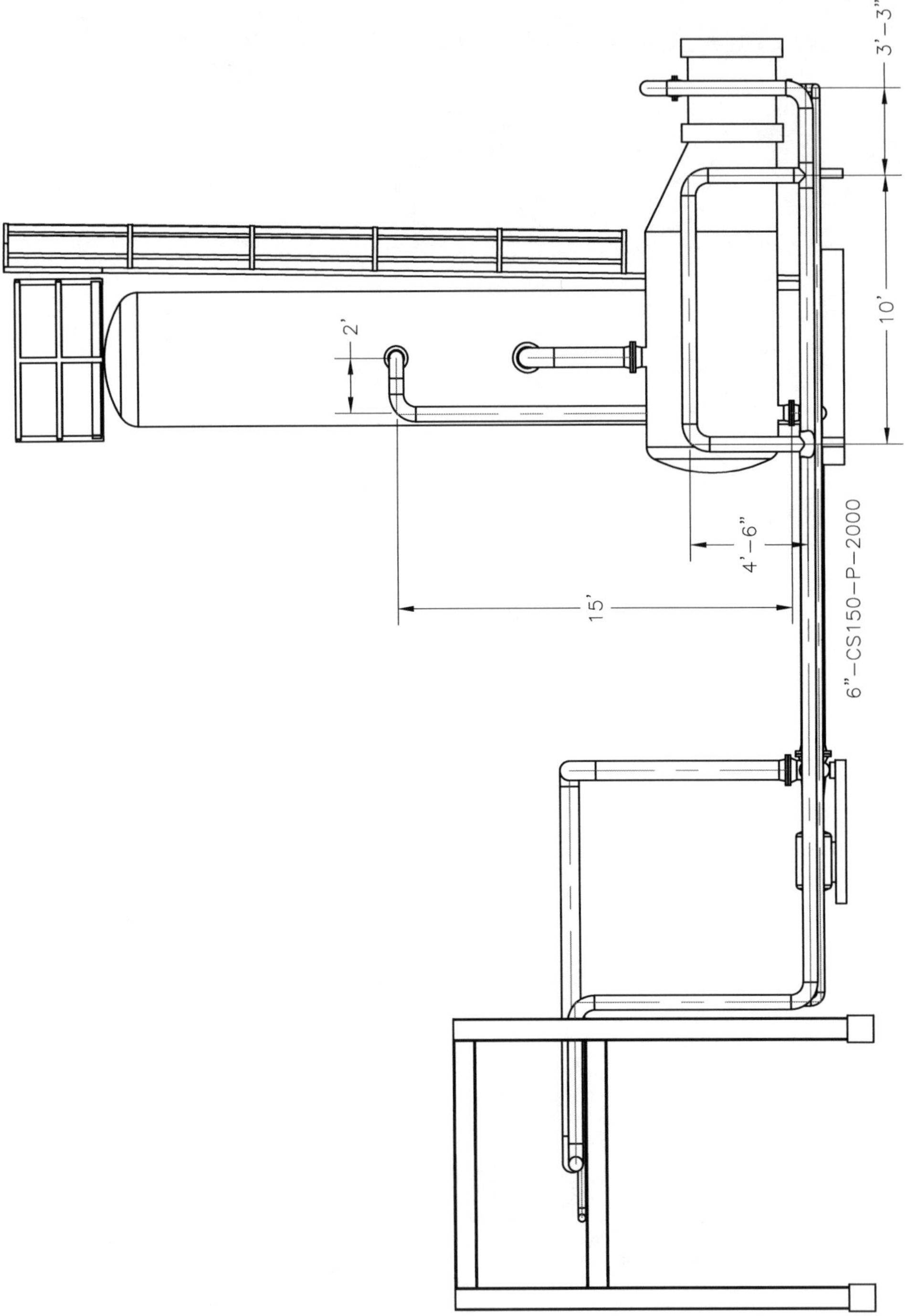

***Figure 6-74** Left view of the model*

Opening a File

1. Double-click on the **AutoCAD Plant 3D 2023 - English** icon; AutoCAD Plant 3D gets started.

2. Select **CADCIM** from the **Current Project** drop-down list in the **PROJECT MANAGER**.

3. Expand the **Plant 3D Drawings** node in the **PROJECT MANAGER** and double-click on **Piping Model**; the selected drawing file is opened.

Selecting the Spec and Pipe Size

In this section, you will select a spec and pipe size.

1. Select **CS150** from the **Spec Selector** drop-down list in the **Part Insertion** panel; the **Dynamic Pipe Spec** in the Tool Palette is loaded with *CS150* piping components.

Inserting the Pipe Rack into the Model

Next, you need to insert the pipe rack created in Tutorial 1 of Chapter 3.

1. When you click on the **Insert** tool from the **Block** panel in the **Insert** tab, a drop-down list is displayed. Select the **Blocks from other drawing**; the **BLOCKS** palette is displayed.

2. Choose the **Browse** button next to the **Filter** drop-down list in the **BLOCKS** palette and then browse to the location *C:\ > Documents > CADCIM > Plant 3D Models*. Next, double-click on the **Pipe_rack.dwg** file; preview of the file is displayed below the **Path** area in the **BLOCKS** palette.

3. Clear the **Insertion point** check box in the **Insertion Options** area if selected. Next, enter **0, 25', 0** in the **X**, **Y**, **Z** edit boxes, respectively.

4. Double click on the preview of the file which is displayed below the **Path** area of the **BLOCKS** palette; the pipe rack is placed in the drawing area at the specified location, as shown in Figure 6-75.

Note

For better visualization, you may hide the line diagram of structure grid of ***Pipe_rack****.*

Connecting the Centrifugal Pumps and Vertical Vessel

In this section, you will connect the pumps and the vertical vessel. First, you need to hide the support at the bottom of the vertical vessel.

1. Select the support of the vertical vessel from the drawing area and then choose the **Hide Selected** button from the **Visibility** panel in the **Home** tab; the support gets hidden.

2. Select the **Route New Line** option from the **Line Number Selector** drop-down list in the **Part Insertion** panel; the **Assign Tag** dialog box is displayed.

3. Enter **1004** in the **Number** edit box in the dialog box.

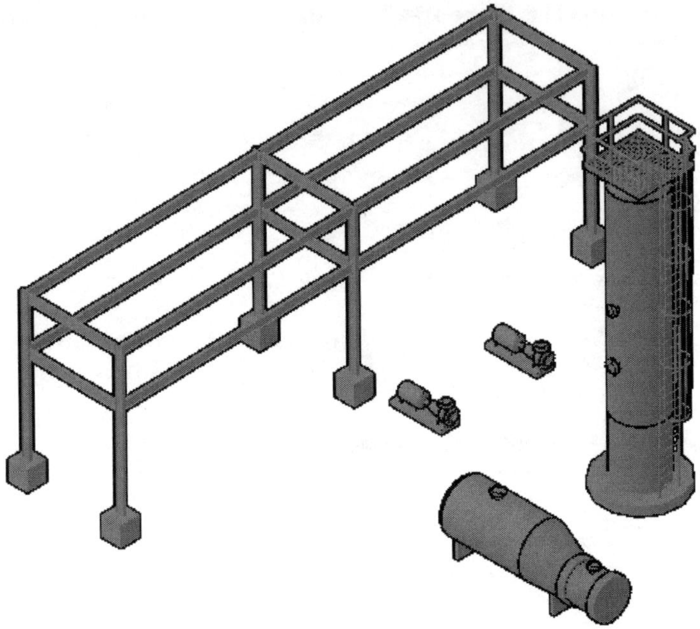

Figure 6-75 *Model after placing the pipe rack*

4. Select the line size as **10"** in the **Size** drop-down list.

5. Choose the **Assign** button; the **Assign Tag** dialog box will be closed and you are prompted to specify the start point of the pipe.

6. Press and hold the SHIFT key and right-click to display a shortcut menu. Choose the **Node** option from the shortcut menu displayed and then select the center point of the inlet nozzle of the left-side pump, as shown in Figure 6-76; you are prompted to specify the next point.

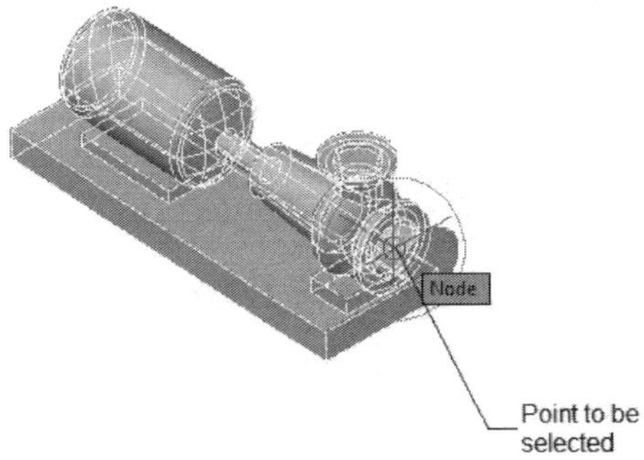

Figure 6-76 *Selecting the center point of the nozzle*

Note

*You can switch to the **Wireframe** view for easy selection of the end point. To do so, click on the down arrow of the **Visual Styles** drop-down list in the **View** panel and then select the **3dWireframe** option.*

7. Move the cursor in a direction which is in-line with the nozzle. Next, enter **4'** at the command prompt to specify the end point; you are prompted to specify the next point of the pipe.

8. Press and hold the SHIFT key and right-click to display a shortcut menu. Choose the **Node** option from the shortcut menu and then select the center point of the inlet nozzle of the right-side pump; the inlet nozzles are connected to the pump, as shown in Figure 6-77.

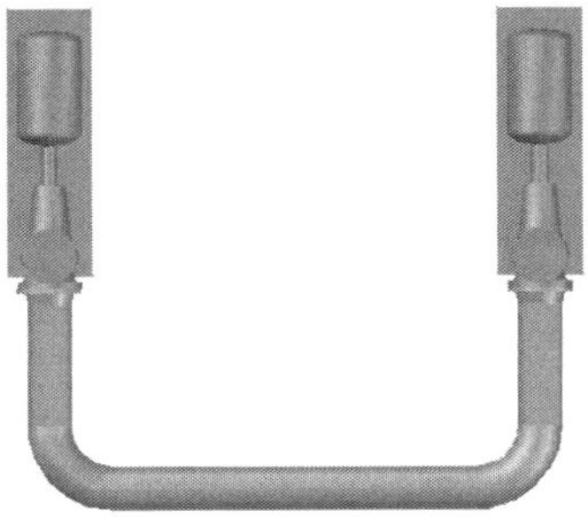

Figure 6-77 *The inlet nozzles connected to the pump*

9. Rotate the model using the **Orbit** tool such that the bottom nozzle of the vertical vessel is visible.

10. Choose the **Toggle Cutback Elbows** button from the **Part Insertion** panel. This creates an elbow whose angle is different from the angle of elbow available in the spec.

11. Click on the vertical vessel to display the plus (+) symbol on the bottom nozzle.

12. Click on the plus (+) symbol displayed and then press and hold the SHIFT key and right-click to display a shortcut menu. Choose the **Midpoint** option from the shortcut menu and then select the mid point of the pipe joining the two pumps, as shown in Figure 6-78; solutions are generated to connect the vessel.

13. Choose the **Next** option from the command prompt to display the required solution, as shown in Figure 6-79.

14. Next, choose the **Accept** option from the command prompt to accept the solution. Also, choose the **Show All** button from the **Visibility** panel of the **Home** tab to unhide the support of the vertical vessel.

Note

*Make sure that the **Toggle Pipe Bends** button is deactivated. If this button is activated then in place of an elbow a pipe bend will be created.*

Next, you need to connect the outlet nozzles of the centrifugal pumps.

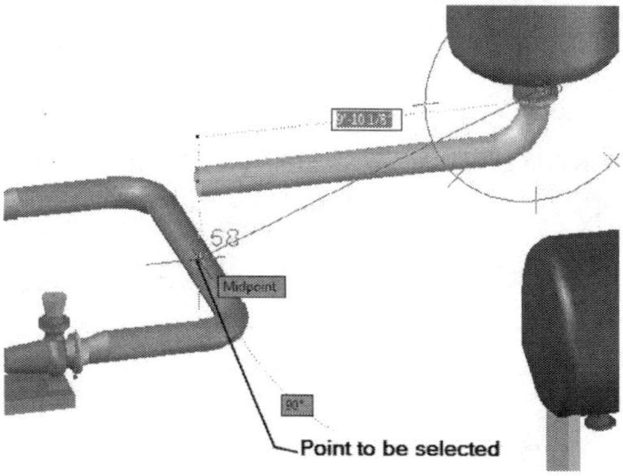

Figure 6-78 *Select the center point of the pipe*

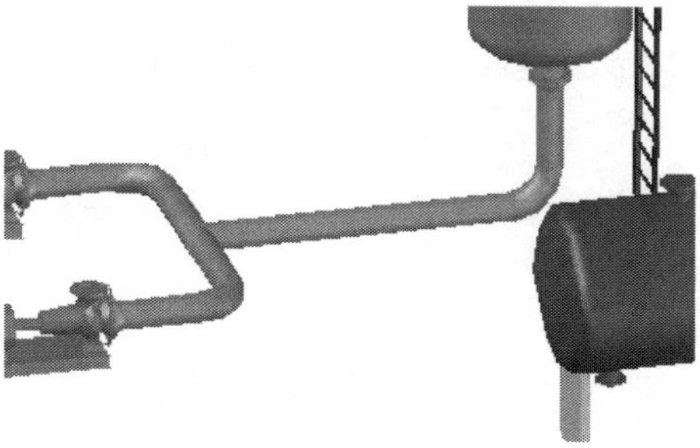

Figure 6-79 *Required solution for connecting the pipes*

Routing Pipes from the Outlet Nozzles of the Pumps

1. Select the **Route New line** option from the **Line Number Selector** drop-down list in the **Part Insertion** panel; the **Assign Tag** dialog box is displayed.

2. Enter **1012** in the **Number** edit box.

3. Select the line size **8”** from the **Size** drop-down list.

4. Choose the **Assign** button; the **Assign Tag** dialog box will be closed and you are prompted to specify the start point of the pipe.

5. Press and hold the SHIFT key and right-click to display a shortcut menu. Then, choose the **Node** option from it.

6. Next, select the center point of the nozzle on the left pump, as shown in Figure 6-80; the pipe gets connected to it and you are prompted to specify the next point.

7. Move the cursor upward and enter **100”** at the command prompt; a vertical pipe is created.

8. Next, press and hold the SHIFT key and right-click to display a shortcut menu. Then, choose the **Node** option from it. Next, select the outlet nozzle of the right-side pump; the outlet nozzles gets connected, as shown in Figure 6-81.

9. Click on the horizontal pipe of the newly created pipe connection; a plus (+) symbol is displayed on it.

10. Select the plus (+) symbol and then move the cursor toward pipe rack

.

11. Change the orientation plane, as shown in Figure 6-82, by entering the **Plane** option at the Command prompt.

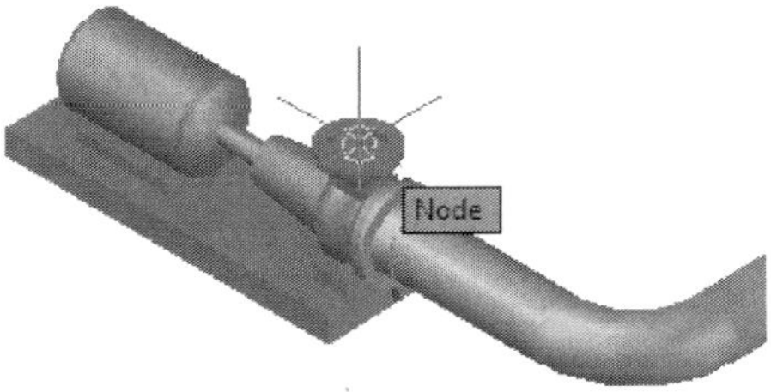

Figure 6-80 *Selecting the center point outlet nozzle of the left pump*

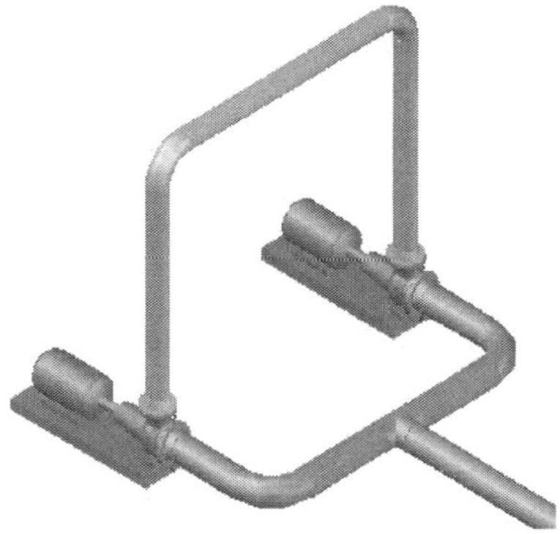

Figure 6-81 *Pumps after connecting the outlet nozzles*

12. Next, move the pipe toward the pipe rack, type **14'** at the command prompt, and then press ENTER to create a pipe passing toward the pipe rack, as shown Figure 6-82.

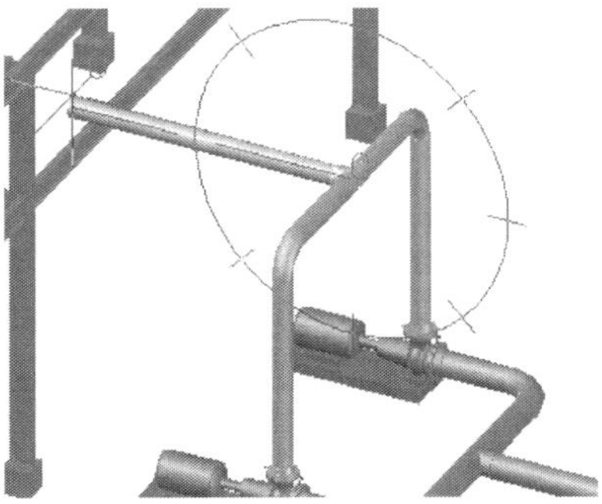

Figure 6-82 *Pipe extending from the right elbow*

13. Next, change the orientation plane and move the cursor toward right, as shown in Figure 6-83.

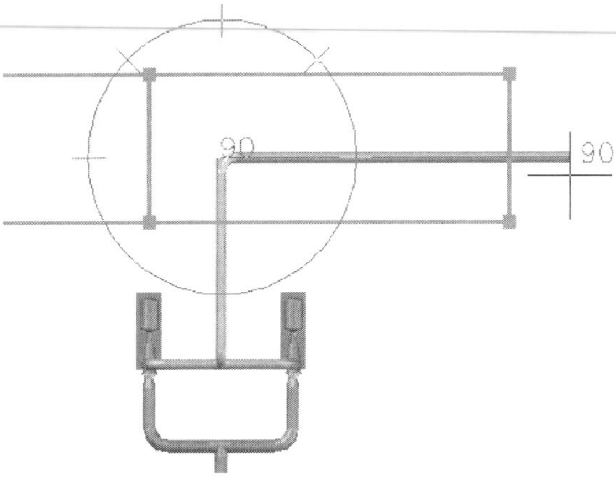

Figure 6-83 *Pipe passing toward the pipe rack*

14. Type **20'** at the command prompt and then press ENTER; a pipe is created upto the right end of the pipe rack. Next, press ESC to exit the **Route Pipe** tool.

Connecting the Vertical Vessel and the Reboiler

In this section, you will connect the vertical vessel and the reboiler using the **Route Pipe** tool.

1. Select the **Route New line** option from the **Line Number Selector** drop-down list in the **Part Insertion** panel; the **Assign Tag** dialog box is displayed.

2. Enter **1007** in the **Number** edit box.

3. Specify the line size as **8"** from the **Size** drop-down list.

4. Choose the **Assign** button; the **Assign Tag** dialog box is closed and you are prompted to specify the start point of the pipe.

5. Press and hold the SHIFT key and right-click to display a shortcut menu. Then, choose the **Node** option and specify the start point of the pipe on the nozzle of the vertical vessel, as shown in Figure 6-84; you are prompted to specify the next point.

6. Press and hold the SHIFT key and right-click to display a shortcut menu. Then, choose the **Node** option.

7. Select the center point of the nozzle on the reboiler, as shown in Figure 6-85; the pipe connection is created between the two selected nozzles, as shown in Figure 6-86.

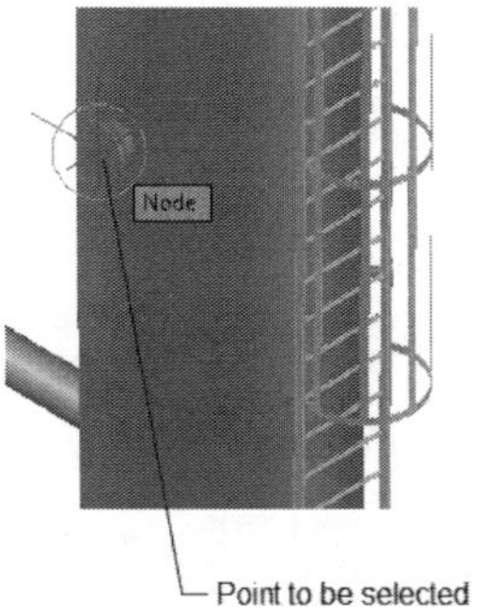

Figure 6-84 *Point to be selected to specify the start point of the pipe*

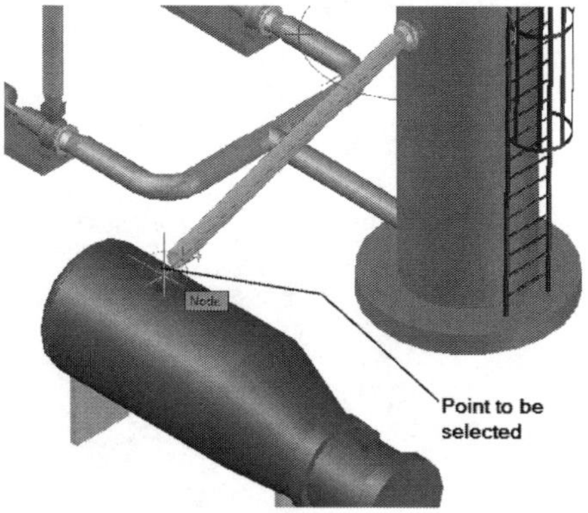

Figure 6-85 *Point to be selected to specify the end point of the pipe*

Note

*Make sure that the **Toggle Pipe Bends** button available in the **Part Insertion** panel is not chosen.*

8. Next, choose the **Accept** option from the command prompt to accept the solution.

Next, you need to connect the nozzle located at the bottom of the reboiler to the nozzle on the vertical vessel.

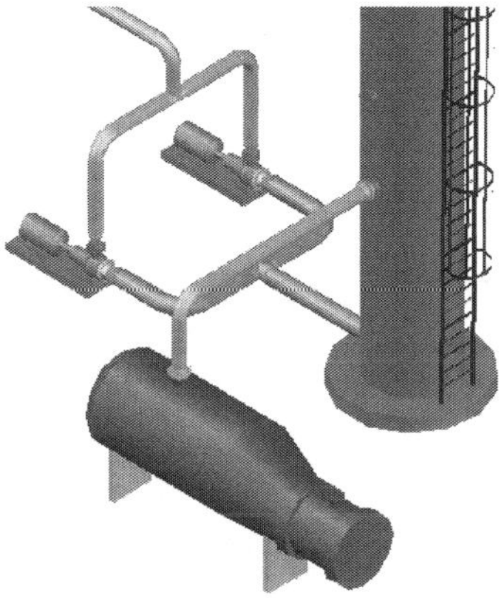

Figure 6-86 *Pipe connection created between the selected nozzles*

9. Rotate the model using the **Orbit** tool such that the nozzle at the bottom of the reboiler is visible.

10. Select the **Route New line** option from the **Line Number Selector** drop-down list in the **Part Insertion** panel; the **Assign Tag** dialog box is displayed.

11. Enter **1006** in the **Number** edit box.

12. Specify the line size as **6"** from the **Size** drop-down list.

13. Choose the **Assign** button; the **Assign Tag** dialog box is closed and you are prompted to specify the start point of the pipe.

14. Select the center point of the nozzle located at the bottom of the reboiler, refer to Figure 6-87.

    Note that to select the nozzle center point you may need to choose the **Node** option from the **Object Snap** shortcut menu.

15. Next, move the cursor toward the vertical vessel, type **3'** at the Command prompt, and then press ENTER.

16. Next, invoke the **Object Snap** shortcut menu and choose the **Node** option from it.

17. Select the center point of the nozzle, as shown in Figure 6-88; solutions are generated to create a connection between the two selected nozzles.

Note

*Make sure that the **Toggle Pipe Bends** button from the **Part Insertion** panel is not chosen.*

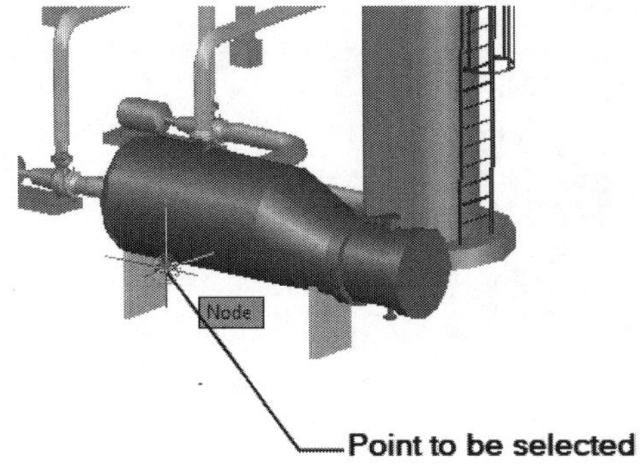

Figure 6-87 Point to be selected

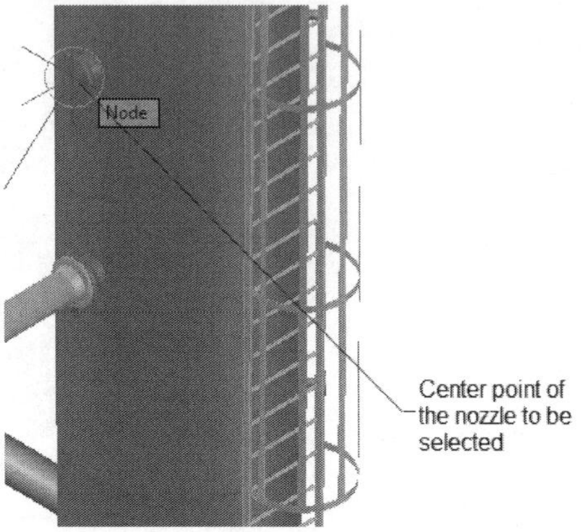

Figure 6-88 Point to be selected

18. Choose the **Next** option from the command prompt until the required solution is displayed, as shown in Figure 6-89.

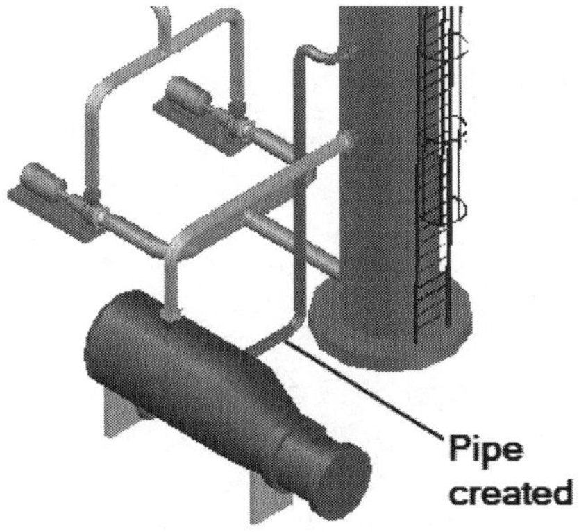

Figure 6-89 The solution to be selected to create the pipe connection

19. Choose the **Accept** option to accept the required solution.

Routing Pipes from the Other Nozzles of the Reboiler

Next, you need to route a pipe from the bottom nozzle of the reboiler.

1. Select the **Route New line** option from the **Line Number Selector** drop-down list in the **Part Insertion** panel; the **Assign Tag** dialog box is displayed.

2. Enter **2002** in the **Number** edit box.

3. Select the line size as **3"** from the **Size** drop-down list.

4. Choose the **Assign** button; the **Assign Tag** dialog box is closed and you are prompted to specify the start point of the pipe.

5. Select the bottom nozzle of the reboiler, as shown in Figure 6-90. Next, move the cursor downward.

6. Type **1'** at the command prompt and then press ENTER.

7. Move the cursor toward left, enter **3'** at the command prompt, and then press ENTER.

8. Change the orientation plane and move the cursor toward the pipe rack. Enter **34'** at the command prompt and then press ENTER.

9. Change the orientation plane, if required, by using the **Plane** option from the command prompt and move the cursor upward. Next, enter **110"** at the command prompt and press ENTER.

10. Move the cursor toward the pipe rack, enter **8'** at the command prompt, and then press ENTER.

11. Change the orientation plane, if required, and then move the cursor toward left upto the end of the pipe rack.

12. Next, click to specify the end point. Figure 6-91 shows the pipe routed from the bottom nozzle of the reboiler.

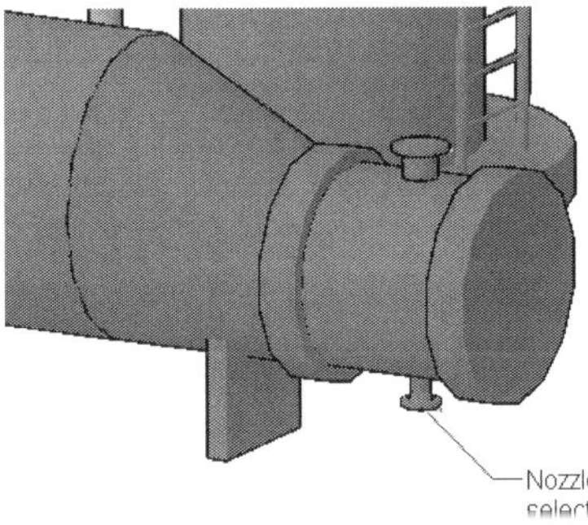

Figure 6-90 *Nozzle to be selected*

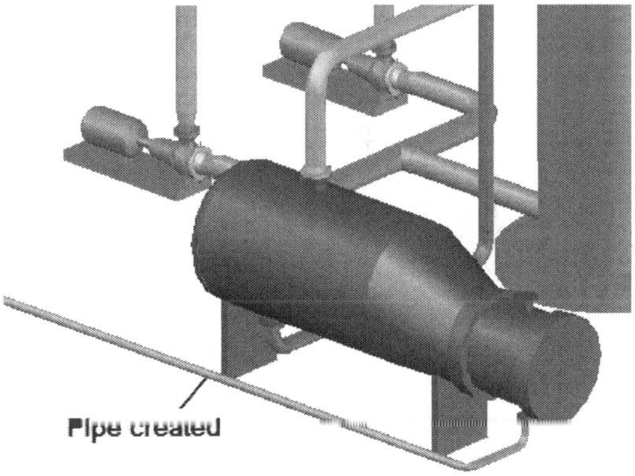

Figure 6-91 *Pipe routed from the bottom nozzle of the reboiler*

Routing Pipe from Another Nozzle of the Reboiler

Next, you need to create other pipes connecting the reboiler.

1. Select the **Route New line** option from the **Line Number Selector** drop-down list in the **Part Insertion** panel; the **Assign Tag** dialog box is displayed.

2. Enter **2000** in the **Number** edit box.

3. Select the line size **6"** from the **Size** drop-down list.

4. Choose the **Assign** button; the **Assign Tag** dialog box is closed and you are prompted to specify the start point of the pipe.

5. Specify the start point on the nozzle of the reboiler, as shown in Figure 6-92.

6. Next, move the cursor toward left, type **5'** at command prompt, and press ENTER.

7. Next, move the cursor downward, type **6'** at the command prompt, and press ENTER.

8. Next, choose the **Plane** option from the command prompt to change the orientation plane, as shown in Figure 6-93.

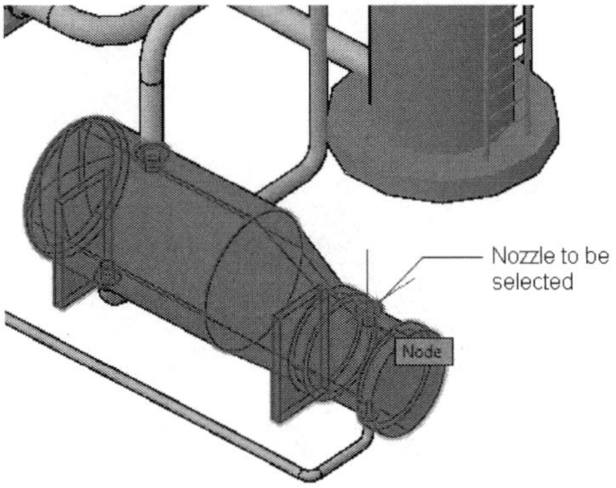

Figure 6-92 *Point to be selected*

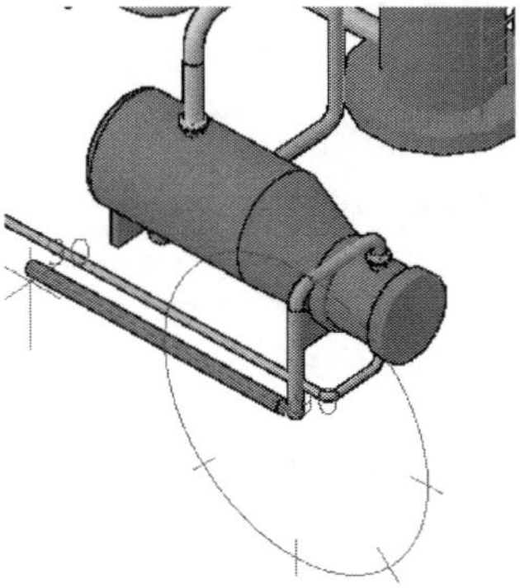

Figure 6-93 *Orientation plane to be set*

9. Move the cursor toward the pipe rack, enter **34**' at the command prompt, and press ENTER.

10. Change the orientation plane, if required, and move the cursor upward. Next, enter **108"** at the command prompt and press ENTER.

11. Move the cursor toward the pipe rack, enter **6'** at the command prompt, and then press ENTER.

12. Change the orientation plane, if required, and move the cursor toward left upto the end of the pipe rack. Next, click to specify the end point.

    Next, you need to create a tee branch on the previously created pipe.

13. Select **Tee, BV,, (CS150)** from the **TOOL PALETTES**, refer to Figure 6-94; the tee gets attached to the cursor and you are prompted to specify the insertion point.

    The insertion point is to be specified at a precise distance from the end point of the pipe.

14. Invoke the **Object Snap** shortcut menu and choose the **From** option. Next, select the end point of the previously created pipe, as shown in Figure 6-95.

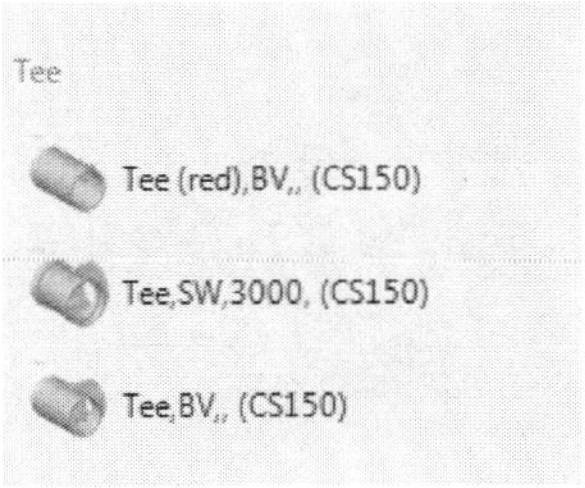

Figure 6-94 *Selecting* ***Tee,BV,, (CS150)*** *from the Tool Palette*

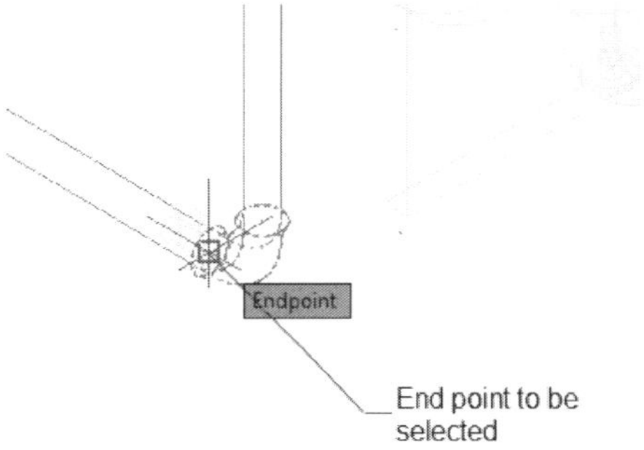

Figure 6-95 *Selecting the end point of the pipe*

15. Next, move the cursor over the pipe and enter **4'** at the command prompt. Next, press ENTER; the tee is placed at the specified point, as shown in Figure 6-96. Also, you are prompted to specify the rotation angle.

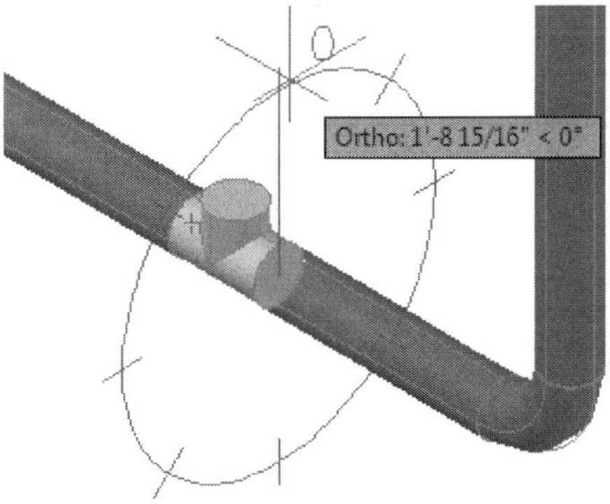

Figure 6-96 *Tee placed at the specified point*

16. Press ENTER to accept the default value. Press ESC to exit the tool.

17. Click on the tee to display a plus (+) symbol on it.

18. Click on the plus (+) symbol and move the cursor upward.

19. Enter **4'** at the command prompt and press ENTER.

20. Change the orientation plane and move the cursor toward the pipe rack.

21. Enter **10'** at the command prompt and press ENTER.

22. Move the cursor downward and select the **STub-in** option from the command prompt.

23. Select the center point of the horizontal pipe, as shown in Figure 6-97; pipe gets created, as shown in Figure 6-98.

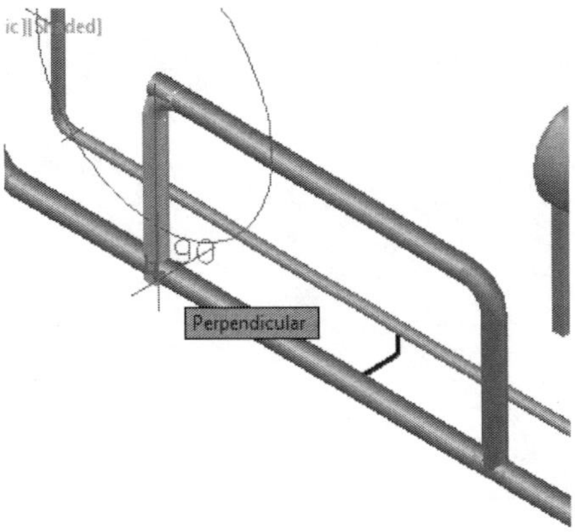

Figure 6-97 Point to be selected on the horizontal pipe

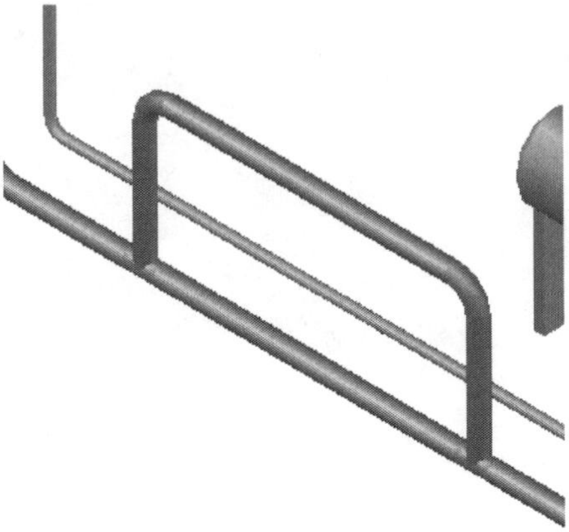

Figure 6-98 Required solution

Adding the Reinforcing Pad

1. Select the vertical pipe, as shown in Figure 6-99.

2. Right-click on the selected pipe and choose the **Add Reinforcing Pad** option from the shortcut menu displayed; the reinforcing pad is added to the stub-in pipe, as shown in Figure 6-100.

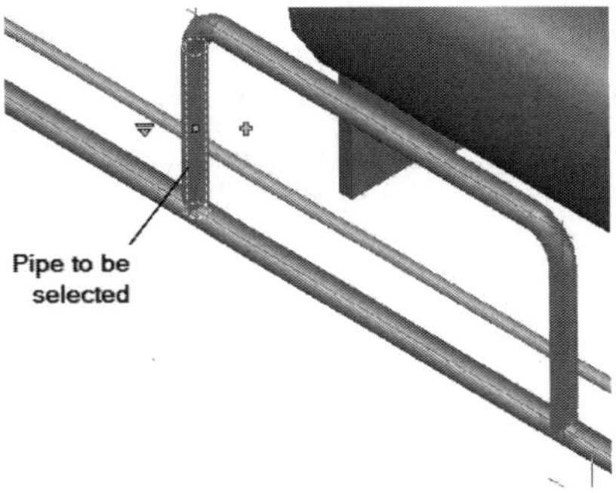

Figure 6-99 Pipe to be selected

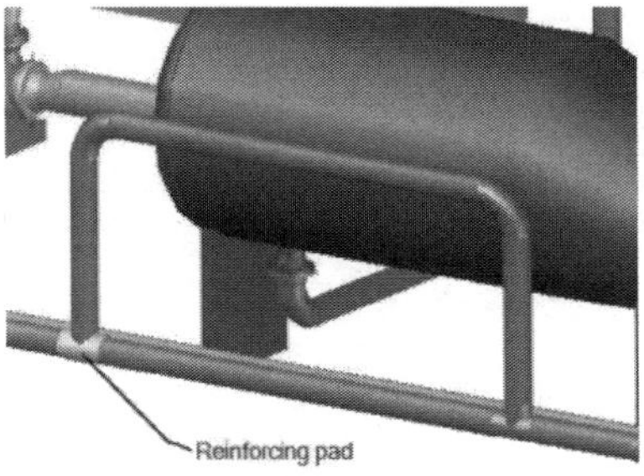

Figure 6-100 Reinforcing pad created

The piping system after adding all the pipes is displayed, as shown in Figure 6-101.

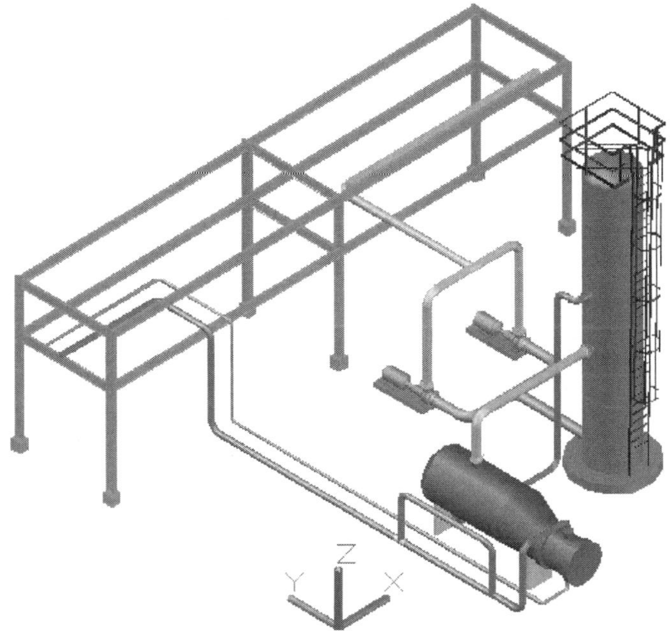

Figure 6-101 Model for Tutorial 1

3. Choose the **Save** button from the **Application** menu and then the **Close** button to close the file.

Self-Evaluation Test

Answer the following questions and then compare them to those given at the end of this chapter:

1. The _________ edit box is used to specify the angular increments through which cursor will snap.
2. The **Spec Viewer** contains the _________ rollout and the _________ rollout.
3. You can insert a part from the **Spec Viewer** by using the _________ button.
4. You can route a pipe with a new line number assigned to it using the _________ option.
5. You can change the position of the route line using the _________ drop-down list.
6. The _________ tool is used to convert a line into a pipe.
7. The _________ button is used to control the display of tick marks on the compass.

8. Before creating a 3D piping model, you need to select a piping spec. (T/F)

9. Solutions are generated for connecting two open ports. (T/F)

10. Compass helps you to rotate a pipe at a precise angle. (T/F)

Review Questions

Answer the following questions:

1. You can select the pipe size from the __________ drop-down list in the **Part Insertion** panel.

2. Using the __________ button, you can create a new tool palette with parts available in the selected spec.

3. You can change the compass settings by using the options in the __________ panel.

4. You can turn on/off the display of tick marks on the compass by using the __________ button.

5. With the __________ mode turned on, you can create elbow at angles other than 90 or 45 degrees.

6. You can create a rolled elbow (twisted elbow) by specifying rotation angles in two planes which are __________ to each other.

7. You can convert a straight pipe into a sloped pipe by invoking the __________ dialog box.

8. On selecting a pipe, the __________ is displayed at the mid point of the pipe.

9. Using the __________, you can connect an AutoCAD Plant 3D pipe to piping components created in other Autodesk applications such as AutoCAD MEP, Civil 3D, and so on.

10. You can change the size of the pipe while routing. (T/F)

Answers to Self-Evaluation Test

1. **Snap Increments**, **2.** **Spec Sheet**, **Part Sizes**, **3.** **Insert in Model**, **4.** **Route New line**, **5.** **Set Routing Line**, **6.** **Line to Pipe**, **7.** **Toggle Tick Marks**, **8.** T, **9.** T, **10.** T

Chapter 7

Adding Valves, Fittings, and Pipe Supports

Learning Objectives

After completing this chapter, you will be able to:

- *Place valves and fittings*
- *Map P&ID objects to Plant 3D objects*
- *Add supports to pipes*
- *Insulate a pipe*
- *Modify pipe components*
- *Validate a 3D model*

INTRODUCTION

In this chapter, you will learn to add valves, fittings, and pipe supports in a 3D model. In addition, you will learn to modify a pipe component and validate a 3D model.

ADDING VALVES AND FITTINGS

In AutoCAD Plant 3D, you can add valves and fittings to a pipe by using different methods. All these methods are discussed next.

Adding Valves and Fittings to a Pipe Using the Spec Sheet

To add a valve or a fitting to a pipe from a spec sheet, invoke the **PIPE SPEC VIEWER** by choosing the **Spec Viewer** button from the **Part Insertion** panel in the **Home** tab. Next, double-click on the required valve in the **Spec Sheet** present in **PIPE SPEC VIEWER**; the valve will be attached to the cursor. You can press the CTRL key to change the port attached to the cursor. Next, you need to specify the insertion point of the valve, refer to Figure 7-1. Make sure that the **Object Snap** is turned on for easy selection of the point. Next, you need to specify the rotation of the valve by using the compass or enter **0** degree at the command prompt to specify the zero degree rotation, refer to Figure 7-2.

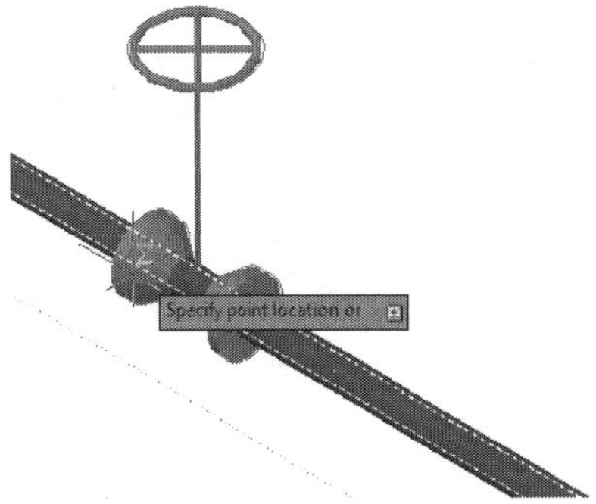

Figure 7-1 *Specifying the insertion point of the valve*

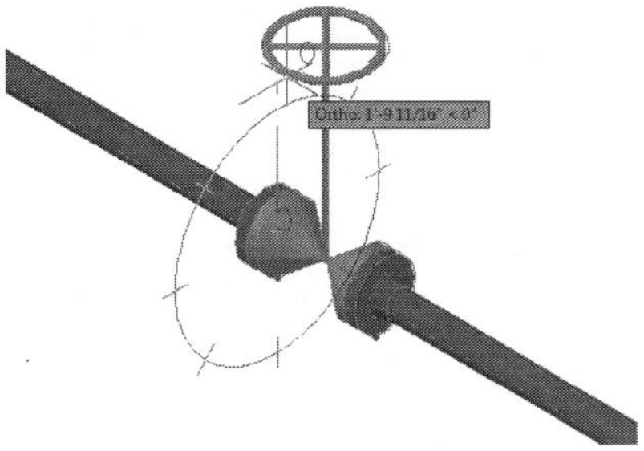

Figure 7-2 *Specifying the rotation of the valve*

You can also place the valves at a precise distance from a point. To do so, press and hold the SHIFT key and then right-click in the drawing area to display a shortcut menu. Next, choose the **From** option from the shortcut menu displayed, refer to Figure 7-3; you are prompted to specify a point from which the distance will be calculated. Press and hold the SHIFT key and right-click to invoke the shortcut menu. Choose the **Node** option from the shortcut menu and then select the end point of the pipe. Next, move the cursor over the pipe, refer to Figure 7-4, and enter a distance at the command prompt; the insertion point of the valve will be specified. Next, you need to specify the rotation angle of the valve.

You can also place a valve or a fitting from the **TOOL PALETTES** located at the right-side of the drawing area. To do so, choose the **Dynamic Pipe Spec** tab from the **TOOL PALETTES** and select the required valve or fitting; the selected valve or fitting will be attached to the cursor. Next, place the valve or fitting in the model.

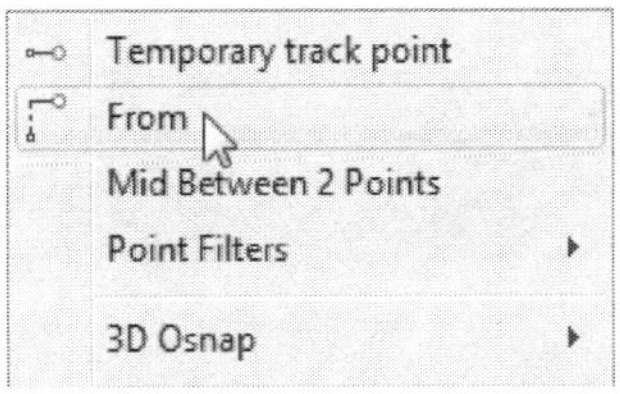

Figure 7-3 Choosing the ***From*** *option from the shortcut menu displayed*

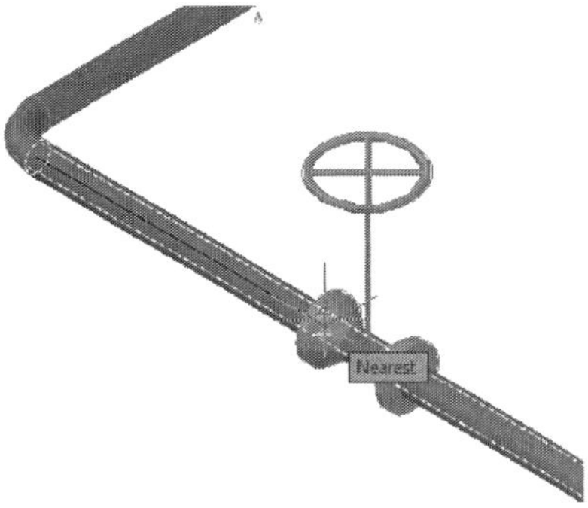

Figure 7-4 Placing the valve at a precise distance from the end point of the pipe

Adding Valves and Fittings Using a P&ID

To place a valve using a P&ID, choose the **P&ID Line List** button from the **Part Insertion** panel; the **P&ID LINE LIST** palette will be displayed, refer to Figure 7-5. In this palette, select **P&ID** from the drop-down list located at the top; all the line numbers present in the selected P&ID are displayed. The components present in the P&ID are grouped under the displayed line numbers. Next, expand the respective pipe line number from the tree view and select the valve available under the pipe line number. Next, choose the **Place** button from the **P&ID LINE LIST** palette; the selected component will be attached to the cursor and you will be prompted to specify the insertion point. Specify the insertion point on the pipe.

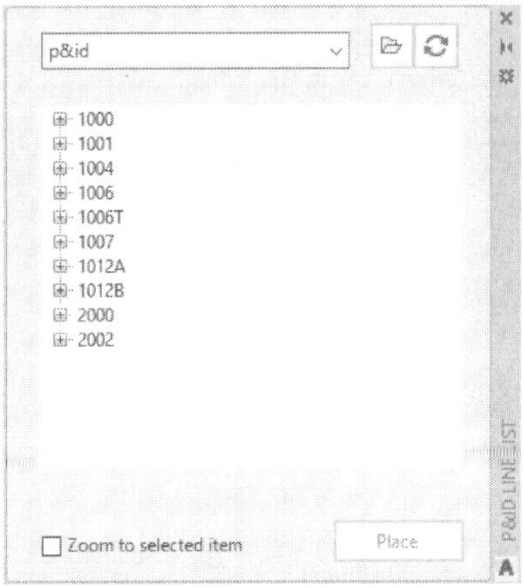

Figure 7-5 The ***P&ID LINE LIST*** *palette*

Note that the component to be placed from a P&ID to plant 3D model should be mapped with the corresponding Plant 3D component. You will learn more about mapping P&ID components with a 3D model later in this chapter.

If the selected component is available in the tree of the **P&ID LINE LIST** palette without the spec, then on choosing the **Place** button, the **Select Size and Spec** dialog box will be displayed, as shown in Figure 7-6. In this dialog box, select the spec and size from the **Spec** and **Size** list boxes, respectively. Next, choose the **Select** button to place the valve.

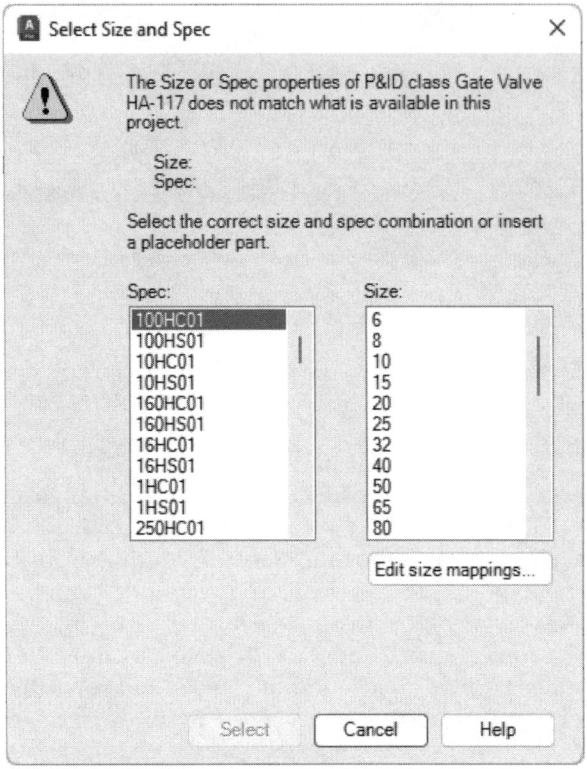

Figure 7-6 The ***Select Size and Spec*** *dialog box*

Placing Valves and Fittings while Routing a Pipe

You can also place a valve or a fitting while routing a pipe. To do so, invoke the **Route Pipe** tool and specify the start point. Next, choose the **pipeFitting** option from the command prompt; the **Part Placement** palette will be displayed, as shown in Figure 7-7.

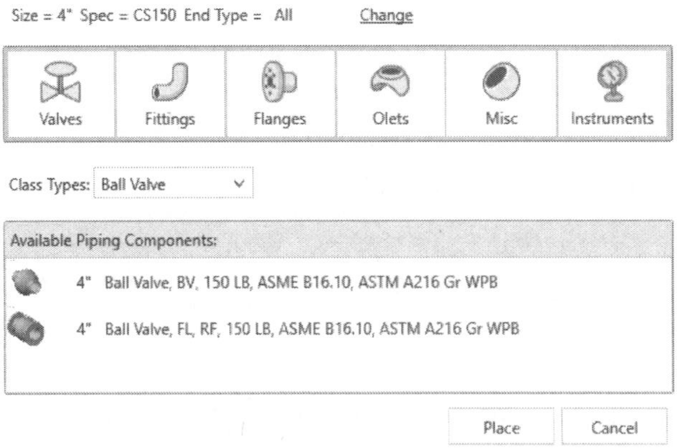

Figure 7-7 The ***Part Placement*** *palette*

To place a valve, choose the **Valves** button from the **Part Placement** palette. Next, select an option from the **Class Types** drop-down list; the available valves will be displayed. Select the required valve from the **Available Piping Components** area and choose the **Place** button; the selected valve will be attached to the end of the pipe, as shown in Figure 7-8. Next, click to place the valve; the compass will be displayed and you will be prompted to specify the valve rotation angle. Specify the rotation angle and then press ENTER. Similarly, you can place other pipe components such as fittings, flanges, caps, and so on.

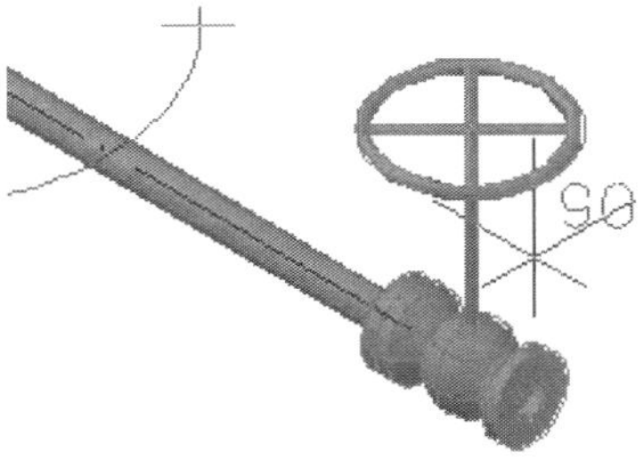

Figure 7-8 *Valve attached to the end of the pipe*

Placing Custom Parts

You can place a custom part or instrument into a 3D model. To do so, choose the **Custom Part** tool from the **Part Insertion** panel; the **CUSTOM PARTS BUILDER** palette will be displayed, as shown in Figure 7-9. In this palette, choose the **Plant 3D Shape** button from the **Graphics** area and select the part type from the **Part Type** drop-down list. Next, choose the **Shape Browser** button from the **Graphics** area; the **Plant 3D Shape Browser** dialog box will be displayed, as shown in Figure 7-10. Note that you need to expand the **Plant 3D Shape Browser** dialog box to view all the options in it. To do so, move the cursor on any of the corners; a double sided arrow will be displayed. Next, drag the cursor diagonally to expand the palette to the desired size. Select the required shape from the **Graphics** area and choose the **OK** button; the preview of the selected shape will be displayed in the **CUSTOM PARTS BUILDER** palette. Select the **Permanent** option from the **Custom Part type** drop-down list in the **Part Properties** rollout. Also, specify the size and other properties in the **CUSTOM PARTS BUILDER** palette. Next, choose the **Insert in Model** button from the **CUSTOM PARTS BUILDER** palette and place the part in the model.

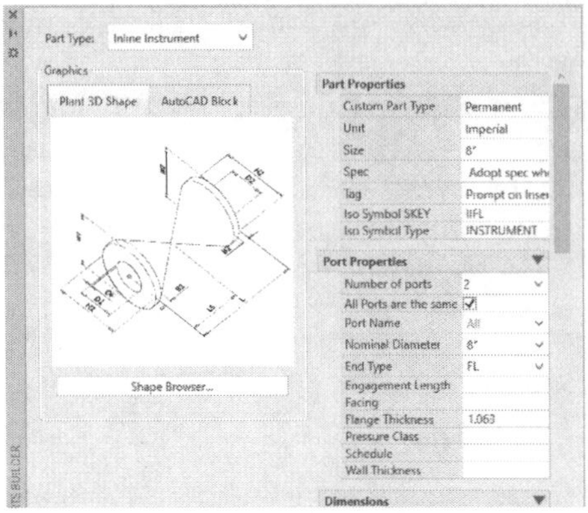

Figure 7-9 *Partial view of the* ***CUSTOM PARTS BUILDER*** *palette*

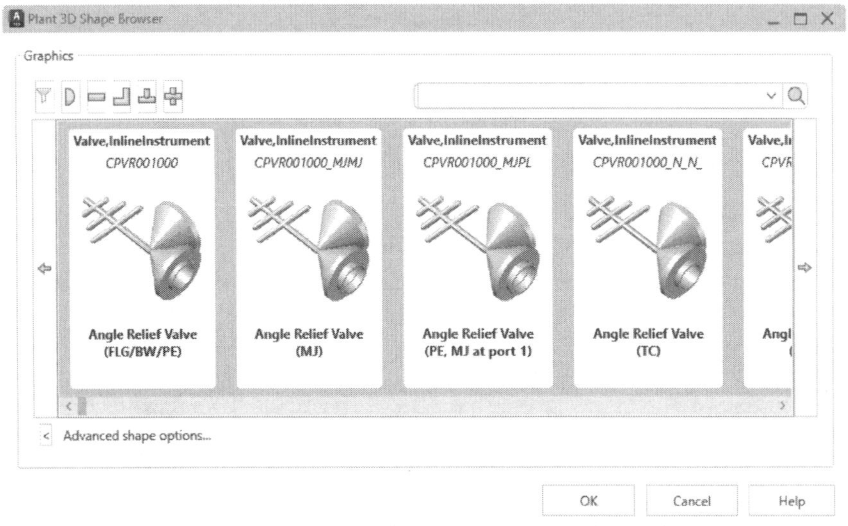

Figure 7-10 *The* ***Plant 3D Shape Browser*** *dialog box*

Mapping a P&ID Object to a Plant 3D Object

Before placing an object from a P&ID into a Plant 3D model, it has to be mapped to the corresponding Plant 3D object. This is a prerequisite to place a P&ID object. Most of the P&ID objects are mapped to the Plant 3D object by default. However, you need to map a newly created P&ID object to the Plant 3D object. For example, when you place a control valve in P&ID, you need to select the required valve body and actuator. A new object type in P&ID will be created. Now, you need to map it to a Plant 3D object. To do so, follow the steps given next.

1. Choose the **Project Setup** tool from the **Project** drop-down in the **Project** panel of the **Home** tab from the **Ribbon**; the **Project Setup** dialog box will be displayed.

2. In this dialog box, expand the **Plant 3D DWG Settings** node and select the **P&ID Object Mapping** option; the **P&ID Object Mapping** area will be displayed on the right side in the dialog box, refer to Figure 7-11.

3. In the **P&ID Object Mapping** area, expand the **P&ID Classes** tree and then select the required class; the properties of the corresponding Plant 3D object will be displayed in the **Plant 3D Classes** area.

4. Choose the **Add** button from the **Plant 3D Classes** area; the **Select Plant 3D Class Mapping** dialog box will be displayed.

5. In this dialog box, expand the **Plant 3D Classes** tree and then select the Plant 3D class. Note that the class should be same as that you have selected from the **P&ID Classes** tree. For example, if you have selected **Control Valve** from **Engineering Items > Instrumentation > Inline Instruments** node in the **P&ID Classes** tree, you need to select the **Valve** class from **Piping and Equipment > Pipe Run Component** in the **Plant 3D Classes** tree.

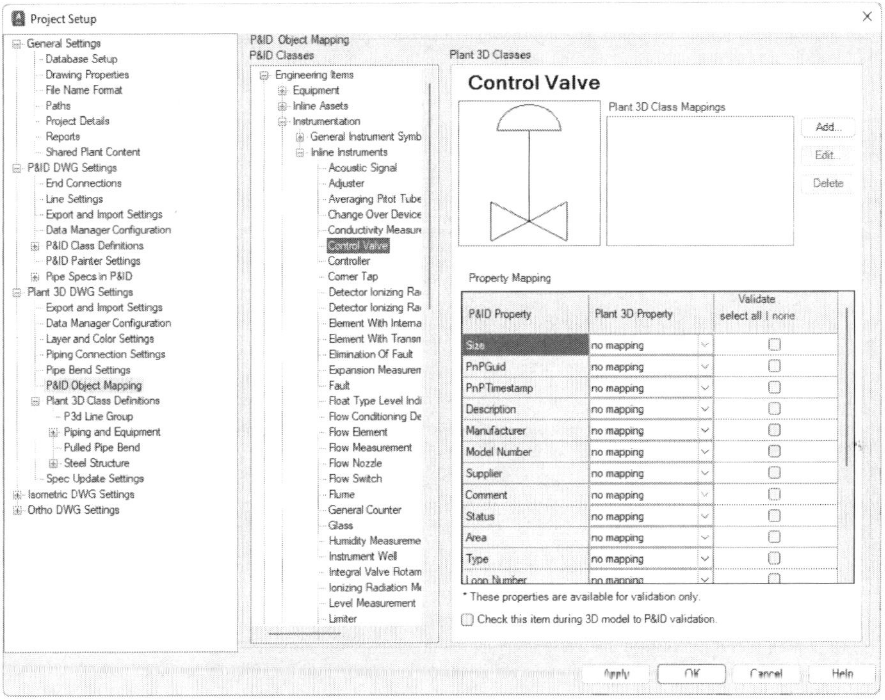

*Figure 7-11 The **Project Setup** dialog box*

6. Next, select the required check boxes from the **Map to one or more specific subtypes of this class** list box in the **Properties** area, refer to Figure 7-12. If you want to select all the subtypes, then select the **Map to all subtypes of this class** check box.

7. Choose the **OK** button; the **Select Plant 3D Class Mapping** dialog box will be closed and the **Project Setup** dialog box will be displayed.

   Next, you need to set the properties that are to be mapped.

8. In the **Project Setup** dialog box, select appropriate values from the drop-down lists located next to the properties in the **Property Mapping** table of the **Plant 3D Classes** area. For

example, you need to select the **Actuator Type** option from the drop-down list next to the **Actuator Type** property.

9. Next, select the required check boxes in the **Validate** column in the **Property Mapping** table. The corresponding properties will be checked while running the validation.

10. Choose the **OK** button to close the **Project Setup** dialog box.

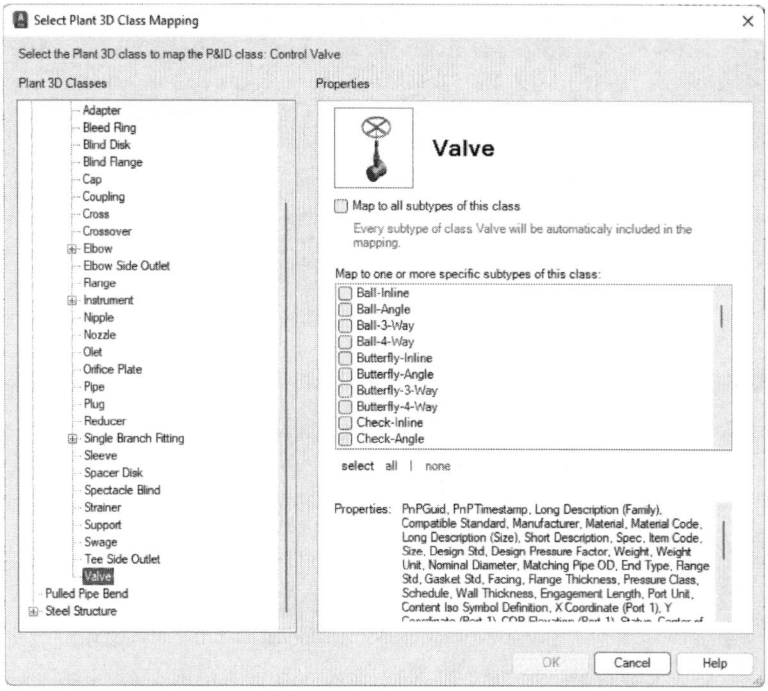

*Figure 7-12 The **Select Plant 3D Class Mapping** dialog box*

ADDING PIPE SUPPORTS

To add a piping support, choose the **Create** button from the **Pipe Supports** panel in the **Home** tab; the **Add Pipe Support** dialog box will be displayed, as shown in Figure 7-13.

Next, select a pipe support from the **Graphics** area. You can filter the display of pipe supports by using the filter buttons present at the top left of the dialog box. For example, to display only the base supports, choose the **Base Supports** button. You can further filter base supports by using the search bar present at the top right of the dialog box. To do so, click on the down arrow present in the search bar; a drop-down list will be displayed with four options: **Bolted Supports**, **Clamped Supports**, **Roller Supports**, **Spring supports**. Select the required option from the drop-down list; the pipe supports will filter as per the option selected from the drop-down list. After selecting the required pipe support, choose the **OK** button; the selected pipe support will get attached to the cursor and you will be prompted to select an insertion point. Select a point on the pipe where you want to place the support. Make sure that the **Object Snap** is turned on for easy selection of the point. Next, press ENTER to exit the tool. Figure 7-14 shows a Side Clamped Stanchion placed as a pipe support.

You can also add a pipe from the **Pipe Supports Spec** tab. To do so, choose the **Pipe Supports Spec** tab from the **TOOL PALETTES** present at the right-side of the drawing window; the pipe supports will be displayed, as shown in Figure 7-15. Scroll through the **TOOL PALETTES** and select the required pipe support from it.

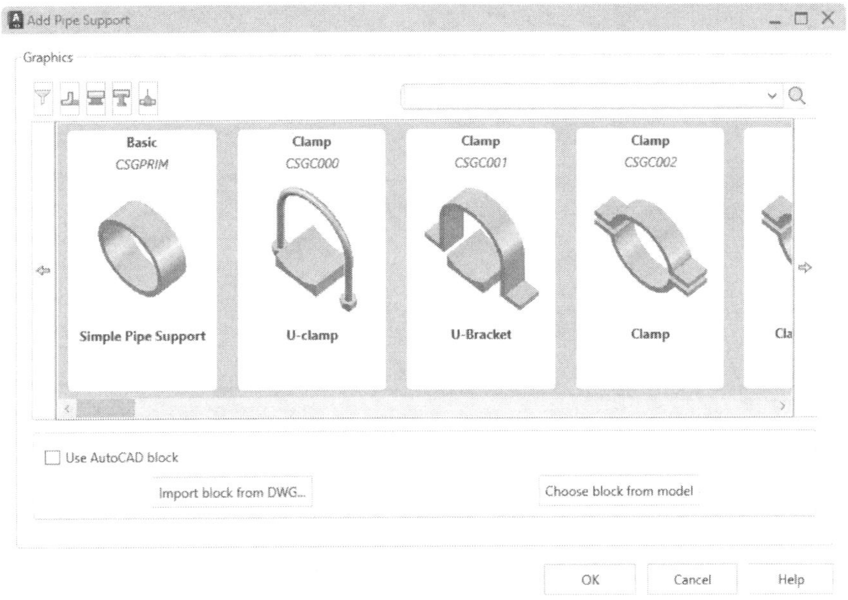

Figure 7-13 *The* ***Add Pipe Support*** *dialog box*

Figure 7-14 A Side Clamped Stanchion placed as a pipe support

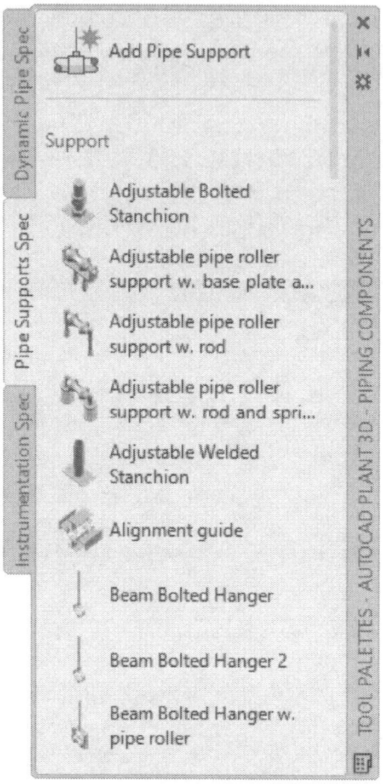

Figure 7-15 Pipe supports displayed in the ***TOOL PALETTES***

Adding a Dummy Leg

To add a dummy leg to a pipe, invoke the **Add Pipe Support** dialog box and then choose the **Dummy legs and General Supports** button from the top left of the dialog box. Next, scroll through the **Graphics** area and select the **Dummy Legs** from it. Choose the **OK** button from the dialog box; the dummy leg will be attached to the cursor and you will be prompted to specify the insertion point. Select an elbow connection point from the drawing area, refer to Figure 7-16. For easy selection, press the SHIFT key and right-click. Now, choose the **Node** option from the shortcut menu displayed.

Figure 7-16 A Dummy leg placed at the elbow

Adding a Hanger and Connecting it to a Structural Member

To add a hanger to a pipe, invoke the **Add Pipe Support** dialog box and then choose the **Hangers** button from it to display hangers. Next, select a hanger and choose the **OK** button. Place the hanger at the required location and press ENTER. After placing the hanger, you can connect it to an existing structural member located at the top of the pipe. To do so, select the hanger that you have placed on the pipe; the **Change Support Elevation** grip will be displayed, as shown in Figure 7-17. Click on this grip, drag and then snap to the point on the structural member located above the pipe, as shown in Figure 7-18; the hanger will be connected to the structural member. You can also specify an elevation value. To do so, press the TAB key and enter a required value in the dynamic input. Next, press ENTER.

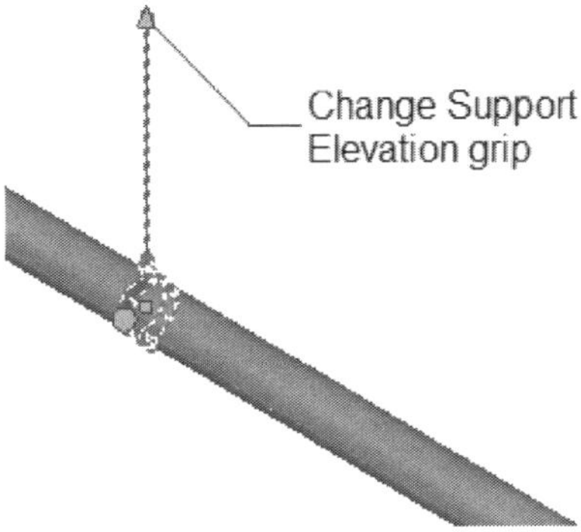

Figure 7-17 *The **Change Support Elevation** grip*

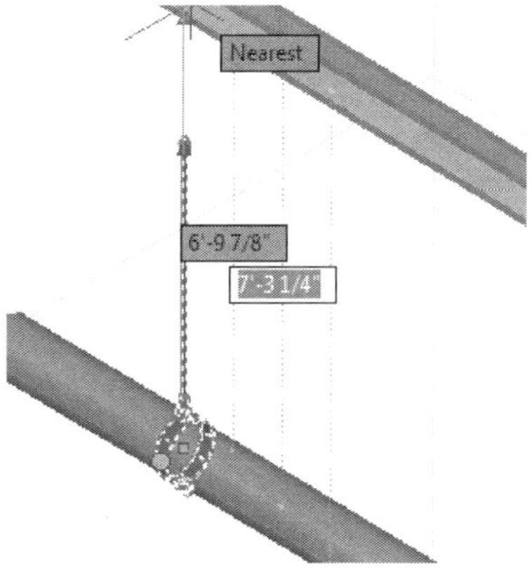

Figure 7-18 *Dragging the **Change Support Elevation** grip*

Modifying the Pipe Supports

To modify a pipe support, right-click on it; a shortcut menu will be displayed. Choose the **Properties** option from the shortcut menu; the **PROPERTIES** palette will be displayed. In the **PROPERTIES** palette, scroll down to the **Dimensions** sub-rollout and edit the dimensions. The dimension details will be shown in the **Preview** window of the **Part Geometry** rollout, as shown in Figure 7-19. You can click in the **Preview** window in order to expand the view. After modifying the dimensions, press ENTER; the pipe support will be modified in the drawing area. Figure 7-20 shows a spring hanger with default dimensions and Figure 7-21 shows the hanger after modifying the dimensions.

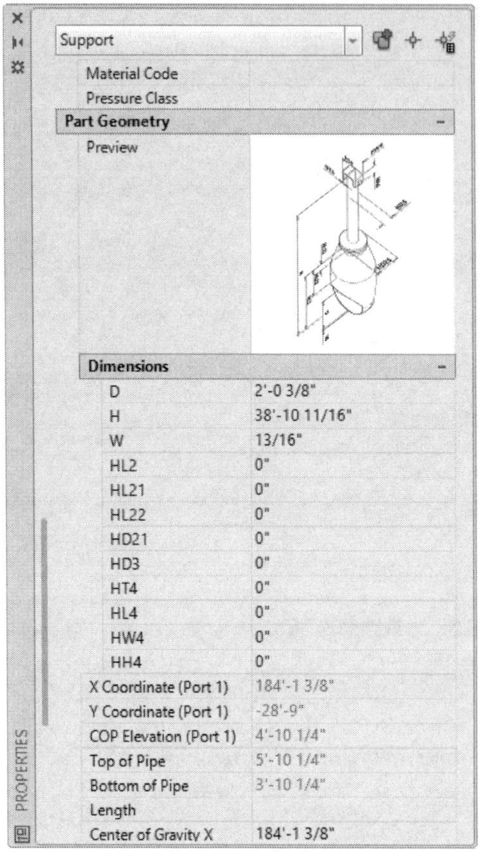

Figure 7-19 *The **Part Geometry** rollout of the **PROPERTIES** palette*

Figure 7-20 *A spring hanger with default dimensions values*

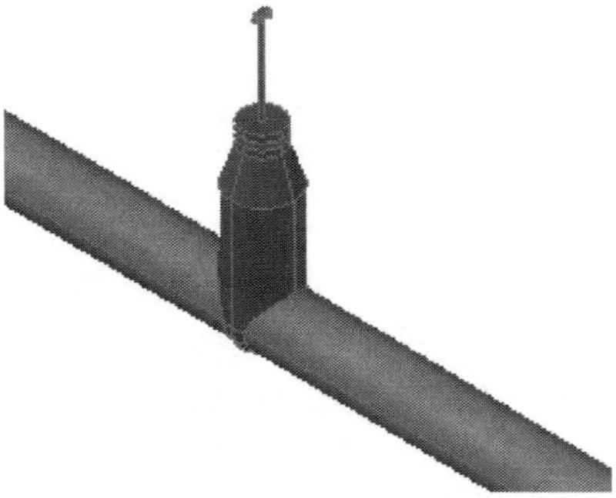

Figure 7-21 *The spring hanger after modifying dimension values*

Copying and Moving a Pipe Support

In AutoCAD Plant 3D, you can copy and move an existing pipe support from one location to another and also maintain the contact with the point of support. To do so, choose the **Toggle Lock Point of Support mode** button from the **Pipe Supports** panel in the **Home** tab; the pipe support will be locked to the point of support. Next, select the pipe support to be copied or moved; the **Move Part** grip will be displayed, as shown in Figure 7-22. Select the **Move Part** grip and move the pipe support. If you want to copy the pipe support, enter **P** at the command prompt and then copy it to the desired location; the pipe support will be copied. Figure 7-23 shows the original part and copied part.

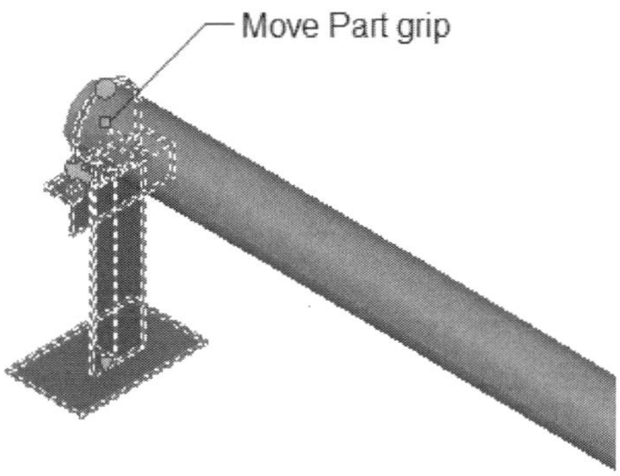

Figure 7-22 *The* ***Move Part*** *grip displayed on the pipe support*

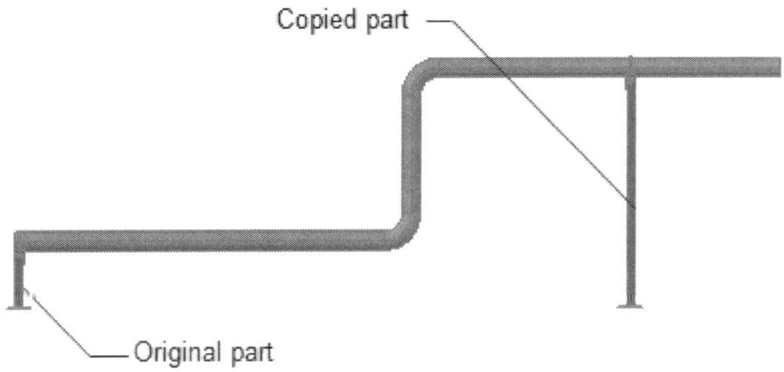

Figure 7-23 *The original pipe support and the copied one*

The **Toggle Lock Point of Support mode** button is much more useful when you are placing a pipe support for a sloped pipe. If you move a pipe support with the **Toggle Lock Point of Support mode** button chosen, height of the pipe support gets updated automatically, refer to Figure 7-24.

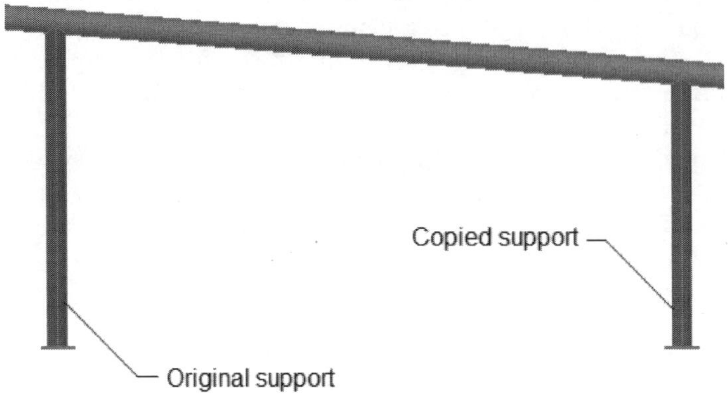

***Figure 7-24** The original pipe support and the copied one*

Connecting Two Pipe Supports

To connect two pipe supports, choose the **Supports on Supports** button from the **Pipe Supports** panel in the **Home** tab; you will be prompted to select pipe supports to connect them together. Select two or more pipe supports to be connected and then press ENTER; the selected pipe supports will connect to each other.

Converting Solids into Pipe Supports

To convert a solid model into pipe support, you need to create a solid model by using tools that are available in the **Modeling** tab. To do so, create a solid model supporting a pipe. Next, choose the **Convert Supports** tool from the **Pipe Supports** panel; you will be prompted to select the solid model. Select the solid model and press ENTER; you will be prompted to specify an insertion point on the pipe or pipe component. Use object snap options to select a point on the pipe or pipe component. After selecting a point on the pipe, the solid will be converted into a pipe support, as shown in the Figure 7-25.

***Figure 7-25** A solid converted into a pipe support*

Attaching Objects to a Pipe Support

To attach objects to a pipe support, choose the **Attach Supports** tool from the **Pipe Supports** panel; you will be prompted to select a pipe support. Select the pipe support to which the object will be attached; you will be prompted to select objects that are to be added to the pipe support. Select the objects from the drawing area, and then press ENTER; the selected object will be attached to the pipe support and will become an integral part of it.

Detaching Objects From a Support

To detach previously added objects from a support, choose the **Detach Supports** tool; you will be prompted to select the support from which the objects will be detached. Select a pipe support with attached objects; the attached objects will be detached from the pipe support.

INSULATING A PIPE

You can insulate a pipe using the **PROPERTIES** palette. To do so, invoke the **PROPERTIES** palette for the pipe by double-clicking on the pipe. Next, scroll to the **Process Line** sub-rollout in the **Plant 3D** rollout in the **PROPERTIES** palette and select an option from the **Insulation Type** drop-down list, refer to Figure 7-26. Select the insulation thickness from the **Insulation Thickness** drop-down list which is located above the **Insulation Type** drop-down list. After applying the insulation, you need to choose the **Toggle Insulation Display** button from the **Visibility** panel to turn on the insulation display.

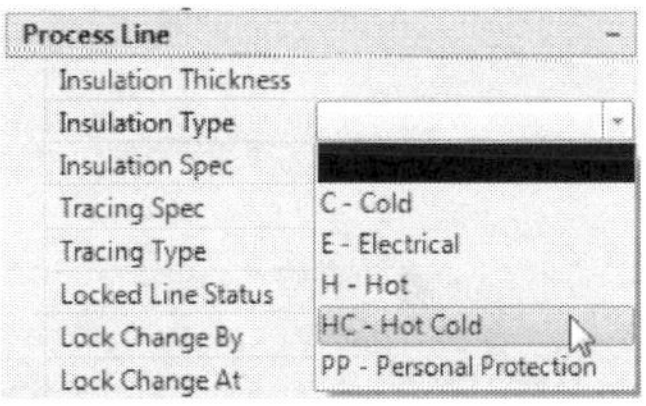

***Figure 7-26** Selecting an option from the **Insulation Type** drop-down list*

MODIFYING THE PIPE COMPONENTS USING GRIPS

You can modify pipes and its components by using the grips which are displayed on selecting the pipe components. The various methods to modify pipes using grips are discussed next.

Substituting a Pipe Component

To substitute a pipe component with a new pipe component, select it from the drawing area; the **Substitute Part** grip will be displayed, refer to Figure 7-27. Click on the **Substitute Part** grip; a flyout will be displayed with a list of substitute components. Choose the required substitute component from the flyout; the existing component will be replaced with the newly selected component.

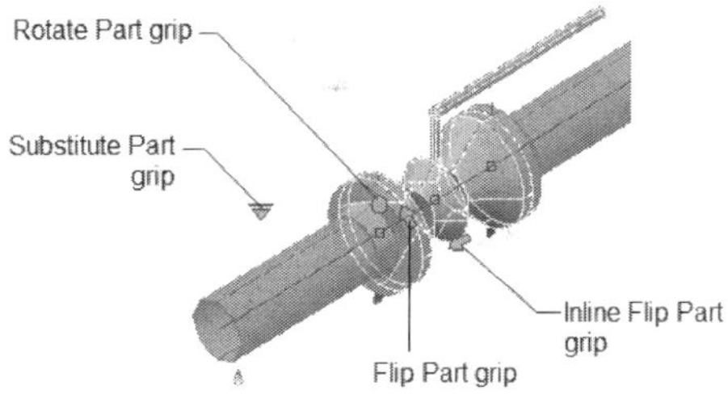

***Figure 7-27** The Part grips displayed on the selected component*

Rotating a Pipe Component

To change the angular orientation of the pipe component, select the pipe component from the drawing area; the **Rotate Part** grip will be displayed, refer to Figure 7-27. Click on this grip; the compass will be invoked. Make sure the **Toggle Compass** button available in the **Compass** panel of the **Home** tab is chosen. Next, use the compass to rotate the pipe component and then press ENTER; the component will be reoriented.

Flipping a Pipe Component

To flip a pipe component, select it from the drawing area to display the **Flip Part** grip, refer to Figure 7-27. Click on the **Flip Part** grip; the pipe component will be flipped.

Flipping a Component Inline with the Pipe

To flip a pipe component inline with the pipe, select the **Inline Flip Part** grip, refer to Figure 7-27. Click on this grip to flip the component in line with the pipe, refer to Figures 7-28 and 7-29.

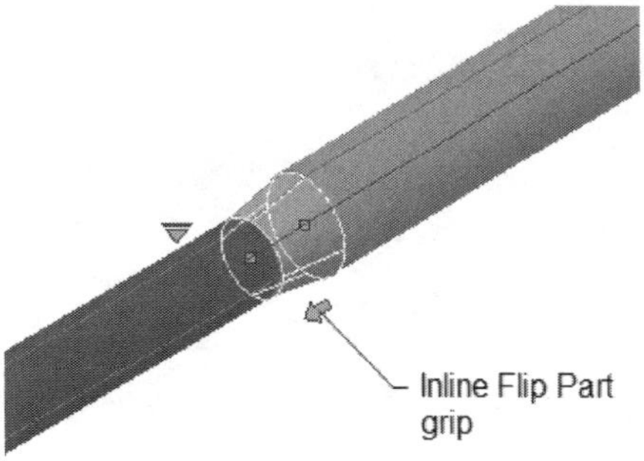

Figure 7-28 The ***Inline Flip Part*** *grip displayed on the pipe component*

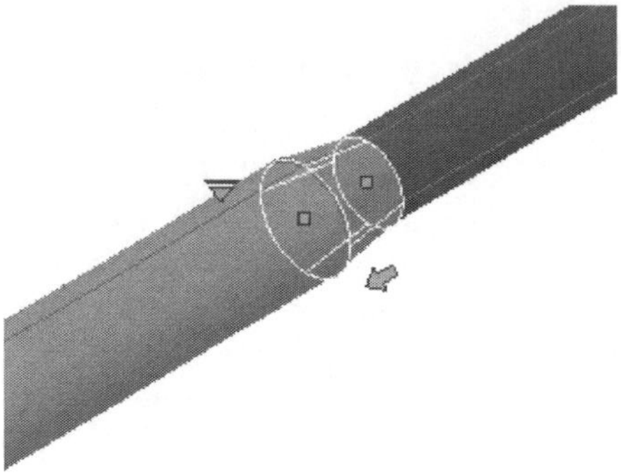

Figure 7-29 The component after flipping it in the inline direction

Changing the Elevation of the Pipe

To change the elevation of a pipe, select it from the drawing area to display the **Change Pipe Elevation** grip, as shown in Figure 7-30. Click on this grip and move the cursor to the required elevation. You can also enter an elevation value in the dynamic input box. Figure 7-31 shows the pipe with the changed elevation.

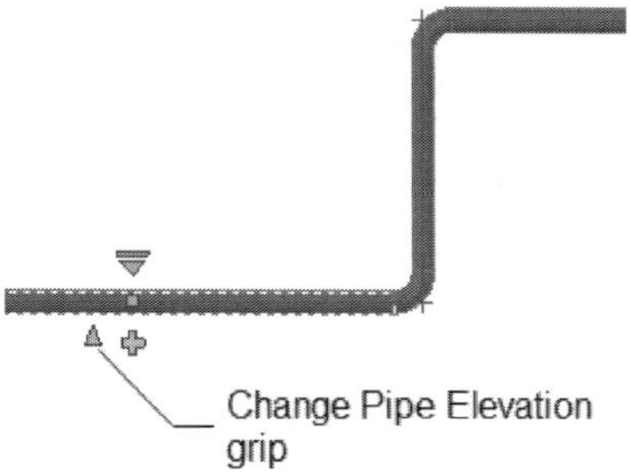

Figure 7-30 The ***Change Pipe Elevation*** *grip displayed on the pipe*

Figure 7-31 The pipe with the changed elevation

Changing the Valve Operator

After placing a valve in a pipe, you may need to changethe valve operator. To do so, select the valve from the drawing area. Next, right-click and choose the **Properties** option from the shortcut menu displayed; the **PROPERTIES** palette will be displayed. In the **PROPERTIES** palette, expand the **Part Properties** rollout and then expand the **Valve Operator** sub-rollout. Next, select the required operator type from the **Operator** drop-down list, as shown in Figure 7-32; the selected valve operator will replace the existing one.

You can also select an operator by invoking the **Override Valve Operator** dialog box. To do so, choose the button next to the **Operator** drop-down list; the **Override Valve Operator** dialog box will be displayed, as shown in Figure 7-33.

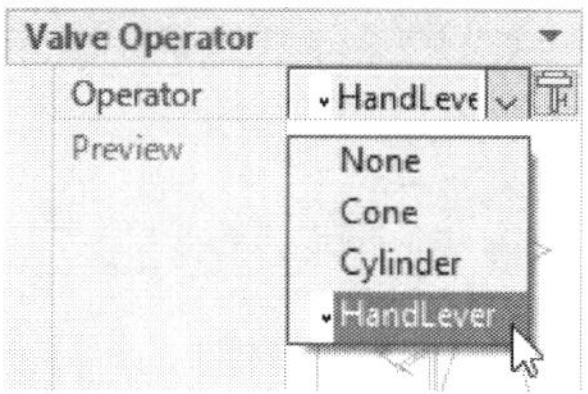

***Figure 7-32** Selecting an option from the **Operator** drop-down list*

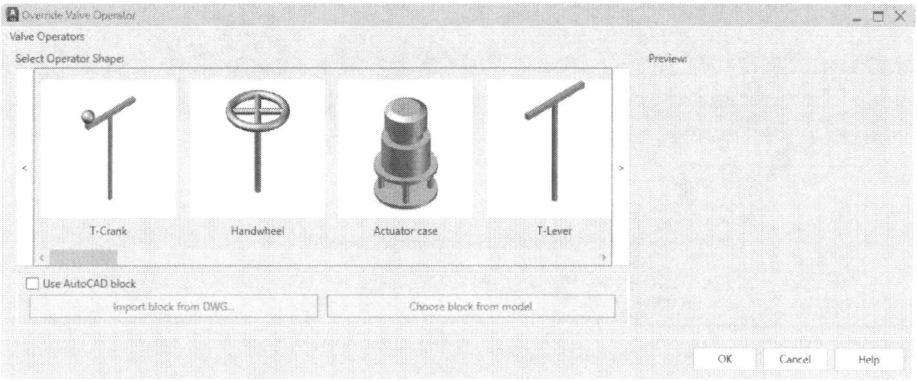

***Figure 7-33** The **Override Valve Operator** dialog box*

Select an operator from the **Select Operator Shape** area and then specify the dimensions in the **Dimensions** area. Next, choose the **OK** button; the new operator will replace the old one. You can also use a block to override an existing operator. To do so, select the **Use AutoCAD block** check box located below the **Select Operator Shape** area. Next, choose the **Import block from DWG** or **Choose block from model** button to import a block or select it from an existing model. Figure 7-34 shows a gate valve with the hand wheel operator and Figure 7-35 shows the valve after changing the operator to T-Crank.

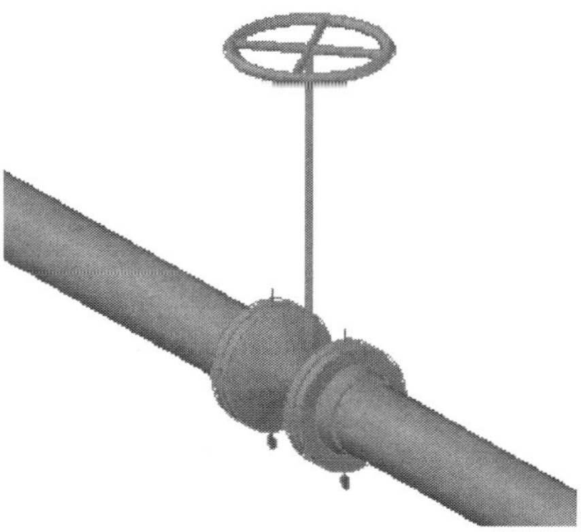

***Figure 7-34** Gate valve with a Hand Wheel operator*

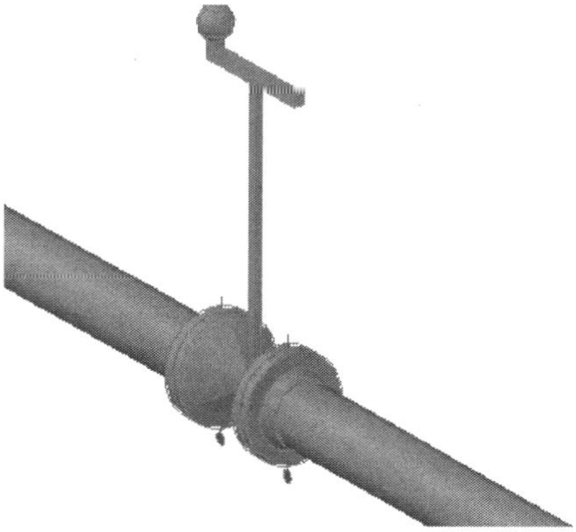

***Figure 7-35** Valve with a T-Crank operator*

VALIDATING A 3D MODEL

In Chapter 2, you learned to validate a P&ID. Similarly, you need to validate a 3D model and correct the errors in it. In addition, you also need to validate a 3D model against the P&ID. It means that you need to make sure that the parts in 3D model match with the corresponding components in the P&ID.

Before starting the validation, you need to specify the error types to be validated. To do so, enter **VALIDATECONFIG** command at the command prompt; the **P&ID Validation Settings** dialog box is displayed, as shown in Figure 7-36. In this dialog box, expand the **3D Piping** node and specify the error types to be validated in the 3D model. Next, you need to specify the types of mismatches to be checked between the 3D model and the P&ID. To do so, expand the **3D Model to P&ID checks** node and select the check boxes of the mismatches to be checked. Next, choose the **OK** button from the **P&ID Validation Settings** dialog box.

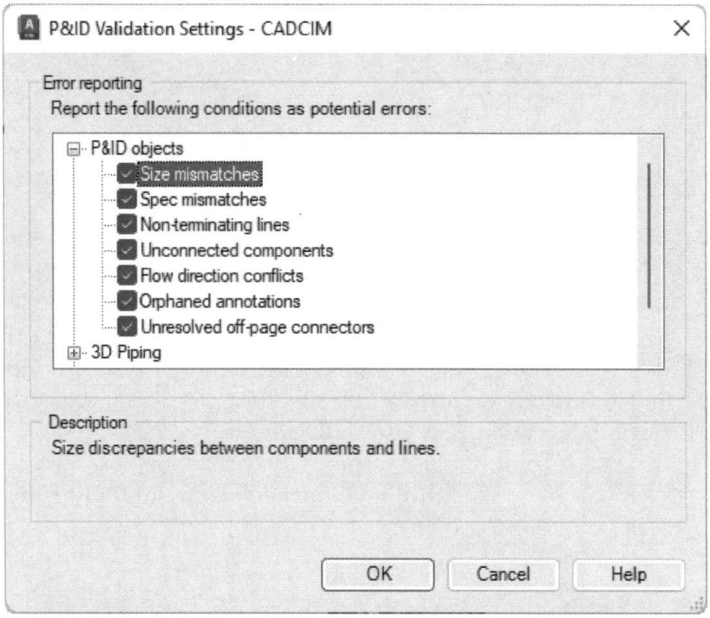

***Figure 7-36** The **P&ID Validation Settings** dialog box*

After specifying the validation settings, expand the **Plant 3D Drawings** node in the **PROJECT MANAGER**. Next, right-click on the drawing file to be validated and choose the **Validate** option from the shortcut menu displayed; the **Validation Progress** message box will be displayed and the drawing will be validated. Also the **VALIDATION SUMMARY** palette will be displayed after the validation process is complete. Select an error from **VALIDATION SUMMARY** palette; the error type and the action to be taken are displayed in the **Details** area. Ignore or correct the error. To ignore the error, select the **Don't display errors marked as ignored** check box.

TUTORIAL

Tutorial 1

In this tutorial, you will open the Piping Model file from the CADCIM project located in the **PROJECT MANAGER** and place valves and fittings in the model space. You can also download this model from the CADCIM website by following the path: *Textbooks > CAD/CAM > Plant 3D > AutoCAD Plant 3D 2023 for Designers > Input Files*. Figures 7-37 through 7-40 show the locations of the valves and valve list. **(Expected time: 45 min)**

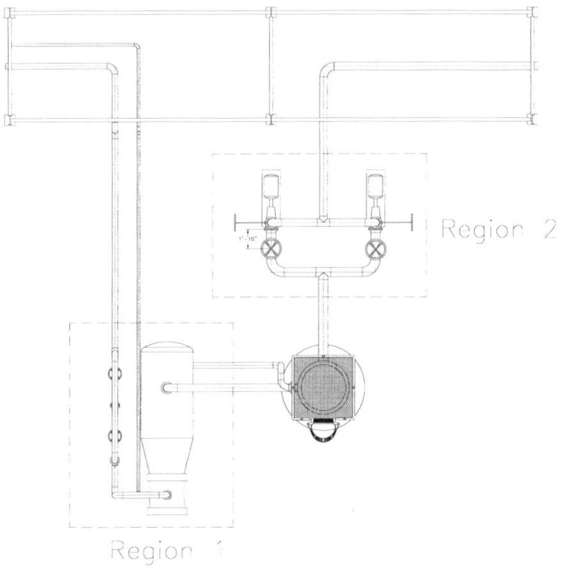

Figure 7-37 *Top view of the model*

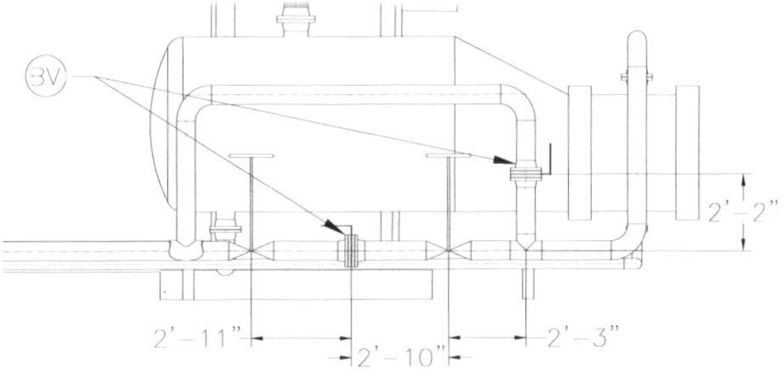

Figure 7-38 *Valves in Region 1*

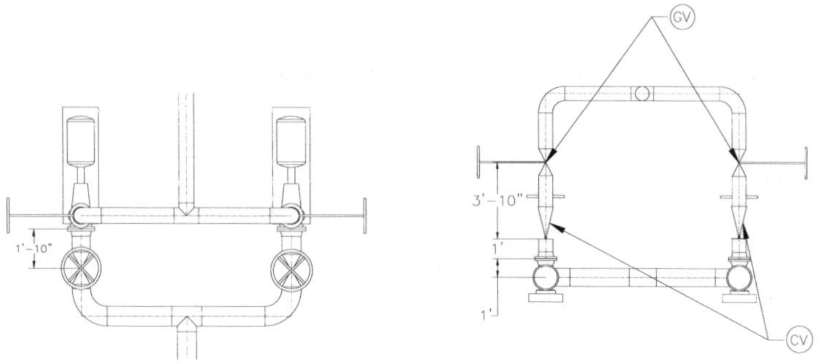

Figure 7-39 *Valves in Region 2*

Valve List			
	Valve Type	Line Number	Actuator Type
BV	Butterfly Valve	2000	Hand Lever
	Butterfly Valve	2000	Hand Lever
CV	Check Valve	1012	
	Check Valve	1012	
GV	Gate Valve	1012	Hand Wheel
	Gate Valve	1012	Hand Wheel
	Gate Valve	1004	Hand Wheel
	Gate Valve	1004	Hand Wheel
	Gate Valve	2000	Hand Wheel
	Gate Valve	2000	Hand Wheel

Figure 7-40 *The valves list*

The following steps are required to complete this tutorial:

a. Open the **Piping Model** from the **PROJECT MANAGER**.
b. Select the required spec and select the pipe size.
c. Place valves at the required locations.
d. Save the file.

Opening the File

1. Double-click on the **AutoCAD Plant 3D 2023 - English** icon; AutoCAD Plant 3D gets started.

2. Select **CADCIM** from the **Current Project** drop-down list in the **PROJECT MANAGER**.

3. Expand the **Plant 3D Drawings** node in the **Project** area and double-click on **Piping Model**; the selected drawing file is opened.

Selecting the Spec and the Pipe Size

In this section, you will select a spec and the pipe size.

1. Select **CS150** from the **Spec Selector** drop-down list in the **Part Insertion** panel; the **Dynamic Pipe Spec** tab in the **TOOL PALETTE** located on the left side of the window is loaded with **CS150** piping components.

Placing Valves Using the P&ID Line List

In this section, you will place valves in the Plant 3D model with reference to the P&ID1 files of the CADCIM project. But before doing so, you need to map the new P&ID objects that you have placed in P&ID.

1. Choose the **Project Setup** tool from the **Project** drop-down in the **Project** panel; the **Project Setup** dialog box is displayed.

2. In this dialog box, expand the **Plant 3D DWG Settings** node and select the **P&ID object Mapping** option; the **P&ID Object Mapping** area is displayed on the right side in the dialog box.

3. In the **P&ID Object Mapping** area, select **Engineering Items > Instrumentation > Inline Instruments > Control Valve** node from the **P&ID Classes** tree; the properties of the corresponding Plant 3D object are displayed in the **Plant 3D Classes** area.

4. Choose the **Add** button from the **Plant 3D Classes** area; the **Select Plant 3D Class Mapping** dialog box is displayed.

5. In this dialog box, select the **Valve** class from **Piping and Equipment > Pipe Run Components** in the **Plant 3D Classes** tree.

6. Select the **Map to all subtypes of this class** check box and choose the **OK** button to close the **Select Plant 3D Class Mapping** dialog box.

7. In the **Project Setup** dialog box, select the ***Actuator Type** option from the drop-down list next to the **Actuator Type** property in the **Property Mapping** table. Similarly, select the ***Valve Body Type** option for the **Body Type** property.

8. Select the **Body Type** and **Actuator Type** check boxes in the **Validate** column of the **Property Mapping** table. Make sure that the **Check this item during 3D model to P&ID validation** check box is selected.

9. Choose the **OK** button to close the **Project Setup** dialog box.

   Next, you need to place a valve using a P&ID.

10. Choose the **P&ID Line List** button from the **Part Insertion** panel of the **Home** tab; the **P&ID LINE LIST** palette is displayed.

11. Select **P&ID1** from the drop-down list located at the top of the window; all the line numbers in this P&ID are listed in the tree view.

12. Select the **Control Valve 01-CV-1002** from **2000 > 6"-CS150-P-2000** in the **P&ID LINE LIST** palette. Figure 7-41 shows the selected pipe in the P&ID.

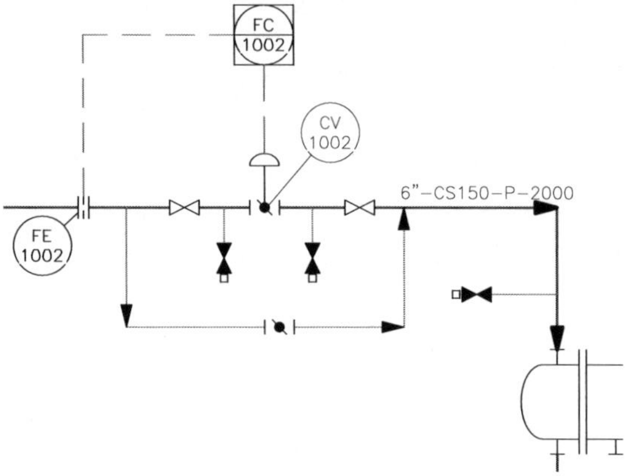

Figure 7-41 *The selected pipe in the P&ID*

13. Choose the **Place** button from the **P&ID LINE LIST** palette; the **Select Size and Spec** dialog box is displayed.

14. In this dialog box, select **CS150** from the **Spec** list box and then select **6"** from the **Size** list box.

15. Choose the **Select** button from the **Select Size and Spec** dialog box; you are prompted to specify the insertion point.

16. Select a point on the pipe connecting the reboiler, refer Figure 7-42; the control valve is placed at the specified point and you are prompted to specify the rotation angle.

17. Type **0** in the command prompt and press ENTER. Figure 7-43 shows the control valve after specifying the rotation angle.

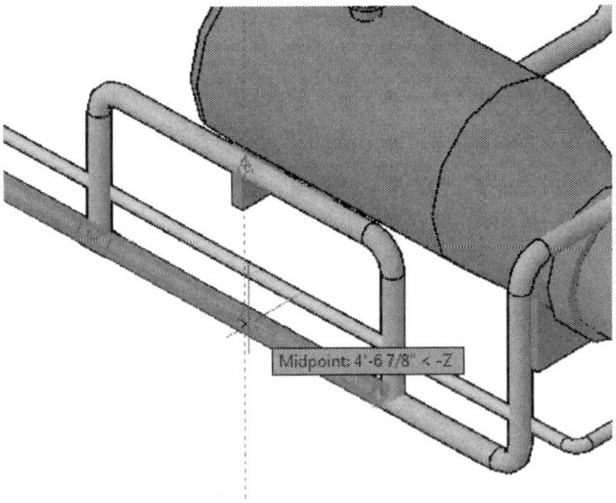

Figure 7-42 *Selecting the midpoint of the pipe connecting the reboiler*

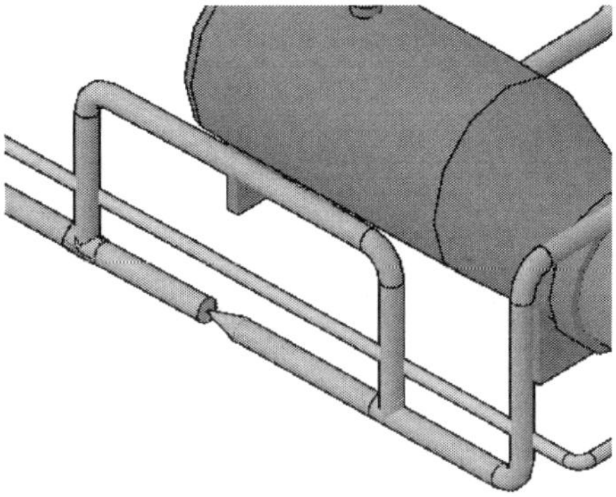

Figure 7-43 *Control valve after specifying the rotation angle*

You may notice that the valve body in the P&ID is different from the one that you have placed. Next, you need to change the valve body.

18. Click on the valve body and select the **Substitute Part** grip displayed on it, refer to Figure 7-44; a flyout is displayed showing the list of parts that can be substituted.

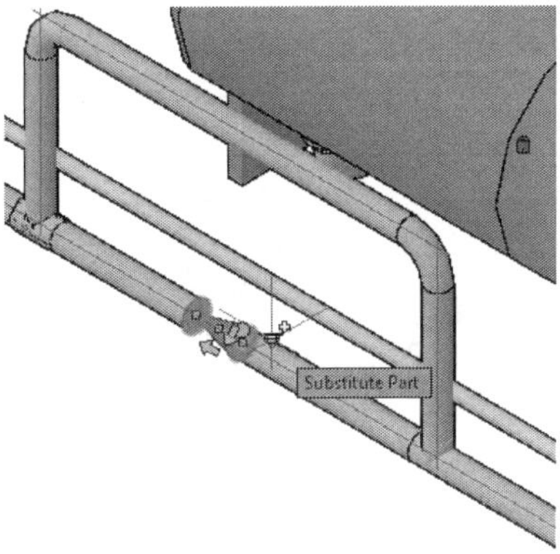

Figure 7-44 *The* ***Substitute Part*** *grip displayed on the valve body*

19. Choose the **6" BUTTERFLY VALVE, OFFSET, 6" ND, LUG, RF, ASME B16.10, 2 1/4" LG** from the flyout, refer to Figure 7-45; the valve body of the control valve is changed to butterfly valve, as shown in Figure 7-46.

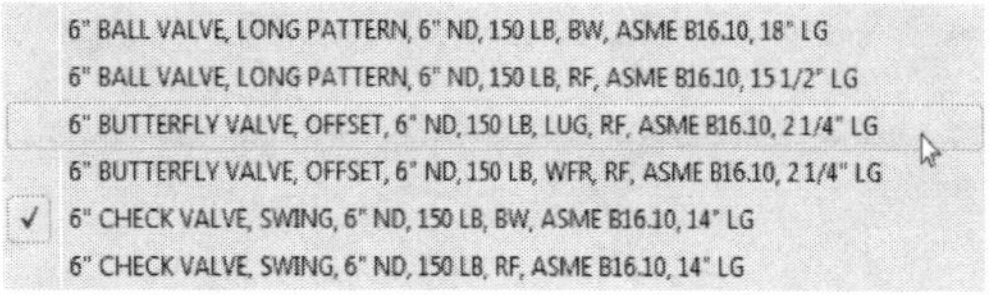

Figure 7-45 *Selecting the butterfly valve from the flyout*

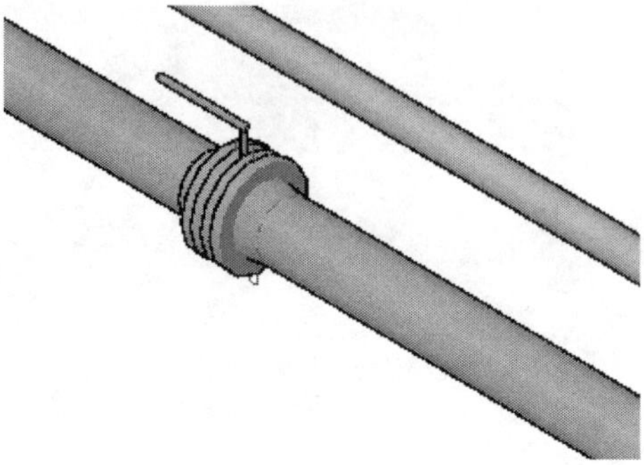

Figure 7-46 *The control valve after changing the valve body to butterfly valve*

Placing Gate Valves

Next, you need to place the gate valves on the same pipe on which control valves were placed, refer to Figure 7-41.

1. Invoke the **P&ID LINE LIST** palette if it is not already available and then select **Gate Valve HA-104** from **2000 > 6"-CS150-P-2000** from this palette.

2. Place a gate valve at the location shown in Figure 7-47.

3. Similarly, select another gate valve **Gate Valve HA-105** from **2000 > 6"-CS150-P-2000** from the **P&ID LINE LIST** palette and then place it on the model, as shown in Figure 7-47.

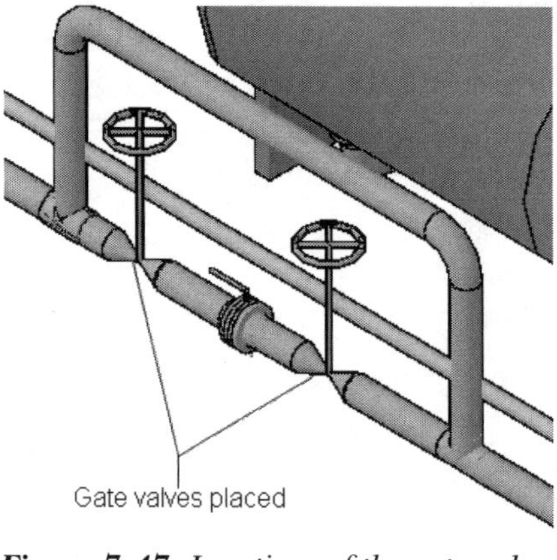

Figure 7-47 *Locations of the gate valves*

4. Select the butterfly valve from the **P&ID LINE LIST** palette, as shown in Figure 7-48, and then place it at the location shown in Figure 7-49. Choose the **Place** button; the **Select Size and Spec** dialog box is displayed.

5. In this dialog box, select the spec as **CS150** and Size as **6"** from the **Spec** and **Size** list boxes, respectively. Next, choose the **Select** button; the **Select 3D Class** window is displayed. In

this window, select the **Valve (Butterfly-Inline)** option from the **3D classes** drop-down list, if not selected by default.

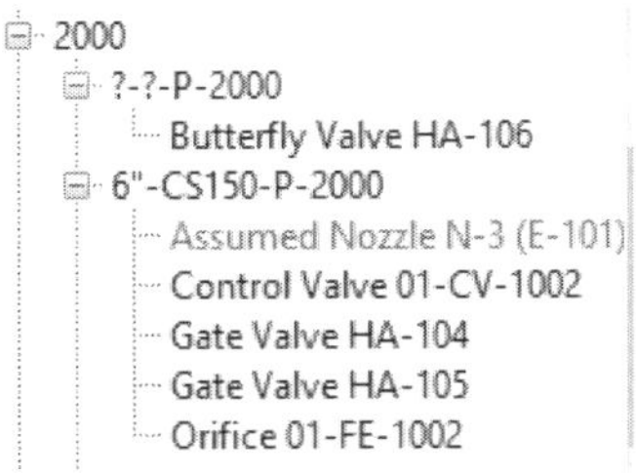

Figure 7-48 *Selecting the Butterfly Valve from the* ***P&ID LINE LIST*** *palette*

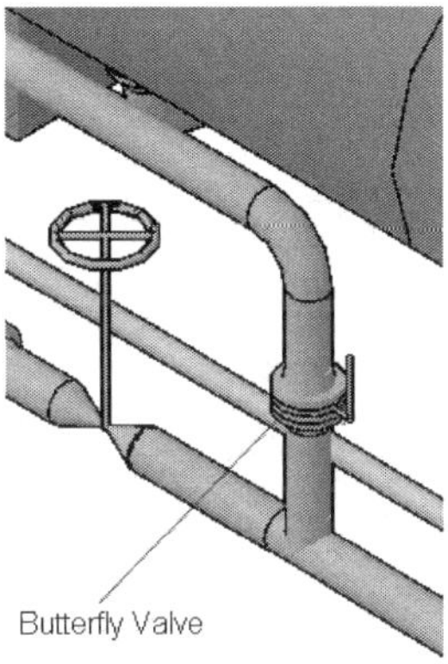

Figure 7-49 *Location of the Butterfly Valve*

Placing Valves on the Pipes Connecting the Pumps

1. Select Gate Valves from the **P&ID LINE LIST** palette, refer to Figure 7-50, and place them on the pipes connecting the pumps, as shown in Figure 7-51.

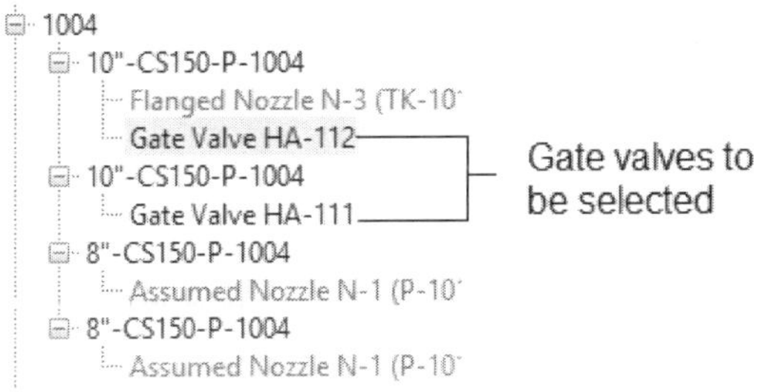

Figure 7-50 *Gate valves to be selected from the* ***P&ID LINE LIST*** *palette*

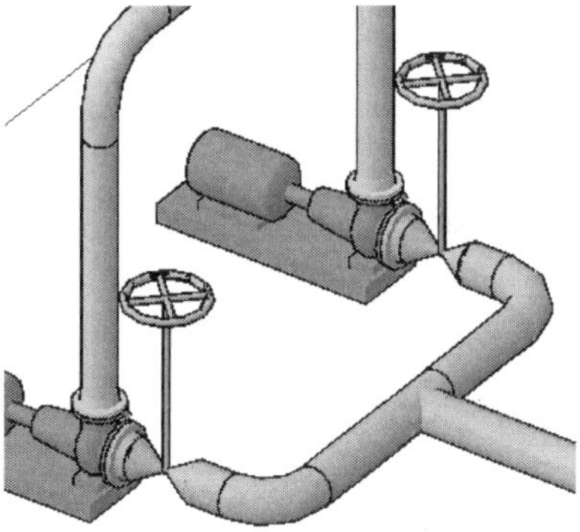

Figure 7-51 *Locations of the gate valves*

2. Select the Check Valves from the **P&ID LINE LIST** palette, refer to Figure 7-52, and place them on the pipes connecting the pumps, as shown in Figure 7-53.

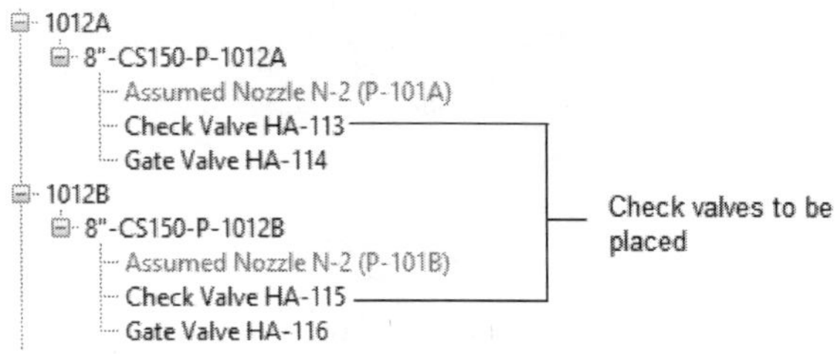

Figure 7-52 *Check valves to be selected from the* ***P&ID LINE LIST*** *palette*

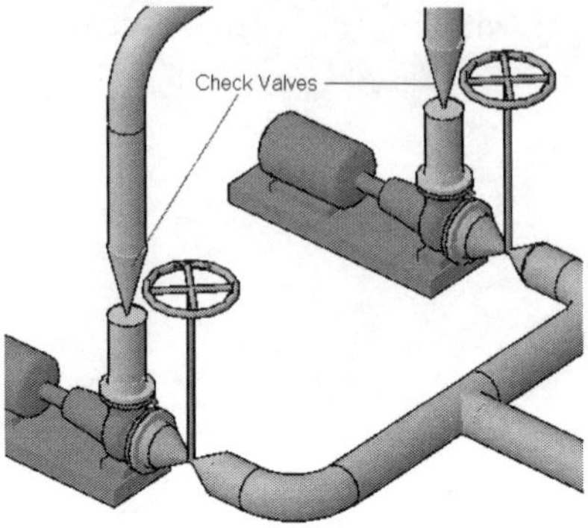

Figure 7-53 *Location of the check valves*

Note

You need to flip the direction of the check valves if they are not placed in the required direction, as shown in Figure 7-53. To do so, select a check valve to display the ***Inline Flip Part*** *grip on it, refer to Figure 7-54. Next, use this grip to flip the direction of the valve. Similarly, flip the direction of the other valve.*

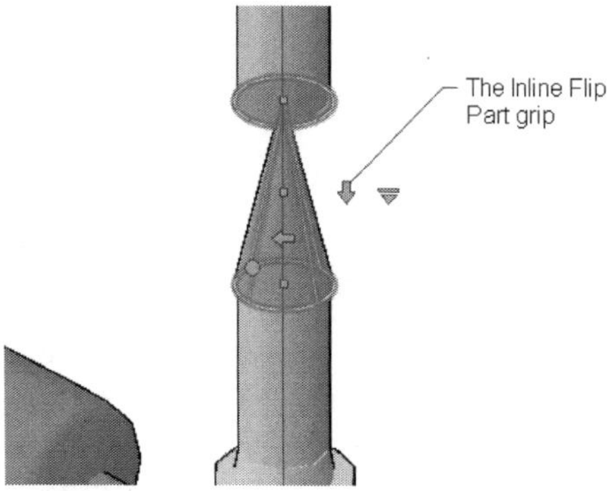

Figure 7-54 *The* ***Inline Flip Part*** *grip displayed*

3. Similarly, select the gate valves from the **P&ID LINE LIST** palette and place them on the pipes connecting the pumps, refer to Figures 7-55 and 7-56.

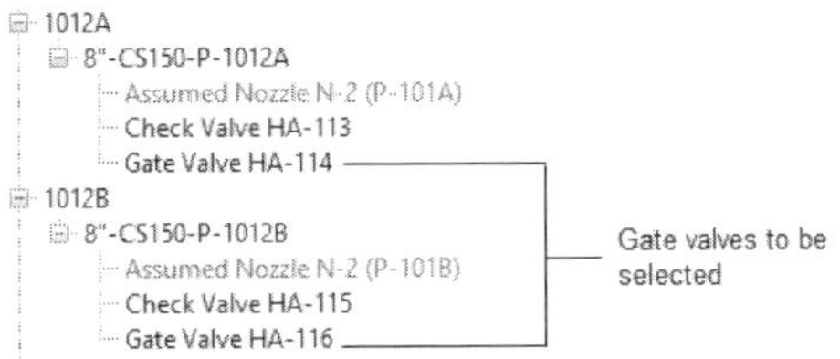

Figure 7-55 *Gate Valves to be selected from the* ***P&ID line list***

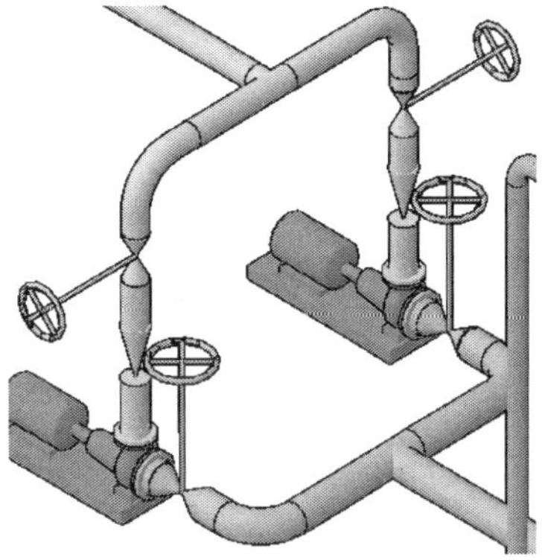

Figure 7-56 *Locations of the gate valves*

4. After placing all the valves, validate the model by choosing the **Validate** option from the shortcut menu displayed on right-clicking on the drawing file in the **PROJECT MANAGER**. Next, choose the **Save** button from the **Application** menu and then the **Close** button to close the file.

Self-Evaluation Test

Answer the following questions and then compare them to those given at the end of this chapter:

1. If the selected component is available in the tree of the **P&ID LINE LIST** palette without the spec, then on choosing the **Place** button, the _________ dialog box will be displayed.

2. The _________ dialog box is used to change the valve operator.

3. To modify a pipe support, you need to invoke its _________ palette.

4. You can maintain the contact between the pipe support and point of support with the _________ button turned on.

5. The _________ tool is used to connect two pipe supports.

6. To insulate a pipe, invoke its _________ palette and select the insulation type from the _________ drop-down list.

7. To substitute a pipe component with a new one, right-click on it and choose the _________ grip.

8. You can modify pipes and their components by using the grips which are displayed on selecting the components. (T/F)

9. You can validate an equipment between a P&ID and a 3D model. (T/F)

10. While routing a pipe, you can place a valve or a fitting on it. (T/F)

Review Questions

Answer the following questions:

1. You can place a custom part into a Plant 3D model using the __________ dialog box.

2. To place a placeholder part, choose the __________ option from the **Custom Part type** drop-down list in the **CUSTOM PART BUILDER** palette.

3. The __________ tool is used to convert a solid model into a pipe support.

4. The __________ tool is used to attach objects to a pipe support.

5. Before placing an object from a P&ID into a Plant 3D model, it has to be mapped to the corresponding Plant 3D object. (T/F)

6. You can insulate a pipe using its **PROPERTIES** palette. (T/F)

7. To route a pipe using a P&ID, the Plant 3D model and the P&ID should be located in the same project. (T/F)

Answers to Self-Evaluation Test

1. **Select Size and Spec**, **2.** **Override Valve Operator**, **3.** **PROPERTIES**, **4.** **Toggle Lock Point of Support mode**, **5.** **Support on Supports**, **6.** **PROPERTIES**, **Insulation Type**, **7.** **Substitute Part**, **8.** T, **9.** T, **10.** T

Chapter 8

Creating Isometric Drawings

Learning Objectives

After completing this chapter, you will be able to:

- *Understand various types of isometric drawings*
- *Create a quick isometric drawing*
- *Create a production isometric drawing*
- *View isometric results*
- *Create iso messages*
- *Export a part component file*
- *Create an iso from a part component file*
- *Lock and unlock lines*
- *Configure the isometric drawing settings*

INTRODUCTION

Isometric drawings are created for easy visualization of the 3D pipe lines which is a challenging task in an orthographic drawing. Isometric drawings are also used for the fabrication of pipes. In this chapter, you will learn to create and modify isometric drawings, Quick Isometric drawings (Check Isometric), and Production (Final) Isometric drawings. Also, you will learn to add iso messages and annotations such as insulation, flow arrows, and so on to the isometric drawings.

ISOMETRIC DRAWING TYPES

The most commonly used isometric drawing types are Check, Stress, Final, and Spool. These types are discussed next.

Check Isometric

The Check isometric drawing is created to check whether all the essential components are represented in the model or not. It also ensures that the final isometric drawing is created without errors.

Stress Isometric

The Stress isometric drawing presents the geometric data which is precise and relevant to the stress analysis of pipes. It is generally used to analyze the stress in pipelines. The pipelines that require stress analysis include high pressure lines, high temperature lines, lines with large pipe size, and so on.

Final Isometric

A Final isometric drawing is the final product document created from the 3D piping model. It is generally produced at the last stage of a project. The Final isometric drawing contains a bill of material (BOM) and is used for carrying out the fabrication and construction process.

Spool Drawings

A Spool drawing is the final isometric drawing separated in individual sections called spools. It is created for shop fabrication.

CREATING A QUICK ISOMETRIC DRAWING

A quick isometric drawing is created to check the piping before creating a production isometric drawing. You can create a Quick isometric drawing by selecting pipes from the drawing area or from the line list in the **PROJECT MANAGER**. To create a Quick isometric drawing, choose the **Quick Iso** tool from the **Iso Creation** panel in the **Isos** tab; you will be prompted to select components to create an isometric drawing. Select a component from the drawing area and press ENTER; the **Create Quick Iso** dialog box will be displayed, as shown in Figure 8-1. Next, choose the **Create** button to create a quick isometric drawing.

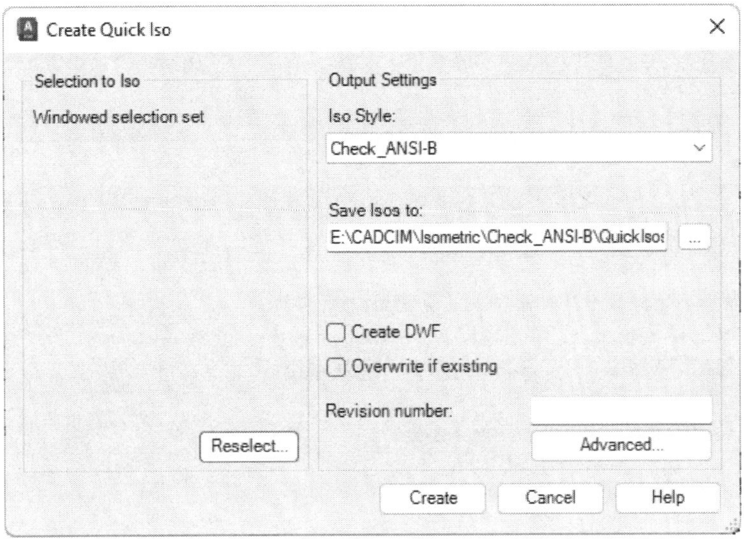

Figure 8-1 *The* ***Create Quick Iso*** *dialog box*

If you want to create a Quick isometric drawing of a particular pipe line, invoke the **Quick Iso** tool and choose the **Line number** option from the command prompt; the **Create Quick Iso** dialog box will be displayed, as shown in Figure 8-2.

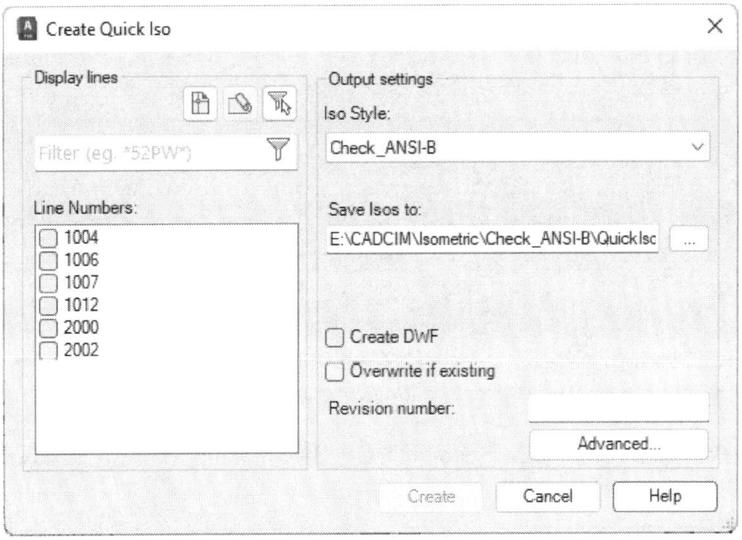

Figure 8-2 *The* ***Create Quick Iso*** *dialog box displayed on choosing the* ***Line number*** *option*

The options in the **Create Quick Iso** dialog box are discussed next.

Selection to Iso

This area will be displayed only when you select components from the drawing area. The **Reselect** button in this area is used to re-select the components from which Iso is to be created.

Display lines

This area will be displayed only when you choose the **Line number** option from the command prompt after invoking the **Quick Iso** tool. The options in this area are discussed next.

Show Line Numbers in current drawing only

On choosing this button, the line numbers of only the current drawing are displayed in the **Line Numbers** list box present in the **Display lines** area.

Show all Line Numbers in Project

On choosing this button, the line numbers from all the drawing files present in the project are displayed.

Show only selected Line Numbers

On choosing this button, only the selected lines are displayed in the **Line Numbers** list box. You can also enter the required line numbers in the Filter edit box present in the **Display lines** area to filter the line number list.

Output settings

This area is used to specify the settings of the isometric output file. The options in this area are discussed next.

Iso Style

This drop-down list is used to specify the style of the isometric drawing. You can select predefined isometric styles from it. You can also modify these styles by using the **Project Setup** dialog box.

Save Isos to

This display box displays the path of the output file. You can change the default location of the output file by choosing the Browse button next to it.

Create DWF

If you select this check box, you can create a Design Web Format (DWF) file from the output file.

Overwrite if existing

This check box is selected to overwrite the existing isometric file. On clearing this check box, the file will be saved with a new name.

Revision number

This edit box is used to enter the revision number. The entered number will be displayed in the title block of the isometric drawing.

Advanced

On choosing this button, the **Advanced Iso Creation Options** dialog box will be displayed. You can use this dialog box to specify advanced settings while creating an isometric drawing.

After specifying the options in the **Create Quick Iso** dialog box, choose the **Create** button from it; the Quick isometric drawing will be created in the background and a message will be displayed at the Status Bar, as shown in Figure 8-3.

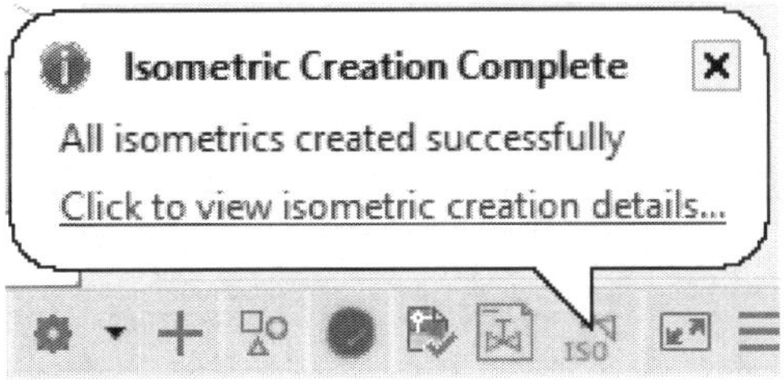

***Figure 8-3** Message displayed at the Status Bar*

CREATING A PRODUCTION ISOMETRIC DRAWING

To create a Production isometric drawing, choose the **Production Iso** tool from the **Iso Creation** panel in the **Isos** tab; the **Create Production Iso** dialog box will be displayed. The options in this dialog box are the same as that in the **Create Quick Iso** dialog box.

In the **Create Production Iso** dialog box, select the lines from the **Line Numbers** list box in the **Display lines** area. Next, select the isometric style from the **Iso Style** drop-down list in the **Output settings** area. You can also select the **Create DWF** check box, if you want to create a Design Web Format (DWF) file from an isometric drawing. If you want to override the existing file with a new file, select the **Overwrite if existing** check box. You can also specify advanced settings by choosing the **Advanced** button. After specifying all the settings, choose the **Create** button; the Production isometric drawing will be created and a message will be displayed at the Status Bar, refer to Figure 8-3.

Viewing Isometric Results

To view the isometric results, double-click on the isometric icon  located in the Status Bar; the **Isometric Creation Results** dialog box will be displayed, as shown in Figure 8-4. To open an isometric file, click on the file path displayed in the dialog box, refer to Figure 8-4; the isometric file will be opened.

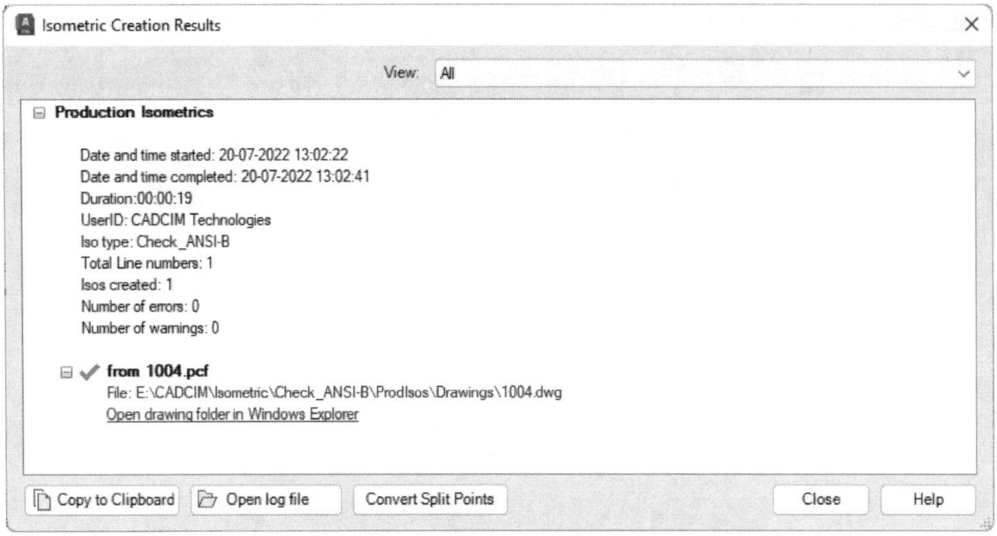

*Figure 8-4 The **Isometric Creation Results** dialog box*

PLACING ISO MESSAGES AND ANNOTATIONS

In AutoCAD Plant 3D, you can place additional information (Iso message) in a 3D model using the **Iso Message** tool. The information will be displayed in the isometric drawing. Also, a sphere symbol will be displayed in the 3D model. To create an Iso message, choose the **Iso Message** tool from the **Iso Annotations** panel in the **Isos** tab; the **Create Iso Message** dialog box will be displayed, as shown in Figure 8-5. Note that this dialog box will be displayed only when you are in workspace where 3D model is displayed. In this dialog box, select the message enclosure from the **Enclose message in** drop-down list; the selected message enclosure will be displayed in the preview area. Next, enter the text in the **Message** edit box. You can select the **Draw dimension to message** check box if you want to place a dimension for locating the message. Next, choose the **OK** button; the dialog box will be closed and you will be prompted to select the insertion point on a pipe or a fitting. After selecting the insertion point, a sphere will be added inside the pipe, as shown in Figure 8-6. You need to switch to the wireframe view to view this sphere.

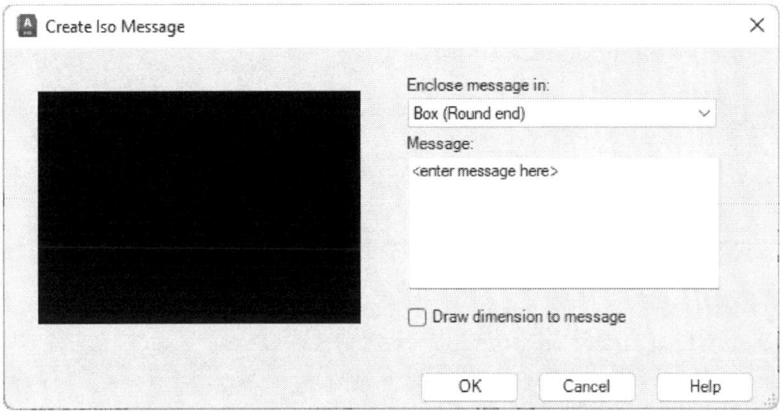

***Figure 8-5** The **Create Iso Message** dialog box*

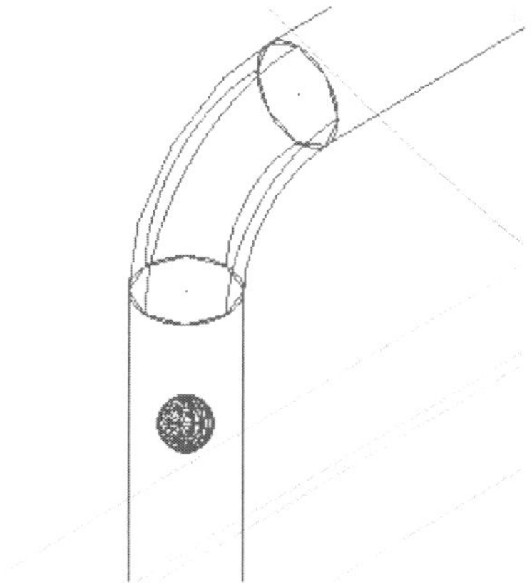

***Figure 8-6** The sphere displayed inside the pipe*

You can place annotation items such as **Floor Symbol**, **Flow Arrow**, **Insulation Symbol**, **Location Point**, **Start Point**, and **Break Point** in a 3D model. These items will be displayed in an isometric drawing. To place an annotation item, select it from the **Iso Annotations** panel in the **Isos** tab; you will be prompted to specify an insertion point on a pipe or a fitting. After specifying the insertion point, a sphere will be displayed inside the pipe. However, if you are placing a **Flow Arrow** symbol, you need to specify the insertion point and then choose the **Accept** or **Reverse** option from the command prompt.

EXPORTING A PIPE COMPONENT FILE

As the Piping Component File (PCF) is the primary input to create an isometric drawing. It is a text file containing information about piping components and routing. A typical PCF contains information of piping components (flanges, valve, and so on), coordinate values, sizes of the components, and shapes to be used to represent the components in an isometric drawing. There will be a separate PCF created for each selected line. Later, you can use this PCF to create an isometric file. To create a PCF, choose the **PCF Export** tool from the **Export** panel in the **Isos** tab; the **Export PCF** dialog box will be displayed, as shown in Figure 8-7. In this dialog box, select the line numbers to be exported from the **Line Numbers** list box and then choose the **Create** button; a separate PCF will be created for each line selected.

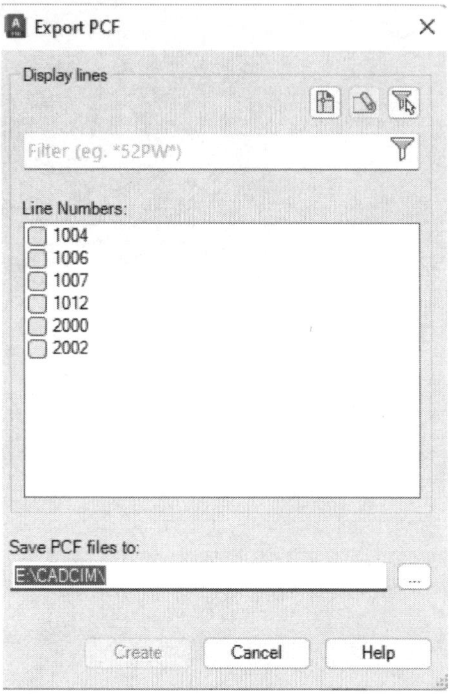

***Figure 8-7** The **Export PCF** dialog box*

Creating an Iso from a Pipe Component File

To create an isometric drawing from a pipe component file, choose the **PCF to Iso** tool from the **Iso Creation** panel in the **Isos** tab; the **Create Iso from PCF** dialog box will be displayed, as shown in Figure 8-8. In this dialog box, choose the **Add** button from the **Display lines** area; the **Pick PCF File** dialog box will be displayed. In this dialog box, browse to the file location and double-click on it; the file name will be displayed in the **PCF files** list box. You can also add more files to the list. After adding the required files, choose the **Create** button; the isometric files will be created.

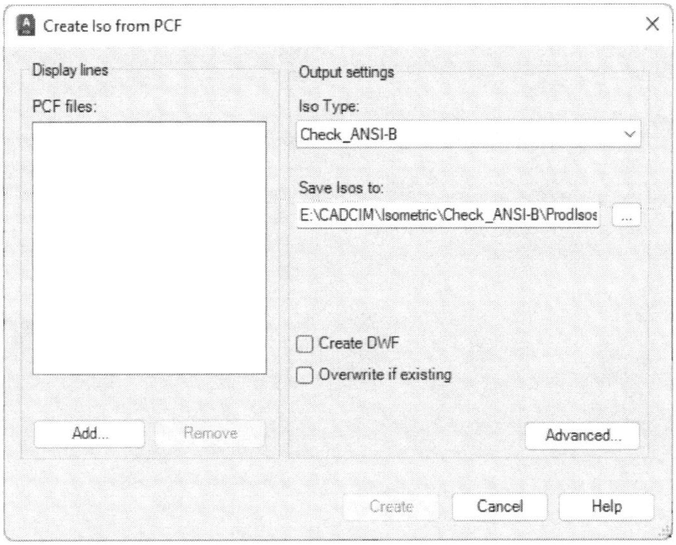

*Figure 8-8 The **Create Iso from PCF** dialog box*

LOCKING A LINE NUMBER

To lock a line, choose the **Isometric DWG** tab in the **PROJECT MANAGER** and then expand the **Isometric Drawings** node; different folders containing isometric drawings are displayed. Expand the required folder to display the line numbers present in the drawing. Next, right-click on the line that you want to lock; a shortcut menu will be displayed. Choose the **Lock Line and Issue** option from the shortcut menu; the line will be locked.

CONFIGURING ISOMETRIC DRAWING SETTINGS

By default, four types of Iso styles are available in AutoCAD Plant 3D: Check, Stress, Final, and Spool styles. If you want more styles, you need to change the old styles or create new ones using the **Project Setup** dialog box. To change an existing Iso style, you need to modify any of the following settings:

a. Iso Style Setup
b. Annotations
c. Dimensions
d. Themes
e. Sloped and Offset Piping
f. Title Block and Display

Configuring Iso Style Settings

To configure Iso style settings, choose the **Project Setup** tool from the **Project** drop-down in the **Project** panel of the **Home** tab; the **Project Setup** dialog box will be displayed. Next, select the **Iso Style Setup** sub-node from the **Isometric DWG Settings** node in the **Project Setup** dialog box; the Iso Style Setup page will be displayed, as shown in Figure 8-9. Using this page, you can create a new Iso style and configure the settings such as drawing format, file naming convention, field weld control, field fit weld makeup, table overflow settings, spool settings, and content paths.

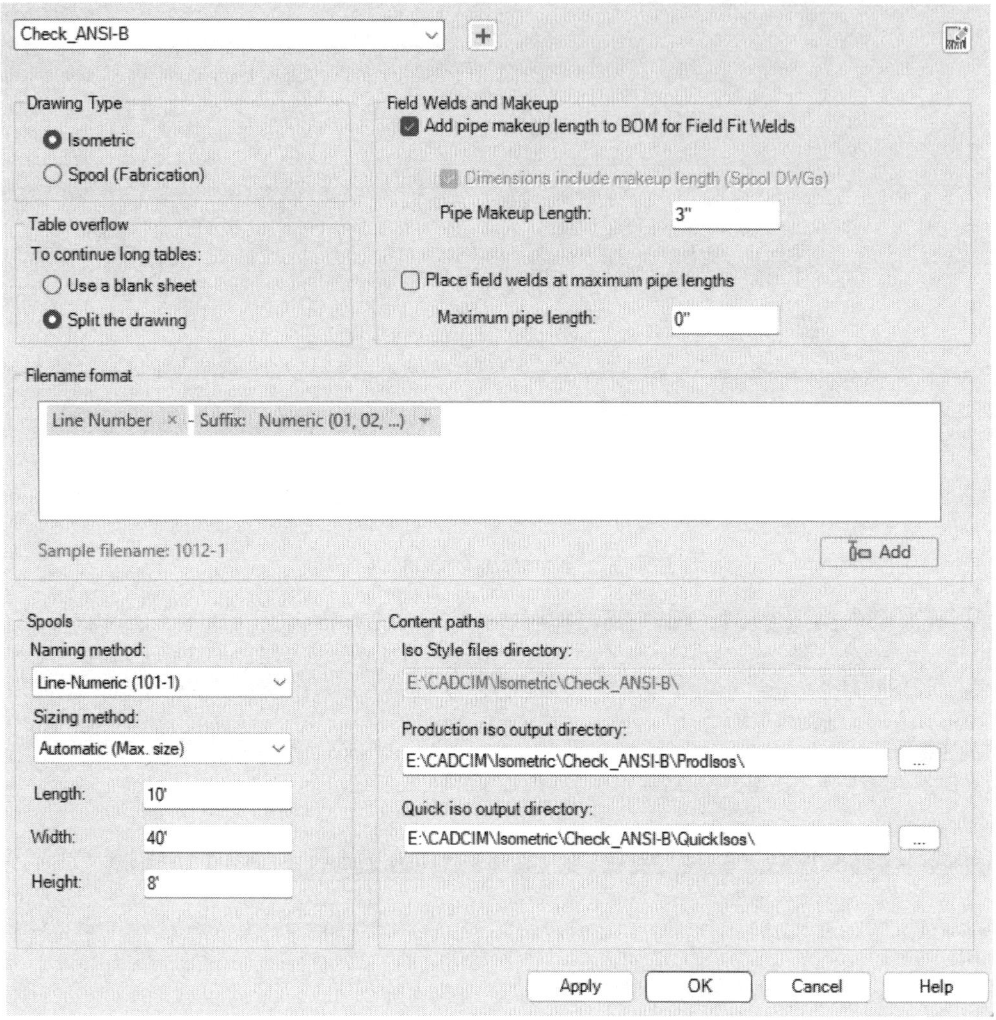

Figure 8-9 *Partial view of the Iso Style Setup page*

Configuring Annotation Settings

To configure the annotation settings, select the **Annotations** sub-node from the **Isometric DWG Settings** node; the Annotations page will be displayed, as shown in Figure 8-10. In this page, you can configure the display of annotations related to bill of materials, spool, welds, valves, cut piece, and connections and continuations.

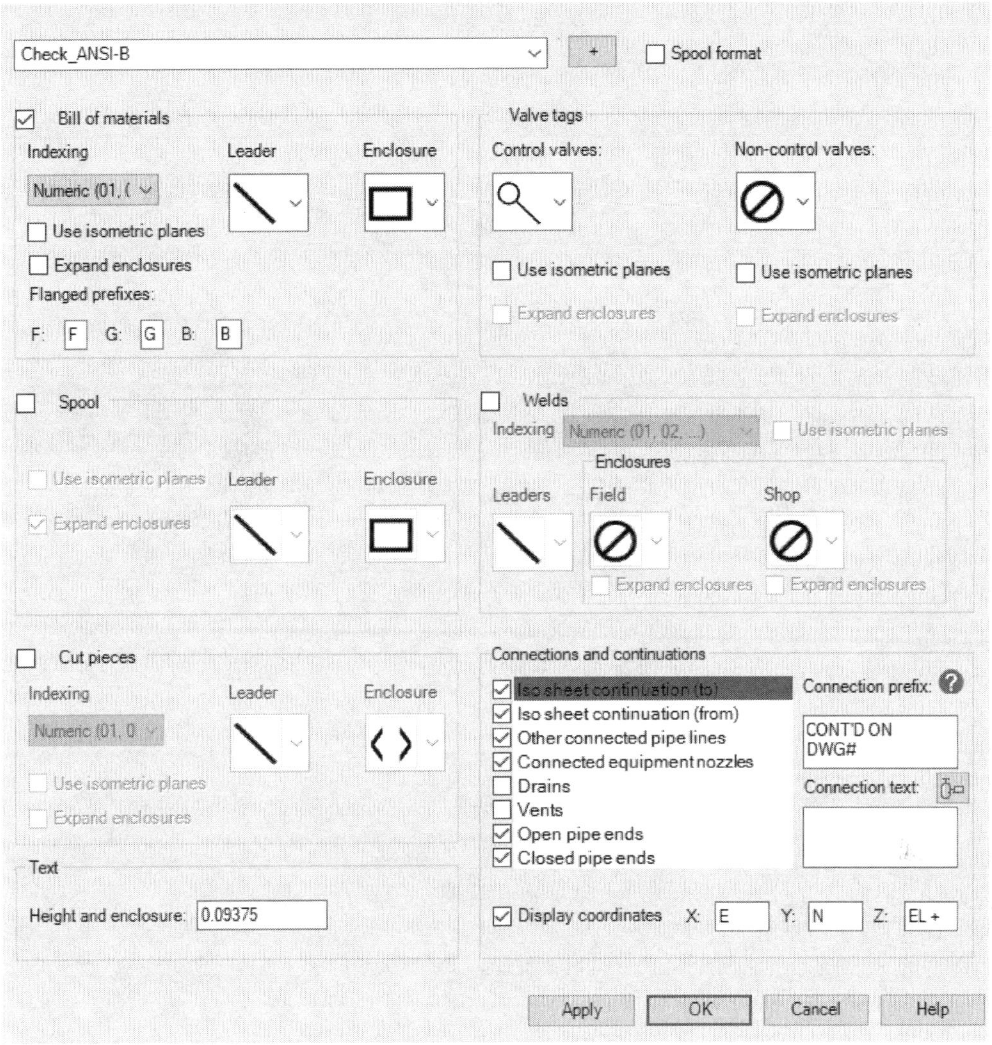

Figure 8-10 Partial view of the Annotations page

Configuring Dimensional Settings

To configure the dimensional settings, select the **Dimensions** sub-node from the **Isometric DWG Settings** node; the Dimensions page will be displayed, as shown in Figure 8-11. In this page, you can define the pipeline dimensions and alternate line settings.

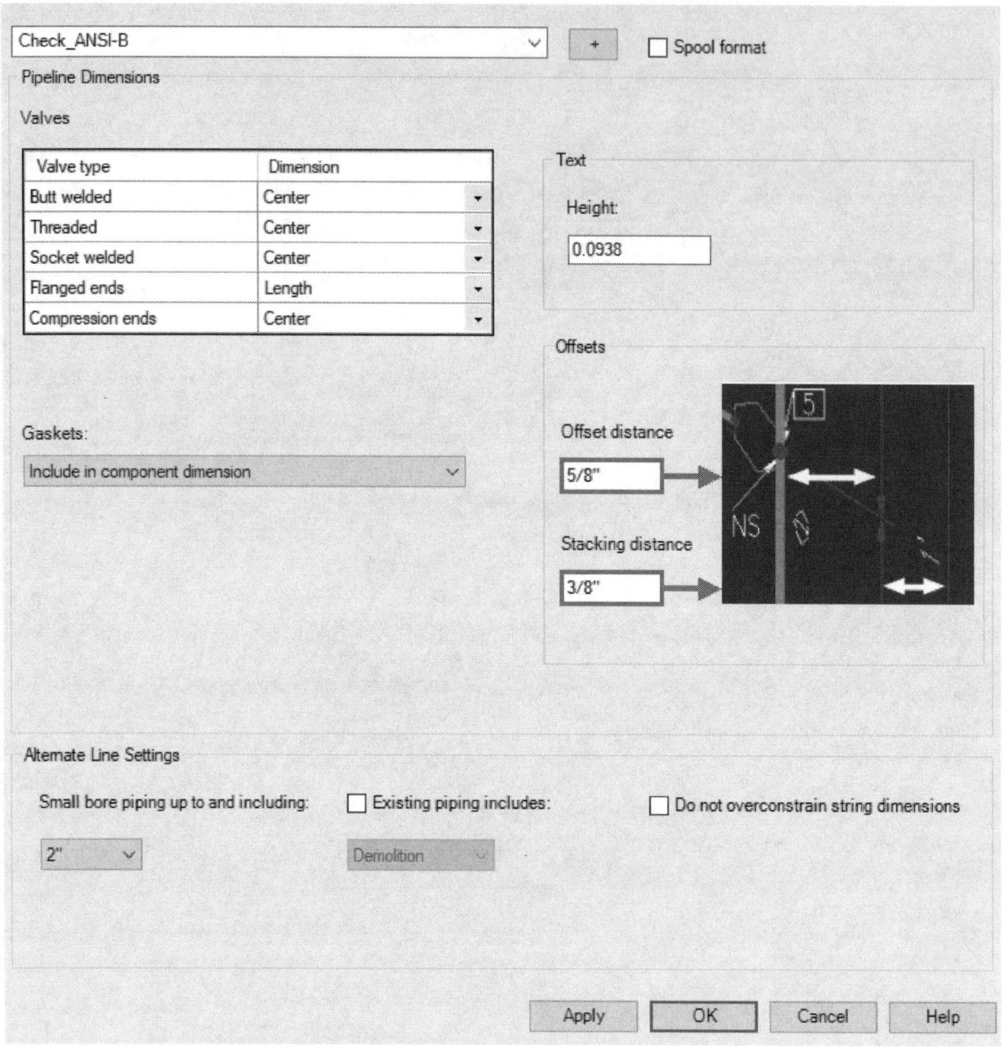

Figure 8-11 *Partial view of the Dimensions page*

Configuring Themes

To configure the themes, select the **Themes** sub-node from the **Isometric DWG Settings** node; the Themes page will be displayed, as shown in Figure 8-12. In this page, you can select the required theme and control the display of dimensioning types such as end to end, string, and locating by using the options from the **Dimension types** area. From this area, you can also control the behavior of end to end, string, and locating type dimensions using the table.

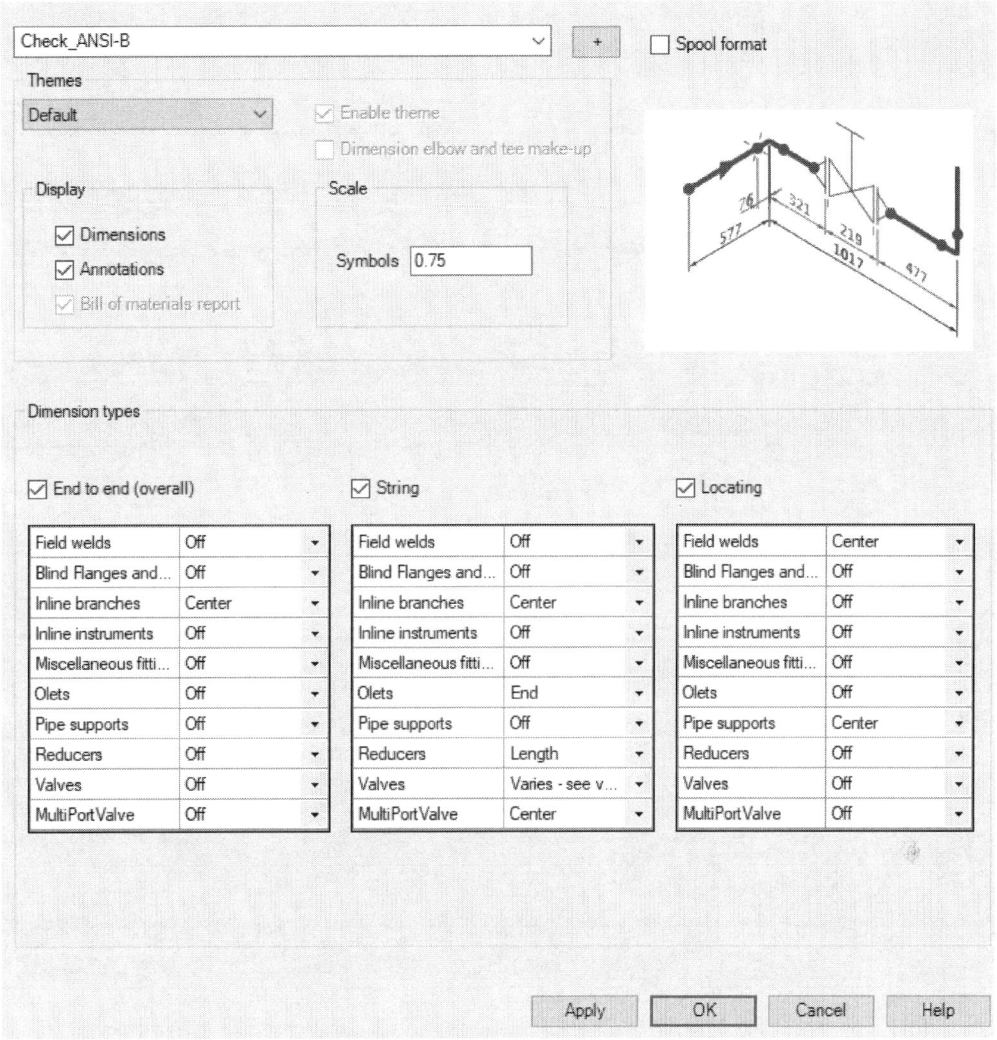

***Figure 8-12** Partial view of the Themes page*

Configuring Sloped and Offset Piping Settings

To configure the sloped and offset piping settings, select the **Sloped and Offset Piping** sub-node from the **Isometric DWG Settings** node; the Sloped and Offset Piping page will be displayed, as shown in Figure 8-13. In this page, you can set the representation of falls, offset piping, and annotations.

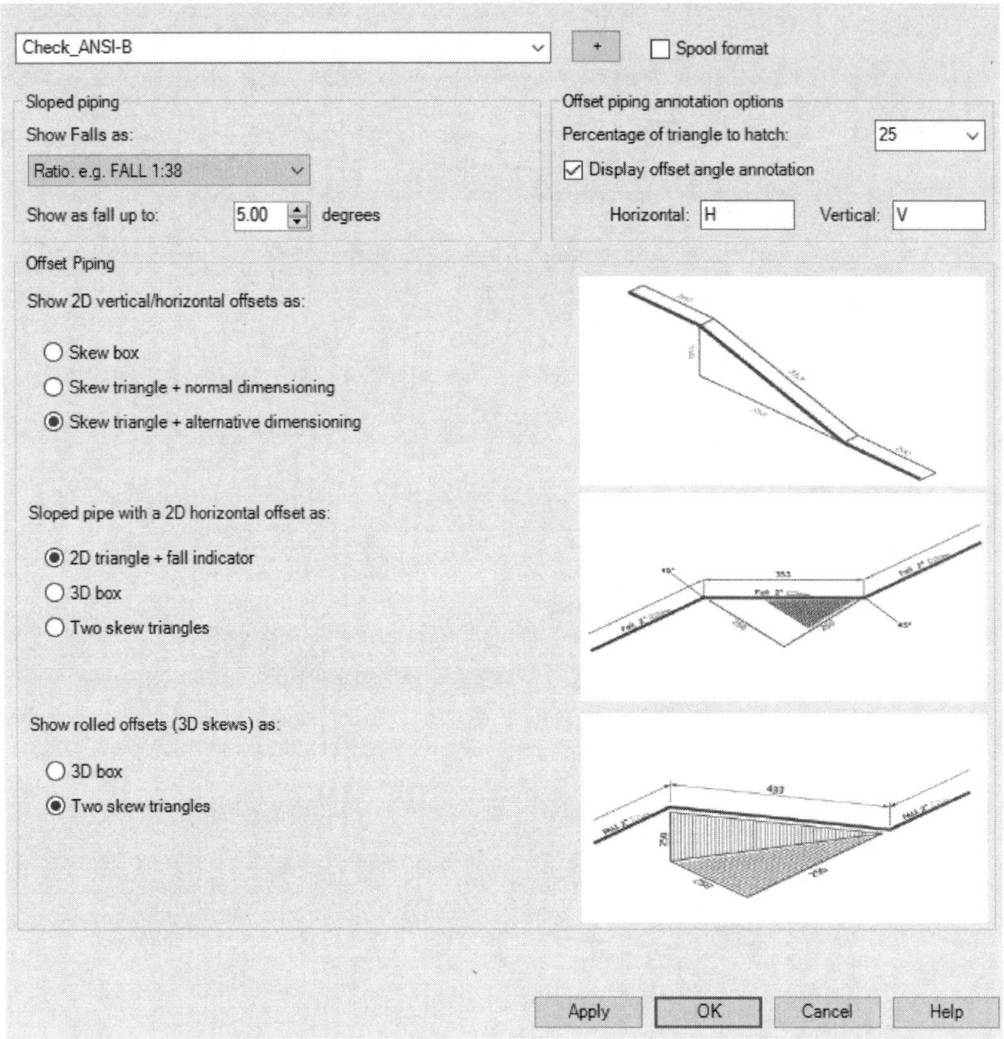

Figure 8-13 Partial view of the Slope and Offset Piping page

Setting the Title Block and Display Properties

To set the title block and other display properties, select the **Title Block & Display** sub-node from the **Isometric DWG Settings** node; the Title block & display page will be displayed, as shown in Figure 8-14. Using this page, you can setup a new title block. You can also modify the display of Isometric symbols, bends, and elbows. To setup a new title block, choose the **Setup Title Block** button; the **Title Block Setup** contextual tab will be displayed in the **Ribbon**, as shown in Figure 8-15. You can use this tab to setup the title block.

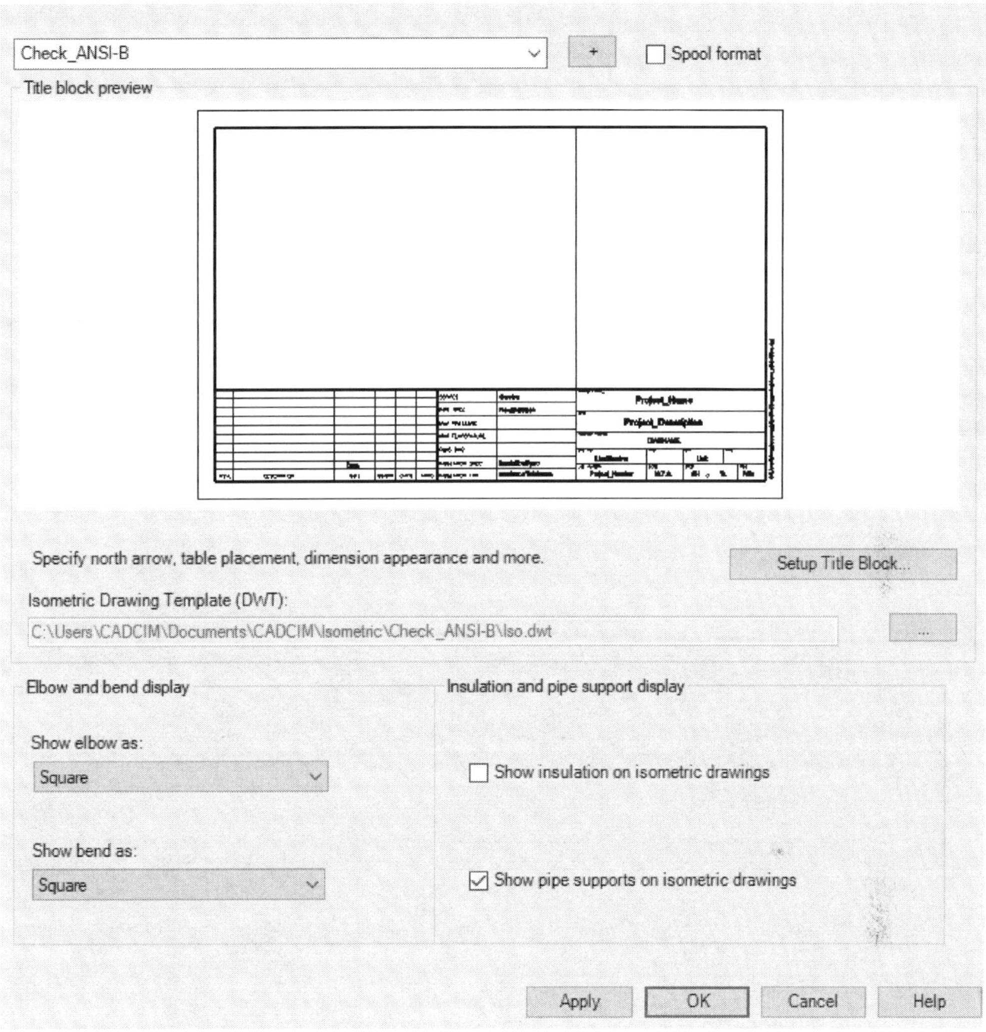

Figure 8-14 *Partial view of the Title block and display page*

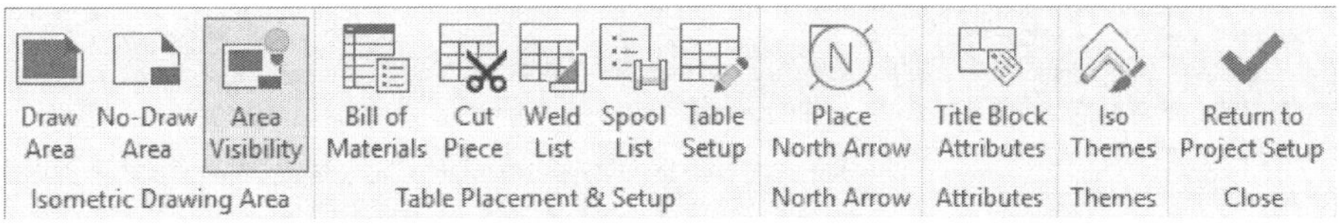

Figure 8-15 *The* ***Title Block Setup*** *contextual tab*

Configuring Bill of Materials in the Title block

Bill of materials is one of the important part of an Isometric drawing. Default style of bill of materials is available in the title block. You can customize the bill of materials by choosing the **Table Setup** tool from the **Table Placement & Setup** panel in the **Title Block Setup** contextual tab. On choosing this tool, the **Table Setup** dialog box will be displayed, as shown in Figure 8-16. You can use this dialog box to customize the bill of materials. You can modify other tables such as **Cut Piece List**, **Weld List**, and **Spool List** by selecting the corresponding option from the **Table type** drop-down list in the dialog box.

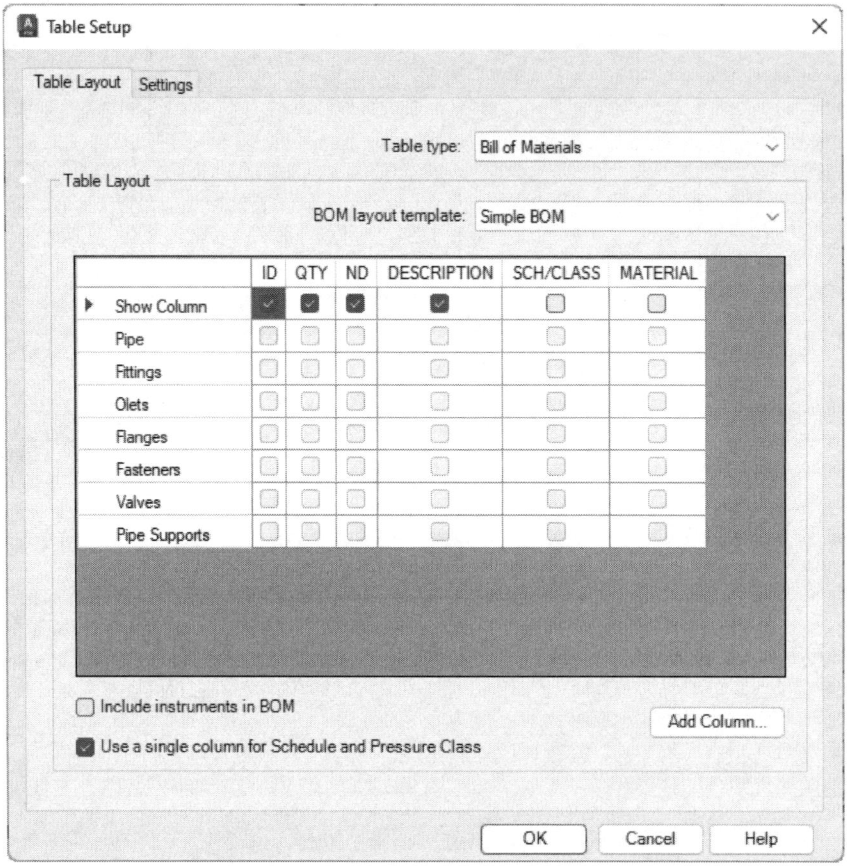

***Figure 8-16** The **Table Setup** dialog box*

TUTORIALS

Tutorial 1

In this tutorial, you will create an isometric drawing of line number 1004 in CADCIM project, as shown in Figure 8-17. **(Expected time: 30 min)**

The following steps are required to complete this tutorial:

a. Open the 3D model file.
b. Check the line number of the pipe for which the isometric drawing is to be created.
c. Create the isometric drawing.
d. Lock the pipe.

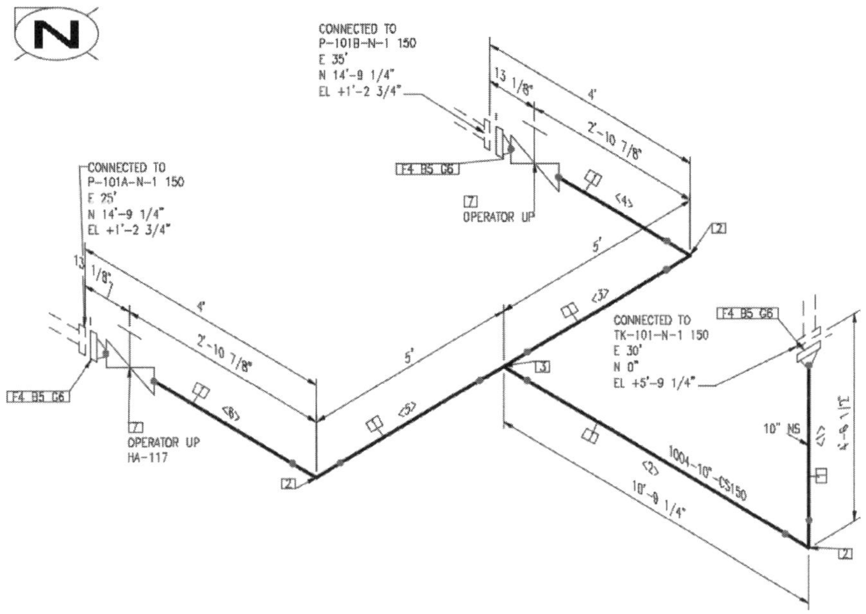

Figure 8-17 Isometric drawing for Tutorial 1

Opening the File

1. Start AutoCAD Plant 3D and open the *Piping Model.dwg* file of the CADCIM project from the **PROJECT MANAGER**.

Checking the Line Number

Next, you need to create the isometric drawing of the pipe connecting the pumps. To do so, you need to check the line number of the pipe.

1. Zoom in the pumps connected to the vertical vessel and hover the cursor on the pipe connecting the inlet nozzles of the pumps, refer to Figure 8-18; the line number of the pipe is displayed.

Creating the Isometric Drawing

1. Choose the **Production Iso** tool from the **Iso Creation** panel of the **Isos** tab; the **Create Production Iso** dialog box is displayed.

2. In this dialog box, select the check box next to the line number **1004** in the **Line Numbers** list box.

3. Select the **Final_ANSI-B** option from the **Iso Style** drop-down list in the **Output settings** area and choose the **Create** button; the isometric drawing creation starts. When a drawing is created, a balloon is displayed at the right end of the Status Bar, as shown in Figure 8-19.

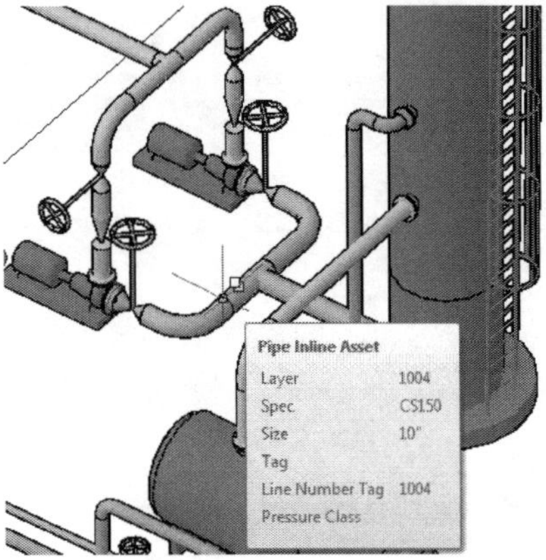

Figure 8-18 Checking the line number of the pipe

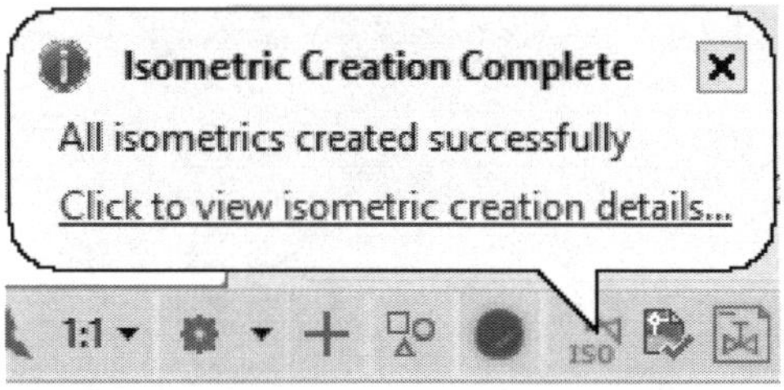

Figure 8-19 Balloon displayed in the Status Bar

4. Click on the **Click to view isometric creation details** link displayed in the balloon; the **Isometric Creation Results** dialog box is displayed, as shown in Figure 8-20.

5. In this dialog box, click on the file path of the isometric drawing, refer to Figure 8-20; the isometric drawing is displayed, as shown in Figure 8-21.

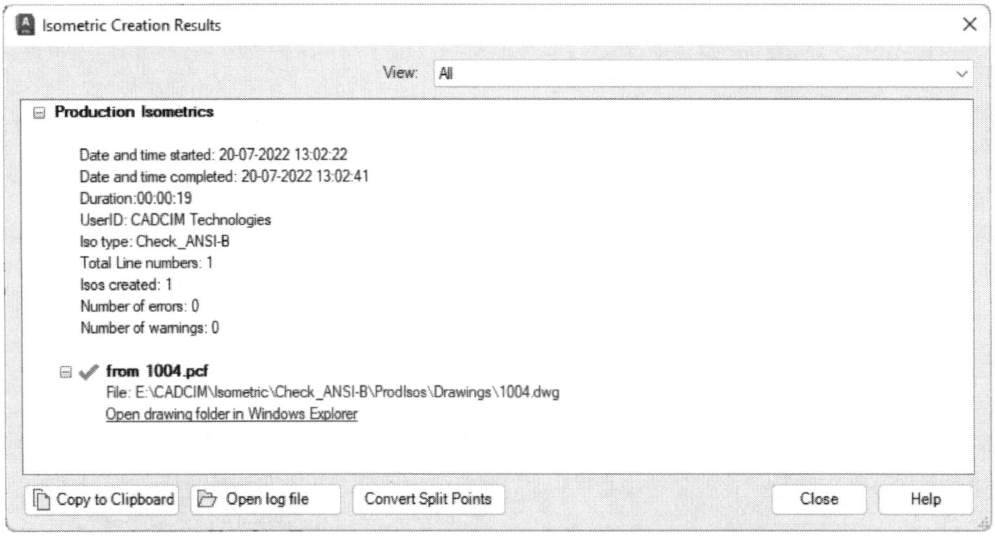

Figure 8-20 Selecting the file path of the isometric drawing

Also, the bill of materials and cut piece list is displayed, refer to Figures 8-22 and 8-23. Notice that the weld symbols in the drawing are displayed as points.

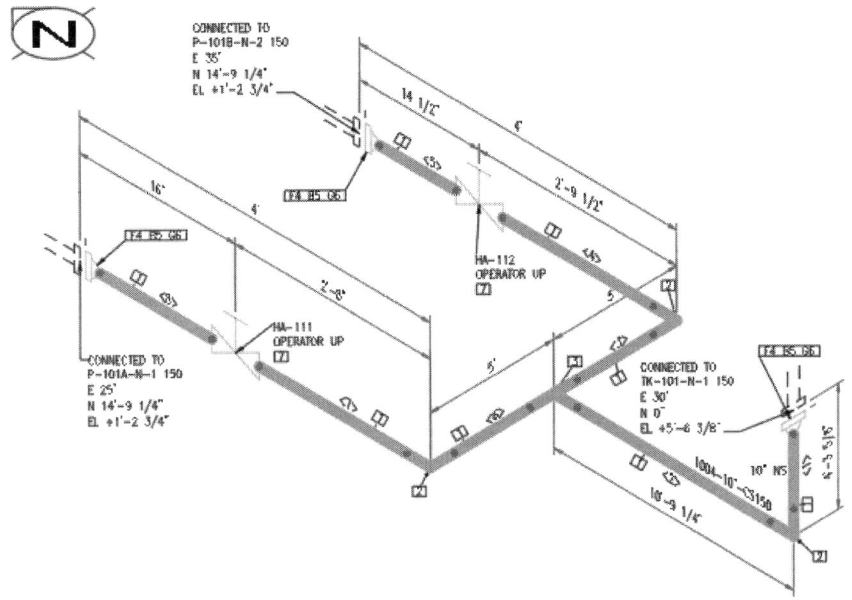

Figure 8-21 *The isometric drawing of line number 1004*

BILL OF MATERIALS				
ID	QTY	ND	SCH/ CLASS	DESCRIPTION
1	19'-5"	10"		PIPE, SEAMLESS, PL, ASME B36.10
2	3	10"		ELL 90 LR, BV, ASME B16.9
3	1	10"		TEE, BV, ASME B16.9
4	3	10"	150	FLANGE WN, RF, 150 LB, ASME B16.5
5	36	7/8"X4 1/2"	150	BOLT SET, RF, 150 LB, STUD BOLT
6	3	10"	150	GASKET, SWG, 1/8" THK, RF, 150 LB, ASME B16.20, CS/PTFE
7	2	10"	150	GATE VALVE, BV, 150 LB, ASME B16.10

Figure 8-22 *The bill of materials of the selected pipe segment*

CUT PIECE LIST				
ID	LENGTH	ND	END1	END2
1	2'-8 1/2"	10"	BEVEL	BEVEL
2	8'-9 3/4"	10"	BEVEL	BEVEL
3	3'-0 1/2"	10"	BEVEL	BEVEL
4	9 1/2"	10"	BEVEL	BEVEL
5	1 7/16"	10"	BEVEL	BEVEL
6	3'-0 1/2"	10"	BEVEL	BEVEL
7	8"	10"	BEVEL	BEVEL
8	2 15/16"	10"	BEVEL	BEVEL

Figure 8-23 *Cut piece list of the selected pipe segment*

Locking the Pipe Line

1. Choose the **Isometric DWG** tab from the **PROJECT MANAGER**; the **Isometric Drawings** node is displayed.

2. Expand the **Isometric Drawings** node and then the **Final ANSI-B** sub-node; the line numbers are displayed.

3. Right-click on the 1004 line number and choose the **Lock Line and Issue** option from the shortcut menu displayed, as shown in Figure 8-24; the line is locked and you cannot make changes to the isometric drawing, refer to Figure 8-25.

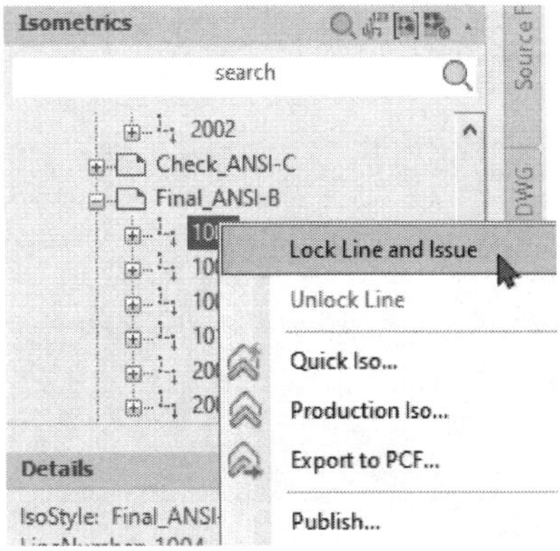

Figure 8-24 *Locking the line*

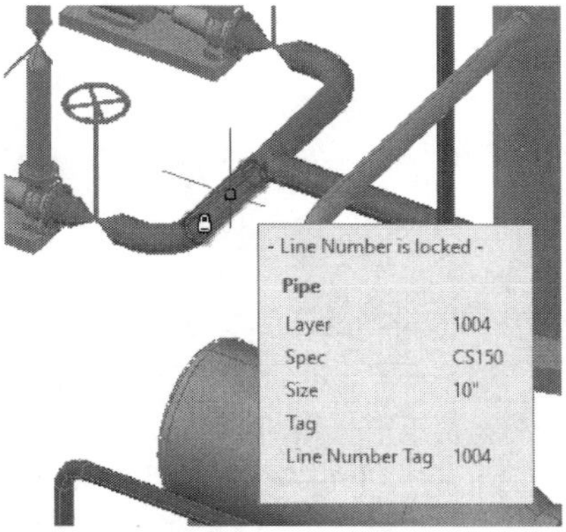

Figure 8-25 *Checking the status of the locked line*

4. Save and close the 3D model and isometric drawing files.

Tutorial 2

In this tutorial, you will create an isometric drawing of line number 1007 in CADCIM project. Later, you will modify the pipe attributes and view changes in the isometric drawing.

(Expected time: 30 min)

The following steps are required to complete this tutorial:

a. Open the 3D model file.
b. Check the line number of the pipe for which the isometric drawing is to be created.
c. Create the isometric drawing.
d. Change the weld type of the pipe in the 3D model.
e. Add insulation symbol to the pipe.
f. Generate the isometric drawing.
g. Lock the pipe.

Creating the Isometric Drawing

1. Open the *Piping Model.dwg* file of the CADCIM project from the **PROJECT MANAGER**.

   Next, you need to create the isometric drawing of the pipe connecting the pumps.

2. Choose the **Isometric DWG** tab from the **PROJECT MANAGER**.

3. Select the line number **1007** from **Isometric Drawings > Check_ANSI-B** node.

4. Right-click on the selected line number and choose the **Quick Iso** option from the shortcut menu displayed; the **Create Quick Iso** dialog box is displayed.

5. Choose the **Create** button from the **Create Quick Iso** dialog box; the isometric drawing creation starts. Also, while the drawing is created, a balloon is displayed at the right end of the Status Bar.

6. Open the isometric drawing created. Figure 8-26 shows the isometric drawing of the selected line.

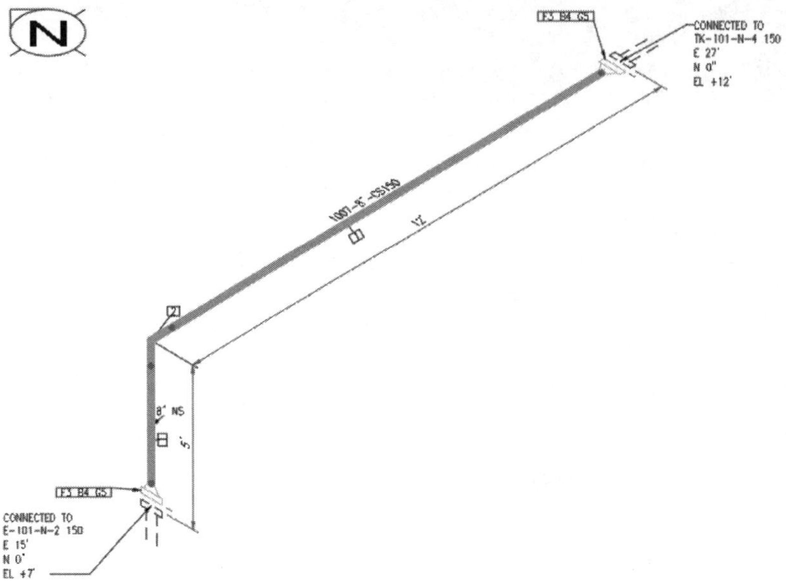

Figure 8-26 *The isometric drawing of line number* ***1007***

Changing the Weld Type in the 3D Model

Next, you need to change the weld type in the 3D model and see the representation in the isometric drawing.

1. Open the 3D model if not already open and then set the view style to **3dWireframe**.

2. Turn on the weld display by choosing the **Toggle Weld Display** button from the **Visibility** panel of the **Home** tab; the pipe connectors are displayed as dots.

3. Select the pipe connector at the elbow by using a selection window, as shown in Figure 8-27; the grips are displayed on it, as shown in Figure 8-28.

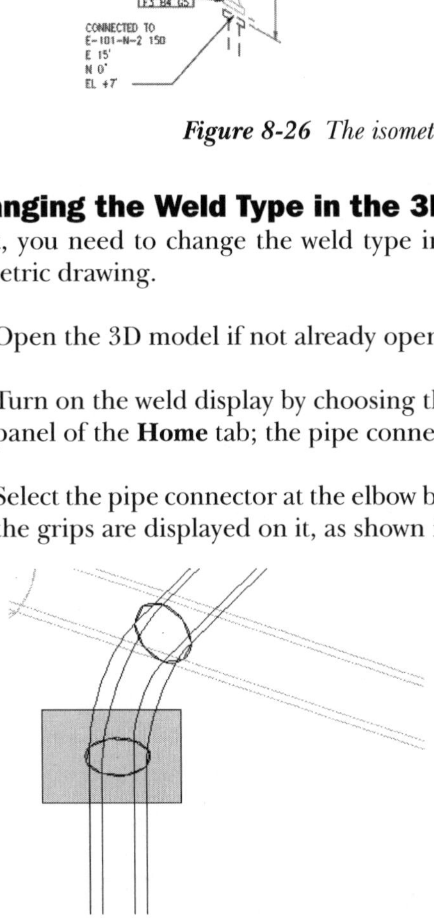

Figure 8-27 *Selecting the pipe connector*

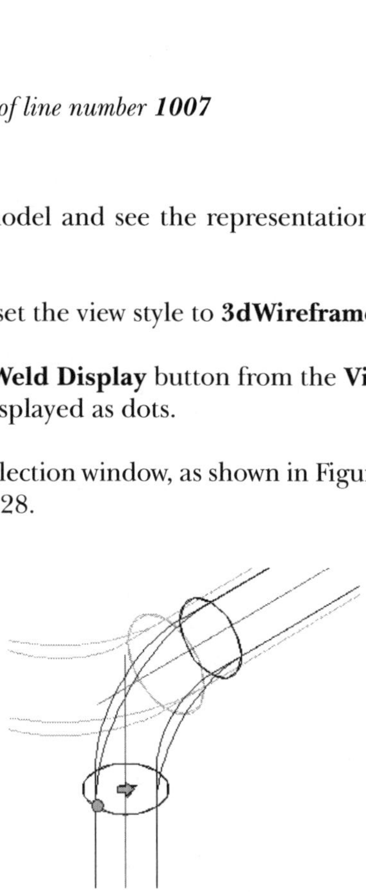

Figure 8-28 *Grips displayed on the pipe connector*

4. Right-click on any one of the grips and choose the **Properties** option from the shortcut menu displayed; the **PROPERTIES** palette is displayed.

   Next, you need to change the welding type in the **PROPERTIES** palette.

5. In this palette, scroll down to the **General** sub-rollout under the **Plant 3D** rollout and select the **FIELD** option from the **Shop/Field** drop-down list, refer to Figure 8-29; the weld type is changed to field weld.

6. Next, close the **PROPERTIES** palette of the connector to exit from it.

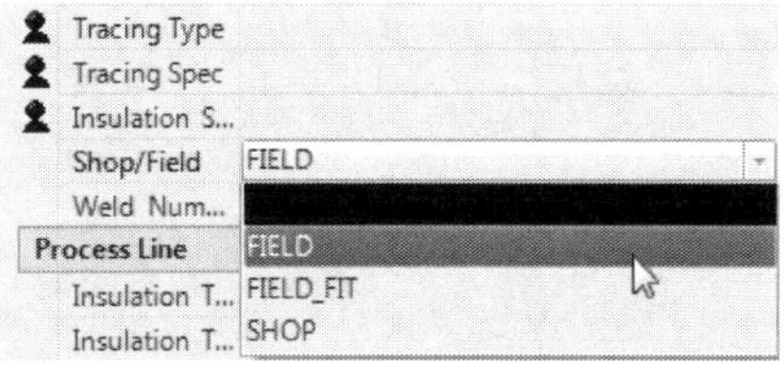

Figure 8-29 *Changing the weld type*

Adding an Insulation Symbol to the Isometric Drawing

1. Select the pipe connecting the reboiler and the vertical vessel and then right-click on it; a shortcut menu is displayed.

2. Choose **Add to Selection > Entire Line Number** from the shortcut menu, refer to Figure 8-30; the entire line number gets selected.

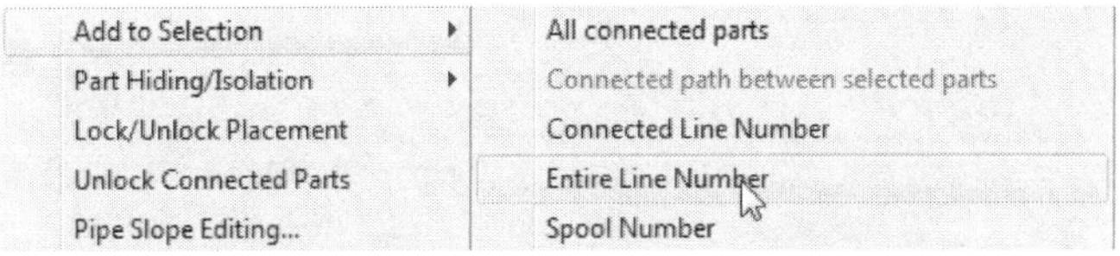

Figure 8-30 *Choosing the* ***Entire Line Number*** *option*

3. Invoke the **PROPERTIES** palette of the pipe and select **0.5-1/2"** from the **Insulation Thickness** drop-down list in the **Process Line** sub-rollout under the **Plant 3D** rollout.

4. Select the **HC - Hot Cold** option from the **Insulation Type** drop-down list, as shown in Figure 8-31.

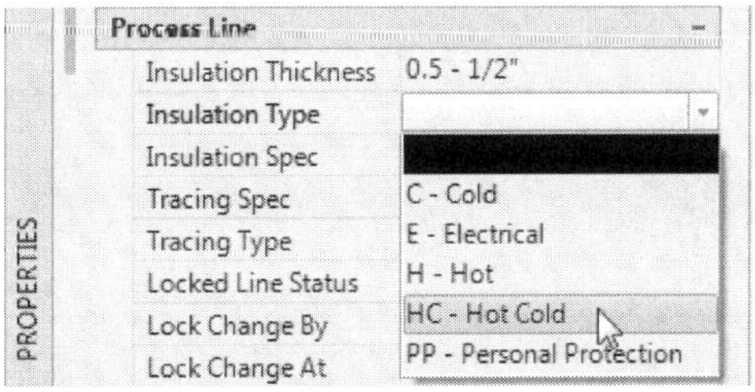

Figure 8-31 *Selecting the insulation type from the* ***PROPERTIES*** *palette*

Next, you need to mark the insulation symbol on the line.

5. Choose the **Insulation Symbol** button from the **Iso Annotations** panel of the **Isos** tab and select the insertion point on the pipe, as shown in Figure 8-32. Notice that a sphere is displayed inside the pipe, as shown in Figure 8-33.

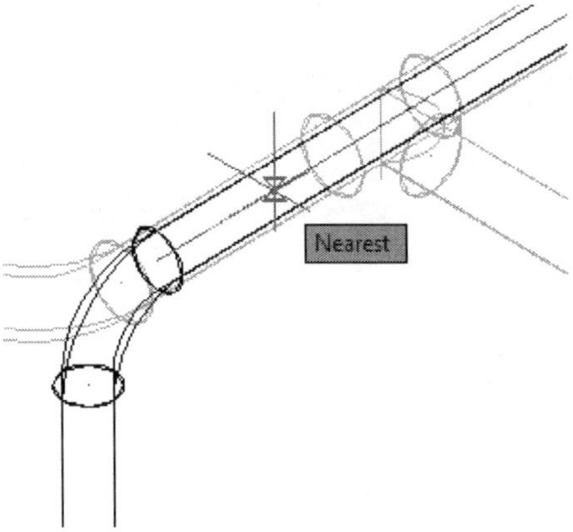

Figure 8-32 *Specifying the insertion point of the insulation symbol*

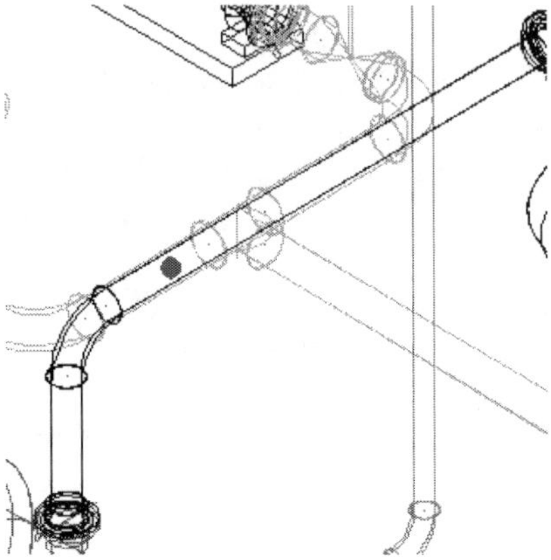

Figure 8-33 *Sphere displayed at the selected point*

6. Choose the **Save** button from the **Quick Access Toolbar** to save the model.

7. Invoke the **Create Production Iso** dialog box and select the line number **1007**. Next, choose the **Create** button; the isometric drawing is created, as shown in Figure 8-34.

   Notice that the weld representation is changed in the isometric drawing. Also, an insulation symbol is displayed, refer to Figure 8-34.

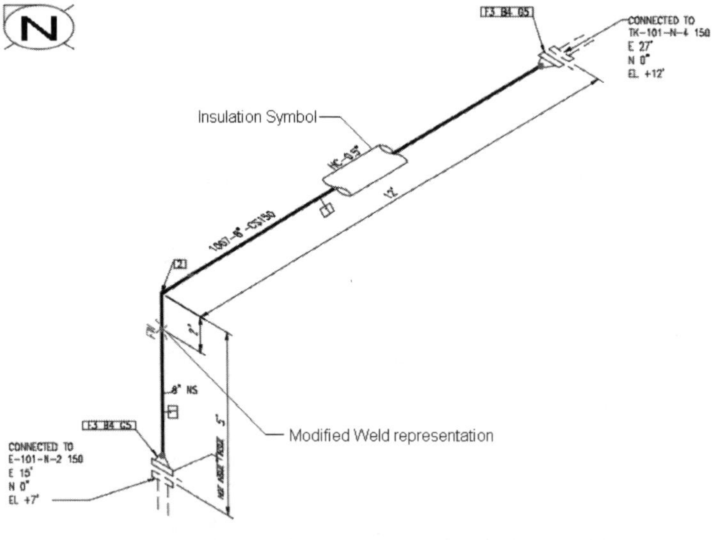

Figure 8-34 *The isometric drawing with the modified weld representation and the insulation symbol*

8. Lock the line number.

9. Save and close the 3D model and isometric drawing files.

Tutorial 3

In this tutorial, you will create a new Iso style based on the Final_ANSI-B style in CADCIM project. Next, you will customize this style and create an isometric drawing, as shown in Figure 8-35.

(Expected time: 45 min)

The following steps are required to complete this tutorial:

a. Open the CADCIM project and create a new iso style and an isometric drawing.
b. Customize the iso style and create an isometric drawing.
c. Create a new iso template.
d. Customize the Title Block.
e. Define the Draw area, No-Draw area, BOM list and Cut Piece list area.
f. Save and close the *iso.dwt* file.
g. Create an isometric drawing with the customized iso style.

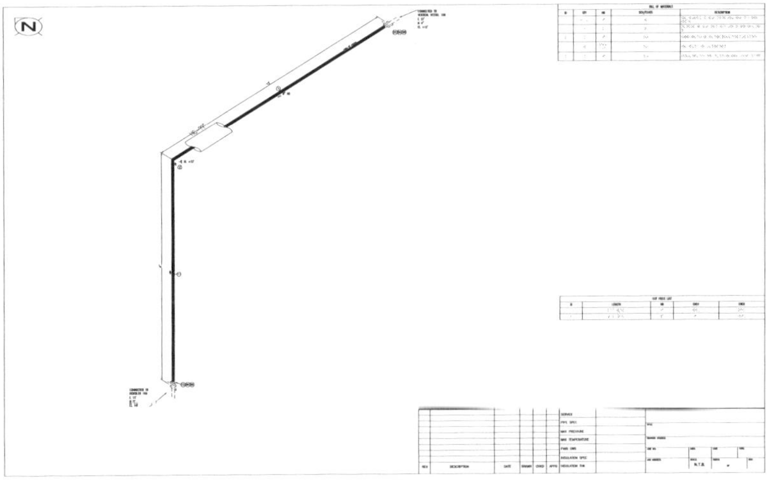

Figure 8-35 *Isometric drawing for Tutorial 3*

Opening the Project and Creating a New Iso Style

1. Start AutoCAD Plant 3D and open the CADCIM project from the **PROJECT MANAGER** and open the 3D model.

2. Choose the **Project Setup** tool from the **Project** drop-down in the **Project** panel; the **Project Setup** dialog box is invoked.

3. In this dialog box, expand the **Isometric DWG settings** node and select **Iso Style Setup**; the Iso Style Setup page is displayed.

4. Choose the **Create a new Iso style based on an existing one** button from the page displayed; the **Create Iso Style** dialog box is displayed.

5. Enter **CADCIM** in the **New style name** edit box and select the **Final_ANSI-B** option from the **Select an existing style** drop-down list.

6. Choose the **Create** button from the **Create Iso Style** dialog box; a new iso style is created and it gets listed in the iso style drop-down list of the **Project Setup** dialog box.

7. Choose the **OK** button from the **Project Setup** dialog box.

8. Create an isometric drawing of the line numbered 1007 using the newly created Iso style. Note that the isometric drawing will be the same as that created in Tutorial 2 of this chapter, as shown in Figure 8-36.

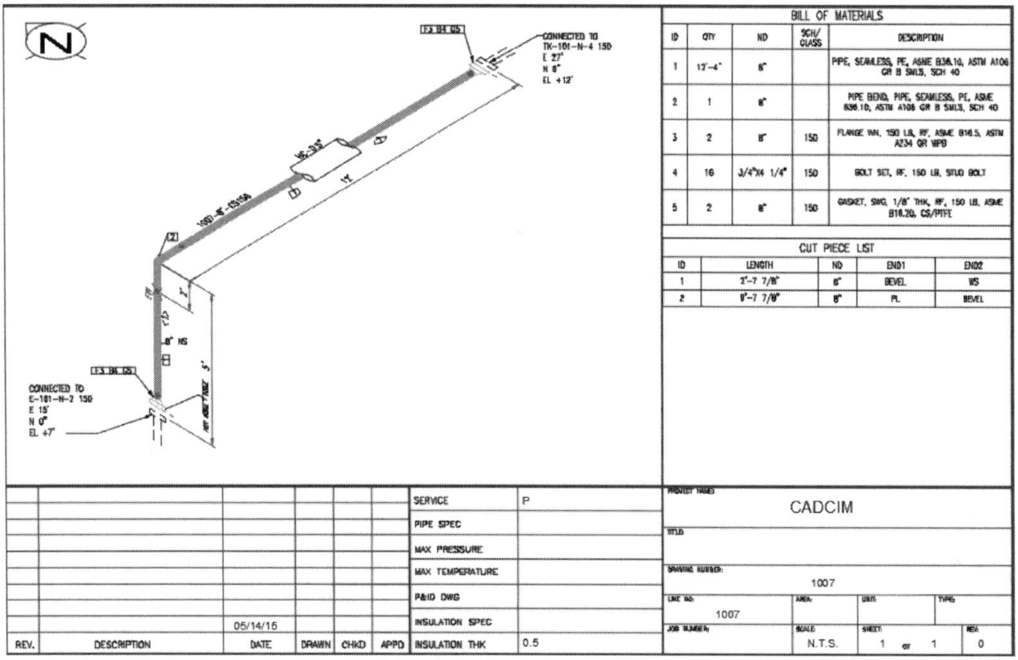

*Figure 8-36 Isometric drawing of the line number **1007** created with default settings*

Customizing the Iso Style

1. Invoke the **Project Setup** dialog box and select **Annotations** from the **Isometric DWG Settings** node; the Annotations page is displayed. Make sure that **CADCIM** is selected in the iso style drop-down list.

2. Select the **Circle** option from the **Enclosure** drop-down list in the **Bill of materials** area, refer to Figure 8-37.

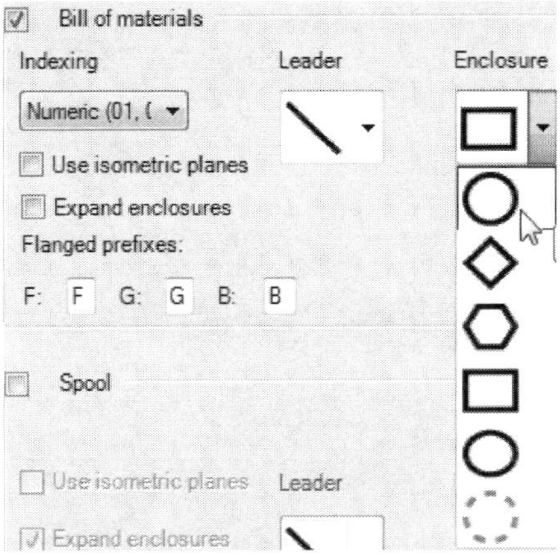

Figure 8-37 *Selecting* ***Circle*** *from the* ***Enclosure*** *drop-down list in the* ***Bill of materials*** *area*

Next, you need to modify the dimensional settings of the iso style.

3. Select **Themes** from the **Isometric DWG Settings** node; the **Themes** and **Dimension types** areas are displayed.

4. Clear the **String** and **Locating** check boxes from the **Dimension types** area and choose the **OK** button.

5. Invoke the **Create Production Iso** dialog box and select **CADCIM** from the **Iso Style** drop-down list.

6. Select the **1007** check box from the **Line Numbers** list and choose the **Create** button.

7. Open the isometric drawing; you will notice that the BOM annotations are enclosed in circles, as shown in Figure 8-38. Also, the string dimensions and locating dimensions are not displayed in the drawing, refer to Figure 8-38.

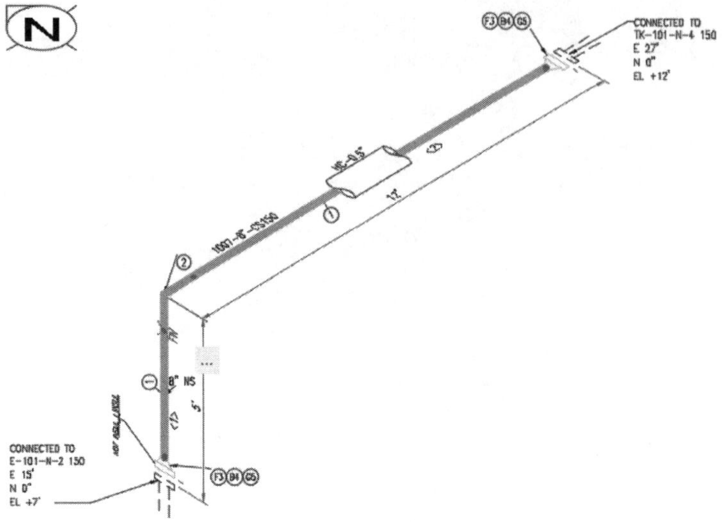

Figure 8-38 *Isometric drawing after modifying the annotation and dimensional settings*

Creating a New Iso Template

In this section, you will create a new template for the isometric drawings.

1. Choose the **Open** button from the **Quick Access Toolbar**; the **Select File** dialog box is displayed.

2. In this dialog box, select the **Drawing Template (*.dwt)** option from the **Files of type** drop-down list.

3. Browse to the location *C:\Users\User_name\Documents\CADCIM\Isometric\CADCIM* and double-click on the **iso.dwt** file; the template file is opened.

4. Select entire geometry from the drawing area and delete it by pressing DELETE.

5. Enter **PURGE** at the Command prompt; the **Purge** dialog box is displayed.

6. Expand the **Blocks** node and select **Title Block** from the **Named Items Not Used** area of the **Purge** dialog box.

7. Choose the **Purge Checked Items** button; the **Purge - Confirm Purge** message box is displayed.

8. Choose **Purge this item** from the message box; the title block is purged.

9. Close the **Purge** dialog box.

Next, you need to insert a new title block into the drawing area.

10. Download the *CADCIM_Iso.dwg* file from *https//www.cadcim.com*. The path of the file is as follows:

    Textbooks > CAD/CAM >AutoCAD Plant 3D > AutoCAD Plant 3D 2023 for Designers > Input Files

11. Save the file at the location *C:\Users\User_name\Documents\CADCIM\Isometric\CADCIM.*

12. Expand the **Insert** drop-down from the **Block** panel of the **Insert** tab and then click on **Blocks from Other Drawings**; the **BLOCKS** palette is displayed.

13. Choose the **Browse** button from the **BLOCKS** palette and open the *CADCIM_Iso.dwg* file.

14. Accept the default settings in the **BLOCKS** palette and double-click on *CADCIM_Iso.dwg* from the preview area.

15. Specify the insertion point as shown in Figure 8-39, the title block is placed in the drawing area.

16. Enter **RENAME** at the Command prompt; the **Rename** dialog box is displayed.

17. Select **Blocks** from the **Named Objects** list box and **CADCIM_Iso** from the **Items** list box.

18. Enter **Title Block** in the **Rename To** edit box and choose the **Rename To** button.

19. Choose the **OK** button from the dialog box to close it.

    Next, you need to set the limits for the drawing area.

20. Enter **LIMITS** at the Command prompt.

21. Press **ENTER** to accept the origin as the first corner.

22. Select the top-right corner of the title block to define the second corner, refer to Figure 8-40.

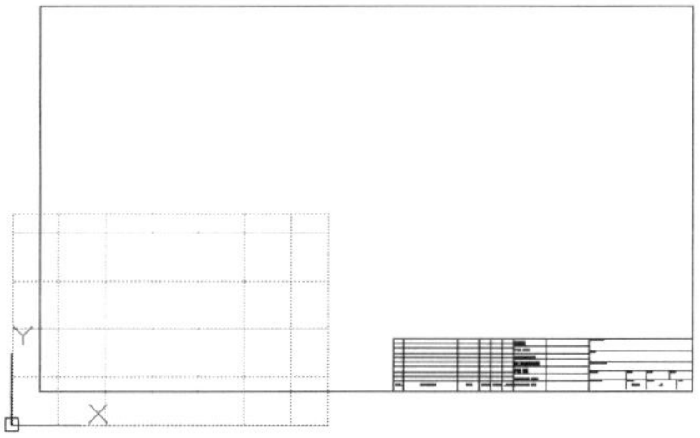

Figure 8-39 *The iso.dwt file after placing the title block*

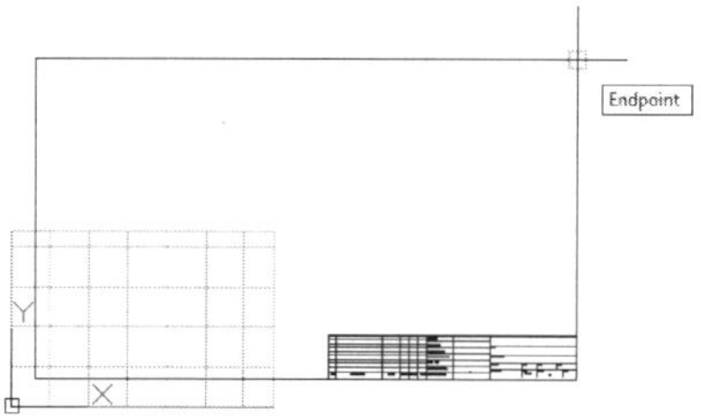

Figure 8-40 *Specifying the second corner of the drawing sheet*

23. Choose the **Save** button from the **Quick Access** Toolbar.

Customizing the Title Block

In this section, you will customize the drawing and the BOM area of the title block.

1. Invoke the **Project Setup** dialog box and select **Title Block and Display** from the **Isometric DWG Settings** node; the Title block & display page is displayed.

2. Make sure that **CADCIM** is selected in the iso style drop-down list.

3. Choose the **Setup Title Block** button from the **Title block preview** area; the **Project Setup** dialog box is closed and the **Title Block Setup** contextual tab is displayed.

4. Choose the **Draw Area** tool from the **Isometric Drawing Area** panel of the **Title block Setup** contextual tab; you are prompted to specify the first corner point of the drawing area.

5. Select the top-left corner of the title block, as shown in Figure 8-41; you are prompted to select the second corner.

6. Select the second corner of the drawing area, as shown in Figure 8-42; the drawing area is defined.

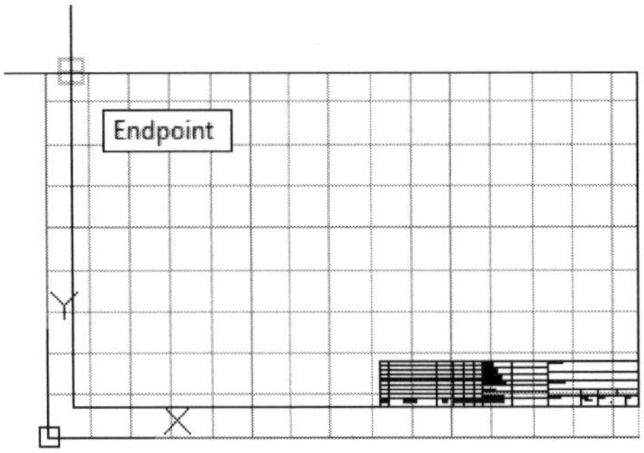

Figure 8-41 *Selecting the first corner of the drawing area*

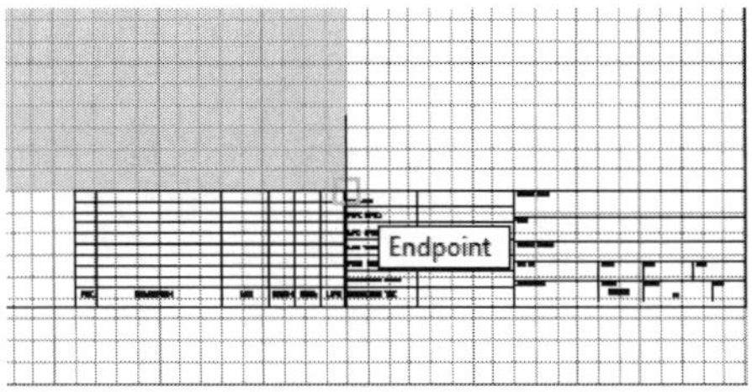

Figure 8-42 *Selecting the second corner of the drawing area*

Next, you need to place the north arrow.

7. Choose the **Place North Arrow** tool from the **North Arrow** panel; you are prompted to specify default north arrow direction.

8. Choose the **upper Left** option from the Command prompt and place the north arrow at the top-left corner, as shown in Figure 8-43.

9. Choose the **No-Draw Area** tool from the **Isometric Drawing Area** panel and define the No-draw area, as shown in Figure 8-44.

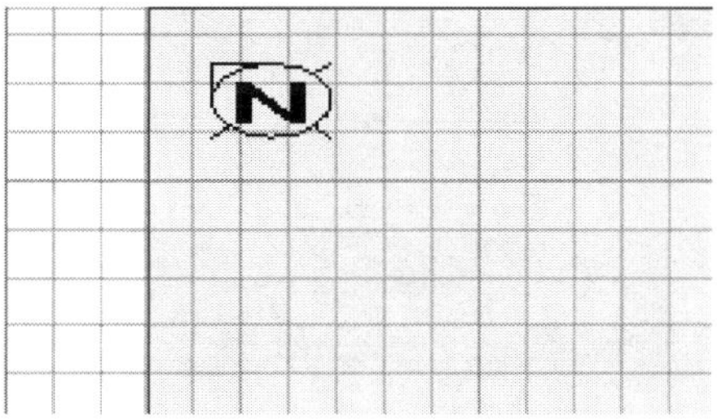

Figure 8-43 Position of the North Arrow

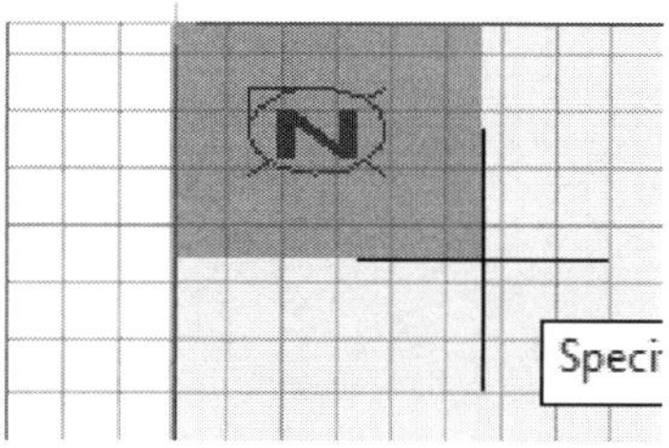

Figure 8-44 Defining the No-draw area

Next, you need to define the BOM list and Cut Piece list area.

10. Choose the **Bill of Materials** tool and define the BOM list area, as shown in Figure 8-45.

11. Choose the **Cut Piece** tool and define the Cut Piece list area, as shown in Figure 8-46.

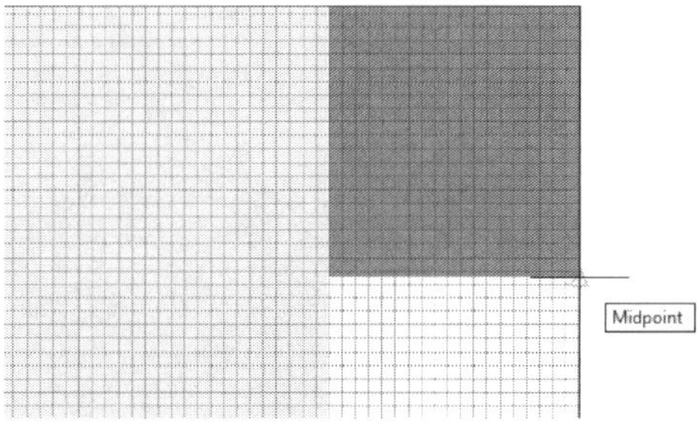

Figure 8-45 Defining the BOM list area

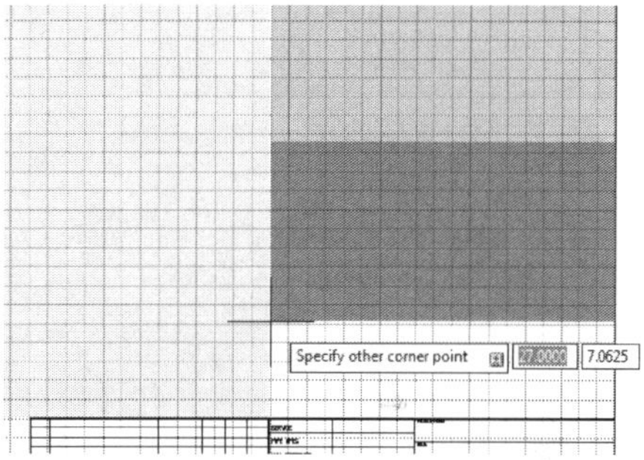

Figure 8-46 Defining the Cut Piece list area

The drawing sheet after configuring the settings is shown in Figure 8-47.

12. Choose the **Return to Project Setup** button from the **Close** panel of the **Title Block Setup** contextual tab; the **Block-Changes Not Saved** message box is displayed.

13. Choose the **Save Changes to "iso.dwt"** option from the message box; the changes made to the *iso.dwt* file are saved. Also, the **Project Setup** dialog box is displayed.

14. Choose the **OK** button from the **Project Setup** dialog box to close it.

15. Create the Production Iso of the line number 1007 and open it. The Isometric drawing is displayed, as shown in Figure 8-48.

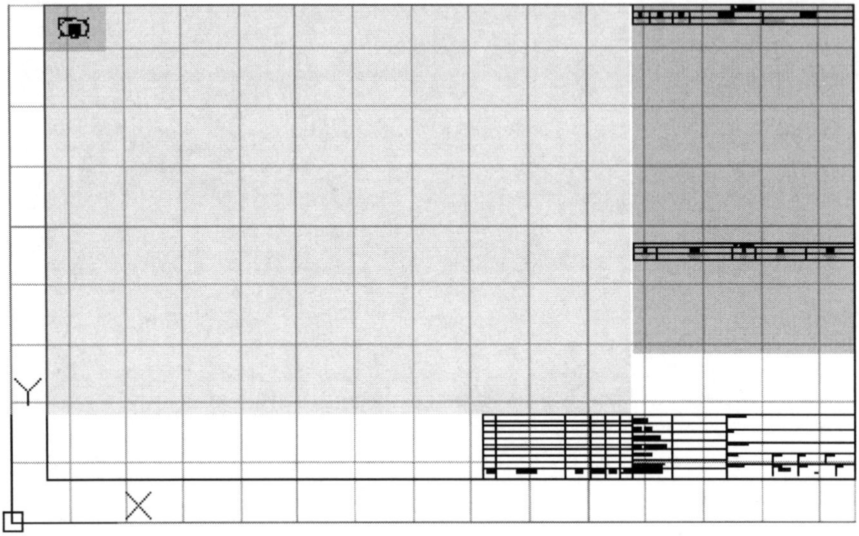

Figure 8-47 The drawing sheet after configuring all the settings

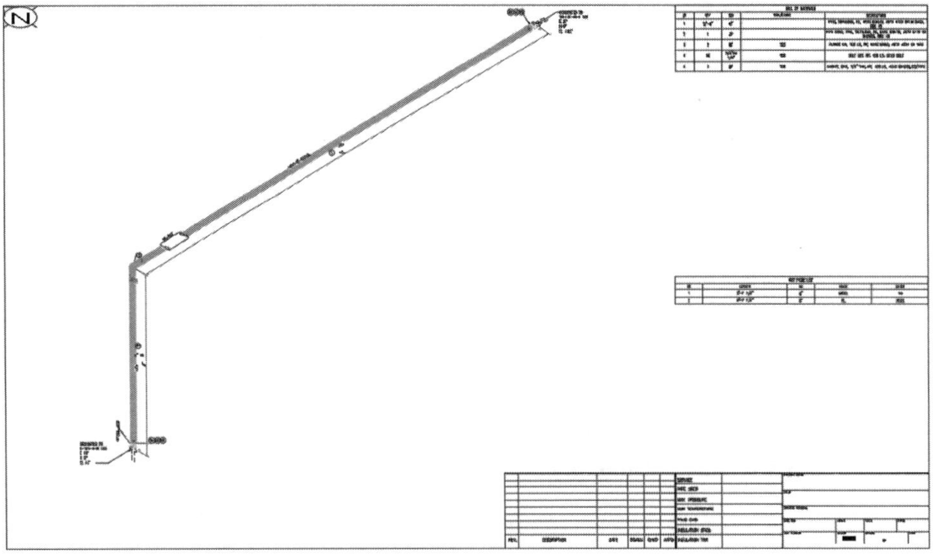

Figure 8-48 The isometric drawing created after modifying the Iso style

16. Save and close all the files.

Self-Evaluation Test

Answer the following questions and then compare them to those given at the end of this chapter:

1. A __________ is created to ensure that the final isometric drawing will be created without errors.

2. The __________ tool is used to define the area in which the isometric drawing will be created.

3. The __________ option is used to lock a line.

4. The __________ drawing will present the geometric data which is precise and relevant to the stress analysis of pipes.

5. The __________ tool is used to create an isometric file from a PCF (Pipe Component File).

6. To create a new iso style, choose the __________ button from the Iso style Setup area in the **Project Setup** dialog box.

7. By default, there are __________ types of Iso styles available in AutoCAD Plant 3D.

8. A balloon will be displayed on the __________ after the isometric drawing is created.

9. You can control the display of dimensions in the __________ page of the **Project Setup** dialog box.

10. A __________ is the final isometric drawing separated in individual sections.

Review Questions

Answer the following questions:

1. The __________ drawing will have a bill of material (BOM) and it is used for carrying out the fabrication and construction process.

2. You can create __________ by selecting pipes either from the drawing area or from the line list.

3. After creating an Iso message, a __________ will be added inside the pipe.

4. You can select predefined isometric styles from the __________ drop down list in the **Create Production Iso** dialog box.

5. The Iso template file is saved as __________.

Answers to Self-Evaluation Test

1. Check Isometric, **2. Draw Area**, **3. Lock Line and Issue**, **4.** Stress isometric, **5. PCF to Iso**, **6. Create a new Iso style based on an existing one**, **7.** four, **8.** Status Bar, **9. Dimensions**, **10.** Spool drawing

Chapter 9

Creating Orthographic Drawings

Learning Objectives

After completing this chapter, you will be able to:

- *Create Orthographic views*
- *Add annotations and dimensions to views*
- *Update Views*
- *Generate Bill of Material*
- *Generate BOM Annotation*

INTRODUCTION

In the previous chapters, you learned to create 3D models and now, you will learn to create their orthographic views. Orthographic views are the two-dimensional (2D) representations of a 3D model. In addition, you will learn to create orthographic and sectional views of a 3D model and then update the views as per the changes made in the 3D model. You will also learn to add dimensions and annotations to the 2D views.

CREATING ORTHOGRAPHIC DRAWINGS

To create a new orthographic drawing, choose the **Orthographic DWG** tab from the **PROJECT MANAGER**; the **Orthos** tree view will be displayed. In this tree view, right-click on the **Orthographic Drawings** node and choose the **New Drawing** option from the shortcut menu displayed, refer to Figure 9-1; the **New DWG** dialog box will be displayed. Next, enter the file name and author name in their respective fields and then choose the **OK** button; a new orthographic drawing will be created.

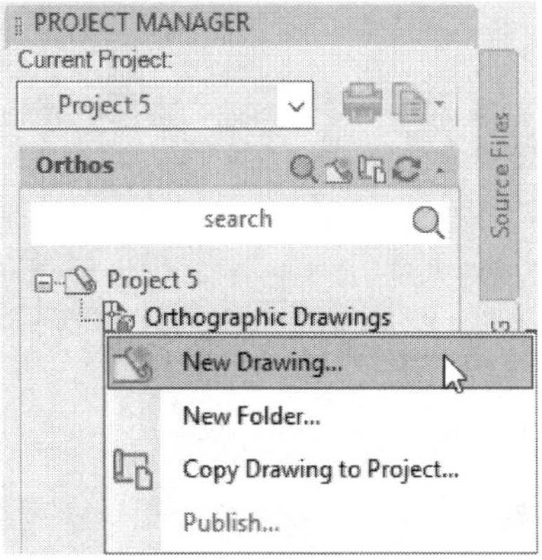

Figure 9-1 *Choosing the* ***New Drawing*** *option*

When you open or create an orthographic drawing, a drawing sheet will be displayed with the **Ortho View** tab chosen by default, refer to Figure 9-2. The tools in this tab are used to perform operations such as editing, updating, annotating the drawing view, and so on. These operations are discussed next.

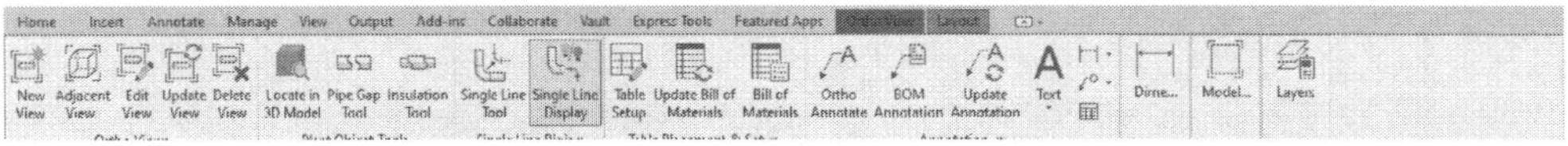

Figure 9-2 *The* ***Ortho View*** *tab*

Generating the First View

To generate the first view of the drawing, choose the **New View** tool from the **Ortho Views** panel of the **Ortho View** tab; the **Select Reference Models** dialog box will be displayed, as shown in Figure 9-3. In this dialog box, select the models to be included from the **Project models** list box. You need to select the check boxes adjacent to the drawing file names to include them in the orthographic view to be created. Next, choose the **OK** button from the dialog box; the **Orthographic View Selection** window will be displayed. Also, the **Ortho Editor** contextual tab will be added to the **Ribbon**, as shown in Figure 9-4. Also, the

OrthoCube is displayed in the drawing area, as shown in Figure 9-5. The options in **Ortho Editor** contextual tab help you create the orthographic views and the OrthoCube helps you to specify the view boundary.

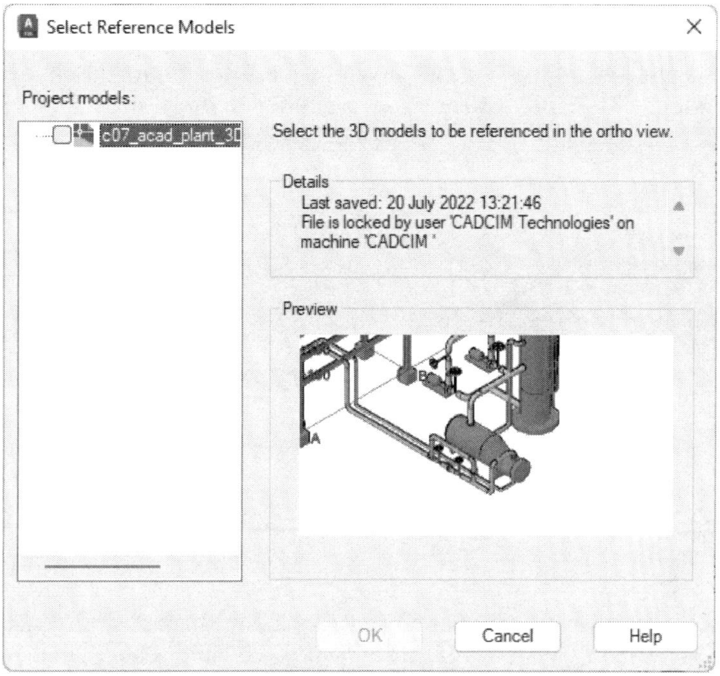

Figure 9-3 *The **Select Reference Models** dialog box*

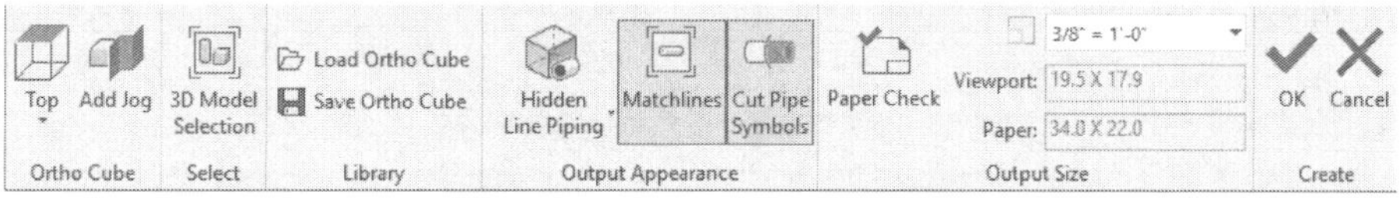

Figure 9-4 *The **Ortho Editor** tab*

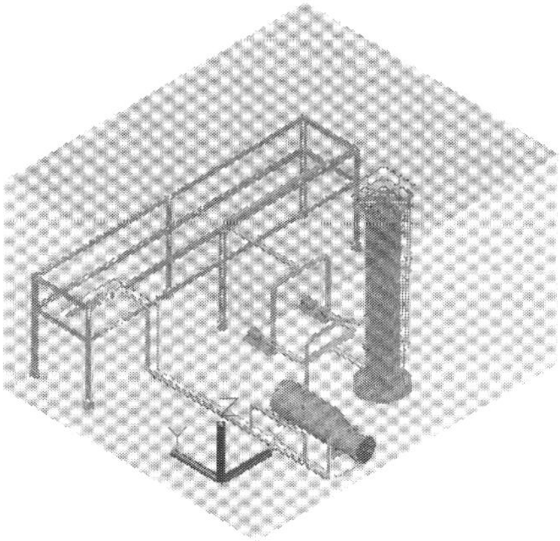

Figure 9-5 *The OrthoCube*

The options in the **Ortho Editor** contextual tab are discussed next.

Ortho Cube

In this panel, the View drop-down is used to select the view to be created. You can select the required orthographic view or isometric view from this drop-down. The **Add Jog** tool is used to add a jog to the OrthoCube. A jog is used to exclude a portion of the model from the view to be created. To add a jog, choose the **Add Jog** tool from the **Ortho Cube** panel and select an edge of the OrthoCube, refer to Figure 9-6; the jog will be added, as shown in Figure 9-7.

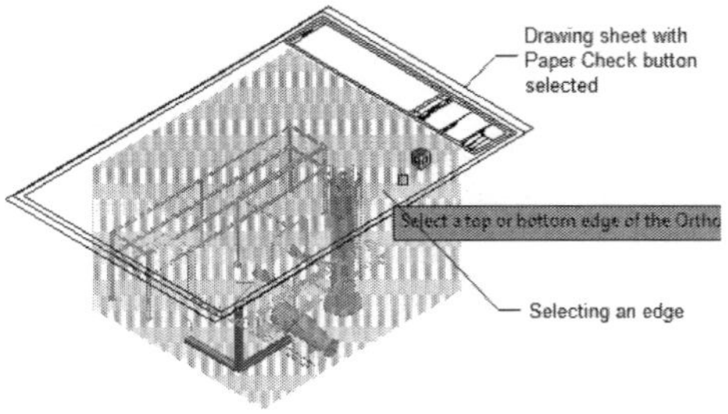

***Figure 9-6** Selecting the edge of the OrthoCube*

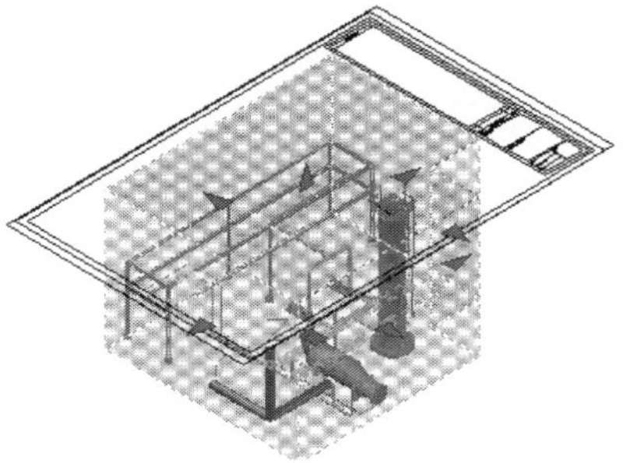

***Figure 9-7** Jog added to the OrthoCube*

Output Size

The **Paper Check** button in this panel is used to display the drawing sheet in the graphics window, refer to Figure 9-6. You can adjust the OrthoCube and scale the view according to the paper size. The **Scale** drop-down list is used to set the scaling factor of the view. The **Viewport** and **Paper** display boxes display the viewport and the paper size, respectively in the drawing. Note that these values are read only.

Select

In this panel, the **3D Model Selection** tool is used to invoke the **Select Reference Models** dialog box. Using this dialog box, you can modify the selection of piping models for creating orthographic views.

Library

Using the tools in this panel, you can save the settings made in the **Ortho View Selection** window. Also, you can load the previously saved settings.

Output Appearance

The tools in this panel are used to set the display of the view outputs. The **Hidden Line Piping** drop-down in this panel is used to set the display style of the orthographic views. The **Hidden Line Piping** tool in this drop-down is used to display the pipes which are hidden, refer to Figure 9-8. The **All Hidden Lines** tool is used to display all the hidden objects, refer to Figure 9-9. If you choose the **No Hidden Lines** tool, the hidden objects will not be displayed, refer to Figure 9-10.

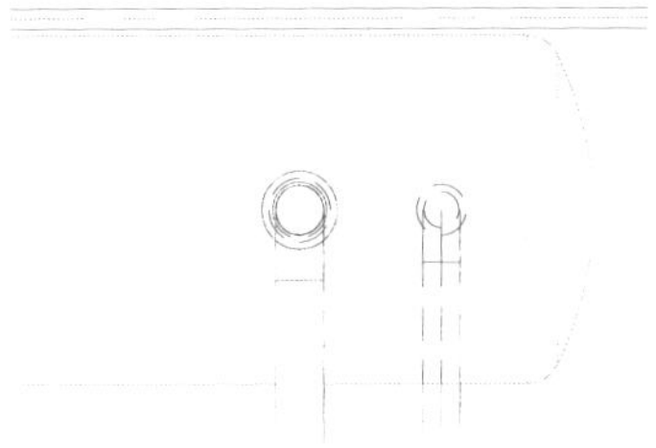

Figure 9-8 *A view with* ***Hidden Line Piping*** *tool chosen*

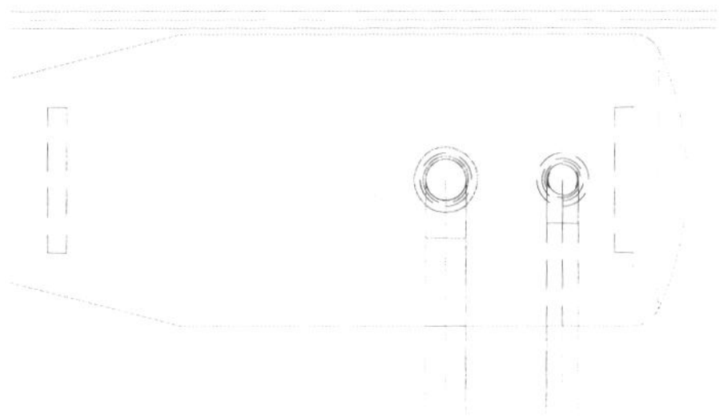

Figure 9-9 *A view with* ***All Hidden Lines*** *tool chosen*

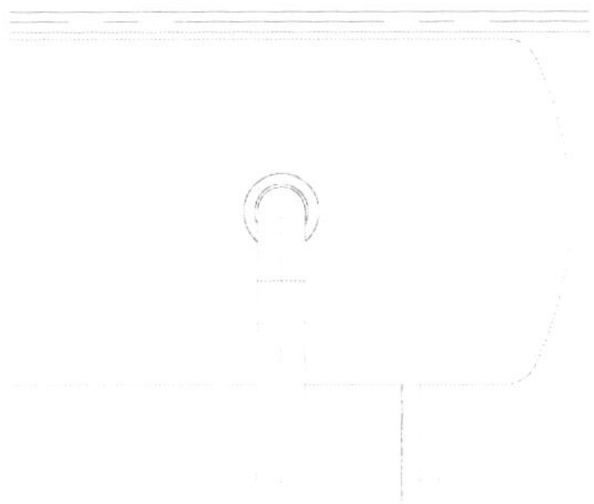

Figure 9-10 *A view with* ***No Hidden Lines*** *tool chosen*

The **Matchlines** button is used to display the matchlines on the boundary of an orthographic view, as shown in Figure 9-11. Note that the matchlines are displayed around the boundary only when you place a Plan (top) view in an ortho drawing.

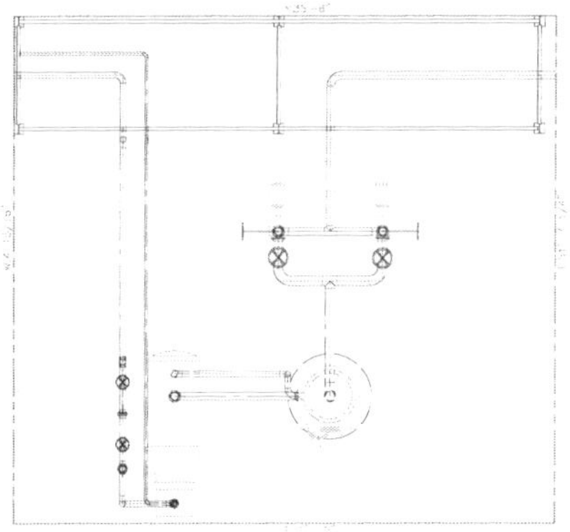

Figure 9-11 *A view with matchlines*

The **Cut Pipe Symbols** button is used to display a cut pipe symbol on the pipe when it is sectioned and only a portion of pipe is displayed in the view.

Create

The **OK** button in this panel is used to accept the settings and generate the orthographic views. You can choose the **Cancel** button to cancel the view generation.

After specifying the required settings in the **Orthographic View Selection** window, choose the **OK** button from the **Create** panel; the viewport will be attached to the cursor and you would need to specify its insertion point in the drawing sheet. At this stage, you can specify the scale value of the viewport by choosing the **Scale** option from the Command prompt. Click to specify the insertion point of the viewport; the **Ortho Generation** message box indicating the Ortho generation progress will be displayed, refer to Figure 9-12. Once the view generation is completed, the **Ortho Generation** message box will be closed and the orthographic view will be placed in the drawing sheet.

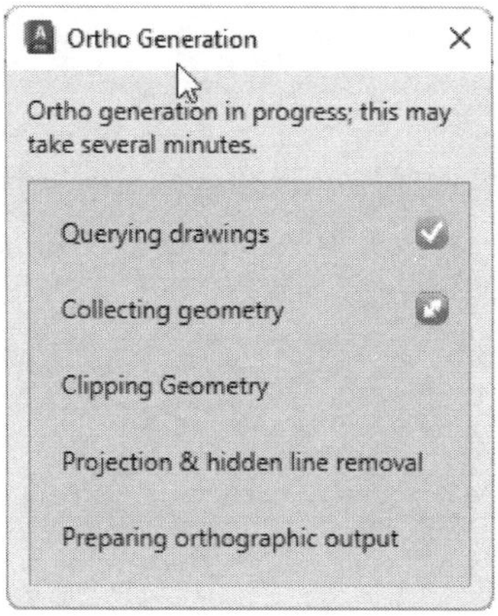

Figure 9-12 The ***Ortho Generation*** *message box*

Creating the Adjacent View

To create an adjacent view, first you need to open the Orthographic drawing by using the **PROJECT MANAGER**. To do so, choose the **Orthographic DWG** tab and then expand the tree view. Next, double-click on the required orthographic drawing to open it. In the orthographic drawing, choose the **Adjacent View** tool from the **Ortho Views** panel of the **Ortho View** tab; you will be prompted to select an orthographic view to create an adjacent view. Select the view frame of an existing orthographic view from the drawing sheet; the **Create an Adjacent View** dialog box will be displayed, as shown in Figure 9-13.

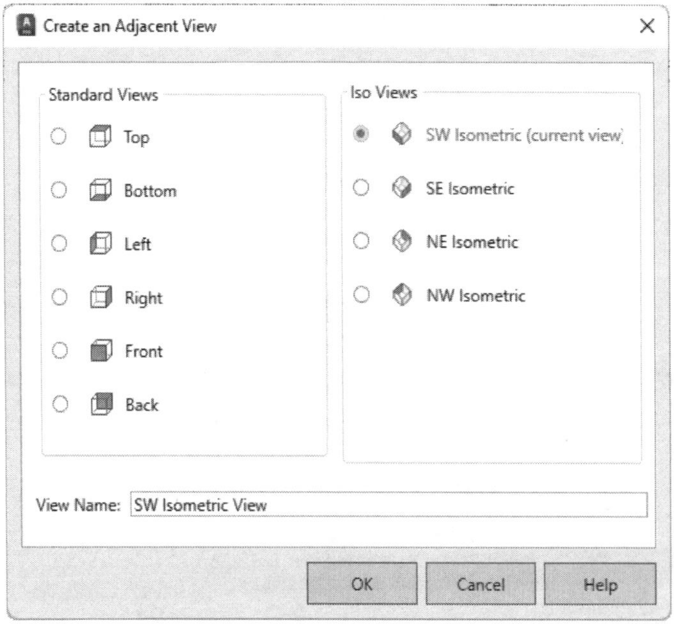

Figure 9-13 The ***Create an Adjacent View*** *dialog box*

Next, select the required orthographic view from this dialog box. Choose the **OK** button; the selected view will be attached to the cursor and you will be prompted to specify its insertion point. At this stage, you can scale or rotate the view by choosing the **Scale** or **Rotate** option from the command prompt. Next, place the view depending upon its orientation with the existing view.

Adding Annotations to the Drawing

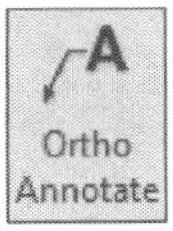

To add annotations to the drawing, choose the **Ortho Annotate** tool from the **Annotation** panel of the **Ortho View** tab; you will be prompted to select the component to annotate. Select the component to be annotated from any of the orthographic views in the drawing sheet; you will be prompted to specify the annotation style. Press ENTER if you want to select the annotation displayed by default. Otherwise, you can specify any annotation style at the command prompt. If you want to know the available annotation styles, enter **?** at the command prompt and then press ENTER; a list of styles will be displayed at the command prompt. After specifying the annotation style, press ENTER; you will be prompted to specify the position of the annotation. Position the annotation anywhere near the component.

You can also add annotation to the drawing by right-clicking in the drawing area and choosing the required annotation from the **Ortho Annotate** sub menu, refer to Figure 9-14. The **Ortho Annotate** sub menu contains the options to annotate pipes, steel structures, equipment, instruments, and so on. You can select the required annotation from this sub menu and select the component to be annotated from the orthographic view. For example, to annotate the coordinate position of a pipe, choose the **Piping Coordinate [Position]** option from the **Ortho Annotate** sub menu and select the pipe to be annotated from the orthographic view.

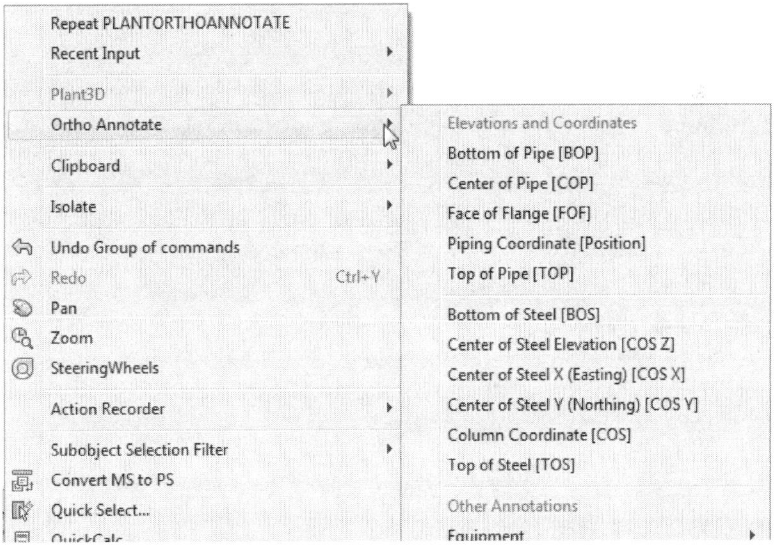

Figure 9-14 *Partial view of the* ***Ortho Annotate*** *sub menu*

Note

You can update an annotation as per the changes made in the 3D model. To do so, choose the ***Update Annotation*** *tool from the* ***Annotation*** *panel of the* ***Ortho View*** *tab; the annotation gets updated.*

Adding Dimensions to the Drawing

You can add dimensions to a drawing. For example, to add a linear dimension to the drawing, choose the **Linear** tool from the **Dimension** drop-down of the **Annotation** panel of the **Ortho View** tab; you will be prompted to specify the first extension line origin. Specify the origin point of the dimension; you will be prompted to specify the second extension line origin. After specifying the second extension line origin, the dimension will be attached to the cursor and you will be prompted to specify the location of the dimension line. Move the cursor and click to specify the location of the dimension.

Locating a Component in the 3D Model

To locate a component in the 3D model, choose the **Locate in 3D model** tool from the **Plant Object Tools** panel of the **Ortho View** tab; you will be prompted to select a component to locate from the 3D model. Select the component from the orthographic view; the 3D model will be opened and it will zoom to the selected component.

Editing a Drawing View

To edit a drawing view, choose the **Edit View** tool from the **Ortho Views** panel of the **Ortho View** tab; you will be prompted to select a viewport. On selecting the required viewport from the drawing sheet, the **Ortho View Selection** window will be displayed. Use the options in the **Ortho Editor** contextual tab of this window to edit the view settings. You can also use the OrthoCube to redefine the boundaries of the view. After modifying all the settings, choose the **OK** button from the **Create** panel of the **Ortho Editor** contextual tab; the selected view will be edited.

Updating a View

While designing a pipe, a lot of changes are made in a 3D model of a project. You need to update the drawing views when changes have been made in the 3D model. To do so, choose the **Update View** tool from the **Ortho Views** panel in the **Ortho View** contextual tab, and then select the view to be updated; the view will be updated. Note that the dimensions will not get updated along with the view. You need to manually update the dimensions.

Adding Gaps to Pipes

You can add gaps to a pipe to display the pipe hidden behind it. To add a gap, choose **Pipe Gap Tool** tool from the **Plant Object Tools** panel and select the pipe; a gap will be added to the pipe. Next, choose the **Update View** tool and update the view; the hidden pipe will become visible.

Generating Bill of Material

You can generate the bill of materials for a drawing. To generate a bill of material for the piping components, choose the **Bill of Materials** tool from the **Table Placement & Setup** panel of the **Ortho View** tab; you will be prompted to select the viewport. Select the required viewport; you will prompted to specify the first corner point of the table. Click in the drawing area to specify the first corner point; you will be prompted to specify the second corner point. Click in the drawing area to specify the second corner point; the bill of material table will be generated in the drawing area. Note that if you make any changes in the 3D model then to update those changes in bill of material table, you need to choose the **Update Bill of Materials** tool from the **Table Placement & Setup** panel of the **Ortho View** tab. You can also define the settings for the bill of material table by using the **Table Setup** tool available in the **Table Placement & Setup** panel of the **Ortho View** tab.

Displaying Pipes as Single Lines

In AutoCAD Plant 3D 2023, you can display the piping in a line representation in orthographic drawings according to different criteria. To work with ortho single-line piping, choose the **Single Line Tool** from the **Single Line Piping** panel of the **Ortho View** tab; you will be prompted to select an orthographic view to manage single line piping display. Next, select an orthographic view; you will be prompted to select the piping to dispaly as single line. Also, different options will be displayed in the command line which are discussed next.

Allpiping

This option allows you to display all the pipes contained by the orthographic view in a single line. After selecting the orthographic view, choose the **Allpiping** option and then choose the **Update** option.

byLinenumber

This option allows you to display the pipelines in a single line representation based on the desired line number. After selecting the orthographic view, choose the **byLinenumber** option and then select the lines that you want to display as single lines. Next, select the **Update** option.

bySize

This option allows you to display the pipelines in a single line representation based on the size of the pipelines. After selecting the orthographic view, choose the **bySize** option; you will be prompted to specify the largest pipe size. Enter the largest pipe size; you will be prompted to select the smallest pipe size. Enter the smallest pipe size. Next, select the Update option.

byproperty

This option allows you to display the pipelines in a single line representation based on the specified property of the pipe.

byselection

This option allows you to display the pipelines in a single line representation based on the selection of particular objects in the orthographic view.

Double line piping

This option allows you to convert a single line piping representation into double line piping representation.

Creating BOM Annotation

After generating bill of materials, you can annotate them in the drawing views by using the **BOM Annotation** tool. To do so, choose the **BOM Annotation** tool from the **Annotation** panel of the **Ortho View** tab; you will be prompted to select the BOM data row or entire table. In the BOM table, select the row whose annotation tag you want to show; you will be prompted to specify location where you need to place the annotation. Click in the drawing area to specify the location for annotation. You can also annotate whole BOM table by choosing the **Table** option from the command prompt. You can also change the viewport by using the **changeView** option from the command prompt.

TUTORIAL

Tutorial 1

In this tutorial, you will open the *Piping Model.dwg* file of the CADCIM project located in the **PROJECT MANAGER** and create orthographic views of the 3D model. You can also download this model from *www.cadcim.com* by following path: *Textbooks > CAD/CAM > AutoCAD Plant 3D > AutoCAD Plant 3D 2023 for Designers > Input Files*. You will then add dimensions and annotations to the views created. Also, you will modify the 3D model and update the orthographic views.

(Expected time: 30 min)

The following steps are required to complete this tutorial:

a. Open the *Piping Model.dwg* file from the **PROJECT MANAGER**.
b. Create orthographic drawing and configure the settings.
c. Create the adjacent views.
d. Add dimensions and annotations to the views.
e. Save the drawing file.
f. Locate a pipe in the 3D model and modify it.
g. Validate the views and then update them.
h. Manually update the dimension.

Opening the File

1. Double-click on the **AutoCAD Plant 3D 2023 - English** icon; AutoCAD Plant 3D gets started.

2. Select **CADCIM** from the **Current Project** drop-down list in the **PROJECT MANAGER**.

3. Expand the **Plant 3D Drawings** node in the **Project** area and double-click on **Piping Model**; the selected drawing file is opened.

Creating the Orthographic Drawing and Configuring the Settings

1. Choose the **Create Ortho View** tool from the **Ortho Views** panel of the **Home** tab; the **Select Orthographic Drawing** dialog box is displayed.

2. Choose the **Create New** button from this dialog box; the **New DWG** dialog box is displayed.

3. In this dialog box, enter **OrthoViews** in the **File name** edit box and then choose the **OK** button; the **Ortho Editor** contextual tab is displayed along with the OrthoCube in the drawing area.

4. Select the OrthoCube to display the grips on it. Next, use the grips and modify the OrthoCube, as shown in Figure 9-15. Make sure that all the components are covered under the OrthoCube.

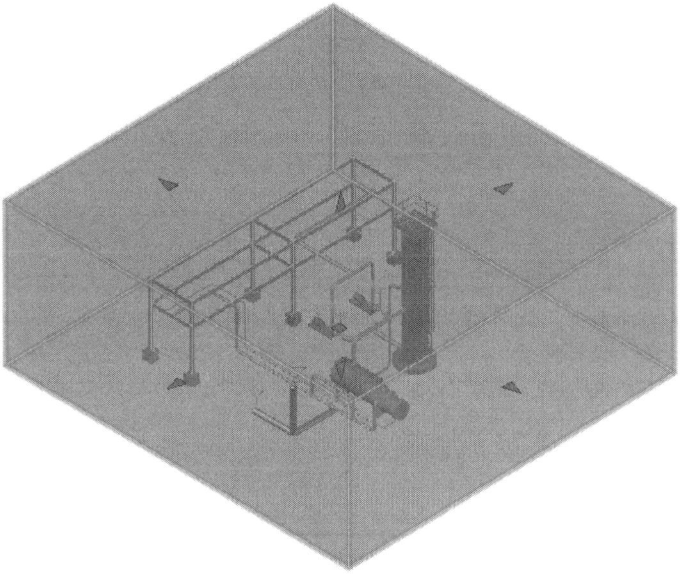

Figure 9-15 *The modified OrthoCube*

Next, you need to specify the orientation of the first view.

5. Select the **Top** option from the **View** drop-down list in the **Ortho Cube** panel if not selected.

6. Select **1:100** from the **Scale** drop-down list in the **Output Size** panel.

   Next, you need to save the configured settings.

7. Choose the **Save Ortho Cube** tool from the **Library** panel; the **Save View** dialog box is displayed.

8. In this dialog box, enter **Top View** in the **View Name** edit box and choose the **OK** button; the view settings are saved. Now, these settings can be used further.

9. Choose the **OK** button from the **Create** panel of the **Ortho Editor** contextual tab; the view is attached to the cursor and you are prompted to specify the insertion point of the view.

10. Place the view at the top right of the layout. Figure 9-16 shows the top view of the model.

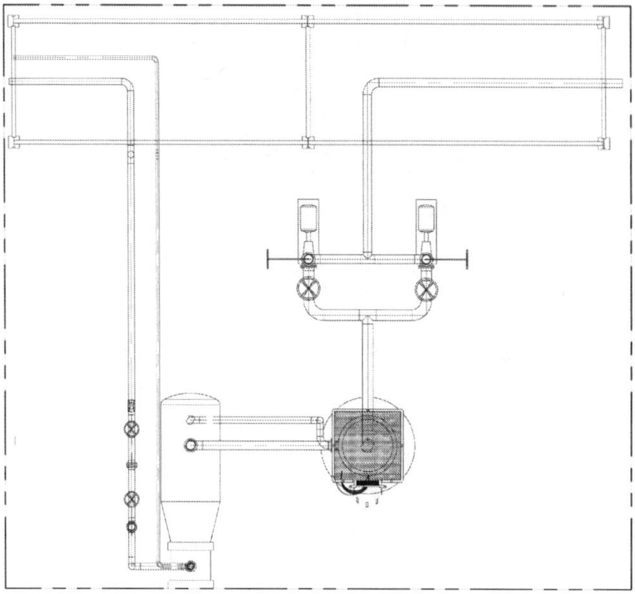

Figure 9-16 *Top view of the 3D model*

Creating the Adjacent Views

In this section, you will create the front and left views.

1. Choose the **Adjacent View** tool from the **Ortho Views** panel; you are prompted to select the orthographic view to create an adjacent view.

2. Select the view frame of the Top view; the **Create an Adjacent View** dialog box is displayed.

3. In this dialog box, choose the **Front** radio button and then choose the **OK** button; the front view is attached to the cursor and you are prompted to specify insertion point for the view.

4. Place the view below the top view, as shown in Figure 9-17. Note that you need to use object tracking to position the front view precisely with respect to the top view.

5. Next, invoke the **Adjacent View** tool and create the left view of the model, refer to Figure 9-17. Note that you need to select the front view to create this view.

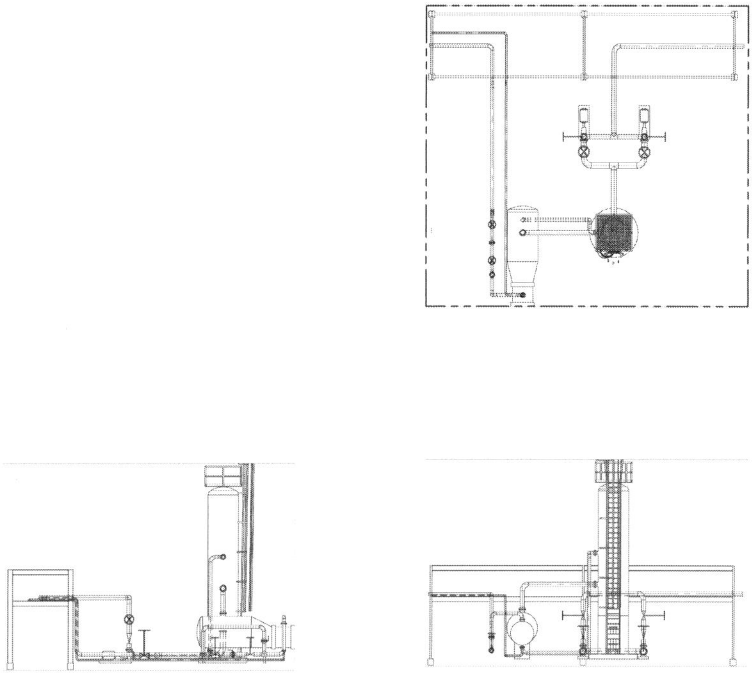

Figure 9-17 *Orthographic views of the 3D model*

Adding Annotations and Dimensions

In this section, you will add annotations and dimensions to the view.

1. Choose the **Ortho Annotate** tool from the **Annotation** panel; you are prompted to select a component to annotate.
2. Zoom in to the Top view and then select the reboiler from the view; you are prompted to specify the annotation style.
3. Enter the **Equipment Annotation [Tag]** in the Command prompt and press ENTER; you are prompted to specify the location for placing the annotation.
4. Place the annotation below the reboiler, as shown in Figure 9-18.
5. Select the vertical thin pipe line connecting the reboiler and enter **Full Line Number Callout** in the command prompt; the annotation is attached to the cursor. Place the annotation near the pipe line, as shown in Figure 9-19.

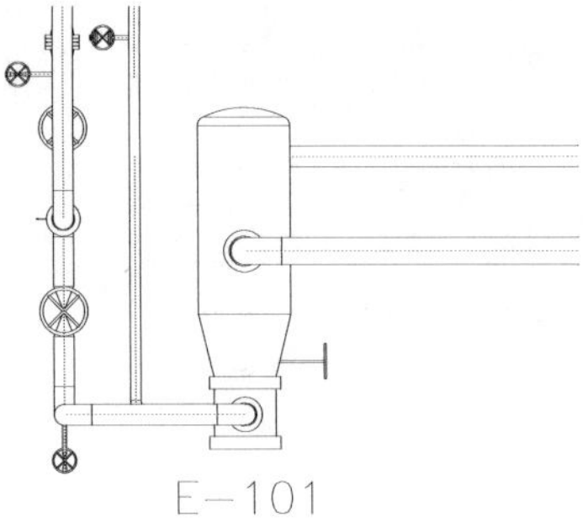

Figure 9-18 The annotation placed below the reboiler

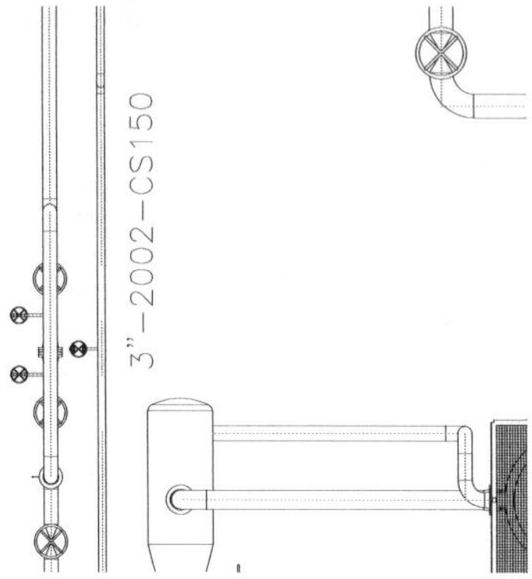

Figure 9-19 The pipe being annotated

Next, you need to create dimensions.

6. Choose the **Linear** tool from the **Dimension** drop-down of the **Dimensions** panel; you are prompted to specify the first extension line origin.
7. Zoom in on the reboiler in the Front view and select the start point of the pipe connecting the bottom nozzle of the reboiler, as shown in Figure 9-20.
8. Next, specify the second extension line origin, refer to Figure 9-20; the dimension line is attached to the cursor and you are prompted to specify its location.
9. Place the dimension line, as shown in Figure 9-21. Note that this dimension will be used later in the tutorial to explain the **Update Views** option.

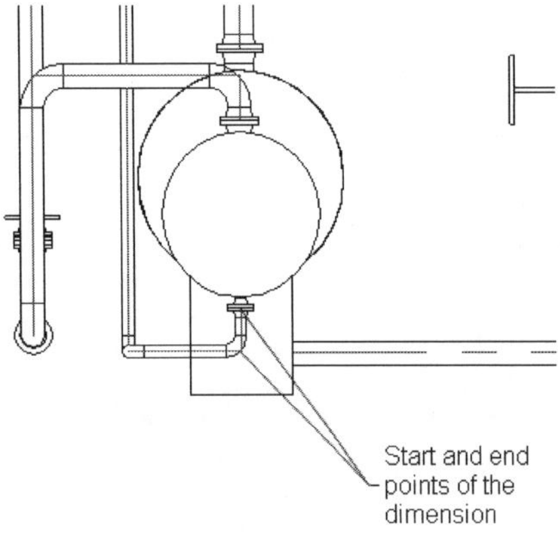

Figure 9-20 The origin points of the dimension extension lines

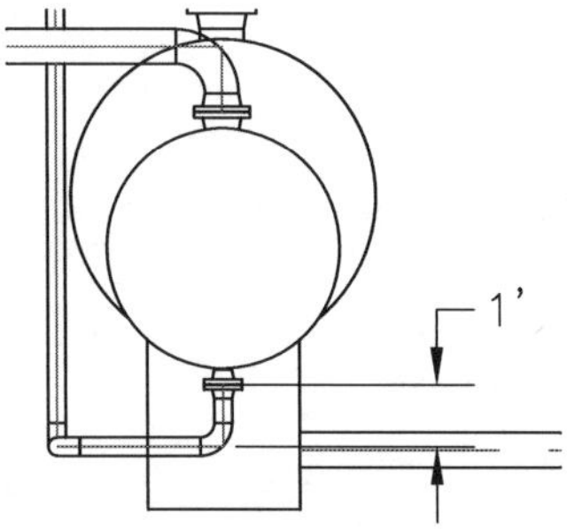

Figure 9-21 Location of the dimension

10. Choose the **Save** button from the **Quick Access Toolbar**; the drawing file is saved.

Updating the 3D Model

In this section, you will update the 3D model.

1. Choose the **Locate in 3D Model** tool from the **Plant Object Tools** panel; you are prompted to select a component to locate in the 3D model.

2. Select the horizontal pipe, as shown in Figure 9-22; it is located in the 3D model.

3. Select the highlighted pipe in the 3D model; the **Move Part** grip is displayed, as shown in Figure 9-23.

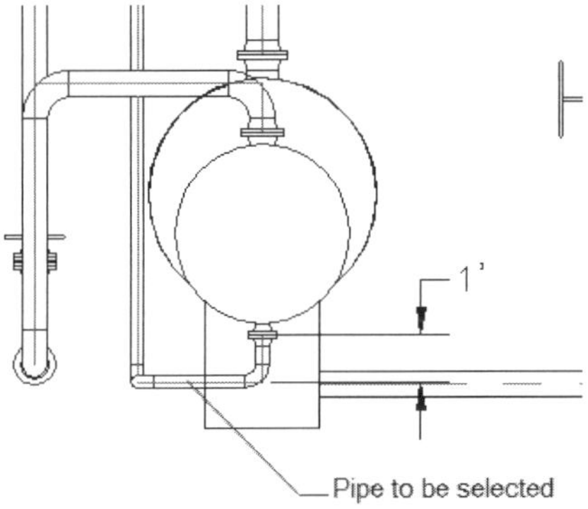

Figure 9-22 The pipe to be selected in the 3D model

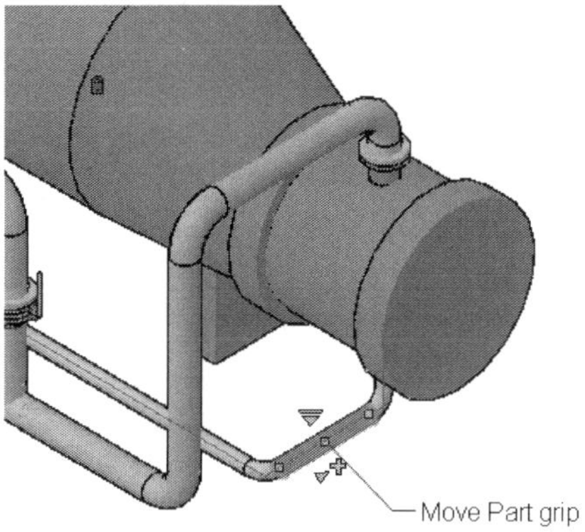

***Figure 9-23** The **Move Part** grip displayed on the pipe*

4. Select the **Move Part** grip and move the cursor downward. Next, enter **6"** at the Command prompt; the pipe is modified, as shown in Figure 9-24.

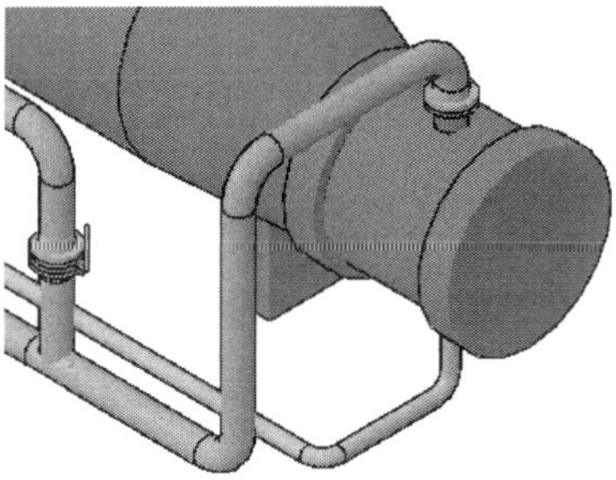

Figure 9-24 The pipe after modification

5. Choose the **Save** button in the **Quick Access Toolbar**.

Validating the Views and then Updating them

1. Right-click on **OrthoViews** under the **Orthographic Drawings** node in the **Orthographic DWG** tab of the **PROJECT MANAGER** and choose the **Validate Views** option, refer to

Figure 9-25; all the views turn red and a message is displayed showing that the referenced models have changed.

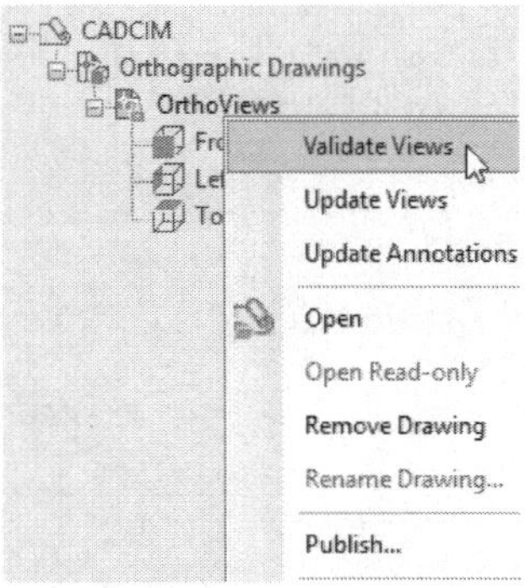

Figure 9-25 *Choosing the **Validate Views** option*

2. Right-click on **OrthoViews** in the **PROJECT MANAGER** and choose the **Update Views** option; the views are updated.

   You will notice that while updating the views, the dimension was not updated, refer to Figure 9-26. Therefore, you need to manually update the dimension.

3. Select the origin point of the second extension line of the dimension and snap to the endpoint of the pipe, as shown in Figure 9-27; the dimension value is modified.

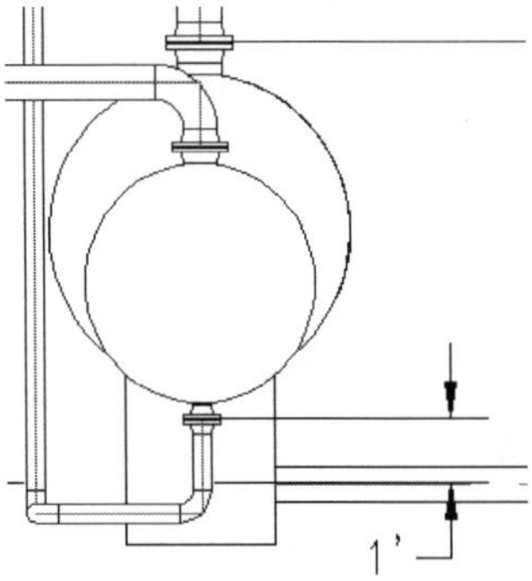

Figure 9-26 *The dimension not updated in the drawing view*

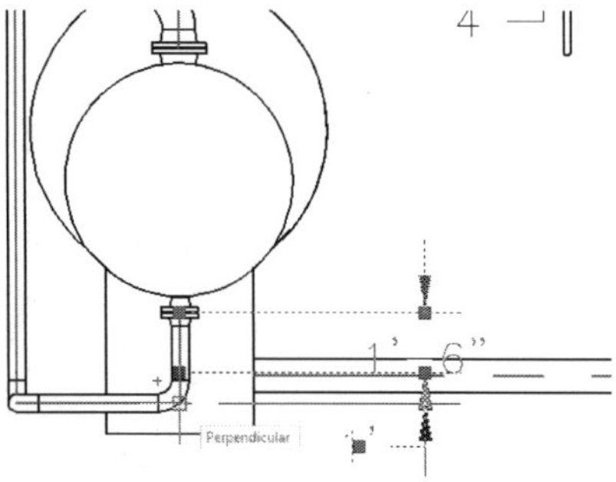

Figure 9-27 *Modifying the dimension*

4. Choose **Close > All Drawings** from the **Application** menu; the **AutoCAD** message box is displayed.

5. Choose the **NO** button in this message box to omit the changes made to the files.

Self-Evaluation Test

Answer the following questions and then compare them to those given at the end of this chapter:

1. In the _________ window, you need to use the _________ to specify the boundaries of the view.
2. You can save the orthographic view settings by choosing the _________ button.
3. The tools in the _________ tab are used to perform operations such as editing, updating, and annotating the drawing view.
4. You can annotate an equipment or a part by using the _________ button.
5. You can use the _________ tool to add gaps to a pipe.

Review Questions

Answer the following questions:

1. You can use the _________ tool to locate a component in a 3D model.
2. On choosing the _________ button, dimensions do not get updated along with the view.
3. You can update the bill of material after making changes in 3D model by choosing the _________ tool.
4. You can also include all the 3D model files present in the current project, while creating the orthographic view. (T/F)
5. You can also scale the drawing view at the time of its placement. (T/F)

Answers to Self-Evaluation Test

1. Orthographic View Selection, OrthoCube, **2. Save Ortho Cube**, **3. Ortho View**, **4. Ortho Annotate**, **5. Pipe Gap**

Chapter 10

Managing Data and Creating Reports

Learning Objectives

After completing this chapter, you will be able to:

- *Understand the usage of the Data Manager*
- *View project and drawing data*
- *Create reports*
- *Export project data*
- *Import project data*

INTRODUCTION

In this chapter, you will learn to use the **DATA MANAGER** to view the information in P&IDs and Plant 3D drawings and to export and import the data. In addition to that, you will be able to filter data in the **DATA MANAGER** to view a particular information.

Also, you will learn to generate, export, and import reports using the **DATA MANAGER**.

DATA MANAGER

The **DATA MANAGER** is a location where the entire project data remains available. You can view, modify, export, and import this data using the **DATA MANAGER**. In addition, you can arrange the information within the **DATA MANAGER** into different forms and then generate reports. To display the **DATA MANAGER**, choose the **Data Manager** tool from the **Project** panel of the **Home** tab. Figure 10-1 shows a typical **DATA MANAGER** and its components.

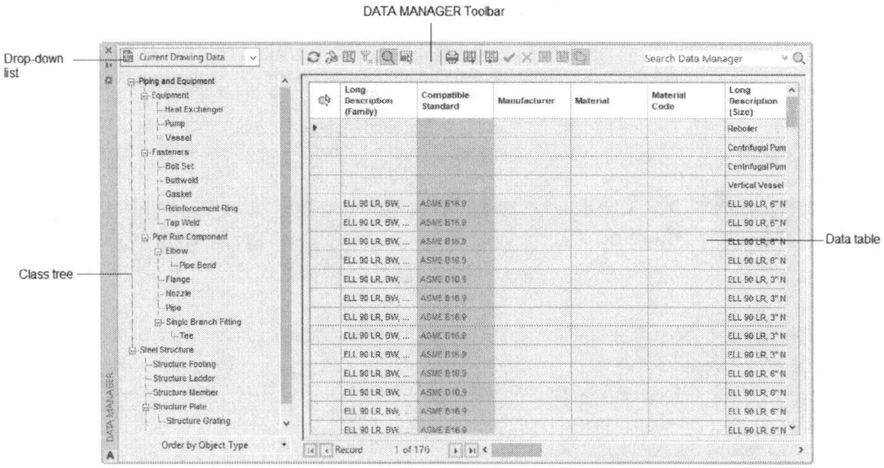

***Figure 10-1** The DATA MANAGER*

The components of the **DATA MANAGER** are discussed next.

Drop-down List

It is used to display different data types such as **Current Drawing Data**, **Plant 3D Project Data** or **P&ID Project Data**, and **Project Reports**.

Class Tree

The Class tree displays the data related to a particular component of the drawing.

DATA MANAGER Toolbar

Using the DATA MANAGER toolbar, you can perform various tasks such as refreshing, exporting, importing, data printing, and so on.

Data Table

It is the place where the project data is arranged in rows and columns.

VIEWING DATA IN THE DATA MANAGER

In the **DATA MANAGER**, you can view the project data related to Plant 3D and P&ID drawings. You can also view the project reports. To view Plant 3D data, first you need to open a Plant 3D model. Next, invoke the **DATA MANAGER** and select the **Plant 3D Project Data** option from the drop-down list available at the top left corner; the Plant 3D project data will be displayed, as shown in Figure 10-2. Next, select the required class from the Class tree present on the left of the **DATA MANAGER**; the data related to the selected component will be displayed in the data table.

To view the P&ID data, first you need to open a P&ID file. Next, invoke the **DATA MANAGER** and select the **P&ID Project Data** option from the drop-down list in the **DATA MANAGER**; the P&ID project Data will be displayed in the **DATA MANAGER**, as shown in Figure 10-3.

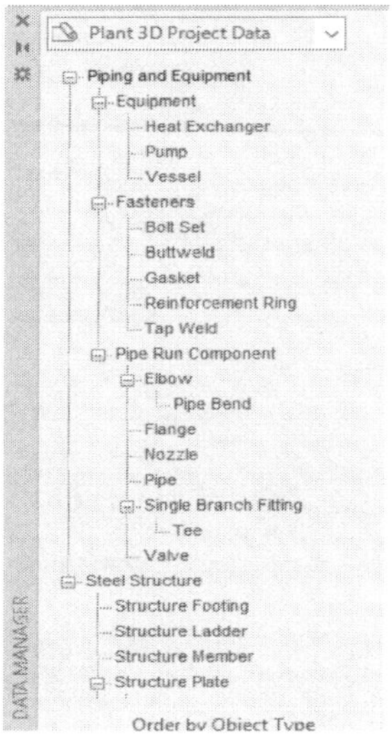

Figure 10-2 *The Plant 3D project data*

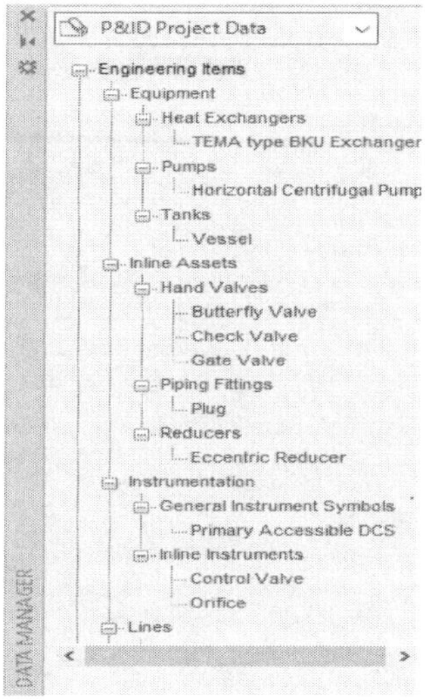

Figure 10-3 *The P&ID project data*

To view the data of the current drawing file, select the **Current Drawing Data** option from the drop-down list available in the **DATA MANAGER**. Next, select the required class from the Class tree; the data related to the current drawing file is displayed.

MODIFYING THE DISPLAY OF DATA

You can modify the display of data by using the options available in the **DATA MANAGER**. These options are discussed next.

Displaying the Data by Object Type and Area

You can display the data based on object type or area. To do so, select the **Order by Object Type** or **Order by Area** option from the drop-down list located at the bottom of the Class tree, refer to Figure 10-4; the data displayed will be modified based on the option selected. Figure 10-5 and 10-6 show the data based on the object type and area, respectively.

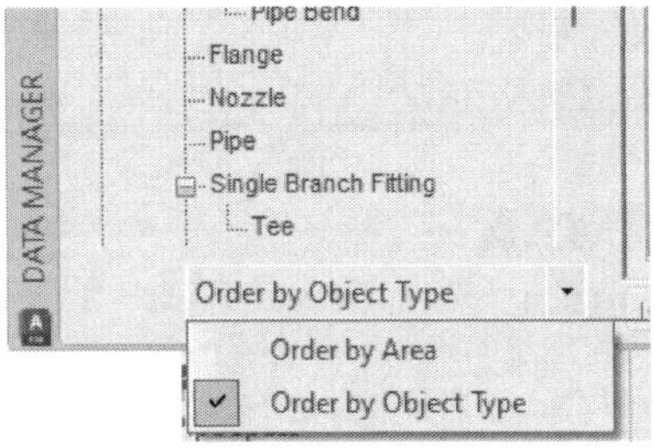

***Figure 10-4** The drop-down list located at the bottom of the Class tree*

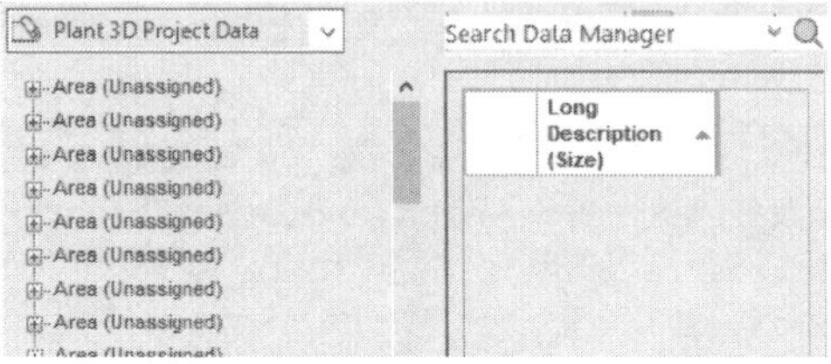

***Figure 10-5** The project data displayed based on area*

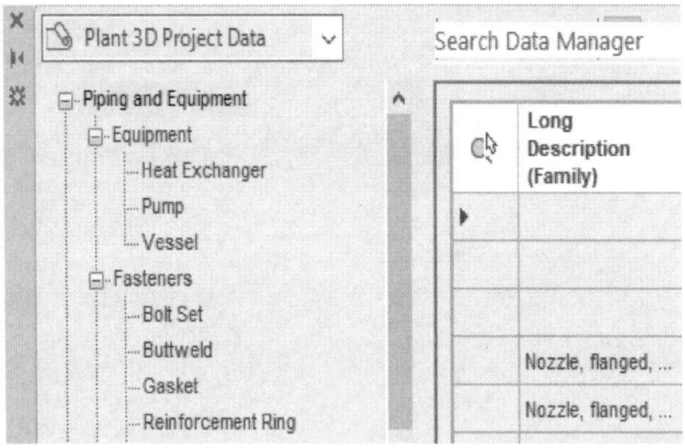

***Figure 10-6** The project data displayed based on the object type*

Note

*The **Order by Object Type** or **Order by Area** options in **P&ID Project Data** will be available by right-clicking in the blank area available at the bottom of the Class tree.*

Tip

*You can modify the display of nodes in the Class tree. To do so, right-click in the Class tree and choose the **Show All Nodes** or **Show Only Nodes with Content** option; the Class tree will be modified depending upon the chosen option.*

ZOOMING TO PLANT OBJECTS

You can zoom to an object in the drawing area by using the **DATA MANAGER**. Also, you can locate the object in the **DATA MANAGER** by selecting it from the drawing area. Various methods to zoom and scroll through objects in the **DATA MANAGER** and to locate them in the drawing area are discussed next.

Locating an Object in the Drawing Area

You can locate an object in the drawing area using the **DATA MANAGER**. To do so, select the **Current Drawing Data** option from the drop-down list located at the top left corner of the **DATA MANAGER**. Next, select the required component from the tree; the related data will be displayed in the data table. In the data table, click on the header row of the object, refer to Figure 10-7; the object will be located in the drawing area.

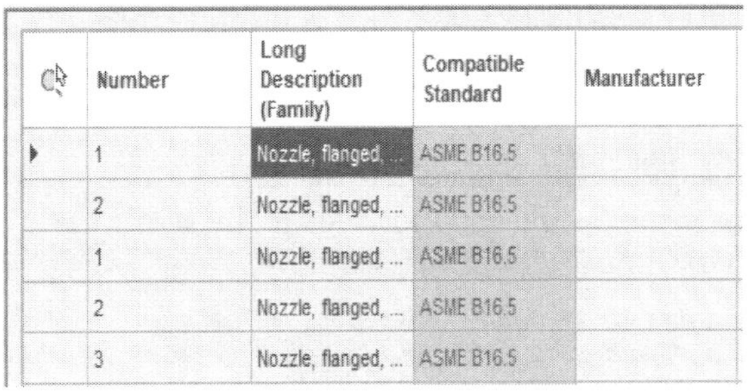

Figure 10-7 *Clicking on the header row*

Scrolling through the Data in the DATA MANAGER

You can scroll through the objects in the **DATA MANAGER** and locate them in the drawing area using the arrow buttons available at the bottom of the data table, refer to Figure 10-8. On choosing the right arrow button, the next record will be highlighted in the data table. Also, the object will be located in the drawing area. On choosing the left arrow button, the previous record will be highlighted in the data table.

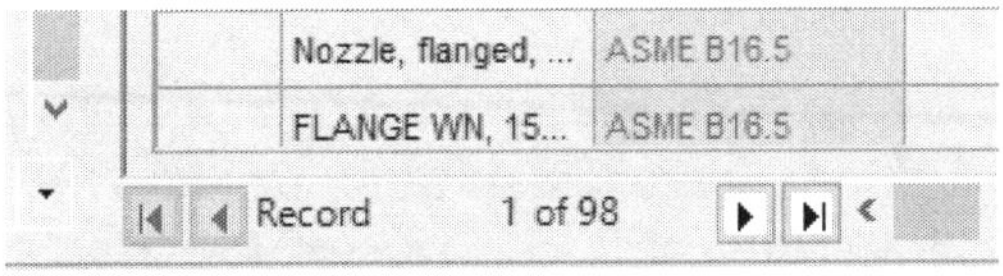

Figure 10-8 *Arrow buttons located at the bottom of the data table*

> **Tip**
> *You can highlight the first or last object of a data table by using the arrow buttons with a vertical line.*

EDITING DATA IN THE DATA MANAGER

You can edit the data in the **DATA MANAGER** manually. To do so, open the **DATA MANAGER** and double-click on the cell whose data needs to be modified; a drop-down list will be displayed or the corresponding cell will turn into an edit box. You can now specify the new value by using the drop-down list or by entering it manually in the edit box. You can also specify the value in the edit box by copying it from another cell. However, note that you cannot change the value of cells displayed with shaded background.

PLACING ANNOTATIONS IN A P&ID USING THE DATA MANAGER

You can place annotations in a P&ID using the **DATA MANAGER**. To do so, open the P&ID and then invoke the **DATA MANAGER**. In the **DATA MANAGER**, select the **Current Drawing Data** option from the drop-down list located at the top of the Class tree; the data related to the current drawing file will be displayed. Next, select the required class from the Class tree; the data related to the selected class will be displayed in the data table. Click on the required cell in the Data table and drag the cursor into the drawing area; the object corresponding to the selected cell will be highlighted in the P&ID and you will be prompted to specify the location of the annotation. Specify the location of the annotation near the highlighted object.

FILTERING THE INFORMATION IN THE DATA TABLE

You can filter the data in the data table so that you can view only the required data and export it. The various methods to filter the information in the data table are discussed next.

Viewing Only the Selected Record in the Data Table

To view only the selected item in the data table, right-click on it and choose the **Filter By Selection** option from the shortcut menu displayed; all the records except the selected one will be hidden.

Viewing All the Records in the Data Table Except the Selected one

To view all the items in the data table except the selected one, right-click on it and choose the **Filter Excluding Selection** option from the shortcut menu displayed; all the records except the selected one will be displayed.

Viewing the Data of the Objects Selected in the Drawing Area

You can view the data of the objects selected in the drawing area. To do so, select the object from the drawing area, as shown in Figure 10-9. Next choose the **Show Selected Items** button from the **DATA MANAGER** toolbar, as shown in Figure 10-10; the object will be located in the data table, refer to Figure 10-11.

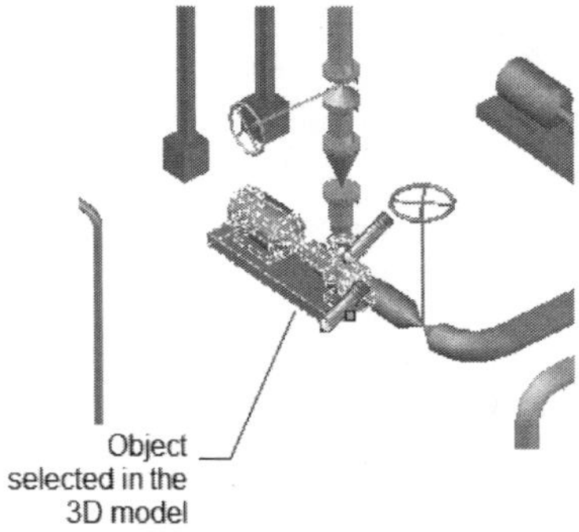

Figure 10-9 Object selected in the 3D model

*Figure 10-10 Choosing the **Show Selected Items** button*

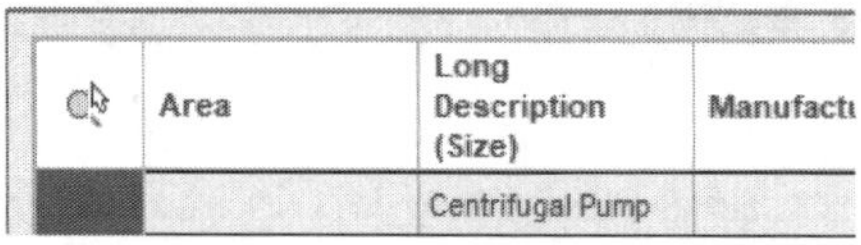

	Area	Long Description (Size)	Manufactu
		Centrifugal Pump	

*Figure 10-11 Object located in the **DATA MANAGER***

To remove all the filters, choose the **Remove Filter** button from the **DATA MANAGER** toolbar; all the filters will be removed from the Data table.

EXPORTING DATA FROM THE DATA MANAGER

To export the data from the **DATA MANAGER**, follow the steps given next.

1. Select a node from the Class tree and choose the **Export** button from the **DATA MANAGER** toolbar; the **Export Data** dialog box will be displayed, as shown in Figure 10-12.

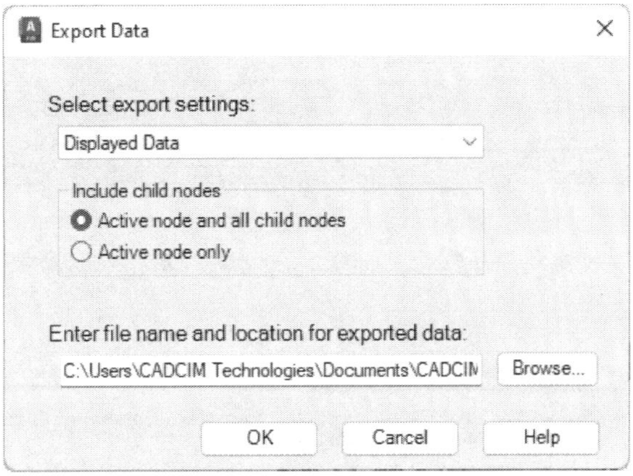

Figure 10-12 *The* ***Export Data*** *dialog box*

2. In this dialog box, select an option from the **Select export settings** drop-down list. Next, select the required radio button from the **Include child nodes** area. If you select the **Active node and all child nodes** radio button, the node selected from the Class tree and all the sub-nodes under it will be included while exporting the data. For example, if you select the **Equipment** node from the Class tree; all the sub-nodes under it will be included and separate worksheets will be created from each node. If you select the **Active node only** radio button, only the selected node will be included while exporting the data.

3. Choose the **Browse** button from the dialog box; the **Export To** dialog box will be displayed.

4. In this dialog box, specify the location of the exported file, enter its name and select the file type from the **Files of type** drop-down list.

5. Choose the **Save** button and then the **OK** button; the data will be exported in the file format selected in the **Files of type** drop-down list.

IMPORTING DATA TO THE DATA MANAGER

To import data to the **DATA MANAGER**, follow the steps given next.

1. Select the node from the Class tree to which the data is to be imported and choose the **Import** button from the **DATA MANAGER** toolbar; the **AutoCAD Plant** message box will be displayed showing that **A log file for the accept and reject will be created in the same**

folder as the import file. Choose the **OK** button from the message box; the **Import From** dialog box will be displayed.

2. In this dialog box, select the required option from the **Files of type** drop-down list and browse to the location of the file from which you want to import the data.

3. Select the file and choose the **Open** button; the **Import Data** dialog box will be displayed, as shown in Figure 10-13.

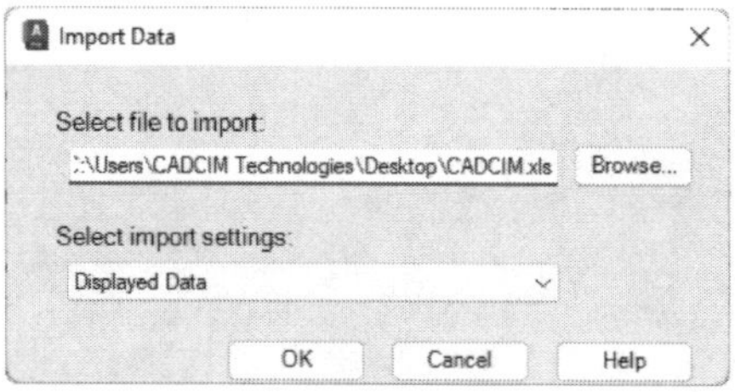

*Figure 10-13 The **Import Data** dialog box*

4. Choose the **OK** button; the data will be imported to the data table.

Accepting or Rejecting Changes in the Imported Data

The imported data may undergo some modifications. After importing the data, you need to accept or reject the changes. To do so, first select the node from the Class tree to which the data is to be imported. You will notice that the changes made in the data are highlighted with yellow background. Choose the **Accept** button to accept the changes one by one or choose the **Accept All** button to accept all the changes at a time from the **DATA MANAGER** toolbar. Similarly, you can reject the changes by choosing the **Reject** or **Reject All** button.

VIEWING REPORTS IN THE DATA MANAGER

You can view, export and import Project Reports using the **DATA MANAGER**. To view Project Reports, click on the **Reports** button in the **PROJECT MANAGER**; a flyout will be displayed. Choose the **Reports** option from the flyout; the project report data will be displayed in the **DATA MANAGER**, as shown in Figure 10-14. In this section, you will learn how to import and export the project reports.

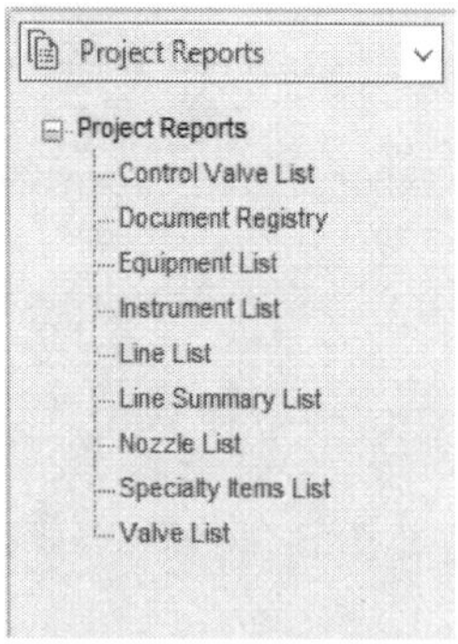

Figure 10-14 The project report data

Exporting the Project Reports

To export Project Reports, first make sure that the **Project Reports** node is selected in the **DATA MANAGER**. Next, choose the **Export** button from the **DATA MANAGER**; the **Export Report Data** dialog box will be displayed, as shown in Figure 10-15. In this dialog box, select the check boxes from the **Reports** list box. Next, choose the **Browse** button; the **Export To** dialog box will be displayed. Select the file format from the **Files of type** drop-down list and specify the location of the file. Next, choose the **Save** button and then the **Export** button; the project report will be exported.

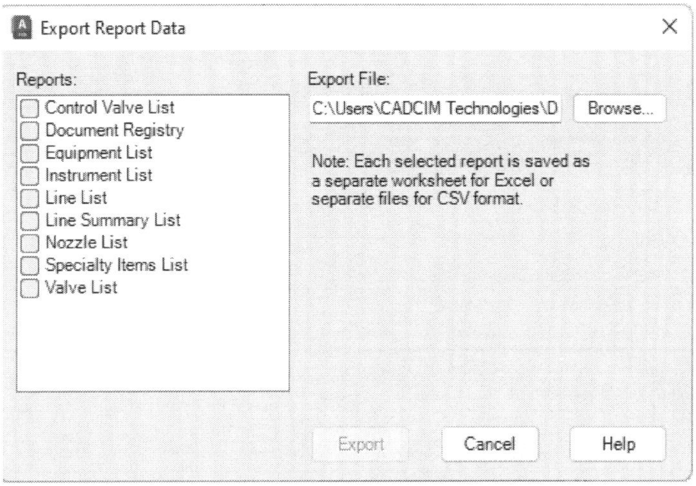

***Figure 10-15** The **Export Report Data** dialog box*

Importing the Project Reports

To import Project Reports, select the **Project Reports** node in the Class tree and choose the **Import** button; the **AutoCAD Plant** message box will be displayed alongwith the message **A log file for accept and reject will be created in the same folder as the import file**. Choose the **OK** button; the **Import From** dialog box will be displayed. Select the type of file from which you want to import the data (*.xls,.xlsx,.csv*) and browse to the file location. Select the file and choose the **Open** button; the **Project Report Selection** dialog box will be displayed, as shown in Figure 10-16. In this dialog box, select the report to import from the **Select project report to import** drop-down list and choose the **OK** button; the selected report will be imported.

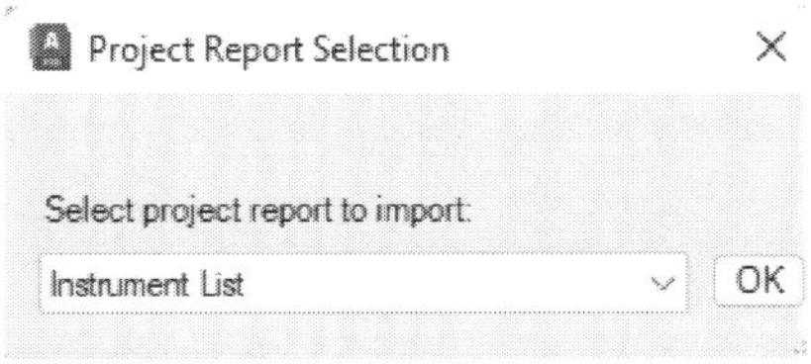

***Figure 10-16** The **Project Report Selection** dialog box*

WORKING WITH THE REPORT CREATOR

Autodesk AutoCAD Plant Report Creator is an additional application which comes along with AutoCAD Plant 3D. It is used to generate reports without starting the AutoCAD Plant 3D application. To start this application, double-click on the **AutoCAD Plant Report Creator 2023 - English** icon from your desktop screen; the **Autodesk AutoCAD Plant Report Creator** with the **Settings** message box will be displayed, as shown in Figure 10-17. Note that the **Settings** message box will be displayed automatically if you are opening the Autodesk AutoCAD Plant Report Creator application for the first time. You can also invoke the **Settings** message box by choosing the **Settings** button available next to the **Report Configuration** drop-down in the **Autodesk AutoCAD Plant Report Creator** application.

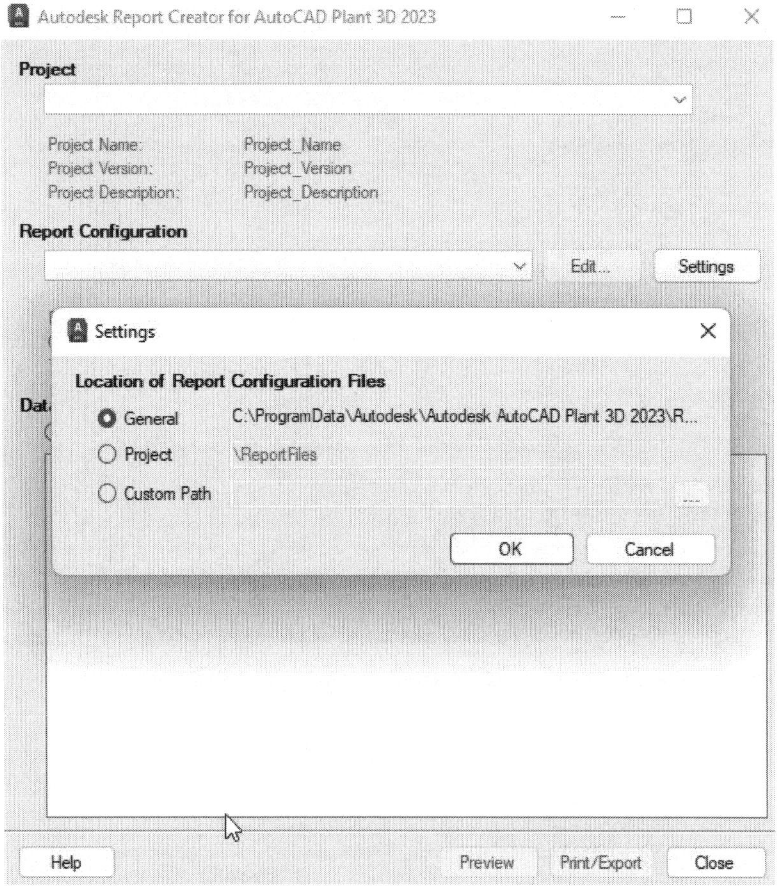

***Figure 10-17** The **Autodesk AutoCAD Plant Report Creator** with the **Settings** message box*

Next, select the **General** radio button from the **Settings** message box and choose the **OK** button. Now, you can start generating reports.

Generating Reports Using the AutoCAD Plant Report Creator

To generate reports, invoke **Autodesk AutoCAD Plant Report Creator** and select the required project available in the **Project** drop-down list. If it is not available in this drop-down list, select the **Open** option; the **Open** dialog box will be displayed. Browse and select the **Project.xml** file. Next, choose the **Open** button; the data configurations related to the selected project will be displayed in **Autodesk AutoCAD Plant Report Creator**. Select the required configuration from the **Report Configuration** drop-down list, refer to Figure 10-18. Select the **Project Data** or **Drawing Data** radio button from the **Data Source** area. If you select the **Drawing Data** radio button, you need to select a drawing or drawings from the **Data Source** list box. Next, choose the **Preview** button; the **Preview** window will be displayed, as shown in Figure 10-19. Close the **Preview** window and choose the **Print/Export** button; the **PDF Export Options** dialog box will be displayed. Choose the **OK** button; the **Export results** window will be displayed, as shown in Figure 10-20. Choose the **OK** button; the report will be exported to a PDF file and placed in the **Reports** folder of the respective project.

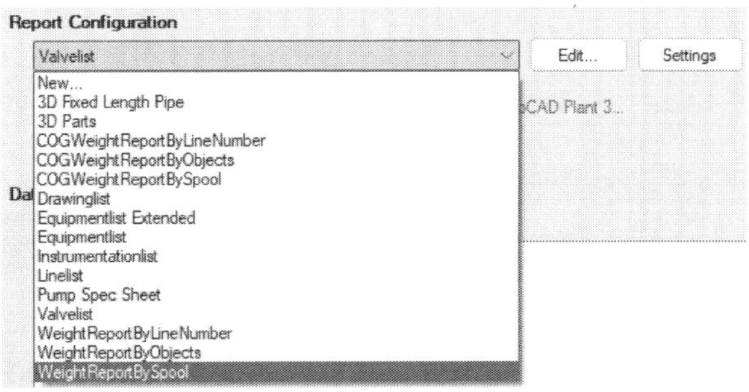

Figure 10-18 *The* ***Report Configuration*** *drop-down list*

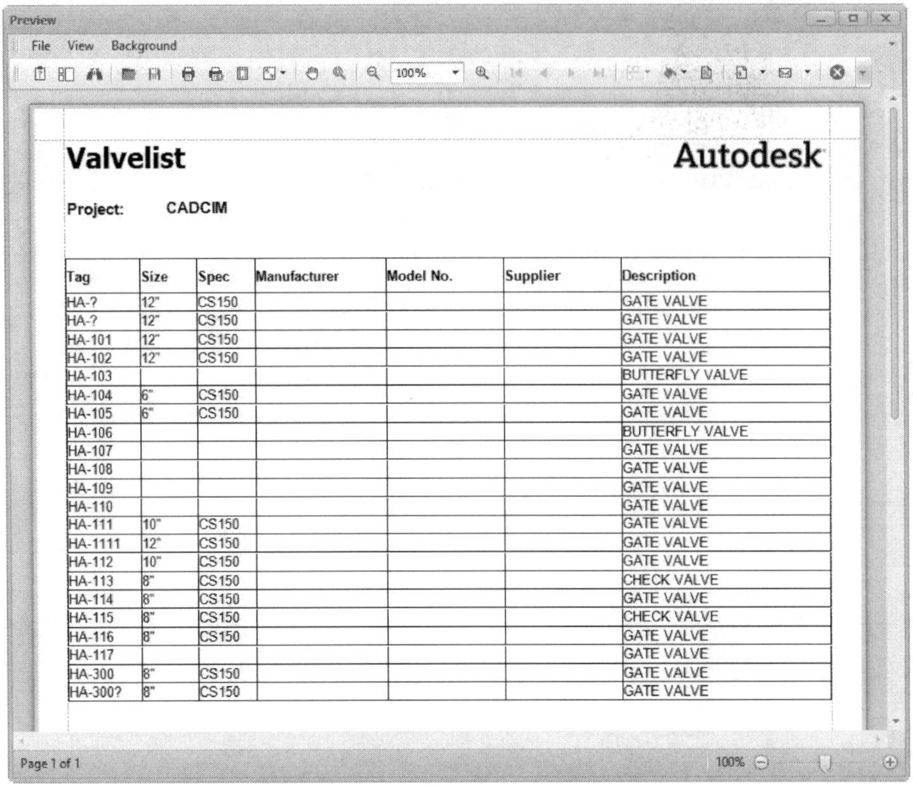

Valvelist

Autodesk

Project: CADCIM

Tag	Size	Spec	Manufacturer	Model No.	Supplier	Description
HA-?	12"	CS150				GATE VALVE
HA-?	12"	CS150				GATE VALVE
HA-101	12"	CS150				GATE VALVE
HA-102	12"	CS150				GATE VALVE
HA-103						BUTTERFLY VALVE
HA-104	6"	CS150				GATE VALVE
HA-105	6"	CS150				GATE VALVE
HA-106						BUTTERFLY VALVE
HA-107						GATE VALVE
HA-108						GATE VALVE
HA-109						GATE VALVE
HA-110						GATE VALVE
HA-111	10"	CS150				GATE VALVE
HA-1111	12"	CS150				GATE VALVE
HA-112	10"	CS150				GATE VALVE
HA-113	8"	CS150				CHECK VALVE
HA-114	8"	CS150				GATE VALVE
HA-115	8"	CS150				CHECK VALVE
HA-116	8"	CS150				GATE VALVE
HA-117						GATE VALVE
HA-300	8"	CS150				GATE VALVE
HA-300?	8"	CS150				GATE VALVE

*Figure 10-19 The **Preview** window*

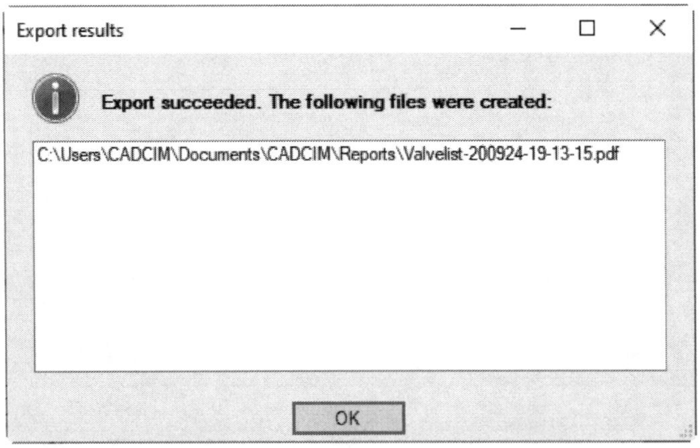

*Figure 10-20 The **Export results** window*

You can also create customized reports. To do so, select the project from the **Project** drop-down list and then select the **New** option from the **Report Configuration** drop-down list; the **New Report Configuration** dialog box will be displayed, as shown in Figure 10-21.

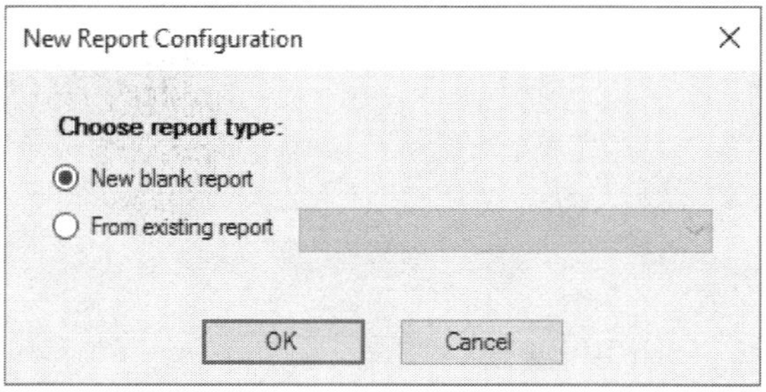

Figure 10-21 The ***New Report Configuration*** *dialog box*

Select the **New blank report** radio button from the dialog box and choose the **OK** button; the **Report Configuration** dialog box will be displayed, refer to Figure 10-22. Specify settings in this dialog box. The procedure to specify the settings in this dialog box is discussed next:

1. Click on the **Edit query** button of the dialog box; the **Query Configuration** dialog box will be displayed.

2. Select the desired radio button from the **Query type** area and then select the class from the **Available Classes** tree. Now, add the required option in the **Included Classes** area by using the **>** button. Next, choose the **OK** button from the dialog box; the **Report Configuration** dialog box will be displayed again.

3. Choose the **Edit report layout** button from the **Report Configuration** dialog box; the **Report Designer** along with the **Report Wizard** will be displayed.

4. Move the required field(s) from the **Available fields** area to the **Fields to display in a report** area from the **Report Wizard**.

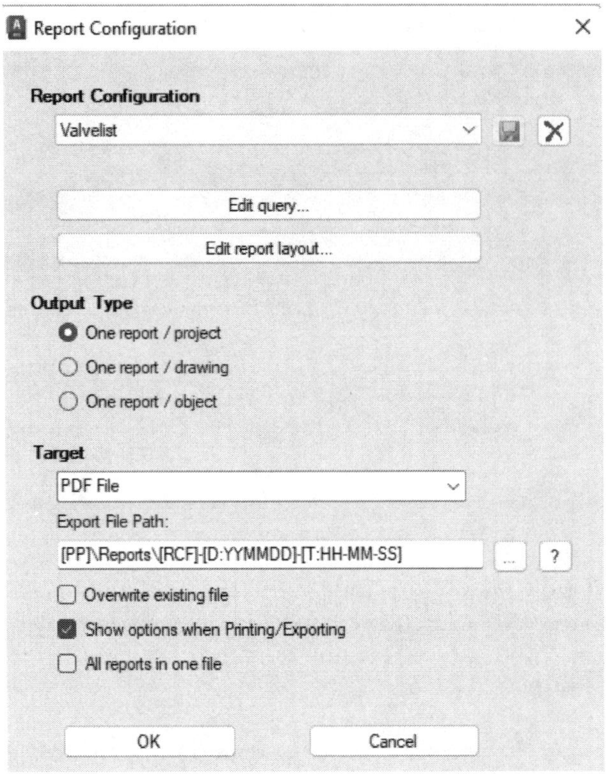

*Figure 10-22 The **Report Configuration** dialog box*

5. Choose the **Next** button and define all the options required for the report and then choose the **Finish** button.

6. Save the report by choosing the **Save** button from the **Report Designer** and close it. Choose the **OK** button from the **Report Configuration** dialog box.

7. Now, you can choose the **Preview** button to see the preview of report or the **Print/Export** button to save or print the report.

TUTORIALS

Tutorial 1

In this tutorial, you will open the *Piping Model.dwg* file from the CADCIM project in the **PROJECT MANAGER**, and filter the data, and export the data from the **DATA MANAGER**. You can also download this model from *www.cadcim.com* by following path: *Textbooks > CAD/CAM > AutoCAD Plant 3D > AutoCAD Plant 3D 2023 for Designers > Input Files*. You will then modify the data, import it and then apply the changes. **(Expected time: 20 min)**

The following steps are required to complete this tutorial:

a. Open the 3D model file.
b. Invoke the **DATA MANAGER**.
c. Filter the data and export it.
d. Open the exported file and modify the data.
e. Import the modified data into the **DATA MANAGER**.

Starting AutoCAD Plant 3D and Opening the Project

1. Start AutoCAD Plant 3D.

2. Select the **CADCIM** option from the **Current Project** drop-down list in the **PROJECT MANAGER**.

3. Open the *Piping Model.dwg* file from the **Plant 3D Drawings** node in the **PROJECT MANAGER**.

Invoking the Data Manager and Viewing the Data

1. Choose the **Data Manager** button from the **Project** panel of the **Home** tab; the **DATA MANAGER** is displayed.

2. In the **DATA MANAGER**, select the **Plant 3D Project Data** option from the drop-down list located at the top-left corner, refer to Figure 10-23; the Plant 3D project data is displayed.

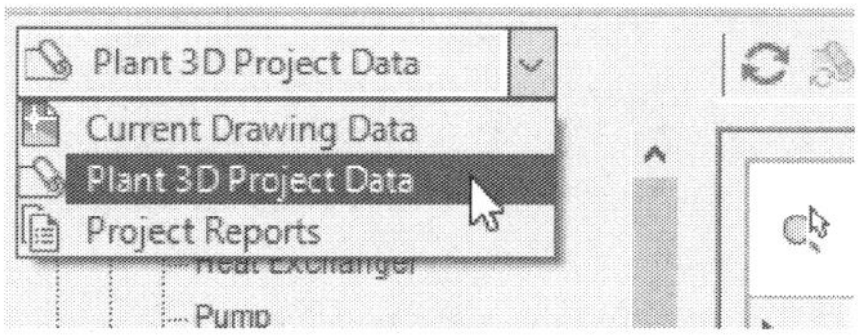

Figure 10-23 *Selecting the* ***Plant 3D Project Data*** *option*

3. Select the **Equipment** node from the Class tree; the data related to equipment is displayed in the Data table.

Filtering and Exporting the Data

1. Right-click on **Centrifugal Pump** from the **Long Description(Size)** column in the Data table and choose the **Filter By Selection** option from the shortcut menu displayed, refer to Figure 10-24; the data related to **Centrifugal Pump** is displayed in the Data table.

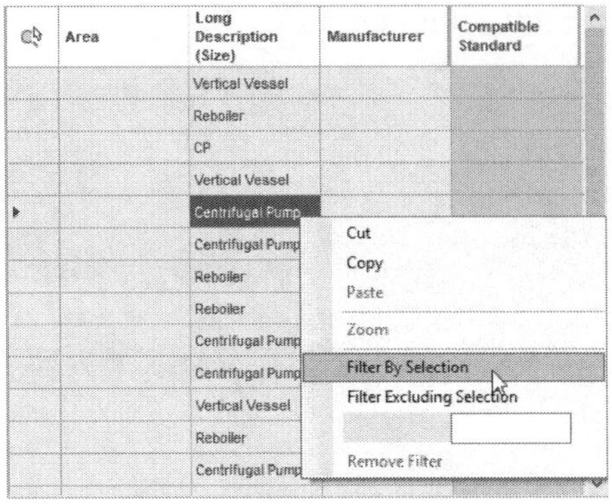

*Figure 10-24 Choosing the **Filter By Selection** option*

2. Choose the **Export** button from the **DATA MANAGER** toolbar, as shown in Figure 10-25; the **Export Data** dialog box is displayed.

*Figure 10-25 Choosing the **Export** button*

3. Select the **Active node only** radio button from the **Include child nodes** area and choose the **Browse** button; the **Export To** dialog box is displayed.

4. In this dialog box, select the **Excel Workbook (*.xlsx)** option from the **Files of type** drop-down list, refer to Figure 10-26, and specify the file location as *C:\Users\User_name\Documents\CADCIM.*

*Figure 10-26 Selecting the **Excel Workbook (*.xlsx)** option*

5. Choose the **Save** button; the **Export To** dialog box is closed.

6. Choose the **OK** button from the **Export Data** dialog box; the data is exported to the Excel file.

Modifying the Exported Data

1. Browse to the location *C:\Users\User_name\Documents\CADCIM* and open the **CADCIM-Equipment.xlsx** file; a spreadsheet is displayed.

2. In the spreadsheet, scroll to the **Spec** column and enter **CS150** in the first cell, refer to Figure 10-27.

AF	AG	AH	AI
Type	PnPID	Spec	Item Code
P	933	CS150	
P	948		

Figure 10-27 *Entering a value in the first cell of the* ***Spec*** *column*

3. Similarly, enter **CS150** in all other cells of the **Spec** column.

4. Save and close the Excel file.

Importing the Data

1. Switch to AutoCAD Plant 3D and open the **DATA MANAGER** if it is closed.

2. In the **DATA MANAGER**, make sure that **Plant 3D Project Data** is selected in the drop-down list located at the top left and Equipment data is displayed in the Data table.

3. Choose the **Import** button from the **DATA MANAGER** toolbar, refer to Figure 10-28; the **AutoCAD Plant** message box is displayed.

Figure 10-28 *Choosing the* ***Import*** *button*

4. Choose the **OK** button; the **Import From** dialog box is displayed.

5. Browse to the location *C:\Users\User_name\Documents\CADCIM* and select the **Excel Workbook (*.xlsx)** option from the **Files of type** drop-down list if it is not selected. Now, double-click on the **CADCIM-Equipment.xlsx** file; the **Import Data** dialog box is displayed.

6. Choose the **OK** button from this dialog box; the data in the Excel file is imported into the **DATA MANAGER**, refer to Figure 10-29. Also, the **Spec** column is highlighted with yellow background.

	Type	PnPID	Spec	Item Code
	P	3299	CS150	
	P	3319	CS150	

Figure 10-29 *The Data table after importing the data*

7. Choose the **Accept All** button from the **DATA MANAGER** toolbar to accept the changes made in the data, refer to Figure 10-30 and then exit the **DATA MANAGER**.

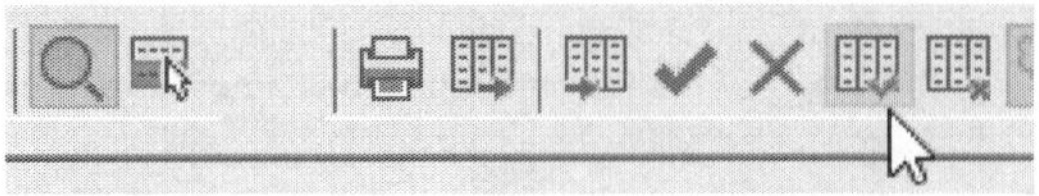

***Figure 10-30** Choosing the **Accept All** button*

8. Choose the **Save** button from the **Application** menu and then the **Close** button to close the file.

Tutorial 2

In this tutorial, you will use **Autodesk AutoCAD Plant Report Creator** to create reports.

(Expected time: 10 min)

The following steps are required to complete this tutorial:

a. Start **Autodesk AutoCAD Plant Report Creator**.
b. Select the required configuration.
c. Select the data source.
d. Export the data.

Starting Autodesk AutoCAD Plant Report Creator and Generating Reports

1. Double-click on the shortcut icon of **AutoCAD Plant Report Creator 2023 - English** in your Desktop; the **Autodesk AutoCAD Plant Report Creator** is displayed.

2. In this dialog box, select the **Open** option from the **Project** drop-down list; the **Open** dialog box is displayed.

3. Browse to the location *C:\Users\User_name\Documents\CADCIM* and double-click on the **Project.xml** file; the data sources related to the **CADCIM** project are displayed in **Autodesk AutoCAD Plant Report Creator**.

4. Choose the **Valvelist** option from the **Report Configuration** drop-down list.

5. Choose the **Project Data** radio button from the **Data Source** area if not already selected.

6. Choose the **Print/Export** button from **Autodesk AutoCAD Plant Report Creator**; the **PDF Export Options** dialog box is displayed.

7. Enter the values in this dialog box, as shown in Figure 10-31 and choose the **OK** button; the report is exported in PDF file format and the **Export results** window is displayed.

8. In this window, double-click on the file path, refer to Figure 10-32; the PDF file containing the valve report is displayed, as shown in Figure 10-33.

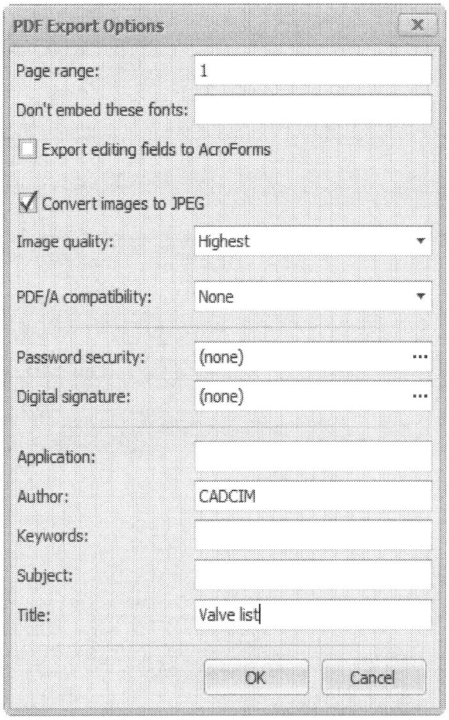

Figure 10-31 *The* ***PDF Export Options*** *dialog box*

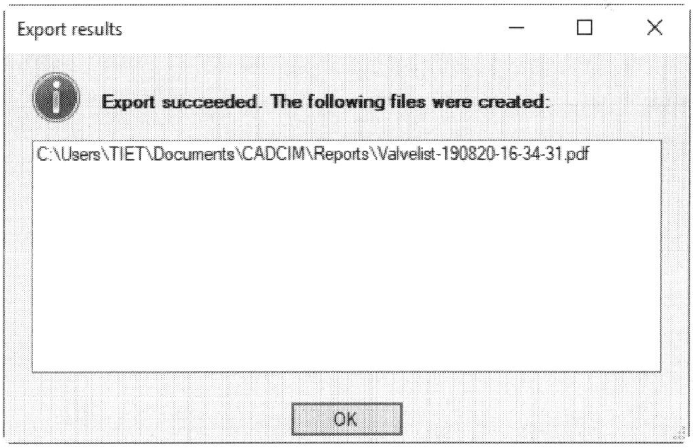

Figure 10-32 *The* ***Export results*** *window showing the file path*

Valvelist Autodesk

Project: CADCIM

Tag	Size	Spec	Manufacturer	Model No.	Supplier	Description
HA-101	12"	CS150				GATE VALVE
HA-102	12"	CS150				GATE VALVE
HA-103						BUTTERFLY VALVE
HA-104	6"	CS150				GATE VALVE
HA-105	6"	CS150				GATE VALVE
HA-106						BUTTERFLY VALVE
HA-107						GATE VALVE
HA-108						GATE VALVE
HA-109						GATE VALVE
HA-110						GATE VALVE
HA-111	10"	CS150				GATE VALVE
HA-112	10"	CS150				GATE VALVE
HA-113	8"	CS150				CHECK VALVE
HA-114	8"	CS150				GATE VALVE
HA-115	8"	CS150				CHECK VALVE
HA-116	8"	CS150				GATE VALVE
HA-117						GATE VALVE

Figure 10-33 *Partial view of the valve report in the PDF file*

9. View the report and close the PDF file.

10. Choose the **OK** button from the **Export results** window and close **Autodesk AutoCAD Plant Report Creator**.

Self-Evaluation Test

Answer the following questions and then compare them to those given at the end of this chapter:

1. The __________ is a location where all project data is located.

2. You can select the __________ option from the drop-down list located at the bottom of the Class tree to display the data based on the area.

3. You can right-click on a selected item in the data table and choose the __________ option from the shortcut menu displayed to hide all other items.

4. You can right-click on a selected item in the data table and choose the __________ option from the shortcut menu displayed to hide it from the data table.

5. You can choose the __________ button from the **DATA MANAGER** to view the data related to the objects selected in the drawing area.

Review Questions

Answer the following questions:

1. If you select the __________ radio button, the node selected from the Class tree and all the sub-nodes under it will be included while exporting the data.

2. You can export the data from the **DATA MANAGER** to an _________ or _________ file.

3. When you make changes to the data, the changes are highlighted with yellow background. (T/F)

4. The **Autodesk AutoCAD Plant Report Creator** can be used to create reports even when the AutoCAD Plant 3D application is not running. (T/F)

5. You can view P&ID data in the **DATA MANAGER** without opening a P&ID file. (T/F)

Answers to Self-Evaluation Test

1. DATA MANAGER, 2. Order by Area, 3. Filter By Selection, 4. Filter Excluding Selection, 5. Show Selected Items

Project

Thermal Power Plant

Project Description

In this project, you will create a steam driven thermal power plant. First, you will create P&ID, plant 3D model, spec file, and pipings and fittings of the plant. Next, you will generate its isometric and orthographic drawings. Finally, you will create and manage the reports of the thermal power plant.

A thermal power plant operates on the basic Rankine cycle. Heat energy obtained by burning coal is converted into electric power with the help of a generator. Before starting with the modeling part, it is better to have understanding of the working of an actual thermal power plant. The stepwise working of a steam driven power plant is given below.

1. In a thermal power plant, the basic purpose is to generate electricity with the help of a generator by running the generator shaft.

2. Generator shaft is run with the help of a turbine. Typically, a multistage steam turbine is used for this purpose. A multistage steam turbine has three units: high pressure turbine, intermediate pressure turbine, and low pressure turbine.

3. In order to run a turbine, you need to subject the inlet of the turbine with high pressure and high temperature steam. This can be done with the help of a boiler. As the steam passes through the blades of the turbine, its temperature and pressure drops towards the outlet. Now, the steam is at low pressure and low temperature. You need to again raise the temperature and pressure to their original state to repeat the process.

4. The first step is to raise the pressure. You can use a compressor for this purpose but compressing of steam is a highly energy utilizing process which will reduce the efficiency of the system. An easy way is to convert the steam into liquid and then boost its pressure. For this purpose, you will introduce a condenser (heat exchanger) unit which sits beneath the low pressure turbine.

5. In the condenser, a stream of cold water flows through the tubes. The steam rejects heat to the cold water and becomes condensed. Now, you can use the pump to increase the pressure of the water. Typically, a multistage centrifugal pump is used for this purpose.

6. The next task is to bring the temperature back to its original state. For this purpose, heat is added at the exit of the pump with the help of a boiler. Thermal power plants generally use the type of a boiler called water tube boiler. Pulverized coal is burnt inside the boiler. The high pressure water ideally flows through an economizer.

7. In the economizer, the water will capture energy from the flue gas. The water flows through a down-comer and then through the water walls where it gets converted into steam. The pure steam is separated at a steam drum.

8. Now the working fluid is back to its original state (high temperature and pressure). This steam can be fed back to the steam turbine and the cycle can be repeated continuously for power production.

9. This power plant working on the basic Rankine cycle will have a very low efficiency (20-25%).

10. You can increase the performance of this system considerably with the help of few techniques.
    Super heating
    Reheating
    Feed water heating

11. In case of super heating, even after converting water into steam, more heat is added to it and with that steam becomes super heated.

12. The higher the temperature of the steam, the more efficient is the cycle. But the turbine material will not withstand the temperature more than 600 degrees. So the super heating is limited to this threshold.

13. The temperature of the steam decreases as it flows along the rows of the turbine blades. Consequently, a great way to increase the efficiency of the power plant is to add more heat after the first turbine stage. This process is known as reheating. This will increase the temperature of the steam again leading to high power output and greater efficiency.

14. The low pressure sides of the turbine are prior to absorb atmospheric air even after providing adequate sealing. The dissolved gases in the atmospheric air will spoil the boiler material over time.

15. To remove these gases, an open feed water heater is introduced. Hot steam from the turbine is mixed into the feed water. Steam bubbles so generated will absorb the dissolved gases. The mixing also preheats the feed water which helps increase the efficiency of the power plant to a greater extent.

16. All these techniques make a power plant to work under an efficiency range of 40-45%.

17. In an actual power plant, the heat addition and heat rejection processes are executed with the help of a cooling tower. The cooling tower supplies the cold liquid at the condenser inlet. The heated up water at the condenser outlet is sprayed in the cooling tower which induces a natural air draft and the sprayed water loses heat. This is how a colder liquid is always provided at the condenser input.

18. At the heat addition side, the burning coal produces many pollutants which can't be released directly into the atmosphere. So, before that the exhaust gases are treated in an electrostatic precipitator.

19. The electrostatic precipitator uses plates with high voltage static electricity to absorb the pollutant particles. Once the exhaust gases are treated, they are released into the atmosphere through a chimney.

After going through the working and the concepts involved in the power plant processes, you can start creating your project in AutoCAD Plant 3D. The following steps are required to complete this project:

1. Start AutoCAD Plant 3D and create a new project

2.Create P&ID of the plant
- Create a new P&ID drawing
- Create custom P&ID symbols
- Place equipment symbols in the drawing sheet
- Connect the equipment with lines

- Add valves to the P&ID
- Validate the drawing
- Save the drawing

3.Create plant 3D model of the plant
- Create and place equipment in the drawing

4.Create spec file for the project
- Add parts to the spec sheet
- Save the spec file

5.Add and modify nozzles
- Add nozzles to the turbine and generator unit
- Add nozzles to the condenser
- Add and modify nozzles of the cooling tower
- Modify nozzles of the vertical inline pump
- Modify nozzles of the centrifugal pump1
- Add and modify nozzles of the low pressure heater
- Add nozzles to the deaerator
- Modify nozzles of the centrifugal pump2
- Add and modify nozzles of the high pressure heater
- Add nozzles to the boiler and boiler drum

6. Connect the equipment

7. Add valves

8. Add structures and supports

9. Add ladders and railings

10. Create isometric drawings
- Create quick isometric drawing for line number 1007
- Create production isometric drawing for line number 3004
- Create production isometric drawing for line number 3006

11. Create orthographic drawings

12. Add annotations and dimensions

13. Create reports and manage data
- Filter and export the data
- Modify the data
- Import the data
- Use Autodesk AutoCAD Plant Report Creator and generate reports

Starting AutoCAD Plant 3D and Creating a New Project

1. Double-click on the **AutoCAD Plant 3D 2023 - English** icon from the desktop of your computer, AutoCAD Plant 3D 2023 starts and the **PROJECT MANAGER** is displayed by default on the left of the screen.

2. In **PROJECT MANAGER**, select the **New Project** option from the drop-down list available in the **Current Project** area; the Project Setup Wizard is displayed.

3. Enter **POWER PLANT** in the **Enter a name for this project** edit box and choose the **Next** button; the **Specify unit settings** page is displayed.

4. Select the **Imperial** radio button, if not selected by default and then choose the **Next** button; the **Specify P&ID settings** page is displayed.

5. Select the **PIP** option from the **Select the P&ID symbology standard to be used** list box, if not selected by default and choose the **Next** button; the **Specify Plant 3D directory settings** page is displayed.

6. Accept the default directory settings and choose the **Next** button; the **Specify database settings** page is displayed.

7. Make sure the **Single User - SQLite local database** radio button is selected and choose the **Next** button; the **Finish** page is displayed.

8. Choose the **Finish** button; you will notice that the **POWER PLANT** project is displayed in the drop-down list available in the **Current Project** area of the **PROJECT MANAGER**. Also, a node named **POWER PLANT** is created in the **Project** area.

Creating P&ID of the Plant

In this section, you will be working with the P&ID of the thermal power plant.

Creating a New P&ID Drawing

1. In the **PROJECT MANAGER**, right-click on the **P&ID Drawings** node in the project tree and then choose the **New Drawing** option; the **New DWG** dialog box is displayed.

2. Enter **P&ID.dwg** in the **File name** edit box and then specify the **PID ANSI D-Color Dependent Plot Styles.dwt** template in the **DWG template** edit box.

3. Choose the **OK** button from the **New DWG** dialog box; the new P&ID drawing file is created with the **TOOL PALETTES** displayed on the right side of the screen.

4. Choose the **Workspace Switching** button on the right-side of the Status Bar; a flyout is displayed. Choose the **PID PIP** option from the flyout; the **P&ID PIP** workspace is invoked and the **TOOL PALETTES** is loaded with the P&ID symbols.

To create the P&ID of the Power Plant, firstly we need to have the list of equipments we are going to use in the plant. The list of equipments is given below.

1. Turbine and Generator Unit
2. Condenser Unit
3. Centrifugal and Inline Pumps
4. LP Heater
5. HP Heater
6. Deaerator
7. Boiler Unit
8. Cooling Tower

Creating Custom P&ID Symbols

You can create custom P&ID symbols of the equipment and add them to the **TOOL PALETTES**. In this project, you need to create custom P&ID symbols for all the equipment except centrifugal pumps. First, you will create a P&ID symbol for the Turbine and Generator unit and then add it to the **TOOL PALETTES**. The stepwise procedure to do so is given next.

1. Create a symbol of the Turbine and Generator unit, as shown in Figure P-1, using the AutoCAD sketching tools.

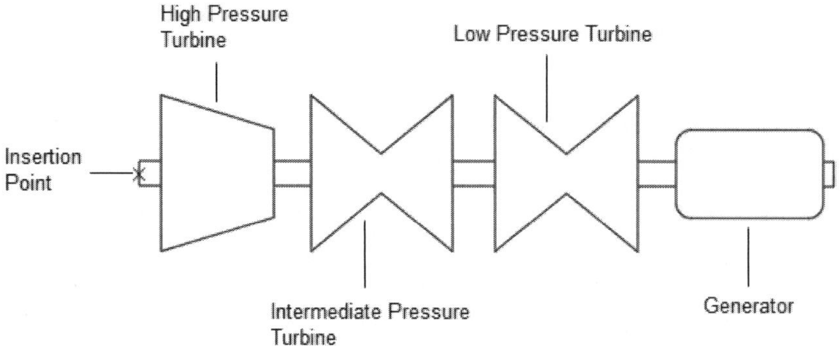

Figure P-1 *Symbol of the Turbine and the Generator Unit*

Note
Assume the dimensions while creating the symbols. You can scale the symbols after inserting them in the drawing sheet as per the requirement.

2. Convert the symbol created into a block named **STEAM TURBINE** and save the drawing. For the insertion point of the block, refer to Figure P-1.

3. Click on the **Equipment** tab in the **TOOL PALETTES - P&ID PIP**. This ensures that the custom symbol will be added to the **Equipment** tab of the **TOOL PALETTES**.

4. Invoke the **Project Setup** dialog box by choosing the **Project Setup** tool from the **Project Manager** drop-down in the **Project** panel of the **Home** tab.

5. In the **Project Setup** dialog box, expand the **P&ID DWG Settings** node by clicking on the + sign. Next, choose **P&ID Class Definitions** > **Engineering Items** > **Equipment**.

6. Click on the **Mechanical Drivers** sub-node; it is highlighted. Next, right-click to display a shortcut menu and choose the **New** option; the **Create Class** dialog box is displayed, refer to Figure P-2.

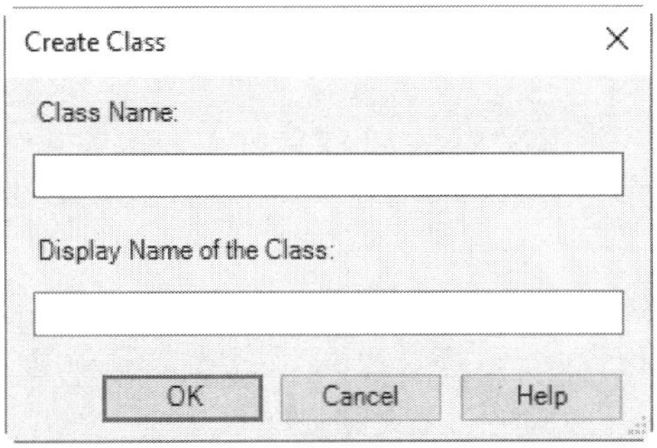

Figure P-2 *The* ***Create Class*** *dialog box*

7. Enter **STEAMTURBINE** in the **Class Name** edit box; the specified name also gets displayed as **STEAMTURBINE** in the **Display Name of the Class** edit box. Next, enter a space after **STEAM** in the **Display Name of the Class** edit box so that the name is displayed as **STEAM TURBINE**. (Note that the class name entered in the **Class Name** edit box should not have any spaces in between the alphabets). Next, choose the **OK** button to exit the dialog box; the specified class name gets listed and highlighted under the **Mechanical Drivers** sub-node. Also, the class name (**STEAM TURBINE**) is displayed next to the **Class settings** area.

8. Choose the **Add Symbols** button from the **Symbol** area; the **Add Symbols - Select Symbols** dialog box will be displayed, refer to Figure P-3.

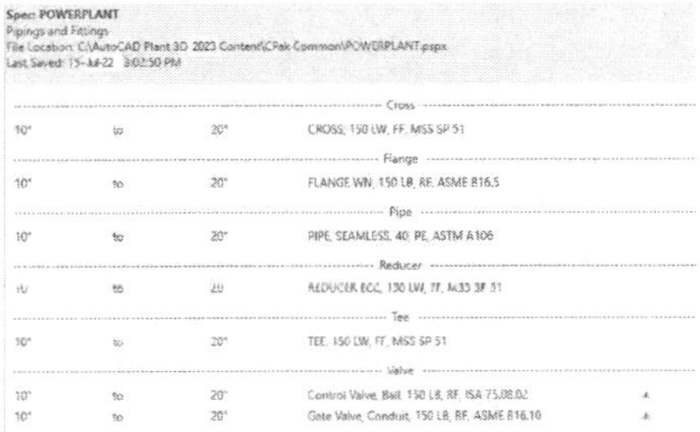

Figure P-3 *The* ***Add Symbols - Select Symbols*** *dialog box*

9. In this dialog box, choose the browse button next to the drop-down list in the **Selected Drawings** area; the **Select Block Drawing** dialog box is displayed. In this dialog box, browse to the file location and open it; the **Add Symbols - Select Symbols** dialog box is displayed again with a list of blocks in the drawing file in the **Available Blocks** list box.

10. Select **STEAM TURBINE** from the **Available Blocks** list box; it is highlighted. Also a preview of the turbine block is displayed in the **Preview** window available below the **Available Blocks** list box.

11. Choose the **Add** button; the block name is displayed in the **Selected Blocks** list box of the dialog box. Choose the **Next** button; the **Add Symbols - Edit Symbol Settings** dialog box is displayed.

12. Enter **STEAM TURBINE** in the **Symbol Name** edit box of the **Symbol Properties** area. Choose the **Yes** option from the **Auto Nozzle** drop-down list and **Flanged Nozzle Style** from the **Auto Nozzle Style** drop-down list of the **Other Properties** area. Next, choose the **Finish** button to exit the **Add Symbols - Edit Symbol Settings** dialog box; the **Project Setup** dialog box is displayed again.

13. Choose the **Add to Tool Palette** button from the **Project Setup** dialog box; the **Create Tool** message box is displayed with a message **'STEAM TURBINE' tool has been added to the current tool palette**. Choose the **OK** button to exit the message box.

14. Choose the **Apply** and then the **OK** button to exit the **Project Setup** dialog box; the **STEAM TURBINE** symbol is added to the **Equipment** tab.

15. Follow the above discussed procedure to add other equipment symbols to the **TOOL PALETTES**. Refer to Figure P-4 for the shapes of the remaining equipment symbols which are to be added to the **TOOL PALETTES**.

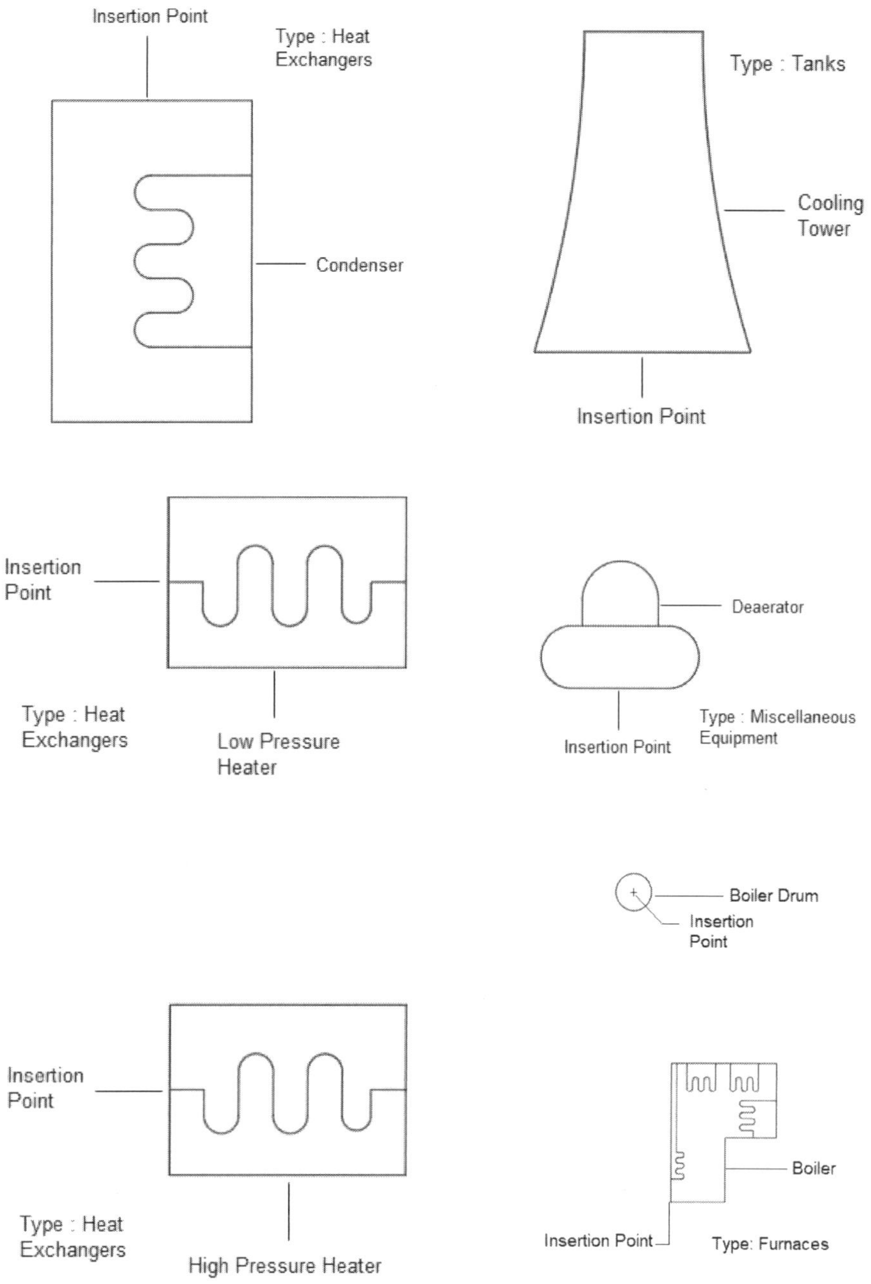

Figure P-4** The Shapes of the symbols to be added to the **TOOL PALETTES

Placing Equipment Symbols in the Drawing Sheet

After adding all the equipment symbols to the **TOOL PALETTES**, now you can start inserting them into the drawing sheet. Figure P-5 shows the final P&ID of the Power Plant (For your reference). The stepwise procedure to add equipment symbols to the drawing sheet is given next.

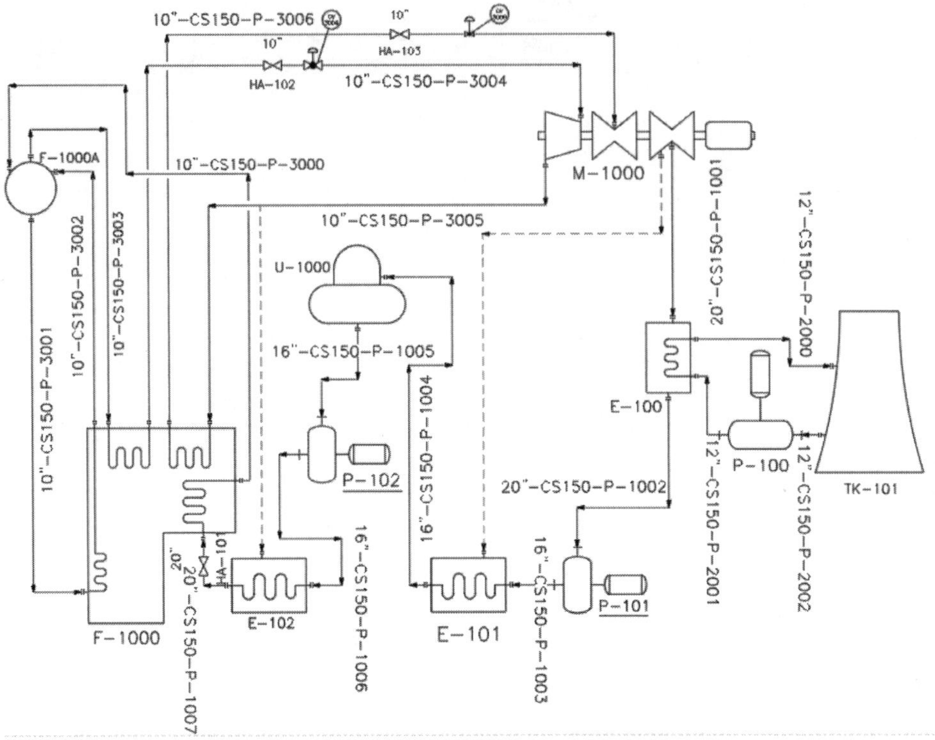

Figure P-5 P&ID of the Plant

1. Choose the **Equipment** tab from the **TOOL PALETTES** if not already chosen.

2. Choose the **STEAM TURBINE** tool from the **Drivers and Agitator/Mixer** area in the Equipment tab of the **TOOL PALETTES - P&ID PIP**; you are prompted to specify an insertion point. Next, specify **(15,16)** as the insertion point of the turbine and press ENTER; the **STEAM TURBINE** is placed at the specified location in the drawing sheet.

Note

You can scale the equipment symbols after inserting them into the P&ID for better visualisation. Make sure while scaling the equipment symbols, you choose their insertion points as the base points.

3. Select the **STEAM TURBINE** from the drawing sheet and right-click; a shortcut menu is displayed. Choose **Annotate > Tag** from the shortcut menu; the tag gets attached to the cursor and you are prompted to specify an annotation position. Drag the cursor and place the annotation tag below the turbine; **M-?** gets displayed on the screen. Next, double-click on the annotation tag; the **Edit Annotation** dialog box is displayed, refer to Figure P-6.

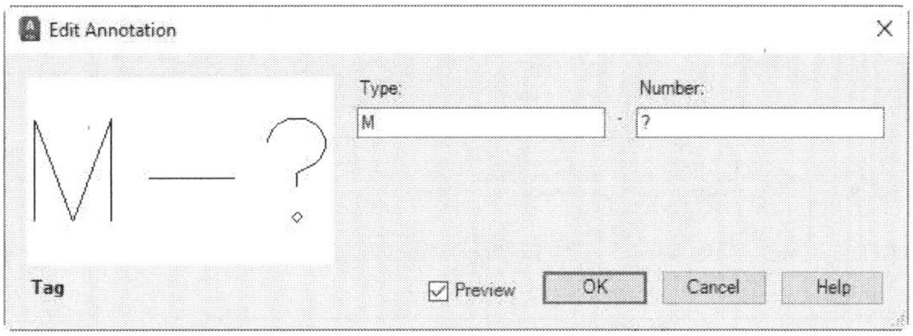

Figure P-6 *The* ***Edit Annotation*** *dialog box*

4. In the **Edit Annotation** dialog box, enter **1000** in the **Number** edit box and then choose the **OK** button; the annotation tag **M-?** gets displayed as **M-1000**, refer to Figure P-5.

5. Choose the **CONDENSER** tool from the **Drivers and Agitator/Mixer** area of the **Equipment** tab; you are prompted to specify the insertion point. Next, specify **(18.5,11)** as the insertion point of the Condenser and press ENTER. The Condenser gets placed at the specified location in the drawing sheet.

6. Follow the procedure discussed above and assign annotation tag **E-100** to the Condenser.

7. Similarly, place other equipment symbols into the drawing sheet and assign annotation tags to them. Table given below shows the equipment names, insertion coordinates, and annotation tags to be assigned to the remaining equipment symbols which are to be added to the drawing sheet.

EQUIPMENT	**INSERTION COORDINATES**	**ANNOTATION TAGS**
Cooling Tower	**(24,7)**	**TK-101**
Vertical Inline Pump (Between Line numbers 2001 and 2002)	**(21,8)**	**P-100**
Horizontal Centrifugal Pump1 (Between Line numbers 1002 and 1003)	**(16,4)**	**P-101**
Low Pressure Heater	**(12,4)**	**E-101**
Deaerator	**(10,11)**	**U-1000**
Horizontal Centrifugal Pump2 (Between Line numbers 1005 and 1006)	**(10,7.5)**	**P-102**
High Pressure Heater	**(6.5,4)**	**E-102**
Boiler	**(2.5,3)**	**F-1000**
Boiler Drum	**(0.9858,14.6250)**	**F-1000A**

Note

The equipment symbols should be placed in the drawing sheet in the same sequence as given in the table.

Connecting the Equipments with lines

1. Choose the **Lines** tab in the **TOOL PALETTES - P&ID PIP**, and then choose the **Primary Line Segment** tool from the **Pipe Lines** area; you are prompted to specify the start point. Specify the start point at the bottom of the Low Pressure Turbine and connect it to the top of the Condenser, refer to Figure P-5. You will notice that the nozzles are added at the start and end points of the line that lies at the bottom of the Low Pressure Turbine and top of the Condenser.

2. Select the previously created line and right-click; a shortcut menu is displayed. Choose the **Assign Tag** option from the shortcut menu; the **Assign Tag** dialog box is displayed.

3. In the **Assign Tag** dialog box, select **20"** from the **Size** drop-down list, **CS150 - 150#Carbon Steel** from the Spec drop-down list, **P - GENERAL PROCESS** from the **Pipe Line Group Service** drop-down list, and enter **1001** in the **Pipe Line Group Line Number** edit box. Next, select the **Place annotation after assigning tag** check box from the **Existing Pipe Line Segments area** and then choose the **Assign** button; you are prompted to select an annotation position. Drag the cursor and place the annotation at the suitable location, refer to Figure P-5.

4. Again, choose the **Primary Line Segment** tool from the **Pipe Lines** area of the **Lines** tab and connect the bottom nozzle of the Condenser to the top nozzle of the Centrifugal Pump1 and assign it the tag **20"- CS150-P-1002**, refer to Figure P-5.

5. Invoke the **Primary Line Segment** tool and connect the left nozzle of the Centrifugal Pump1 and right nozzle of the Low Pressure Heater tagged **E-101** and assign it the tag **16"- CS150-P-1003**, refer to Figure P-5.

6. Similarly, connect the other equipment using the **Primary Line Segment** tool, refer to Figure P-5. Table given below shows the details of the equipment to be connected and the tags to be assigned to the lines connecting them and Figure P-7 shows the Boiler with its labeled components for your reference.

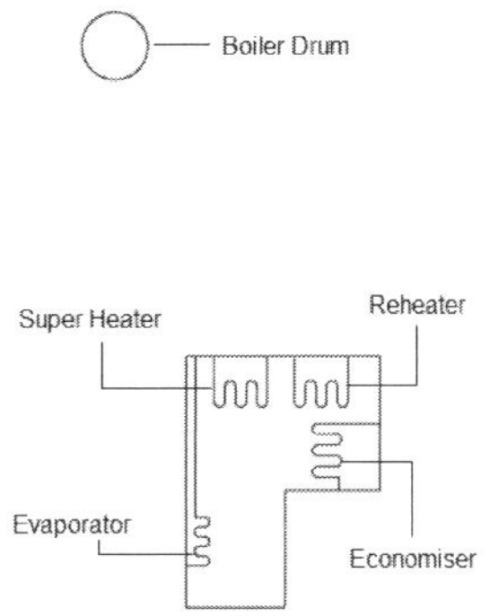

Figure P-7 *Boiler with its labelled components*

EQUIPMENTS TO BE CONNECTED	TAG
Left of LP Heater and right of Deaerator	**16"-CS150-P-1004**
Bottom of Deaerator and top nozzle of Centrifugal Pump2	**16"-CS150-P-1005**
Left nozzle of Centrifugal Pump2 and right of HP Heater tagged E-102	**16"-CS150-P-1006**
Left of HP Heater and bottom of Boiler (Economiser)	**20"-CS150-P-1007**
Top right of Condenser and top of Cooling Tower	**12"-CS150-P-2000**
Bottom of Cooling Tower and inlet nozzle of Vertical Inline Pump	**12"-CS150-P-2002**
Outlet nozzle of Vertical Inline Pump and bottom right of Condenser	**12"-CS150-P-2001**
Top right of Economiser and top left of Boiler Drum	**10"-CS150-P-3000**
Bottom of Boiler Drum and bottom left of Evaporator	**10"-CS150-P-3001**
Top of Evaporator and right of Boiler Drum	**10"-CS150-P-3002**
Top of Boiler Drum and left of Super Heater	**10"-CS150-P-3003**
Right of Super Heater and top of High Pressure Turbine	**10"-CS150-P-3004**
Bottom of High Pressure Turbine and right of Reheater	**10"-CS150-P-3005**
Left of Reheater and top of Intermediate Pressure Turbine	**10"-CS150-P-3006**

7. Choose the **Primary Line Segment-Existing** tool from the **Pipe Lines** area of the **Lines** tab and connect the bottom left of the Low Pressure Turbine and the top of the Low Pressure Heater, refer to Figure P-5.

8. Again, choose the **Primary Line Segment-Existing** tool and connect the line tagged **10"-CS150-P-3005** to the top of the High Pressure Heater, refer to Figure P-5.

Adding Valves to the P&ID

1. Select the Valves tab in the **TOOL PALETTES - P&ID PIP**; the **Control Valve** and **Valves** areas is displayed in the **TOOL PALETTES**.

2. Choose the **Gate Valve** tool from the **Valves** area; you are prompted to specify an insertion point. Place the valve on the line tagged **20"-CS150-P-1007** at the location shown in Figure P-5.

3. Similarly, place the Gate Valves on the lines tagged **10"-CS150-P-3004** and **10"-CS150-P-3006** at the locations shown in Figure P-5.

4. Choose the **Control Valve** tool from the **Control Valve** area; you are prompted as **pick insertion point or [Change body or actuator]**. Choose the **Change body or actuator** option from the command prompt; the **Control Valve Browser** dialog box is displayed. Select **Globe Valve** from the **Select Control Valve Body** list box and **Diaphragm Actuator** from the **Select Control Valve Actuator** list box and then choose the **OK** button. Next, place the valve on the line tagged **10"-CS150-P-3004** at the location shown in Figure P-5; you are prompted to select an annotation position. Drag the cursor and click on the suitable location where you want to place the tag; the **Assign Tag** dialog box is displayed. Enter **3004** in the **Loop Number** edit box and choose the **Assign** button; the tag is attached to the cursor. Place the tag at the location, as shown in Figure P-5.

5. Similarly, place another Control Valve on the line tagged **10"-CS150-P-3006** and assign tag **CV-3006** to it.

After adding all the equipment, pipings, and fittings, the P&ID should look similar to the one shown in Figure P-5.

Validating the Drawing

1. Choose the **Validate Config** button from the **Validate** panel in the **Home** tab; the **P&ID Validation Settings** dialog box is displayed.

2. In this dialog box, expand the **P&ID objects** node and clear the **Unresolved off-page connectors** check box. Similarly, expand the **3D Model to P&ID checks** node and clear all the check boxes under it. Next, choose the **OK** button to close the dialog box.

3. Run validation process by choosing the **Run Validation** tool from the **Validate** panel; the **VALIDATION SUMMARY** palette is displayed with a tree list of errors in the drawing.

4. Right-click on any error in the **VALIDATION SUMMARY** palette and choose Ignore to ignore the error. Similarly, ignore all the errors and exit the **VALIDATION SUMMARY** palette.

Saving the Drawing

1. Choose the **Save** option from the **Application Menu** or choose **File > Save** in the menu bar to save the drawing file.

2. Choose **Close > Current Drawing** from the **Application Menu** to close the drawing file.

CREATING PLANT 3D MODEL OF THE PLANT

After creating the P&ID, you need to create a 3D model of the plant. To do so, you need to create a new Plant 3D drawing. In the **PROJECT MANAGER**, select the **Plant 3D Drawings** node and right-click; a shortcut menu is displayed. Next, choose the **New Drawing** option from the shortcut menu; the **New DWG** dialog box gets displayed. Enter **P3D** in the **File name** edit box and then choose the **OK** button; a new file **P3D** gets created and displayed under the **Plant 3D Drawings** node.

Next, click on the **Workspace Switching** button in the Status Bar; a flyout gets displayed. Choose the **3D Piping** option from the flyout; the 3D Piping environment gets invoked and the **TOOL PALETTES - AUTOCAD PLANT 3D - PIPING COMPONENTS** gets loaded with 3D piping components.

Creating and Placing Equipments in the Drawing

Now, you need to create and place the equipment in the drawing. After creating the P&ID of the plant, you have a clear understanding of the equipment to be used in the plant. A list of equipment to be used in the plant is given below for your reference.

1. Turbine and Generator unit
2. Condenser
3. Vertical Inline Pump
4. Cooling Tower
5. Centrifugal Pump1
6. Low Pressure Heater
7. Deaerator
8. Centrifugal Pump2
9. High Pressure Heater
10. Boiler
11. Boiler Drum

Note

While inserting the equipment in the drawing area, make sure that the ***Dynamic Input*** *is turned off and* ***UCS*** *is set to* ***World****.*

Turbine and Generator unit

First, you will create the Turbine and Generator unit and place it in the drawing. To do so, choose the **Create Equipment** tool from the **Equipment** panel of the **Home** tab; the **Create Equipment** dialog box is displayed. Select the **Mechanical Drivers > New Horizontal Mechanical Drivers** from the Equipment drop-down list. Next, select the **Equipment** tab in the dialog box if it is not already selected.

Next, click on the **Add Shape** button; a flyout is displayed. Choose the **Cylinder** option from the flyout; it is listed in the **Shapes** list box. Select the **Cylinder** from the **Shapes** list box; its default dimensions are displayed in the **Dimensions** area. Next, enter **1'** and **2'** in the **D** and **H** edit boxes, respectively. Click in the **D** edit box; a swatch is displayed on the right. Click on the swatch; a flyout is displayed. Next, choose the **Override mode** option from the flyout.

Note

After entering the values in the various edit boxes in the ***Dimensions*** *area of the* ***Create Equipment*** *dialog box, make sure that you choose the* ***Override mode*** *option in each edit box wherever it is available.*

Again, choose the **Add Shape** button from the **Create Equipment** dialog box and choose the **Pyramid** option from the flyout; it is added to the **Shapes** list box. Select the **Pyramid** from the **Shapes** list box; its default dimensions are listed in the **Dimensions** area. Next, enter **5'**, **3'**, **5'**, and **6'** in the **D1**, **D2**, **W**, and **H** edit boxes, respectively. Make sure that the **Downwards** option is selected from the **Orientation** drop-down list of the **Dimensions** area.

Next, follow the same procedure to add the shapes to the **Shapes** list box. Figure P-8 shows the **Create Equipment** dialog box with list of shapes which are required to create the Turbine and Generator unit.

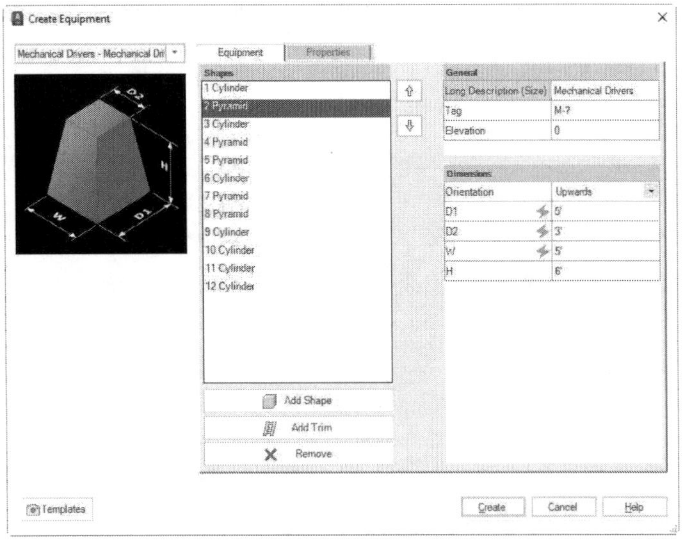

Figure P-8 The ***Create Equipment*** *dialog box with list of Shapes*

Next, select the cylinder numbered 3 in the Shapes list box and change its dimensions to:
D = 1' **H = 2'**

Similarly, for all other shapes, specify the dimension values as given below.

For Pyramid numbered 4
D1 = 7' **D2 = 3'**
W = 5' **H = 5'**
Orientation : Downwards

For Pyramid numbered 5
D1 = 7' **D2 = 3'**
W = 5' **H = 5'**
Orientation : Upwards

For Cylinder numbered 6
D = 1' **H = 2'**

For Pyramid numbered 7
D1 = 8' **D2 = 3'**
W = 5' **H = 5'**
Orientation : Downwards

For Pyramid numbered 8
D1 = 8' **D2 = 3'**
W = 5' **H = 5'**
Orientation : Upwards

For Cylinder numbered 9
D = 1' **H = 4'**

For Cylinder numbered 10
D = 2' **H = 6"**

For Cylinder numbered 11
D = 8' **H = 6'**

For Cylinder numbered 12
D = 2' **H = 6"**

After specifying dimension values for all the shapes, click in the **Tag** edit box in the **General** area, refer to Figure P-8; the **Assign Tag** dialog box is displayed. Enter **1000** in the **Number** edit box and choose the **Assign** button to exit the **Assign Tag** dialog box. You can also assign tag to the equipment using the **PROPERTIES** palette. Next, choose the **Create** button from the **Create Equipment** dialog box; you are prompted to choose an insertion point. Specify the insertion point as **54',-134',7'5"** at the command prompt; the compass tool is displayed at the bottom of the equipment. Set the orientation of the turbine and generator unit to 0 degrees; it is placed at the specified point.

Condenser (Heat Exchanger unit)

Now, you need to create the Heat Exchanger unit which is a vertical Condenser in our case. To do so, choose the **Create Equipment** tool from the **Equipment** panel of the **Home** tab; the **Create Equipment** dialog box is displayed. Next, select the **Heat Exchanger > New Vertical Heat Exchanger** from the Equipment drop-down list. Select the **Equipment** tab in the dialog box if not already selected. Next, follow the same procedure explained before to add shapes to the **Shapes** list box. Refer to Figure P-9 for the list of shapes which are to be added to the **Shapes** list box to create the Condenser unit.

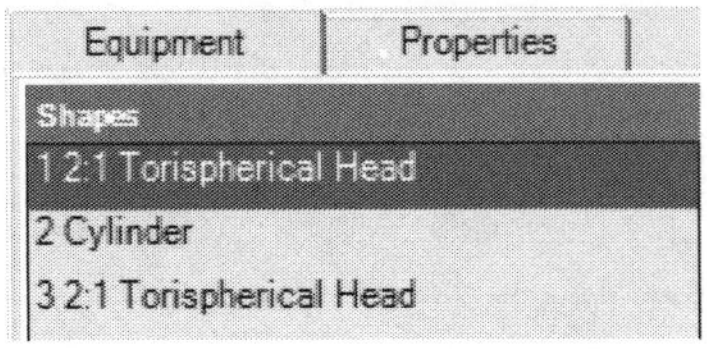

Figure P-9 *The list of shapes to be added to the* ***Shapes*** *area*

Next, select the shapes listed in the **Shapes** area one by one and assign dimension values to them. The dimension values for each shape is given below.

For 2:1 Torispherical Head numbered 1
D = 10' **SF = 0** **T = 1"**

For Cylinder numbered 2
D = 10' **H = 12'**

For 2:1 Torispherical Head numbered 3
D = 10' **SF = 0** **T = 1"**

After specifying dimension values for all the shapes, click in the **Tag** edit box in the **General** area; the **Assign Tag** dialog box is displayed. Enter **100** in the **Number** edit box and choose the **Assign** button to exit the **Assign Tag** dialog box. Next, choose the **Create** button from the **Create Equipment** dialog box; you are prompted to choose an insertion point. Specify the insertion point as **70',-134',-44'** at the command prompt; the compass tool is displayed at the bottom of the equipment. Set the orientation of the Condenser unit to **90** degrees; it is placed at the specified point.

Follow the same procedure to create and place rest of the equipment. The shapes to be added, the dimension values and the tags to be assigned to each equipment are given next.

Cooling Tower

Select the **Tank > Vertical Tank** from the Equipment drop-down list. Refer to Figure P-10 for the shapes to be added, dimension values, and tag to be assigned.

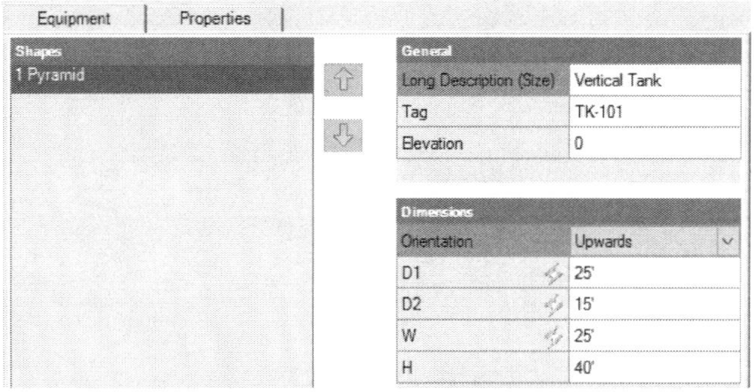

Figure P-10 *The shapes to be added, dimension values, and tag to be assigned to the COOLING TOWER*

Insertion point : **5', -134', -45'**
Compass Orientation : **0 degrees**

Vertical Inline Pump

Select the **Pump > Vertical Inline Pump** from the Equipment drop-down list. Refer to Figure P-11 for the dimension values and tag to be assigned.

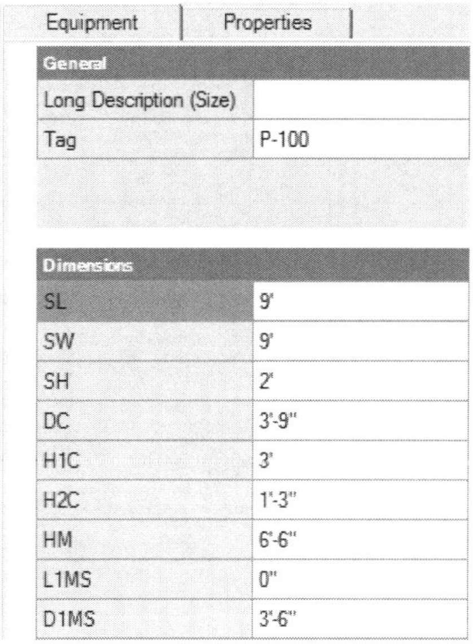

Figure P-11 *The dimension values and the tag to be assigned to the VERTICAL INLINE PUMP*

Insertion point : **40', -134', -40'**
Compass Orientation : **0 degrees**

Centrifugal Pump1

Select the **Pump > Centrifugal Pump** from the Equipment drop-down list. Refer to Figure P-12 for the dimension values and tag to be assigned..

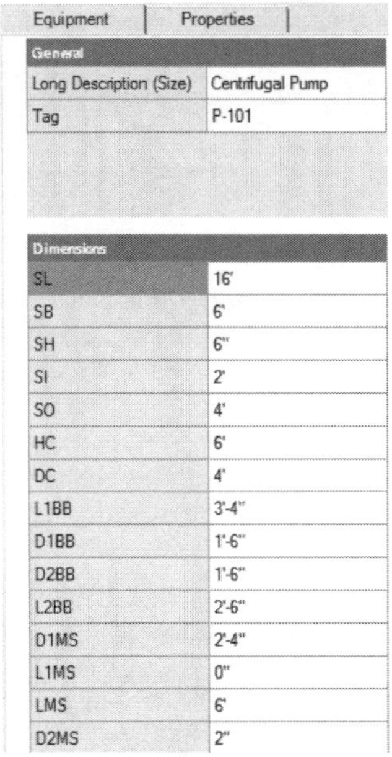

Figure P-12 *The dimension values and the tag to be assigned to the CENTRIFUGAL PUMP1*

Insertion point : **99', -134', -64'**
Compass Orientation : **0 degrees**

Low pressure heater

Select the **Heater > Box Type Heater** from the Equipment drop-down list. Refer to Figure P-13 for shapes to be added and tag to be assigned.

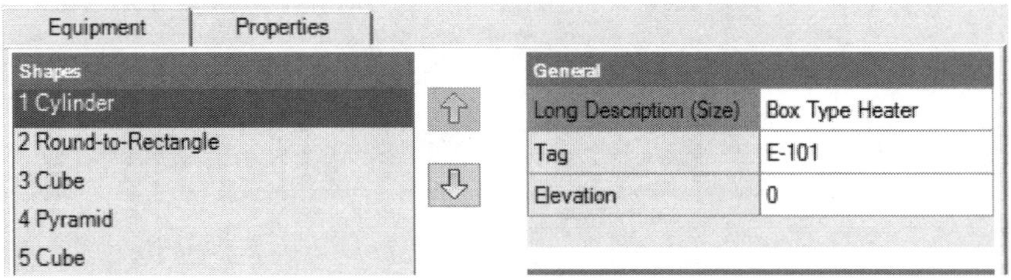

Figure P-13 *The shapes to be added and the tag to be assigned to the LOW PRESSURE HEATER*

The dimension values to be assigned to the shapes listed in the **Shapes** area are given next, refer to Figure P-13.

For Cylinder numbered 1
D = 2'-6" **H = 12'-11"**
For Round-to-Rectangle numbered 2
D1 = 5'-5" **W = 5'-5"** **H = 2'-11"**

D2 = 2'-6" **E = 0"** **A = 0**
Orientation : Upwards

For Cube numbered 3
D = 5'-5" **W = 5'-5"** **H = 4'-7"**

For Pyramid numbered 4
D1 = 12'-11" **D2 = 5'-5"**
W = 12'-11" **H = 2'-11"**
Orientation : Upwards

For Cube numbered 5
D = 12'-11" **W = 12'-11"** **H = 16'-3"**

Insertion point : **117', -173', -64'**
Compass Orientation : **180 degrees**

Deaerator

Select the **Misc Equipment > New Vertical Misc Equipment** from the Equipment drop-down list. Refer to Figure P-14 for shapes to be added and tag to be assigned.

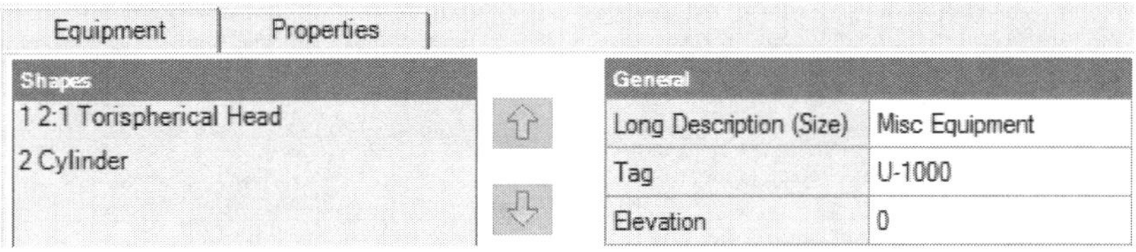

Figure P-14 *The shapes to be added and the tag to be assigned to the DEAERATOR*

The dimension values to be assigned to the various shapes listed in the **Shapes** area are given next.

For 2:1 Torispherical Head numbered 1
D = 10' **SF = 5'** **T = 1'**

For Cylinder numbered 2
D = 15' **H = 8'**

Insertion point : **117', -203', -15'**
Compass Orientation : **0 degrees**

Centrifugal Pump2

Select the **Pump > Centrifugal Pump** from the Equipment drop-down list. Refer to Figure P-15 for the dimension values and tag to be assigned.

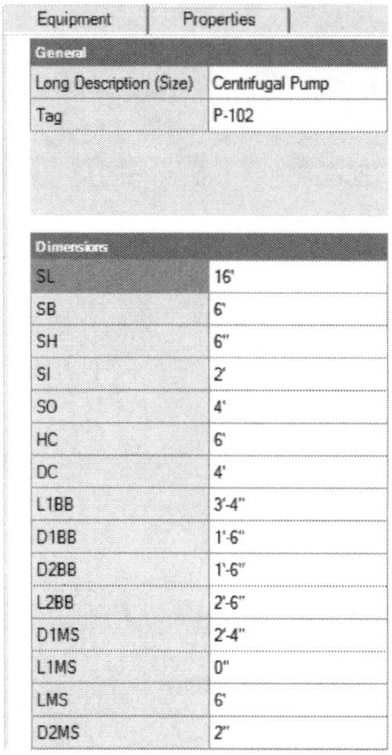
Equipment | Properties

General	
Long Description (Size)	Centrifugal Pump
Tag	P-102

Dimensions	
SL	16'
SB	6'
SH	6"
SI	2'
SO	4'
HC	6'
DC	4'
L1BB	3'-4"
D1BB	1'-6"
D2BB	1'-6"
L2BB	2'-6"
D1MS	2'-4"
L1MS	0"
LMS	6'
D2MS	2"

***Figure P-15** The dimension values and the tag to be assigned to the CENTRIFUGAL PUMP 2*

Insertion point : **69', -203', -64'**
Compass Orientation : **180 degrees**

High Pressure Heater

Select the **Heater > Box Type Heater** from the Equipment drop-down list. Refer to Figure P-16 for shapes to be added and tag to be assigned.

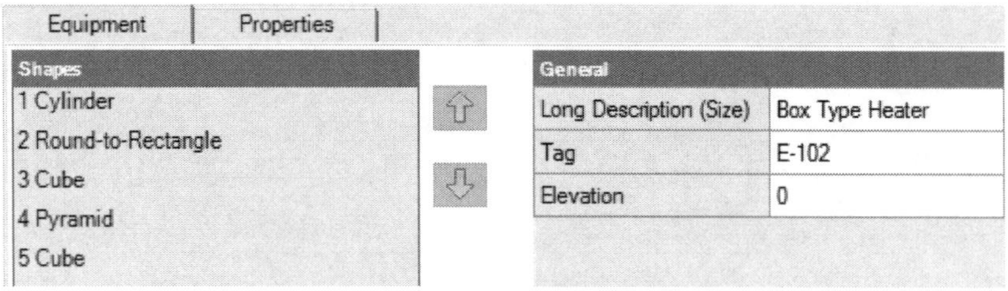
Equipment | Properties

Shapes
1 Cylinder
2 Round-to-Rectangle
3 Cube
4 Pyramid
5 Cube

General	
Long Description (Size)	Box Type Heater
Tag	E-102
Elevation	0

***Figure P-16** The shapes to be added and the tag to be assigned to the HIGH PRESSURE HEATER*

The dimension values to be assigned to the various shapes listed in the **Shapes** area are given next.

For Cylinder numbered 1
D = 2'-11" H = 13'-4"

For Round-to-Rectangle numbered 2
D1 = 5'-10" W = 5'-10" H = 3'-4"
D2 = 2'-11" E = 0" A = 0
Orientation : Upwards

For Cube numbered 3
D = 5'-10" W = 5'-10" H = 5'

For Pyramid numbered 4
D1 = 13'-4" D2 = 5'-10"
W = 13'-4" H = 3'-4"
Orientation : Upwards

For Cube numbered 5
D = 13'-4" W = 13'-4" H = 16'-8"

Insertion point : **33', -203', -64'**
Compass Orientation : **0 degrees**

Boiler

You need a Water Tube Boiler in the plant. So you will create a custom shape for the Boiler and then convert it into the Plant 3D component. Figure P-17 shows the dimensions which are required to create the Boiler.

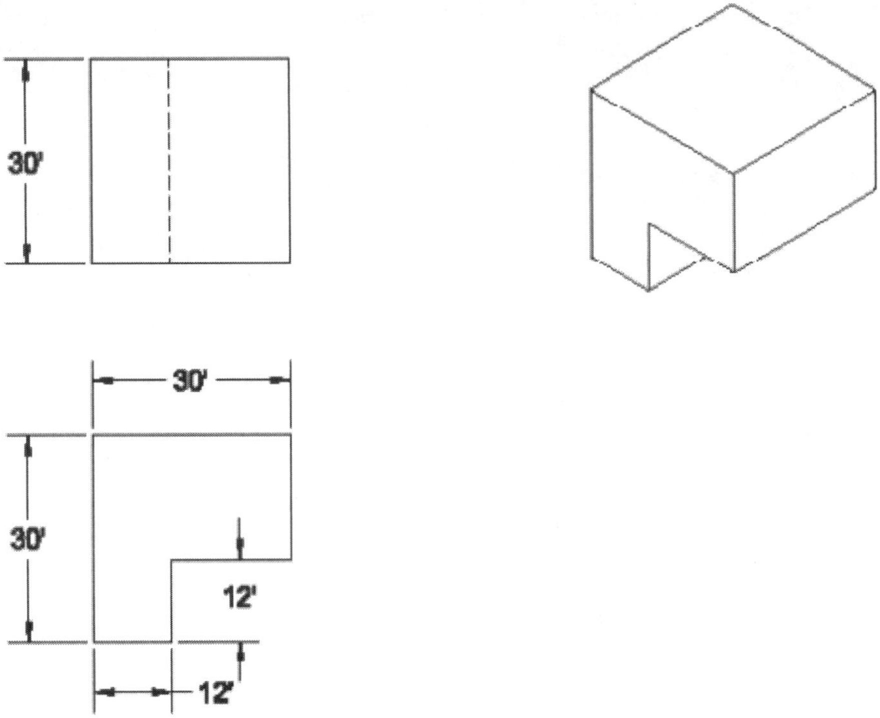

Figure P-17 *Views and dimensions of the Boiler*

Create a 3D model of the boiler, refer to Figure P-17 for dimensions. After creating the 3D model, you need to convert it into a Plant 3D equipment. To do so, choose the **Convert Equipment** tool from the **Equipment** panel of the **Home** tab; you are prompted to select AutoCAD objects to convert. Select the 3D model created and press ENTER; the **Convert to Equipment** dialog box is displayed. Select **Furnace** from the **Select equipment type** list box and then choose the **Select** button; you are prompted to specify the insertion base point. Specify the bottom left vertex of the Boiler as the insertion point, refer to Figure P-18; the **Modify Equipment** dialog box is displayed. Enter **Boiler** in the **Long Description (Size)** edit box. Next, click in the **Tag** edit box; the **Assign Tag** dialog box is displayed. Enter **1000** in the **Number** edit box and then choose the **Assign** button. Next, choose the **Apply** and then the **OK** button. The 3D model is converted into Plant 3D equipment.

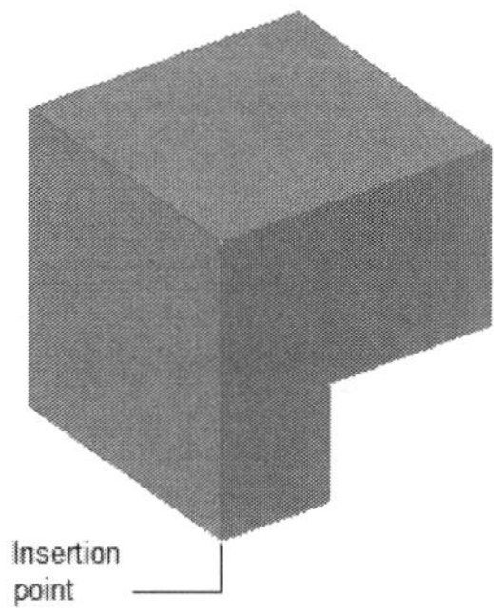

Figure P-18 *3D model of the Boiler with insertion point*

Next, we need to place the Boiler at the required location. Move the Boiler using the insertion point as the base point and palce it at **(-46', -196', -64')**.

Boiler Drum

Select the **Furnace > New Vertical Furnace** from the Equipment drop-down list. Refer to Figure P-19 for shapes to be added, tag, and dimensions to be assigned to the Boiler Drum.

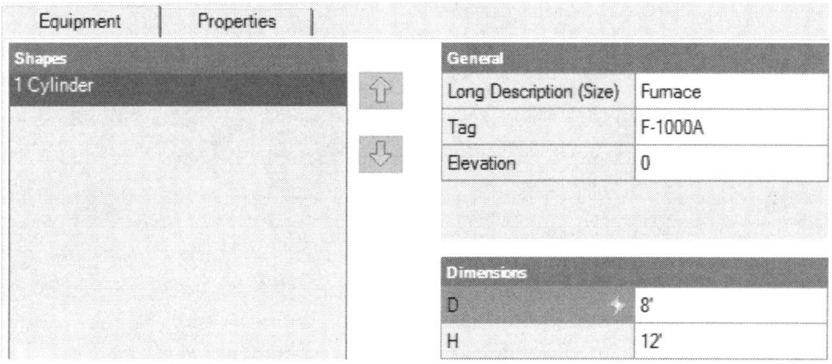

Figure P-19 *Shapes to be added, tag, and dimension values to be assigned to the Boiler Drum*

After assigning all the values to the Boiler Drum in the **Create Equipment** dialog box, choose the **Create** button; you are prompted to specify the insertion point. Specify the insertion point as **(-56', -181', -14')** and orientation angle as **90** degrees, the Boiler Drum is placed at the required location.

After placing all the equipment, the model should look like the one shown in Figure P-20.

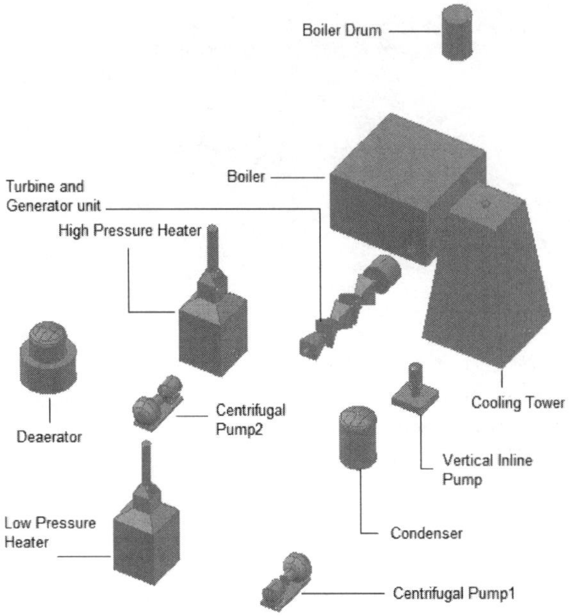

Figure P-20 *The* ***NE Isometric*** *view of the model after placing all the equipments*

Save the drawing by choosing the **Save** tool from the Quick Access Toolbar. Next, choose **Close > Current Drawing** from the **Application Menu**.

Note

While placing the equipment into the plant model, make sure that the orientation of the equipment is exactly similar to the one shown in Figure P-20.

CREATING SPEC FILE FOR THE PROJECT

After creating P&ID and placing equipment in the drawing, you need to connect the equipment. To do so, you need to have a data file containing all pipings, fittings, valves, and other connections required to connect the equipment to run the plant process effectively. The data file is termed as Spec File in AutoCAD Plant 3D. To create the Spec File, double-click on the **AutoCAD Plant 3D Spec Editor 2023 - English** icon on the desktop of your computer; the **Autodesk AutoCAD Plant 3D Spec Editor 2023** window is displayed, as shown in Figure P-21.

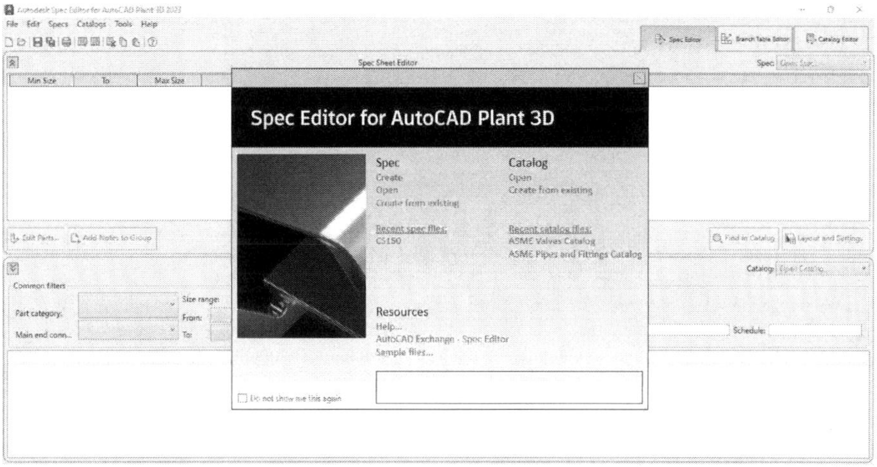

*Figure P-21 The **Autodesk AutoCAD Plant 3D Spec Editor 2023** window*

Choose the **Create** option from the **Spec** area of the **Autodesk AutoCAD Plant 3D 2023 Spec Editor** window; the **Create Spec** dialog box is displayed. Enter **POWER PLANT** in the **New Spec name** edit box and **Pipings and Fittings** in the **Spec description** edit box. Next, choose the **ASME Pipes and Fittings Catalog** option from the **Load catalog** drop-down list and then choose the **Create** button; the **Autodesk AutoCAD Plant 3D Spec Editor 2023** window is displayed with the **Spec Sheet** and **Catalog** areas.

Adding Parts to the Spec Sheet

Now, you need to add parts to the Spec Sheet. To do so, you need to filter the parts in the **Catalog** area under the **Spec Editor** tab by using the filter options available in it. The stepwise procedure to add parts to the Spec Sheet are given next.

1. Select the **Pipe** option from the **Part category** drop-down list in the **Common filters** area and apply the following filters:

   Size range: **10 to 20**
   Units: **in**

2. Select the **PIPE, SEAMLESS** option from the **Short Description** drop-down list in the Catalog Browser to show seamless pipes in it.

3. Select **PIPE, SEAMLESS, 40, PE, ASTM A106** from the Catalog Browser and choose the **Add to Spec** button from the **Spec Sheet** area; the part is added to the spec sheet.

4. Select the **Apply property overrides to parts added to spec** check box from the **Property overrides** area in the **Catalog** area and enter the following values:

Material:	**CS**
Material Code:	**A106**
Schedule:	**100**

5. Next, select the **Fittings** option from the **Part category** drop-down list and apply the following filters:

Size range:	**10 to 20**
Main end connection:	**FL**
Short Description:	**REDUCER ECC**
Pressure Class:	**150**

6. Select **REDUCER ECC, 150LW, FF, MSS SP 51** from the Catalog Browser and choose the **Add to Spec** button from the **Spec Sheet** area; the reducer gets added to the **Spec Sheet**.

   Similarly, add other parts to the **Spec Sheet** area by filtering the Catalog Browser using the following filter options:

Part Category:	**Fittings**
Main end connection:	**FL**
Size range:	**10 to 20**
Pressure Class:	**150**

   Select the following parts from the Catalog Browser:

   TEE, 150 LW, FF, MSS SP 51
   CROSS, 150 LW, FF, MSS SP 51

7. Choose the Flanges option from the Part Category drop-down list and apply the following filters:

Main end connection:	**FL**
Size range:	**10 to 20**
Short Description:	**FLANGE WN**

8. Select **FLANGE WN, 150 LB, RF, ASME B16.5** from the Catalog Browser and choose the **Add to Spec** button; the part will be added to the **Spec Sheet** area.

Next, you need to add valves and actuators to the Spec Sheet.

9. Choose the **Open Catalog** option from the **Catalog** drop-down list available in the area; the **Open** dialog box is displayed. In this dialog box, open the **CPak ASME** folder by double-clicking on it and then open the **ASME Valves Catalog.pcat file**; the **ASME Valves** Catalog is loaded in the **Catalog** area.

10. Select the **Valves** option from the **Part category** drop-down list and apply the following filters:

Size range:	**10 to 20**
Units:	**in**
Pressure Class	**150**

11. Press and hold the CTRL key on the keyboard and select **Control Valve, Ball, 150 LB, RF, ISA 75.08.02** and **Gate Valve, Conduit, 150 LB, RF, ASME B16.10** from the **Catalog Browser**. Next, choose the **Add to Spec** button; the valves are added to the **Spec Sheet** area.

Figure P-22 shows the **Spec Sheet** area after adding parts to it.

Figure P-22 *The* ***Spec Sheet*** *area after adding parts to it*

In this sheet, you can notice that an error symbol is displayed next to the parts with conflicts, refer to Figure P-22. This is because the system is not able to assign part usage priority to parts having the same size. Click on any of the error symbols displayed in the **Part Use Priority** column in the **Spec Sheet**; the **Part Use Priority** dialog box is displayed. Select **10"** size from the **Size Conflicts** list in the dialog box; the parts for the selected size are displayed in the **Spec Part Use Priority** list. Next, move **Gate Valve, Conduit, 150 LB, RF, ASME B16.10** to top in the **Spec Part Use Priority** list and select the **Mark as resolved** check box. Similarly, move **Gate Valve, Conduit, 150 LB, RF, ASME B16.10** to the top and select the **Mark as resolved** check box for all the sizes given in the **Size Conflicts** area. Next, choose the **OK** button; the dialog box is closed and a green dot is displayed in the **Part Use Priority** column, indicating that the conflict is resolved.

Saving the Spec File

1. Choose the Save As tool from the menu bar and specify the location as *C:\Users\User_name\Documents\POWER PLANT\Spec Sheets* to save the file.

2. Choose **File > Exit** from the menu bar to close the spec file.

ADDING AND MODIFYING NOZZLES

Now, you need to modify the existing nozzles and add some new nozzles to the equipment placed in the drawing sheet. To do so, open the **P3D** drawing file under **Plant 3D Drawings** node in **PROJECT MANAGER**.

Before you start adding and modifying nozzles, you need to add the spec file, which was created earlier, to the current project. To do so, select the **Pipe Specs** node from the **Project** area of the **PROJECT MANAGER** and right-click on it; a flyout is displayed. Next, choose the **Copy Specs to Project** option from the flyout; the **Select Files to Copy to Project** dialog box is displayed. Next, browse to the loaction where you have saved the spec file and select it; **POWER PLANT** is displayed in the **File name** edit box. Next, choose the **Open** button; the spec file is added under the **Pipe Specs** node. Next, select **POWER PLANT** from the **Spec Selector** drop-down list in the **Part Insertion** panel of the **Home** tab; the **TOOL PALETTES** is loaded with the parts contained in the spec file. Now, you can start adding and modifying nozzles.

Adding Nozzles to the Turbine and Generator unit

First, you need to add nozzles to the Turbine and Generator unit at the required locations. To do so, select the Turbine and Generator unit from the drawing area; the **Add Nozzle** grip is displayed. Click on the **Add Nozzle** grip; a dialog box for adding or modifying nozzles is displayed with **Change Location** tab chosen. Refer to Figures P-23 and P-24 for the values and parameters to be assigned to the nozzle. Figure P-23 shows the values to be entered in the dialog box for adding or modifying nozzles when the **Change Location** tab is chosen and Figure P-24 shows the values to be entered when the **Change Type** tab is chosen.

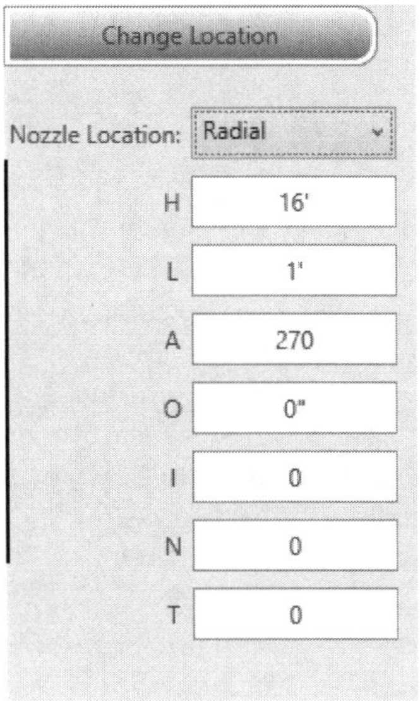

Figure P-23 *Values to be entered when* ***Change Location*** *tab is chosen*

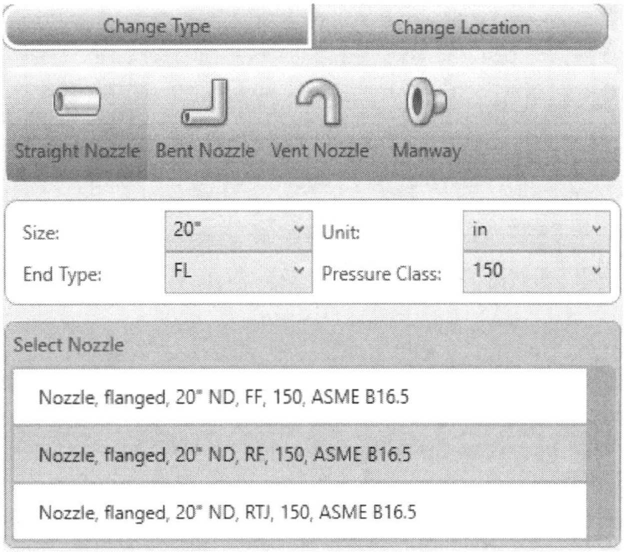

Figure P-24 *Values to be entered when* ***Change Type*** *tab is chosen*

Next, select **Nozzle, flanged, 20" ND, RF, 150, ASME B16.5** from the **Select Nozzle** area by double-clicking on it and then choose the **Close** button; the nozzle is placed at the bottom center of the Low Pressure Turbine.

Next, you need to place a nozzle at the top center of Intermediate Pressure Turbine. Figure P-25 shows the values to be entered in the dialog box for adding or modifying nozzles when the **Change Location** tab is chosen and Figure P-26 shows the values to be entered when the **Change Type** tab is chosen.

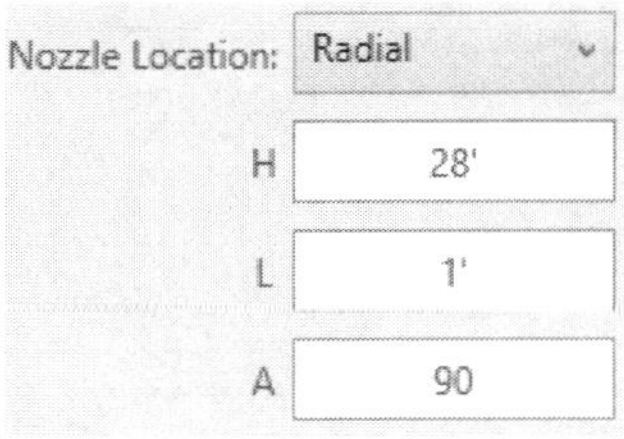

Figure P-25 *Values to be entered when* ***Change Location*** *tab is chosen*

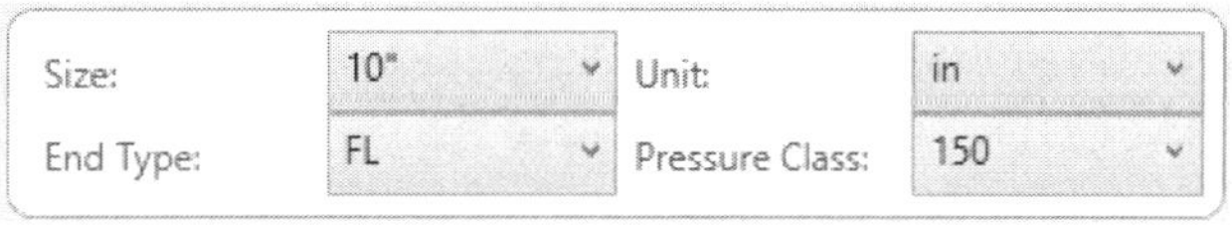

Figure P-26 *Values to be entered when* ***Change Type*** *tab is chosen*

Next, select **Nozzle, flanged, 10" ND, RF, 150, ASME B16.5** from the **Select Nozzle** area by double-clicking on it and then choose the **Close** button; the nozzle is placed at the top center of the Intermediate Pressure Turbine.

Similarly, you can place two more nozzles on the High Pressure Turbine. Figures P-27 and P-28 show the values to be entered in the dialog box for adding or modifying nozzles when the **Change Location** and **Change Type** tabs are chosen, respectively. Using these values, you can place the nozzle at the top of High Pressure Turbine.

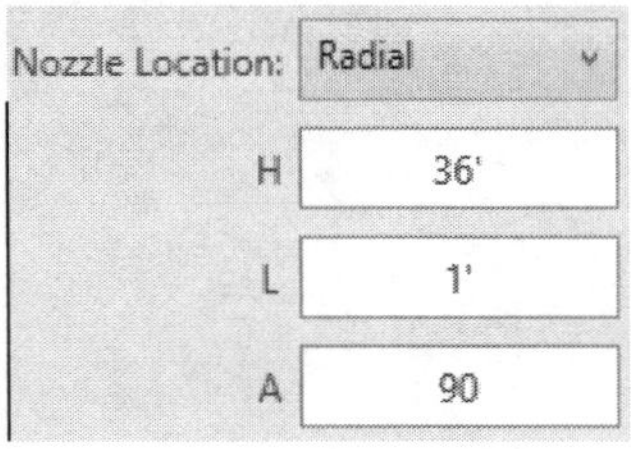

*Figure P-27 Values to be entered when **Change Location** tab is chosen*

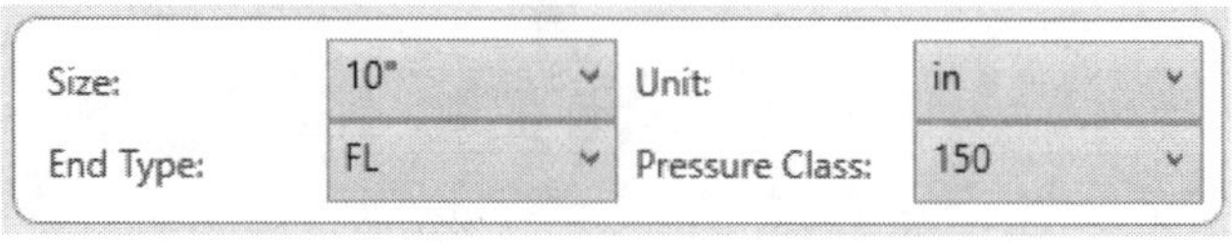

*Figure P-28 Values to be entered when **Change Type** tab is chosen*

After entering all values in the dialog box, select **Nozzle, flanged, 10" ND, RF, 150, ASME B16.5** from the **Select Nozzle** area by double-clicking on it and then choose the **Close** button; the nozzle is placed at the top of the High Pressure Turbine.

Figures P-29 and P-30 show the values to be entered in the dialog box for adding or modifying nozzles when the **Change Location** and **Change Type** tabs are chosen, respectively. Using these values, you can place the nozzle at the bottom of the High Pressure Turbine.

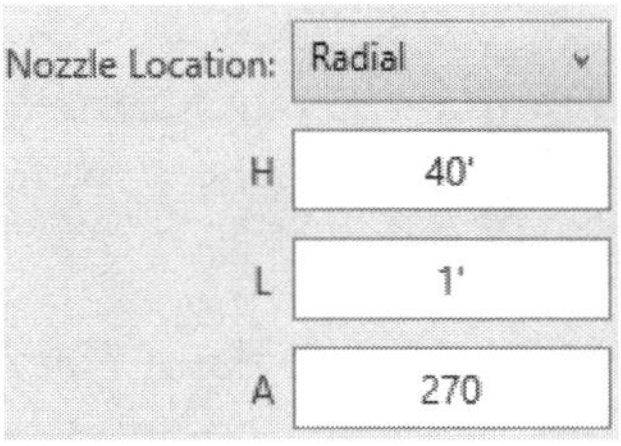

*Figure P-29 Values to be entered when **Change Location** tab is chosen*

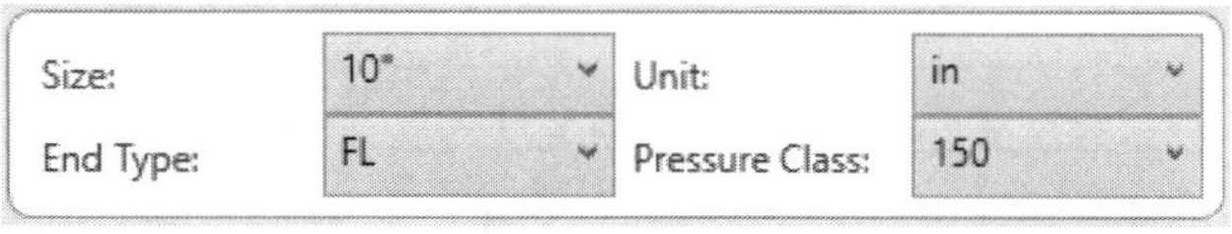

*Figure P-30 Values to be entered when **Change Type** tab is chosen*

Next, choose **Nozzle, flanged, 10" ND, RF, 150, ASME B16.5** from the **Select Nozzle** area by double-clicking on it and then choose the **Close** button; the nozzle is placed at the bottom of the High Pressure Turbine.

Adding Nozzles to the Condenser

Next, you need to add four nozzles to the Condenser (one at the top and three radial nozzles). Figure P-31 shows the Condenser in **Back** view labeled with nozzle numbers. Note that the nozzles are labeled only for your reference.

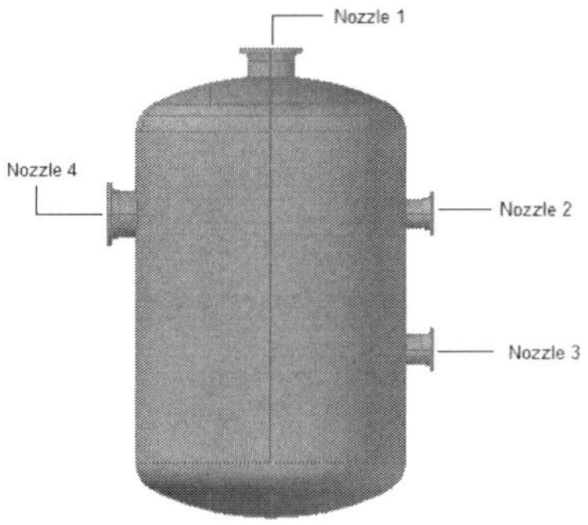

Figure P-31 *Condenser labelled with nozzle numbers (in **Back** view)*

Figures P-32 through P-35 show the values to be entered for nozzles labeled 1 to 4 in the **Change Location** and the **Change Type** tabs, respectively.

For Nozzle 1

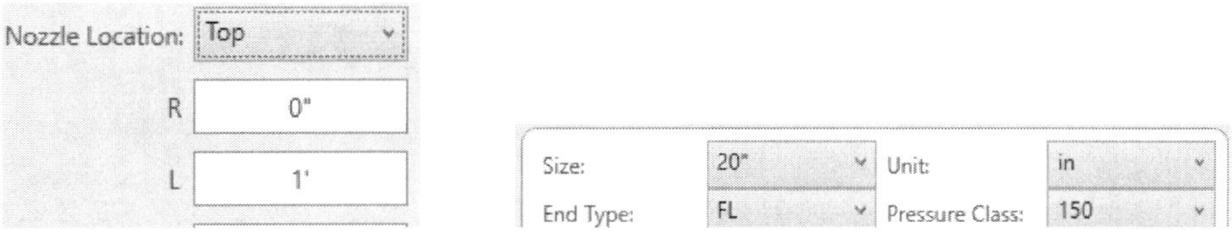

Figure P-32 *Values to be entered under **Change Location** and **Change Type** tabs for Nozzle 1*

After entering all the values under both the tabs, select **Nozzle, flanged, 20" ND, RF, 150, ASME B16.5** from the **Select Nozzle** area of the **Change Type** tab by double-clicking on it and then choose the **Close** button.

For Nozzle 2

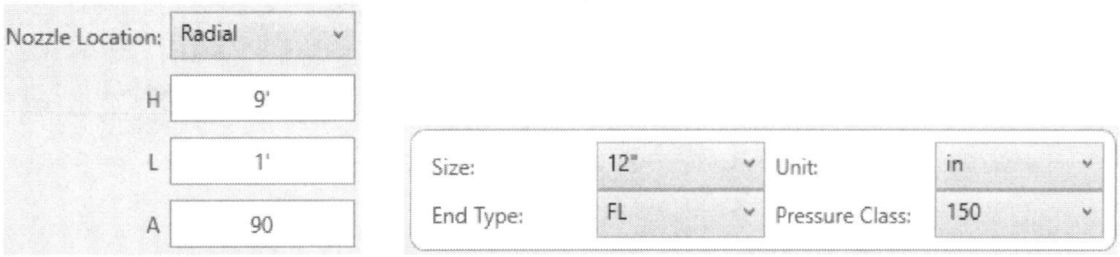

Figure P-33 *Values to be entered under **Change Location** and **Change Type** tabs for Nozzle 2*

Next, select **Nozzle, flanged, 12" ND, RF, 150, ASME B16.5** from the **Select Nozzle** area of the **Change Type** tab by double-clicking on it and then choose the **Close** button.

For Nozzle 3

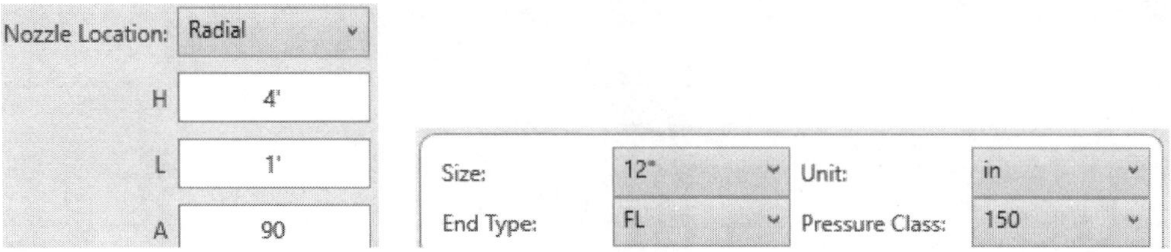

Figure P-34 Values to be entered under ***Change Location*** *and* ***Change Type*** *tabs for Nozzle 3*

Next, select **Nozzle, flanged, 12" ND, RF, 150, ASME B16.5** from the **Select Nozzle** area of the **Change Type** tab by double-clicking on it and then choose the **Close** button.

For Nozzle 4

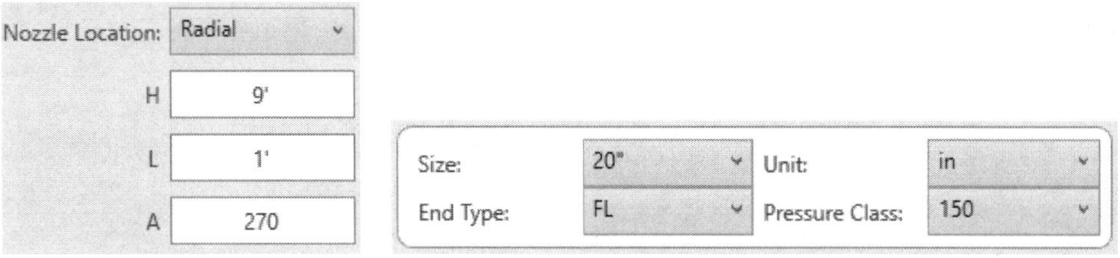

Figure P-35 Values to be entered under ***Change Location*** *and* ***Change Type*** *tabs for Nozzle 4*

Next, select **Nozzle, flanged, 20" ND, RF, 150, ASME B16.5** from the **Select Nozzle** area of the **Change Type** tab by double-clicking on it and then choose the **Close** button.

Adding and Modifying Nozzles of the Cooling Tower

Next, you need to add two nozzles to the Cooling Tower at the required locations and remove the existing ones. First, you will delete the existing nozzles of the Cooling Tower. To do so, press the CTRL key and select all the existing nozzles. Next, press the DELETE key. All the existing nozzles is deleted. Now, you will add nozzles at the required locations. Figure P-36 shows the Cooling Tower labeled with nozzle numbers.

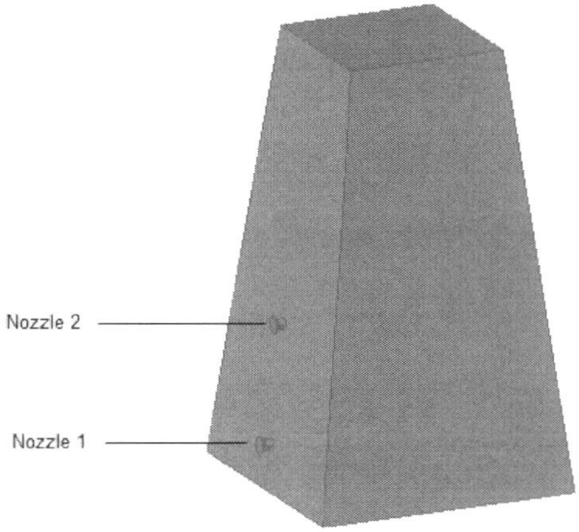

Figure P-36 *Cooling Tower labelled with nozzle numbers*

> **Tip**
> *You can also modify the existing nozzles of the Cooling Tower instead of deleting them.*

Figures P-37 and P-38 show the values to be entered (for nozzles labeled 1 and 2) in the **Change Location** and **Change Type** tabs, respectively.

For Nozzle 1

Figure P-37 *Values to be entered under* ***Change Location*** *and* ***Change Type*** *tabs for Nozzle 1*

After entering all the values under both the tabs, choose **Nozzle, flanged, 12" ND, RF, 150, ASME B16.5** from the **Select Nozzle** area of the **Change Type** tab by double-clicking on it and then choose the **Close** button.

For Nozzle 2

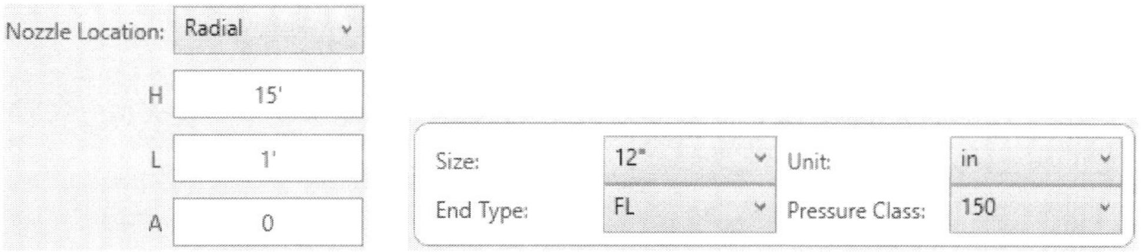

Figure P-38 *Values to be entered under* ***Change Location*** *and* ***Change Type*** *tabs for Nozzle 2*

Next, select **Nozzle, flanged, 12" ND, RF, 150, ASME B16.5** from the **Select Nozzle** area of the **Change Type** tab by double-clicking on it and then choose the **Close** button.

Modifying Nozzles of the Vertical Inline Pump

Now, you need to modify the existing nozzles of the Vertical Inline Pump. To do so, select the Vertical Inline Pump from the drawing area; two pencil grips is displayed on it. Next, click on the pencil grip which is closer to the Cooling Tower; a dialog box for adding or modifying nozzles is displayed. Choose the **Change Location** tab and enter the values shown in Figure P-39. Next, choose the **Change Type** tab and enter the values shown in Figure P-40.

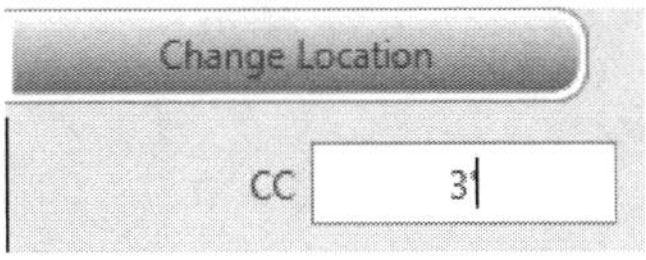

Figure P-39 *Values to be entered when* ***Change Location*** *tab is chosen*

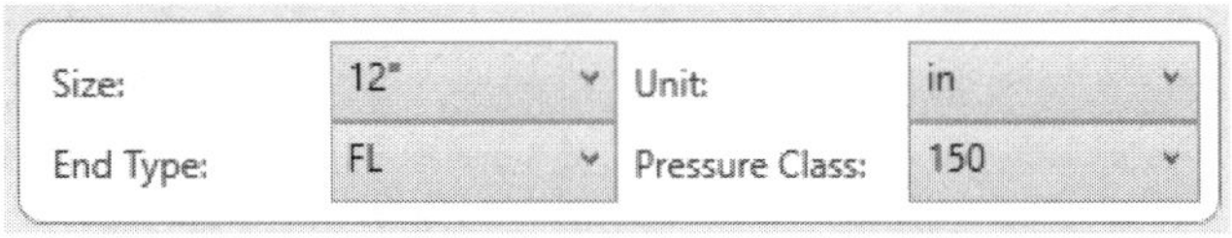

Figure P-40 *Values to be entered when* ***Change Type*** *tab is chosen*

After entering all the values under both the tabs, select **Nozzle, flanged, 12" ND, RF, 150, ASME B16.5** from the **Select Nozzle** area of the **Change Type** tab by double-clicking on it and then choose the **Close** button.

Again, select the Vertical Inline Pump and click on the other pencil grip; a dialog box for adding or modifying nozzles is displayed. Choose the **Change Location** tab and enter the values shown in Figure P-41. Next, choose the **Change Type** tab and enter the values shown in Figure P-41.

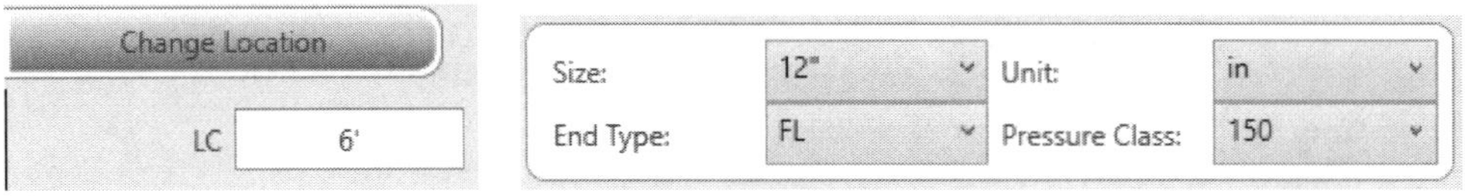

Figure P-41 *Values to be entered when* ***Change Location*** *and* ***Change Type*** *tabs are chosen*

After entering all the values under both the tabs, choose **Nozzle, flanged, 12" ND, RF, 150, ASME B16.5** from the **Select Nozzle** area of the **Change Type** tab by double-clicking on it and then choose the **Close** button. Figure P-42 shows the Vertical Inline Pump after modifying the nozzles.

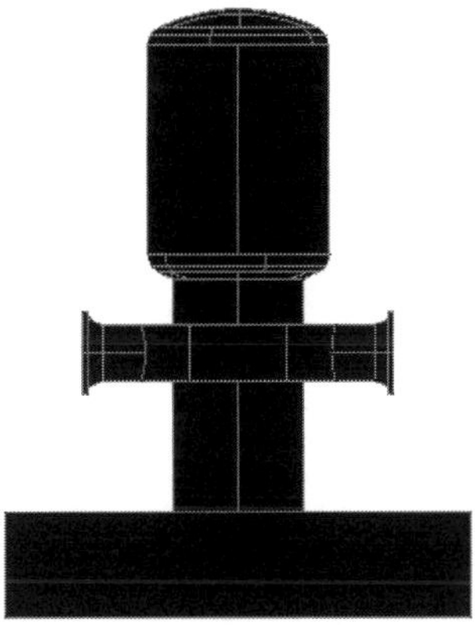

Figure P-42 Vertical Inline Pump after modifying nozzles

Modifying Nozzles of the Centrifugal Pump1

Now, you need to modify the existing nozzles of the Centrifugal Pump1. To do so, select the Centrifugal Pump1 from the drawing area; two pencil grips are displayed on it. Click on the grip which is displayed on the inlet side of the pump; a dialog box for adding or modifying nozzles is displayed. Choose the **Change Location** tab and enter the values shown in Figure P-43. Next, choose the **Change Type** tab and enter the values shown in Figure P-44.

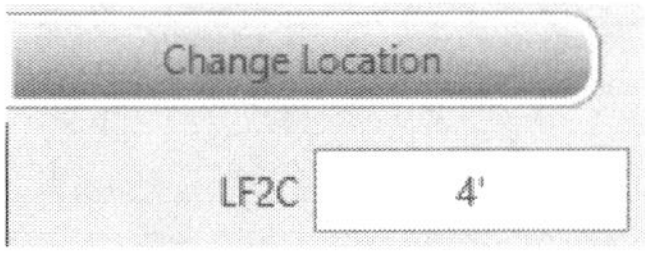

Figure P-43 *Values to be entered when* ***Change Location*** *tab is chosen*

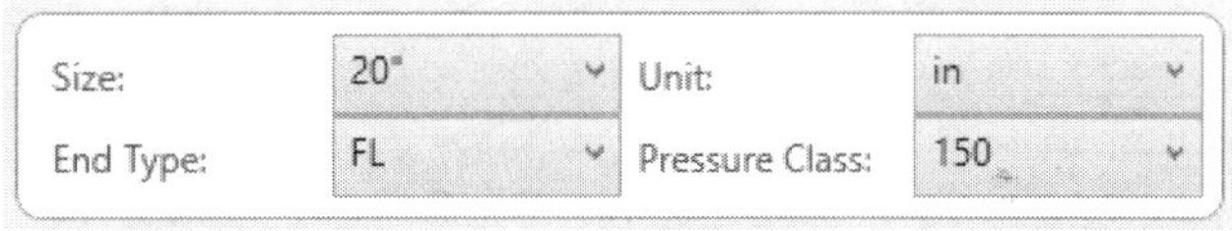

Figure P-44 *Values to be entered when* ***Change Type*** *tab is chosen*

After entering all the values under both the tabs, choose **Nozzle, flanged, 20" ND, RF, 150, ASME B16.5** from the **Select Nozzle** area of the **Change Type** tab by double-clicking on it and then choose the **Close** button.

Similarly, to modify the outlet nozzle of the pump, enter the values shown in Figures P-45 and P-46 under the **Change Location** and **Change Type** tabs, respectively.

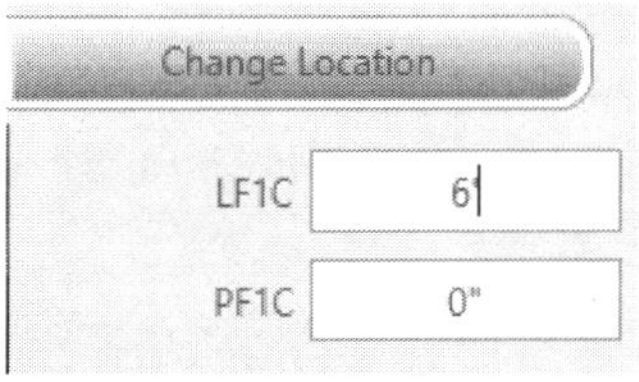

Figure P-45 *Values to be entered when* ***Change Location*** *tab is chosen*

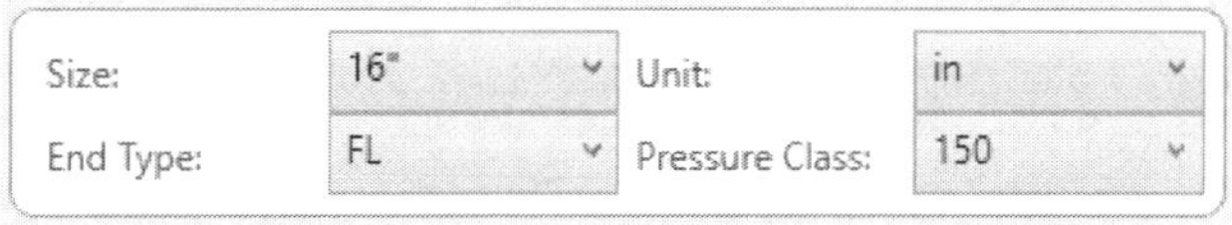

Figure P-46 *Values to be entered when* ***Change Type*** *tab is chosen*

After entering all the values under both the tabs, choose **Nozzle, flanged, 16" ND, RF, 150, ASME B16.5** from the **Select Nozzle** area of the **Change Type** tab by double-clicking on it and then choose the **Close** button.

Adding and Modifying Nozzles of the Low Pressure Heater

Now, you need to delete the existing nozzles of the Low Pressure Heater and add new nozzles to it at the required locations. So, you will first remove the existing nozzles of the Low Pressure Heater. To do so, press the CTRL key and select all the existing nozzles at once and then press the DELETE key.

Next, follow the same procedure to add two nozzles to the Low Pressure Heater. Figures P-47 and P-48 show the values to be entered under the **Change Location** and **Change Type** tabs for the first nozzle (the one facing Centrifugal Pump1).

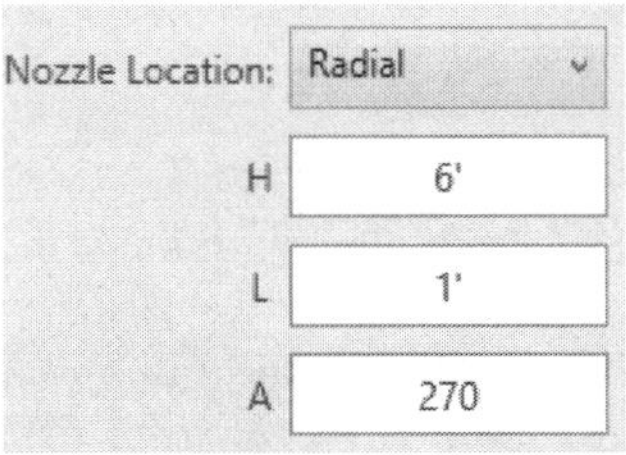

*Figure P-47 Values to be entered when **Change Location** tab is chosen*

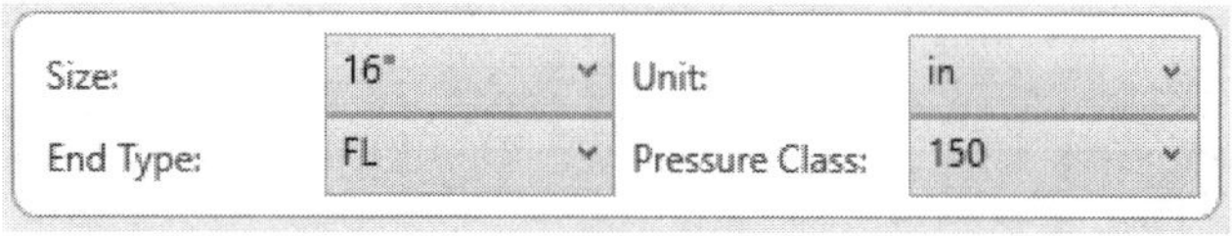

*Figure P-48 Values to be entered when **Change Type** tab is chosen*

After entering all the values under both the tabs, choose **Nozzle, flanged, 16" ND, RF, 150, ASME B16.5** from the **Select Nozzle** area of the **Change Type** tab by double-clicking on it and then choose the **Close** button.

Figures P-49 and P-50 show the values to be entered under the **Change Location** and **Change Type** tabs for the second nozzle (the one facing Deaerator).

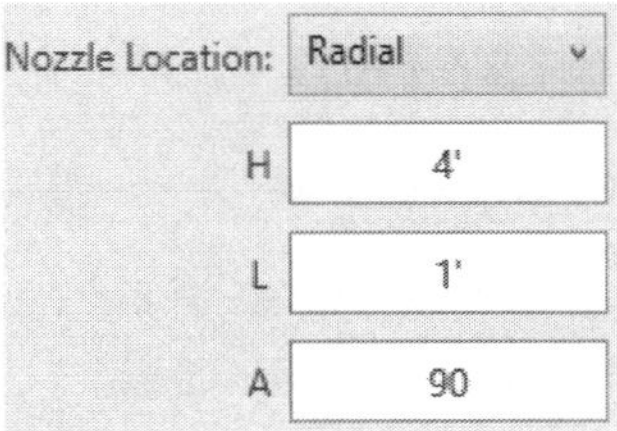

*Figure P-49 Values to be entered when **Change Location** tab is chosen*

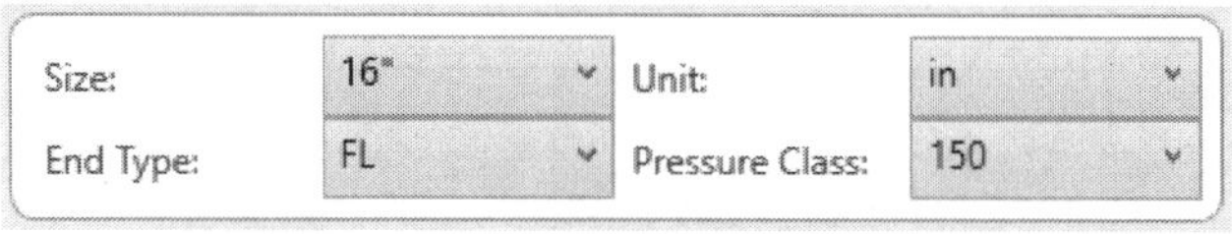

*Figure P-50 Values to be entered when **Change Type** tab is chosen*

After entering all the values under both the tabs, choose **Nozzle, flanged, 16" ND, RF, 150, ASME B16.5** from the **Select Nozzle** area of the **Change Type** tab by double-clicking on it and then choose the **Close** button.

Adding Nozzles to the Deaerator

Now, we need to add two nozzles to the Deaerator. Figures P-51 and P-52 show the values to be entered under the **Change Location** and **Change Type** tabs for the first nozzle (the one facing Low Pressure Heater).

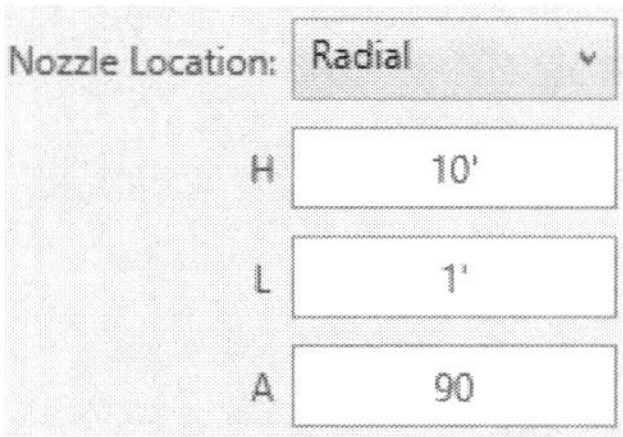

Figure P-51 *Values to be entered when* ***Change Location*** *tab is chosen*

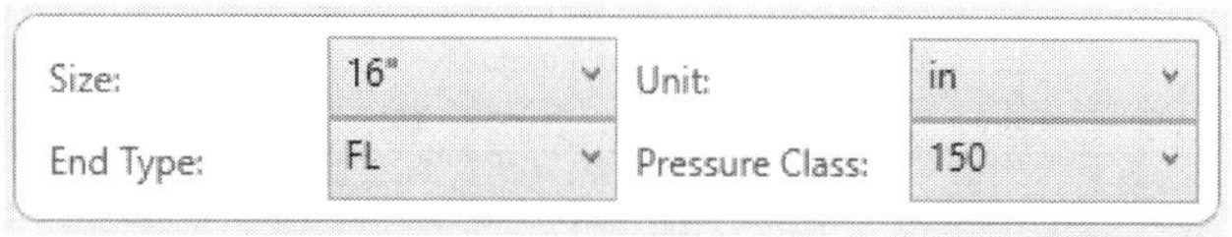

Figure P-52 *Values to be entered when* ***Change Type*** *tab is chosen*

After entering all the values under both the tabs, choose **Nozzle, flanged, 16" ND, RF, 150, ASME B16.5** from the **Select Nozzle** area of the **Change Type** tab by double-clicking on it and then choose the **Close** button.

Figures P-53 and P-54 show the values to be entered under the **Change Location** and **Change Type** tabs for the second nozzle (the one at the bottom center of Deaerator).

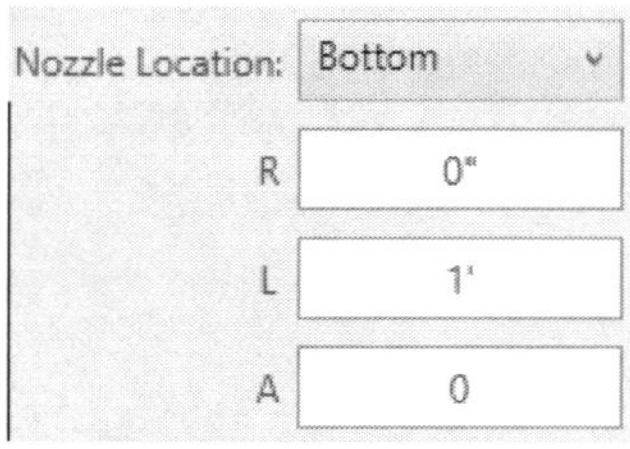

Figure P-53 *Values to be entered when* ***Change Location*** *tab is chosen*

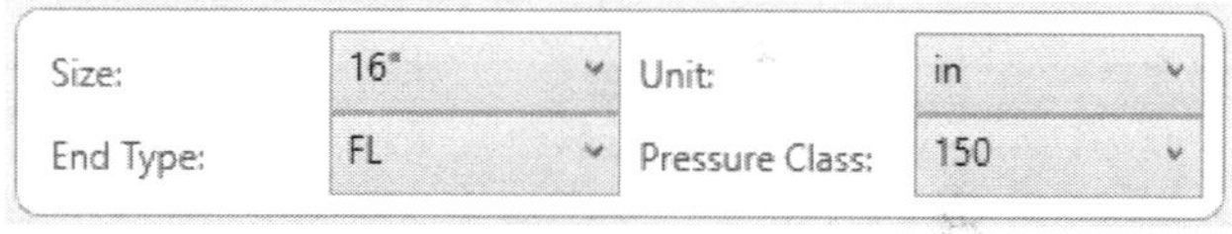

Figure P-54 *Values to be entered when* ***Change Type*** *tab is chosen*

After entering all the values under both the tabs, choose **Nozzle, flanged, 16" ND, RF, 150, ASME B16.5** from the **Select Nozzle** area of the **Change Type** tab by double-clicking on it and then choose the **Close** button.

Modifying Nozzles of the Centrifugal Pump2

Now, you need to modify the existing nozzles of the Centrifugal Pump2. To do so, select the **Centrifugal Pump2** from the drawing area; two pencil grips are displayed on it. Next, click on the grip displayed on the inlet side of the pump; a dialog box for adding or modifying nozzles is displayed. Choose the **Change Location** tab and enter the values shown in Figure P-55. Next, choose the **Change Type** tab and enter the values shown in Figure P-56.

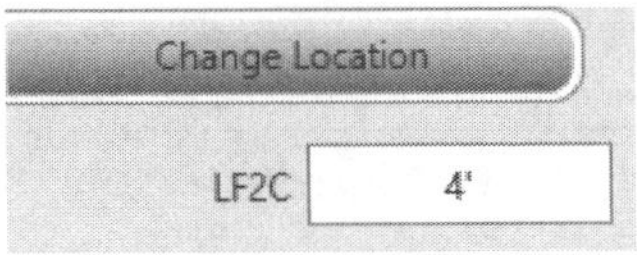

Figure P-55 *Values to be entered when* ***Change Location*** *tab is chosen*

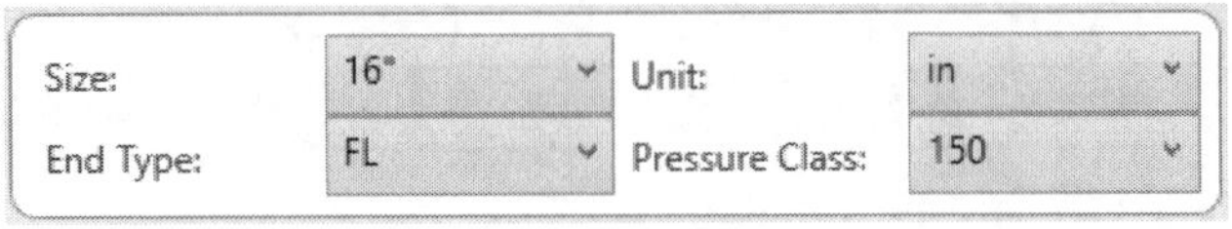

Figure P-56 *Values to be entered when* ***Change Type*** *tab is chosen*

After entering all the values under both the tabs, choose **Nozzle, flanged, 16" ND, RF, 150, ASME B16.5** from the **Select Nozzle** area of the **Change Type** tab by double-clicking on it and then choose the **Close** button.

Similarly, to modify the outlet nozzle of the pump, enter the values shown in Figures P-57 and P-58 under the **Change Location** and **Change Type** tabs, respectively.

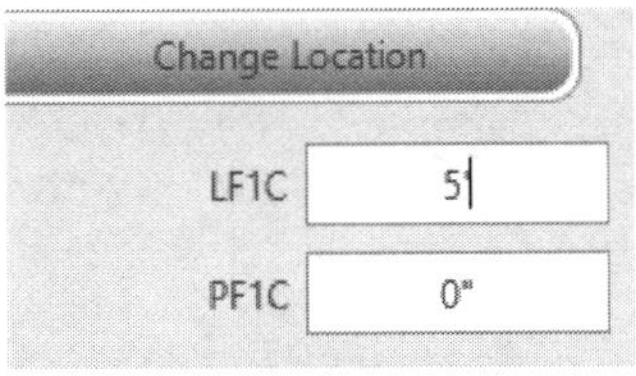

Figure P-57 *Values to be entered when the* ***Change Location*** *tab is chosen*

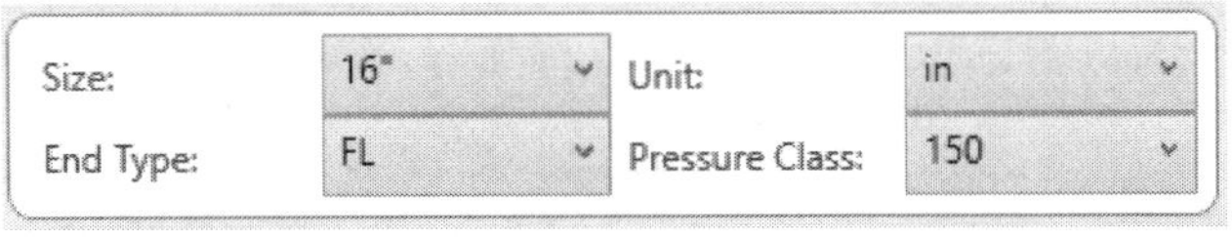

Figure P-58 *Values to be entered when the* ***Change Type*** *tab is chosen*

After entering all the values under both the tabs, choose **Nozzle, flanged, 16" ND, RF, 150, ASME B16.5** from the **Select Nozzle** area of the **Change Type** tab by double-clicking on it and then choose the **Close** button.

Adding and Modifying Nozzles of the High Pressure Heater

Now, you need to delete the existing nozzles of the High Pressure Heater and add new nozzles to it at the required locations. Firstly, you will delete the existing nozzles of the High Pressure Heater. To do so, press the CTRL key and select all the existing nozzles at once and then press the DELETE key.

Next, follow the same procedure explained above to add two nozzles to the High Pressure Heater. Figures P-59 and P-60 show the values to be entered under the **Change Location** and **Change Type** tabs for the first nozzle (the one facing Centrifugal Pump2).

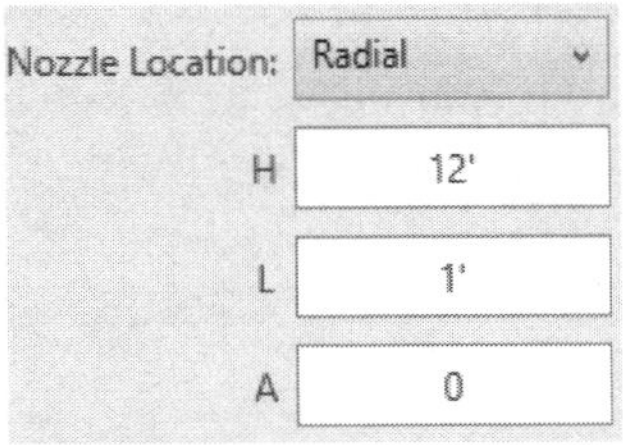

Figure P-59 *Values to be entered when the* ***Change Location*** *tab is chosen*

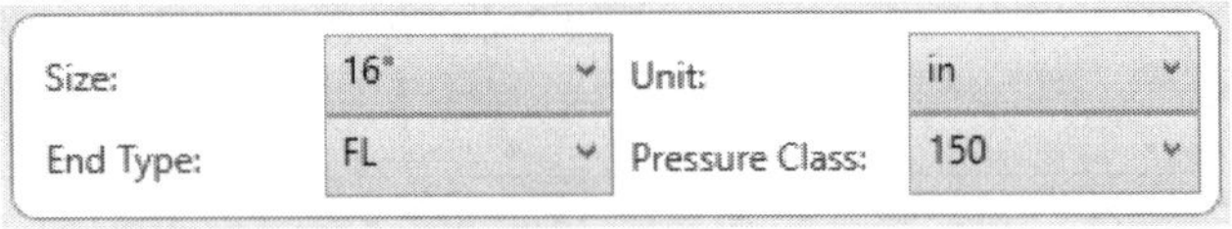

Figure P-60 *Values to be entered when the* ***Change Type*** *tab is chosen*

After entering all the values under both the tabs, choose **Nozzle, flanged, 16" ND, RF, 150, ASME B16.5** from the **Select Nozzle** area of the **Change Type** tab by double-clicking on it and then choose the **Close** button.

Figures P-61 and P-62 show the values to be entered under the **Change Location** and **Change Type** tabs for the second nozzle (the one facing Boiler).

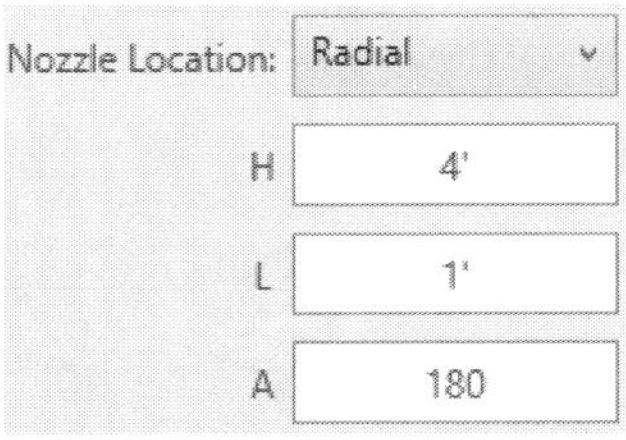

*Figure P-61 Values to be entered when the **Change Location** tab is chosen*

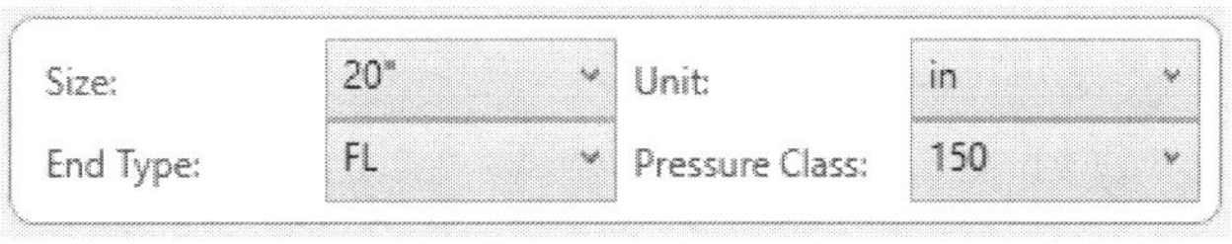

*Figure P-62 Values to be entered when the **Change Type** tab is chosen*

After entering all the values under both the tabs, choose **Nozzle, flanged, 20" ND, RF, 150, ASME B16.5** from the **Select Nozzle** area of the **Change Type** tab by double-clicking on it and then choose the **Close** button.

Adding Nozzles to the Boiler and Boiler Drum

Now, you will place nozzles on the Boiler which is a custom equipment you created earlier. To do so, select the Boiler from the drawing area; the **Add Nozzle** grip is displayed on it. Click on the **Add Nozzle** grip, you are prompted to specify the nozzle center. Press and hold the SHIFT key and then right-click; a flyout is displayed. Next, choose the **From** option from the flyout; you are prompted to specify the base point. Specify the base point at the bottom left vertex of the Boiler, refer to Figure P-63. After specifying the base point, you are prompted to specify the offset value. Enter @**-6',12',0'** at the command prompt; you are prompted to specify the nozzle direction. Move the cursor vertically down and click to specify the nozzle direction; a dialog box for adding or modifying nozzles is displayed. Next, choose the **Change Type** tab and enter the values shown in Figure P-64.

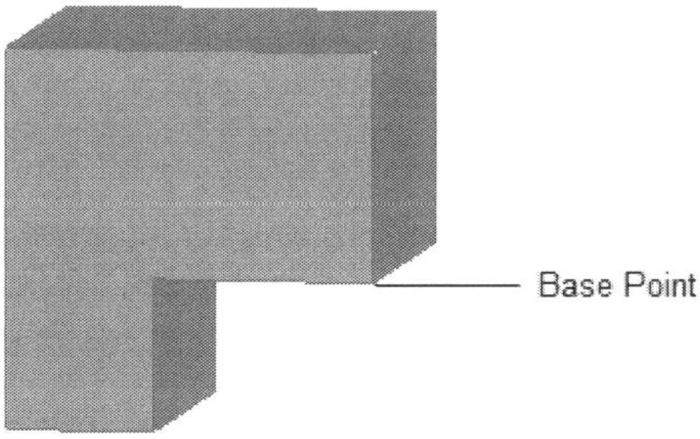

*Figure P-63 Base point to be selected while using **From** snap*

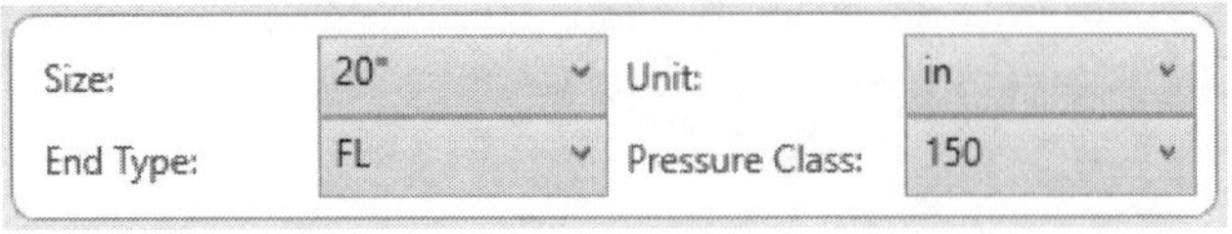

***Figure P-64** Values to be entered when the **Change Type** tab*

After entering all the values under the **Change Type** tab, choose **Nozzle, flanged, 20" ND, RF, 150, ASME B16.5** from the **Select Nozzle** area of the **Change Type** tab by double-clicking on it and then choose the **Close** button.

Now, you have specified the nozzle location. Next, you need to place a flange at the nozzle point. To do so, choose **FLANGE WN, FL, RF, 150, (POWER PLANT)** from the **Flange** area in the **Dynamic Pipe Spec** tab of the **TOOL PALETTES - AUTOCAD PLANT 3D - PIPING COMPONENTS**; you are prompted to specify the insertion point. Press and hold the SHIFT key and right-click, a flyout is displayed. Next, choose the **Node** option from the flyout and place the flange at the node of the nozzle point located earlier; you are prompted to specify the rotation angle. Specify the rotation angle as 0 degrees using the cursor and then press the ESC key; the flange is placed at the specified location, as shown in Figure P-65.

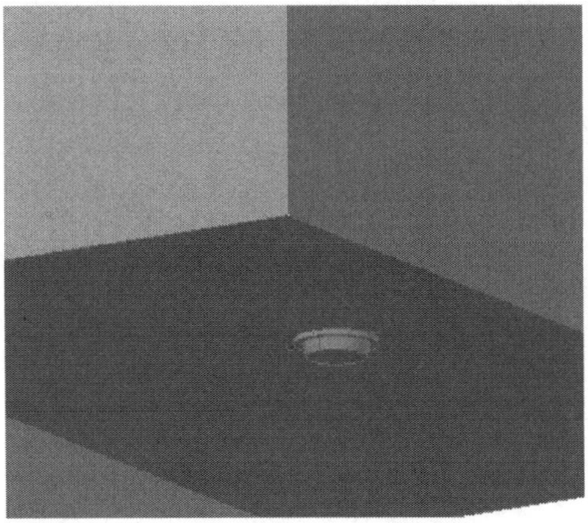

***Figure P-65** Flange placed at the specified location*

Note

*While specifying coordinates for the placement of nozzles on the Boiler unit, make sure that the UCS is set to **World**.*

Now, you need to add nozzles at the top face of the Boiler. Figure P-66 shows the Boiler labeled with nozzle numbers which are required to be placed at its top face.

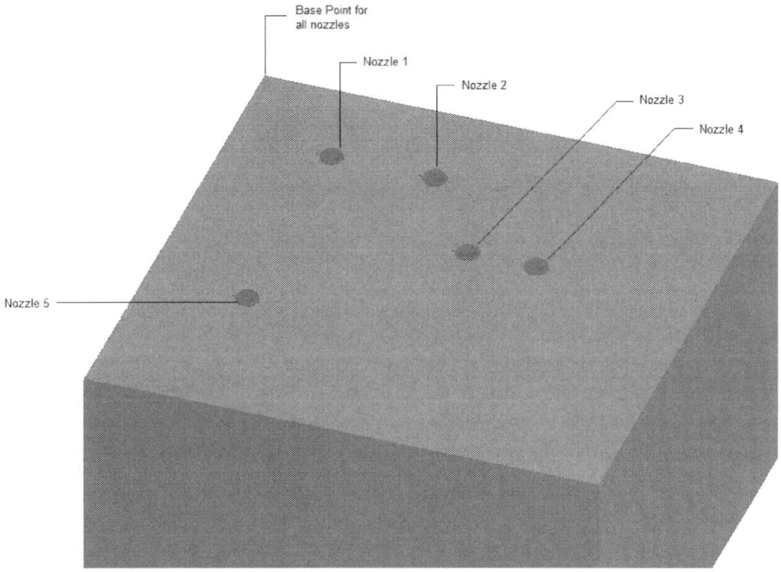

Figure P-66 *Top face of the Boiler with labelled nozzles*

Nozzle 1

To place nozzle 1, select the Boiler from the drawing area; the **Add Nozzle** grip is displayed. Click on the **Add Nozzle** grip; you are prompted to specify the nozzle center. Press and hold the SHIFT key and then right-click; a flyout is displayed. Next, choose the **From** option from the flyout, you are prompted to specify the base point. Specify the base point, refer to Figure P-66; you are prompted to specify the offset value. Enter @**6',-6',0'** at the command prompt, you are prompted to specify the nozzle direction. Move the cursor vertically up and click to specify the nozzle direction; a dialog box for adding or modifying nozzles is displayed. Next, choose the **Change Type** tab and enter the values shown in Figure P-67.

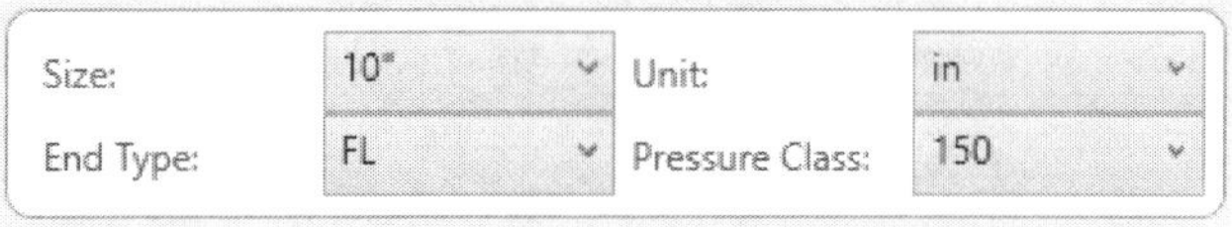

Figure P-67 *Values to be entered in the* ***Change Type*** *tab*

After entering all the values under the **Change Type** tab, choose **Nozzle, flanged, 10" ND, RF, 150, ASME B16.5** from the **Select Nozzle** area of the **Change Type** tab by double-clicking on it and then choose the **Close** button. This step will remain same for rest of the nozzles to be placed on the boiler.

Now, you have specified the nozzle location. Next, you need to place a flange at the nozzle point. To do so, choose the **FLANGE WN, FL, RF, 150, (POWER PLANT)** from the Flange area in the **Dynamic Pipe Spec** tab of the **TOOL PALETTES - AUTOCAD PLANT 3D - PIPING COMPONENTS**; you are prompted to specify the insertion point. Press and hold the SHIFT key and right-click, a flyout is displayed. Next, choose the **Node** option from the flyout and place the flange at the node of the nozzle point located earlier; you are prompted to specify the rotation angle. Specify the rotation angle as **0** degree using the cursor and then press the ESC key; the flange is placed at the required location, refer to Figure P-66.

Similarly, place the other nozzles at the required locations, refer to Figure P-66. Note that the base point will remain same for all the nozzles while using the **From** snap and the nozzle direction is vertically upwards for all the nozzles. The offset values for nozzles numbered 2 to 5 are given next.

Nozzle 2 : **@12',-6',0'**
Nozzle 3 : **@16',-12',0'**
Nozzle 4 : **@20',-12',0'**
Nozzle 5 : **@6',-20',0'**

After specifying the nozzle points, place **FLANGE WN, FL, RF, 150, (POWER PLANT)** at all the nozzle points. After placing the flanges, the top face of Boiler should look similar to the one shown in Figure P-66.

Next, you need to place two nozzles, one at the front and other at the back face of the Boiler. First, you will place the nozzle on the front face. To do so, follow the same procedure as explained before. Refer to Figure P-68 for the base point to be selected while using the **From** snap.

Figure P-68 *Base point to be selected while using* ***From*** *snap*

Offset value for nozzle on front face : **@0',12',8'**
Similarly, place the nozzle on back face of the Boiler. Refer to Figure P-69 for the base point to be selected while using the **From** snap.

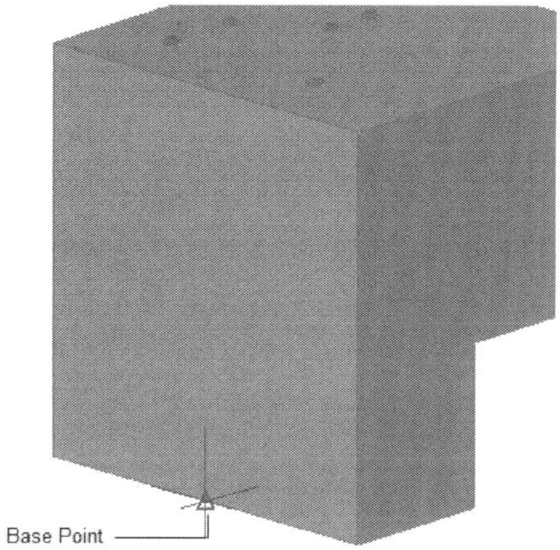

Figure P-69 *Base point to be selected while using* ***From*** *snap*

Offset value for nozzle on back face : **@0,0,12'**

After specifying the nozzle points on the front and back faces, place **FLANGE WN, FL, RF, 150, (POWER PLANT)** at both the nozzle points.

Next, you need to place nozzles on the Boiler Drum. Figure P-70 shows the Boiler Drum labeled with nozzle numbers.

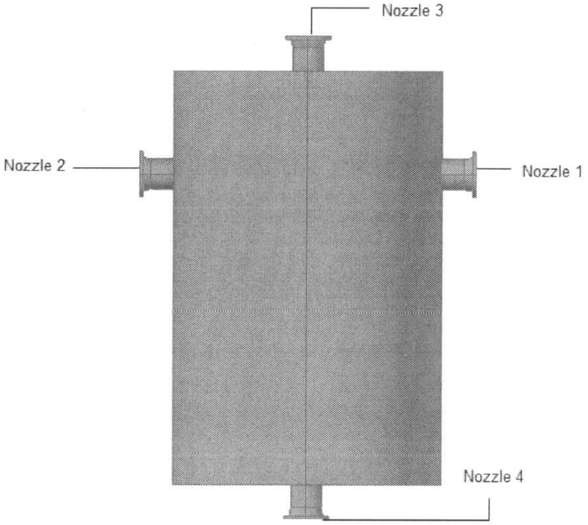

Figure P-70 *Boiler Drum labeled with nozzle numbers*

Figures P-71 through P-74 show the values to be entered (for nozzles labeled 1 to 4) in the **Change Location** and the **Change Type** tabs, respectively.

For Nozzle 1

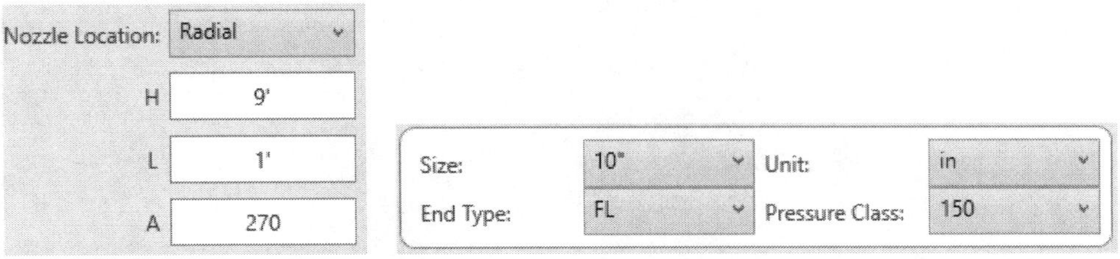

***Figure P-71** Values to be entered under **Change Location** and **Change Type** tabs for Nozzle 1*

After entering all the values under both the tabs, choose **Nozzle, flanged, 10" ND, RF, 150, ASME B16.5** from the **Select Nozzle** area of the **Change Type** tab by double-clicking on it and then choose the **Close** button. This step will remain same for all the nozzles to be placed on the Boiler Drum.

For Nozzle 2

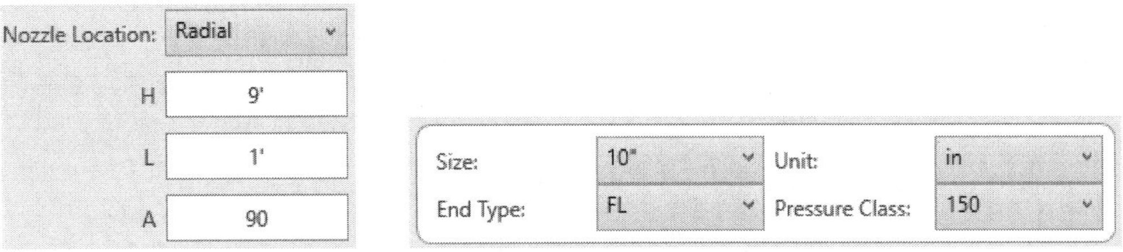

***Figure P-72** Values to be entered under **Change Location** and **Change Type** tabs for Nozzle 2*

For Nozzle 3

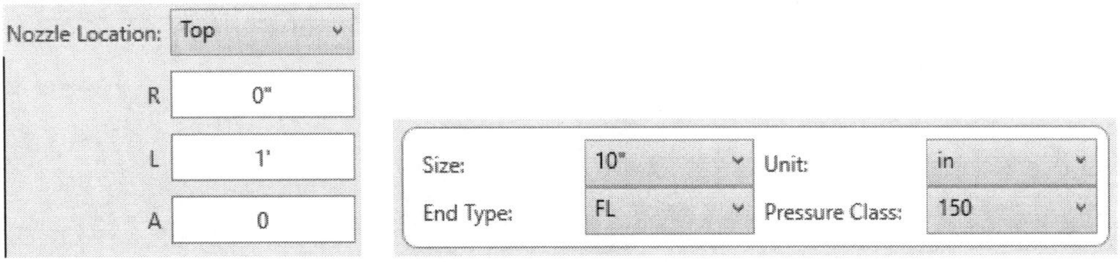

***Figure P-73** Values to be entered under **Change Location** and **Change Type** tabs for Nozzle 3*

For Nozzle 4

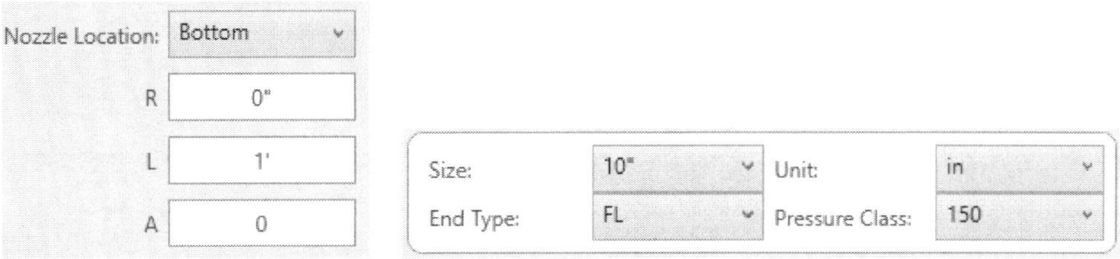

Figure P-74 *Values to be entered under* ***Change Location*** *and* ***Change Type*** *tabs for Nozzle 4*

CONNECTING THE EQUIPMENTS

After placing the nozzles at the required locations, you need to connect the equipment. This can be done by making piping connections between the equipment.

First, you will connect the Low Pressure Turbine with the Condenser. To do so, choose the **Route New Line** option from the **Line Number Selector** drop-down of the **Part Insertion** panel; the **Assign Tag** dialog box is displayed. Enter **1001** in the **Number** edit box in this dialog box. Next, select the line size as **20"** from the **Size** drop-down list. Next, choose the **Assign** button; the **Assign Tag** dialog box is closed and you are prompted to specify the start point of the pipe. Press and hold the SHIFT key and right-click to display a shortcut menu. Choose the **Node** option from the shortcut menu displayed and then select the node of the nozzle located at the bottom of the Low Pressure Turbine; you are prompted to specify the next point. Move the cursor vertically down and select the node of the nozzle located at the top of the Condenser, the pipe is added between the two nozzles, as shown in Figure P-75.

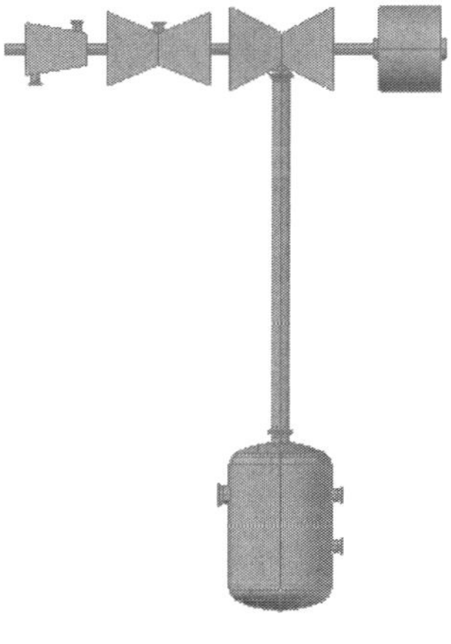

Figure P-75 *Pipe added between bottom nozzle of the Low Pressure Turbine and top nozzle of the Condenser*

Next, you will connect the lower nozzle of the Cooling Tower with the inlet nozzle of the Vertical Inline Pump. To do so, choose the **Route New Line** tool from the **Line Number Selector** drop-down of the **Part Insertion** panel; the **Assign Tag** dialog box is displayed. Enter **2002** in the **Number** edit box in this dialog box. Select the line size as **12"** from the **Size** drop-down list. Next, choose the **Assign** button; the **Assign Tag** dialog box is closed and you are prompted to specify the start point of the pipe. Press and hold the SHIFT key and right-click to display a shortcut menu. Choose the **Node** option from the shortcut menu displayed and then select the node of the lower nozzle of the Cooling Tower; you are prompted to specify the next point. Move the cursor horizontally towards the Vertical Inline Pump and select the node of the inlet nozzle of the Vertical Inline Pump, the pipe is added between the two nozzles as shown in Figure P-76.

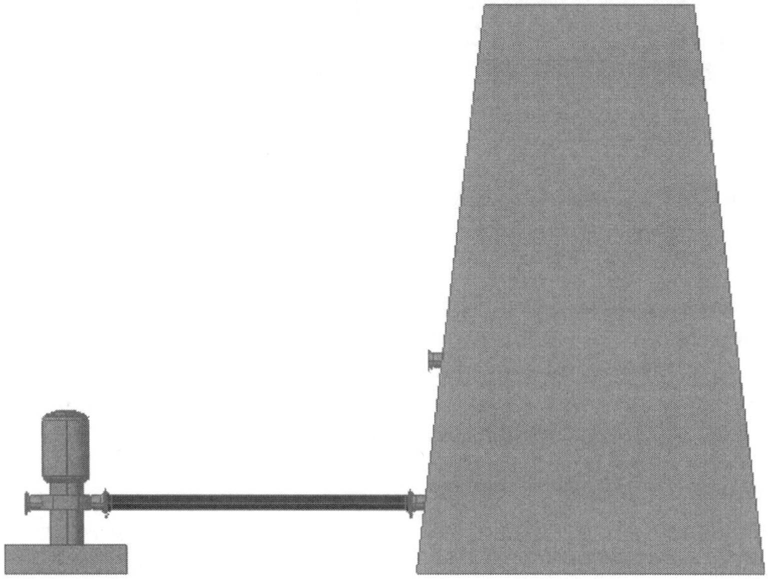

Figure P-76 *Pipe added between lower nozzle of the Cooling Tower and inlet nozzle of the Vertical Inline Pump*

Next, you need to connect the outlet nozzle of the Vertical Inline Pump with the lower radial nozzle of the Condenser. Follow the same procedure explained before to connect the equipment and refer to Figure P-77 for the values to be assigned in the **Assign Tag** dialog box.

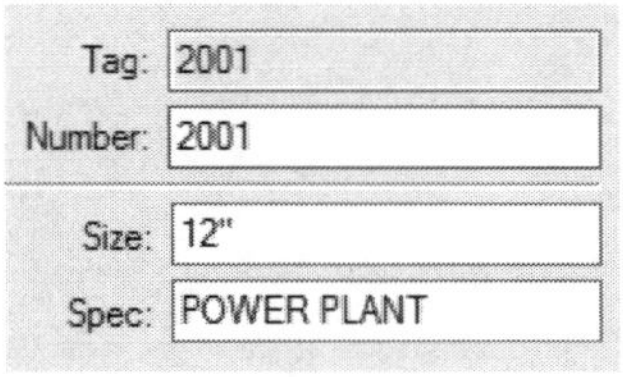

Figure P-77 *Values to be assigned in the* ***Assign Tag*** *dialog box*

Figure P-78 shows the pipe connected between the Vertical Inline Pump and the Condenser.

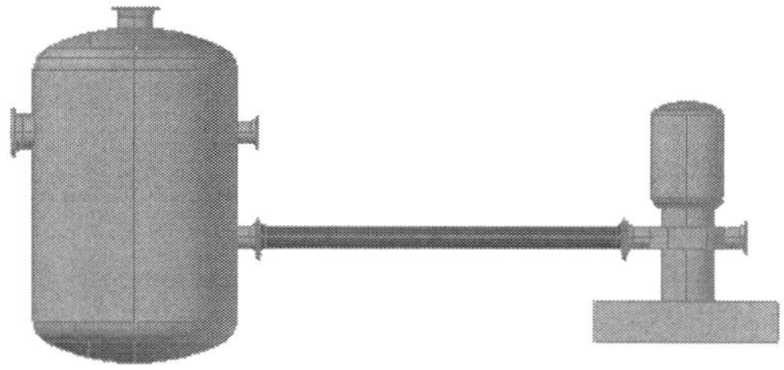

Figure P-78 Pipe added between outlet nozzle of the Vertical Inline Pump and lower radial nozzle of the Condenser for Nozzle 4

Next, you need to add a pipe between the upper radial nozzle of the Condenser with the upper nozzle of the Cooling Tower. To do so, choose the **Route New Line** tool from the **Line Number Selector** drop-down of the **Part Insertion** panel; the **Assign Tag** dialog box is displayed. Enter **2000** in the **Number** edit box in this dialog box. Select the line size as **12"** from the **Size** drop-down list. Next, choose the **Assign** button; the **Assign Tag** dialog box is closed and you are prompted to specify the start point of the pipe. Press and hold the SHIFT key and right-click to display a shortcut menu. Choose the **Node** option from the shortcut menu displayed and then select the node of the upper nozzle of the Condenser; you are prompted to specify the next point. Move the cursor horizontally towards the Cooling Tower and select the node of the upper nozzle of the Cooling Tower. Next, press ENTER to accept the suggested piping solution. Figure P-79 shows the pipe connected between the upper radial nozzle of the Condenser and the upper nozzle of the Cooling Tower.

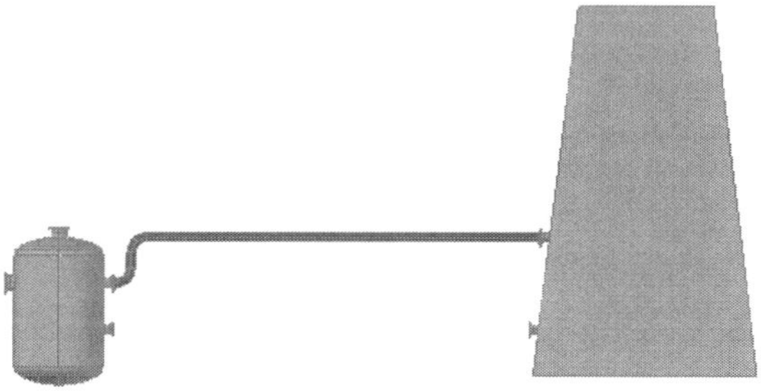

Figure P-79 Pipe added between upper radial nozzle of the Condenser and upper nozzle of the Cooling Tower

Now, you need to connect the radial nozzle of the Condenser facing the Centrifugal Pump1 with the inlet nozzle of the Centrifugal Pump1. To do so, choose the **Route New Line** tool from the **Line Number Selector** drop-down of the **Part Insertion** panel; the **Assign Tag** dialog box is displayed. Enter **1002** in the **Number** edit box in this dialog box. Select the line size as **20"** from the **Size** drop-down list. Next, choose the **Assign** button; the **Assign Tag** dialog box is closed and you are prompted to specify the start point of the pipe. Press and hold the SHIFT key and right-click to display a shortcut menu. Choose the **Node** option from the shortcut menu displayed and then select the node of the radial nozzle of the Condenser; you are prompted to specify the next point. Move the cursor towards the Centrifugal Pump1 and select the node of the inlet nozzle of the Centrifugal Pump1. Next, press ENTER to accept the suggested piping solution. Figure P-80 shows the pipe connected between the radial nozzle of the Condenser and the inlet nozzle of the Centrifugal Pump1.

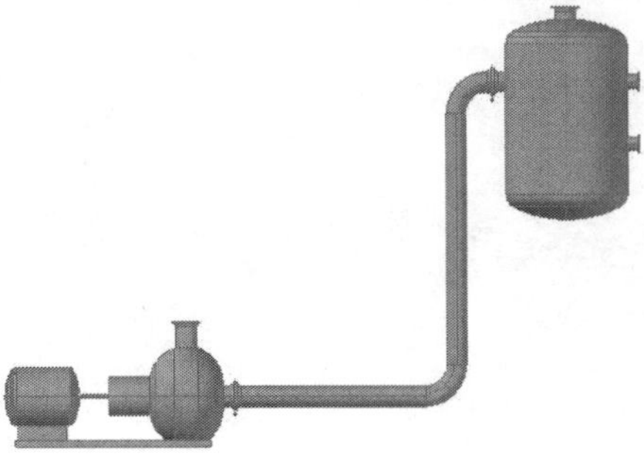

Figure P-80 *Pipe added between upper radial nozzle of the Condenser and inlet nozzle of the Centrifugal Pump1*

Similarly, connect the outlet nozzle of the Centrifugal Pump1 with the inlet nozzle of Low Pressure Heater. Refer to Figure P-81 for the values to be assigned in the **Assign Tag** dialog box.

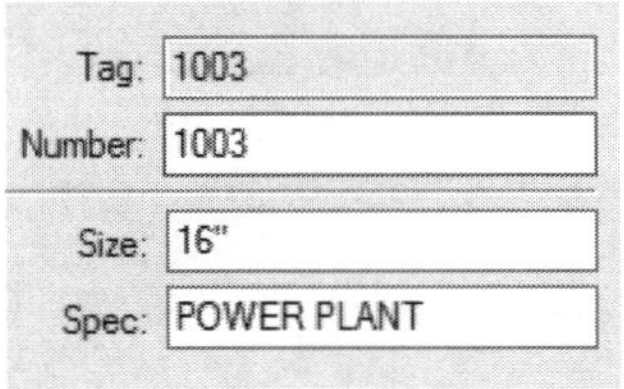

Figure P-81 *Values to be assigned in the* ***Assign Tag*** *dialog box*

Note that you have to click on **Next** option and then choose the **Accept** option in the command prompt when you are prompted to specify the piping solution. Figure P-82 shows the pipe connected between the outlet nozzle of the Centrifugal Pump1 and the inlet nozzle of the Low Pressure Heater.

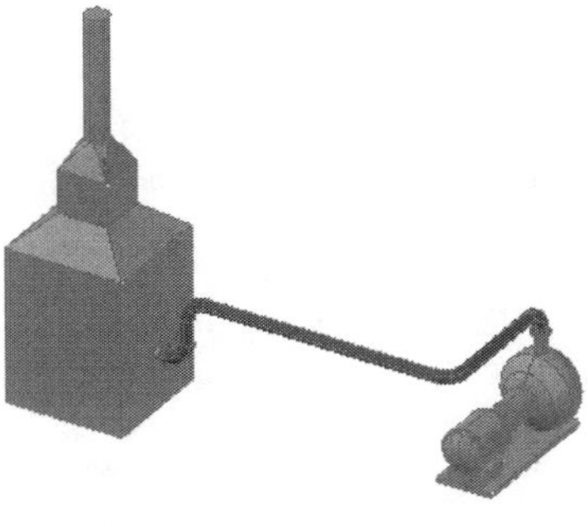

Figure P-82 *Pipe added between outlet nozzle of the Centrifugal Pump1 and inlet nozzle of the Low Pressure Heater*

Similarly, connect the outlet nozzle of the Low Pressure Heater with the inlet nozzle of the Deaerator. Refer to Figure P-83 for the values to be assigned in the **Assign Tag** dialog box. The pipe connected should look like the one shown in Figure P-84.

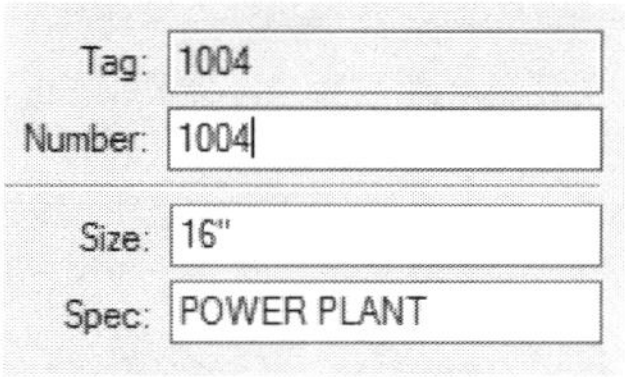

Figure P-83 *Values to be assigned in the* ***Assign Tag*** *dialog box*

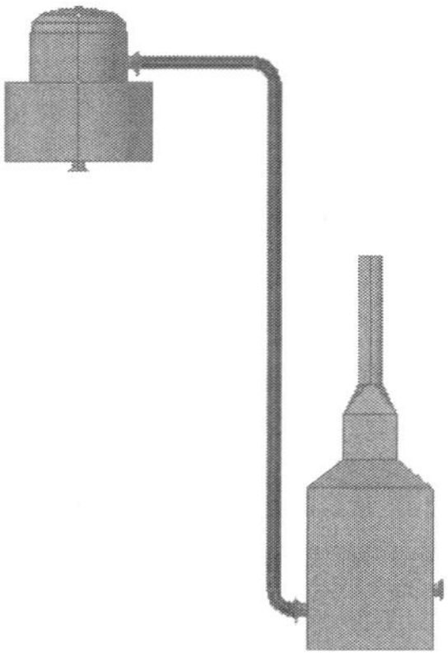

Figure P-84 *Pipe added between outlet nozzle of the Low Pressure Heater and inlet radial nozzle of the Deaerator*

Next, connect the outlet nozzle of the Deaerator with the inlet nozzle of the Centrifugal Pump2. Refer to Figure P-85 for the values to be assigned in the **Assign Tag** dialog box. The pipe connected should look like the one shown in Figure P-86.

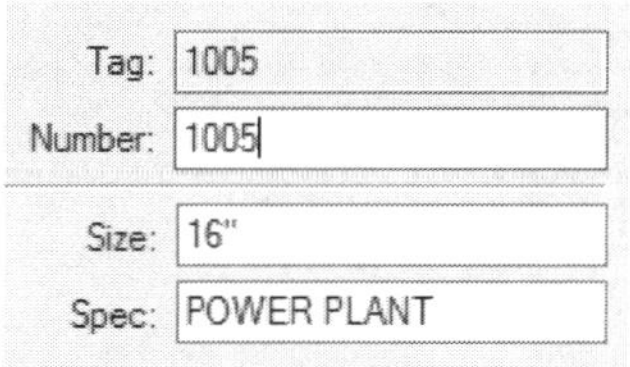

Figure P-85 *Values to be assigned in the* ***Assign Tag*** *dialog box*

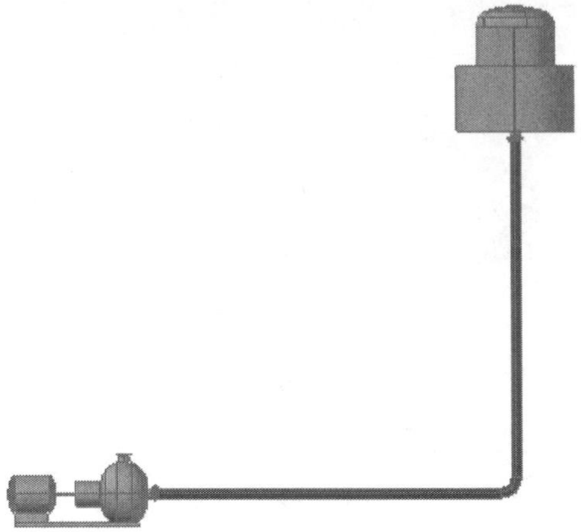

Figure P-86 *Pipe added between outlet nozzle of the Deaerator and inlet nozzle of the Centrifugal Pump2*

Next, connect the outlet nozzle of the Centrifugal Pump2 with the inlet nozzle of the High Pressure Heater. Refer to Figure P-87 for the values to be assigned in the **Assign Tag** dialog box. The pipe connected should look like the one shown in Figure P-88.

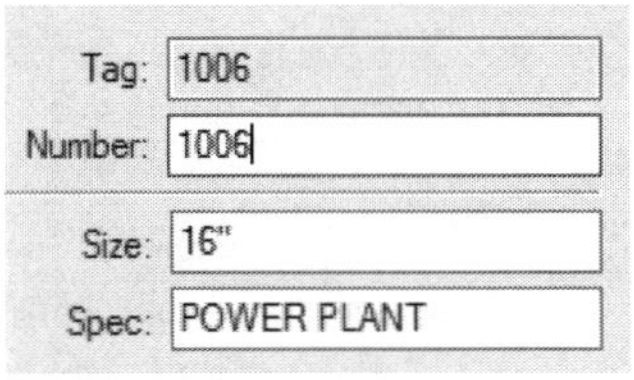

Figure P-87 *Values to be assigned in the* ***Assign Tag*** *dialog box*

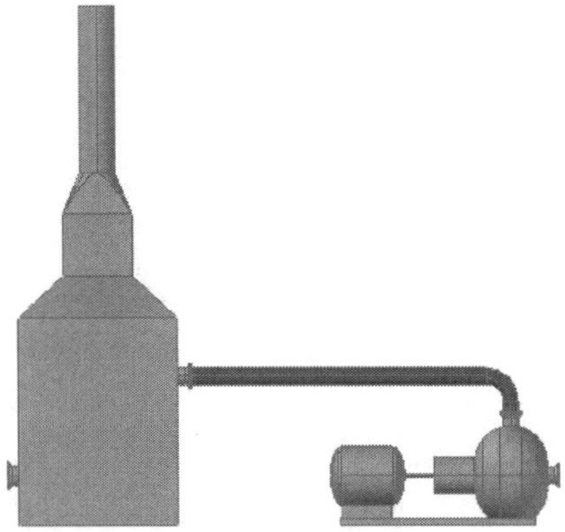

Figure P-88 *Pipe added between outlet nozzle of the Centrifugal Pump2 and inlet nozzle of the High Pressure Heater*

Next, connect the outlet nozzle of the High Pressure Heater with the inlet nozzle of the Economiser unit on the Boiler. Refer to Figure P-89 for the values to be assigned in the **Assign Tag** dialog box. The pipe connected should look like the one shown in Figure P-90.

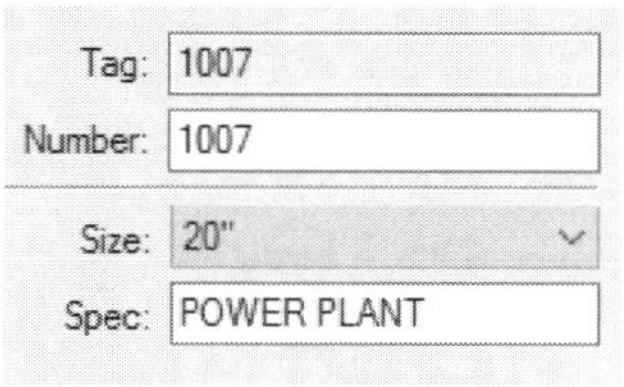

Figure P-89 *Values to be assigned in the* ***Assign Tag*** *dialog box*

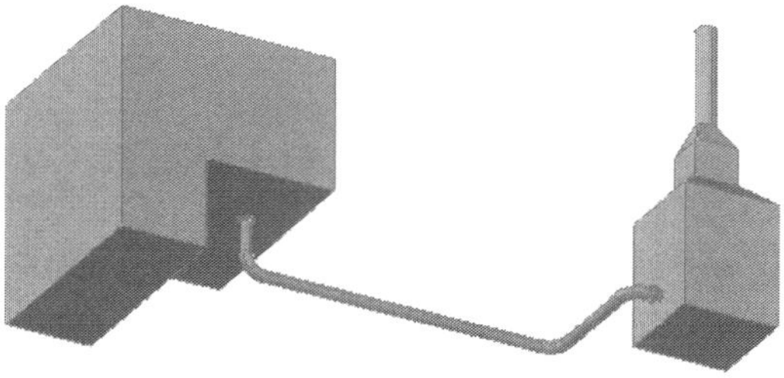

Figure P-90 *Pipe added between outlet nozzle of the High Pressure Heater and inlet nozzle of the Economiser unit on the Boiler*

Now, connect the outlet nozzle of the Economiser unit with the left radial nozzle on the Boiler Drum. To do so, choose the **Route New Line** tool from the **Line Number Selector** drop-down of the **Part Insertion** panel; the **Assign Tag** dialog box is displayed. Enter **3000** in the **Number** edit box in this dialog box. Select the line size as **10"** from the **Size** drop-down list. Next, choose the **Assign** button; the **Assign Tag** dialog box is closed and you are prompted to specify the start point of the pipe. Press and hold the SHIFT key and right-click to display a shortcut menu. Choose the **Node** option from the shortcut menu displayed and then select the node of the outlet nozzle of the Economiser unit; you are prompted to specify the next point. Move the cursor horizontally towards right and enter **4'** at the command prompt; you are prompted to specify the next point. Next, choose the **Plane** option from the command prompt. Move the cursor vertically upwards and enter **25'** at the command prompt. Next, press and hold the SHIFT key and right-click; a shortcut menu is displayed. Choose the **Node** option from the shortcut menu and select the node of the left radial nozzle on the Boiler Drum.

Next, press ENTER to accept the suggested piping solution. The pipe connected should look like the one shown in Figure P-91.

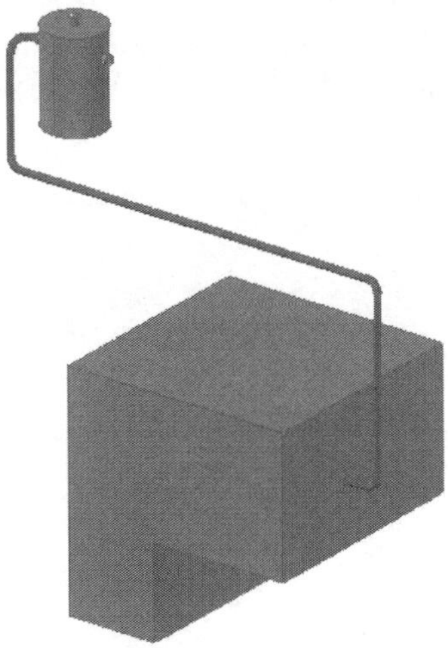

__Figure P-91__ Pipe added between outlet nozzle of the Economiser unit and left radial nozzle of the Boiler Drum

Next, connect the bottom nozzle of the Boiler Drum to the inlet nozzle of the Evaporator unit. Refer to Figure P-92 for the values to be assigned in the **Assign Tag** dialog box. The pipe connected should look like the one shown in Figure P-93.

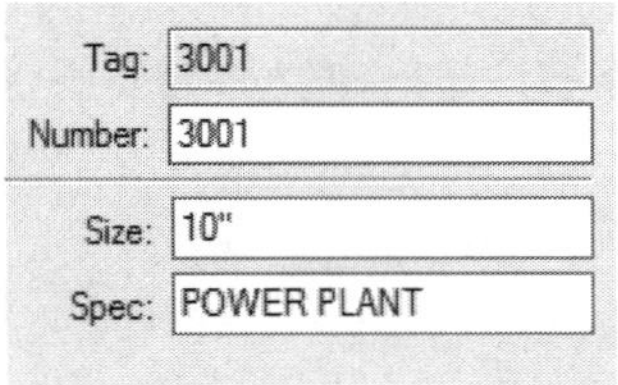

Figure P-92 *Values to be assigned in the* ***Assign Tag*** *dialog box*

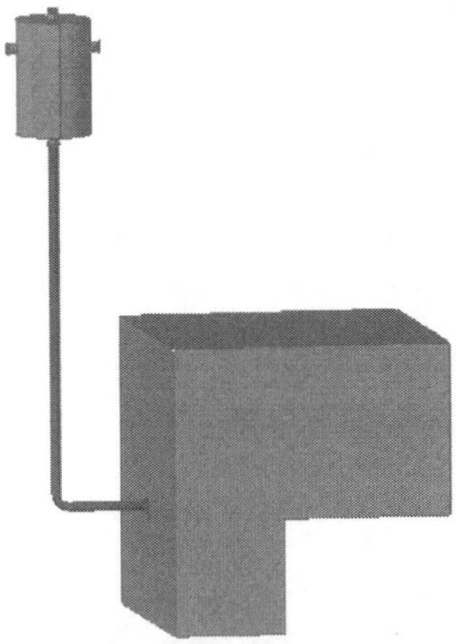

__Figure P-93__ Pipe added between bottom nozzle of the Boiler Drum and inlet nozzle of the Evaporator unit on the Boiler

Next, connect the outlet nozzle of the Evaporator unit with the right radial nozzle on the Boiler Drum. Refer to Figure P-94 for the values to be assigned in the **Assign Tag** dialog box. The pipe connected should look like the one shown in Figure P-95.

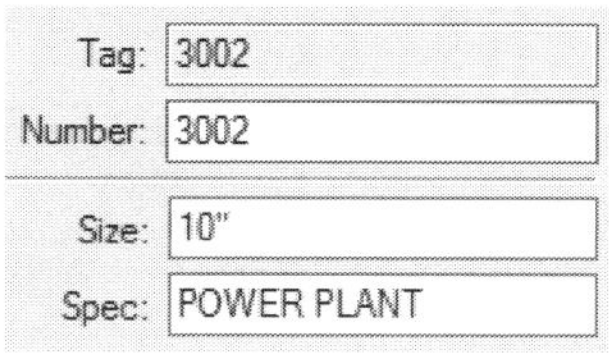

Figure P-94 *Values to be assigned in the* ***Assign Tag*** *dialog box*

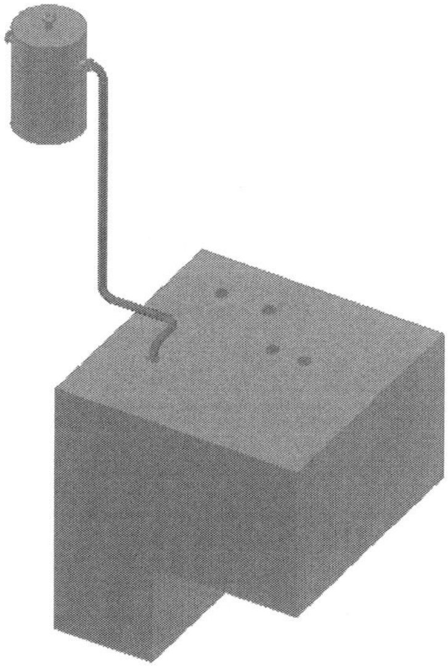

Figure P-95 *Pipe added between outlet nozzle of the Evaporator unit and right radial nozzle of the Boiler Drum*

Note that you have to click on **Next** option and then choose the **Accept** option in the command prompt while you are prompted to specify the piping solution.

Next, connect the top nozzle of the Boiler Drum to the inlet nozzle of the Super Heater. Refer to Figure P-96 for the values to be assigned in the **Assign Tag** dialog box. The pipe connected should look like the one shown in Figure P-97. Note that you have to click on the **Next** option thrice and then the **Accept** option in the command prompt while you are prompted to specify the piping solution.

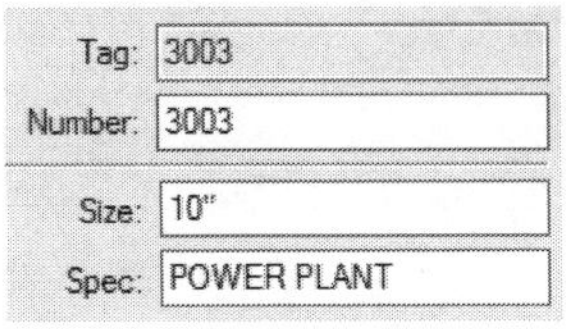

Figure P-96 *Values to be assigned in the* ***Assign Tag*** *dialog box*

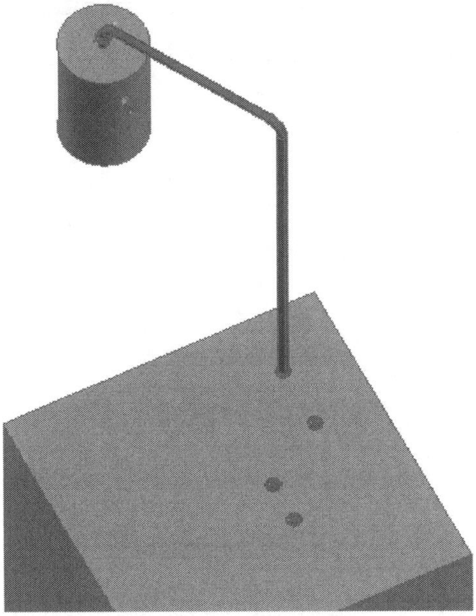

Figure P-97 *Pipe added between top nozzle of the Boiler Drum and inlet nozzle of the Super Heater unit on the Boiler*

Next, connect the outlet nozzle of the Super Heater with the nozzle at top of the High Pressure Turbine. Refer to Figure P-98 for the values to be assigned in the **Assign Tag** dialog box. The pipe connected should look like the one shown in Figure P-99. Note that you have to click on **Next** option twice and then the **Accept** option in the command prompt when you are prompted to specify the piping solution.

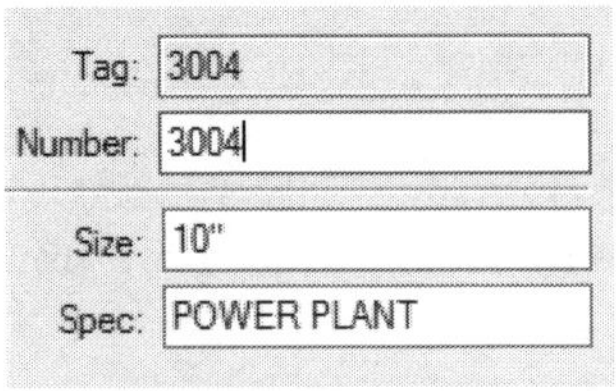

Figure P-98 *Values to be assigned in the* ***Assign Tag*** *dialog box*

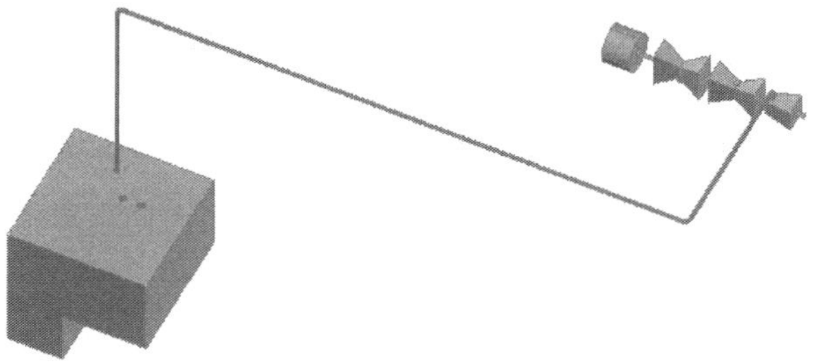

Figure P-99 *Pipe added between outlet nozzle of the Super Heater and top inlet nozzle of the High Pressure Heater*

Next, connect the bottom nozzle of the High Pressure Turbine with the inlet nozzle of the Reheater. Refer to Figure P-100 for the values to be assigned in the **Assign Tag** dialog box. The pipe connected should look like the one shown in Figure P-101. Note that you have to click on the **Next** option thrice and then on the **Accept** option in the command prompt when you are prompted to specify the piping solution.

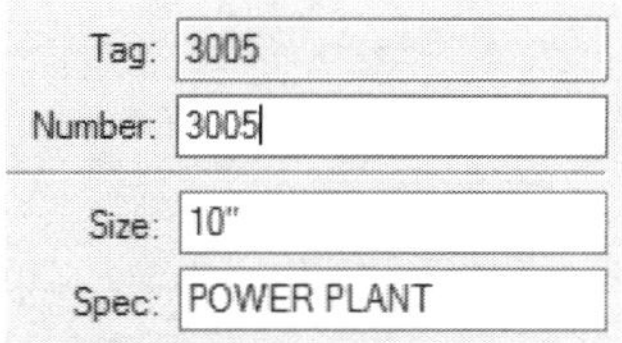

Figure P-100 *Values to be assigned in the* ***Assign Tag*** *dialog box*

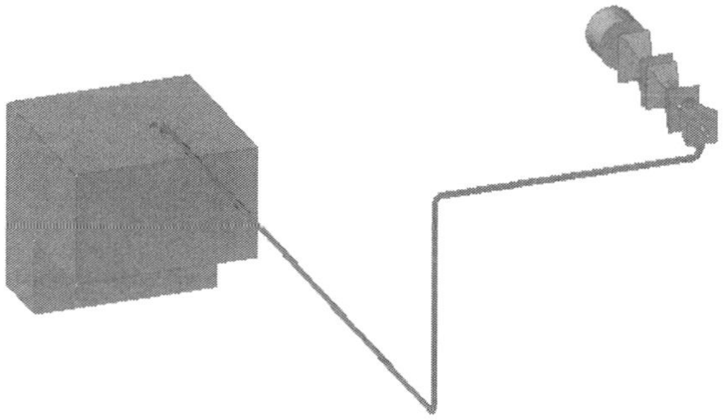

Figure P-101 *Pipe added between bottom outlet nozzle of the High Pressure Turbine and inlet nozzle of the Reheater unit on the Boiler*

Next, connect the outlet nozzle of the Reheater with the nozzle at top center of the Intermediate Pressure Turbine. To do so, choose the **Route New Line** tool from the **Line Number Selector** drop-down of the **Part Insertion** panel; the **Assign Tag** dialog box is displayed. Enter **3006** in the **Number** edit box in this dialog box. Select the line size as **10"** from the **Size** drop-down list. Next, choose the **Assign** button; the **Assign Tag** dialog box is closed and you are prompted to specify the start point of the pipe. Press and hold the SHIFT key and right-click to display a shortcut menu. Choose the **Node** option from the shortcut menu displayed and then select the node of the outlet nozzle of the Reheater unit; you are prompted to specify the next point. Move the cursor vertically upwards and enter **50'** at the command prompt; you are prompted to specify the next point. Next, select the node of the nozzle at top center of the Intermediate Pressure Turbine. In the command prompt, click on the **Next** option once and then choose the **Accept** option. The pipe connected should look like the one shown in Figure P-102.

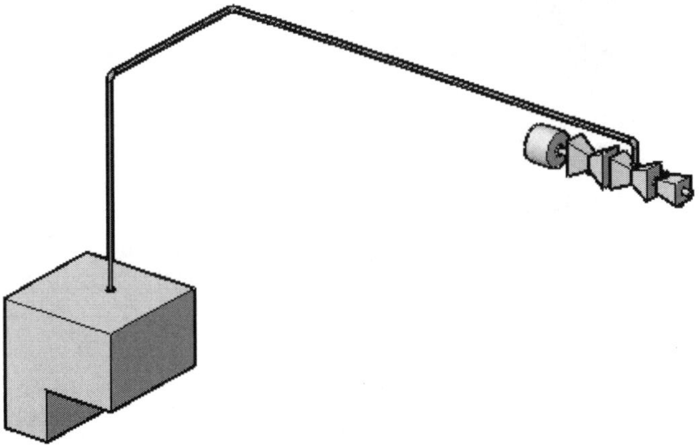

Figure P-102 *Pipe added between outlet nozzle of the Reheater and inlet nozzle of the Intermediate Pressure Turbine*

After adding all the pipings, the plant model should look similar to the one shown in Figure P-103.

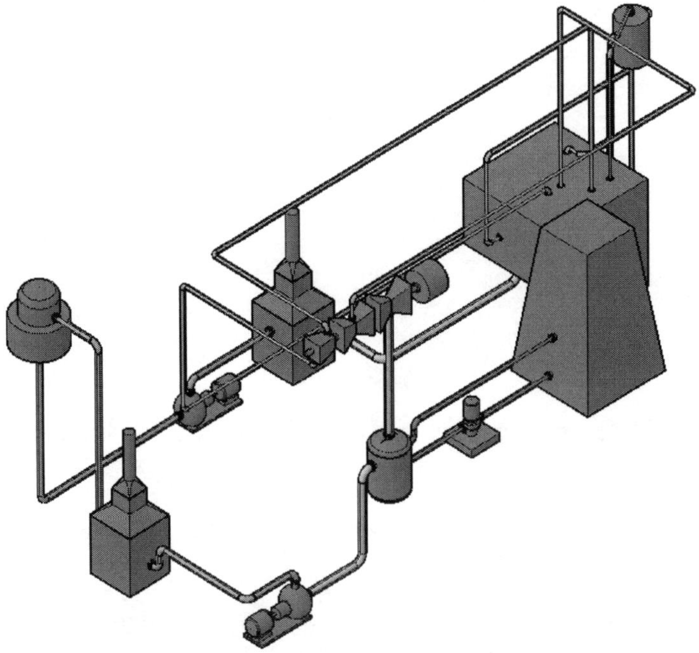

Figure P-103 NE Isometric *view of the plant model after adding pipes*

ADDING VALVES

After making all the piping connections, you need to add valves to the plant model. First, you will add Gate Valves to the model. To do so, choose **Gate Valve, FL, RF, 150** from the **Valve** area in the **Dynamic Pipe Spec** tab of the **TOOL PALETTES**; you are prompted to specify the insertion point. Click on the midpoint of the pipe connecting the High Pressure Heater and the Boiler, refer to Figure P-104; you are prompted to specify the rotation angle. Enter 0 at the command prompt and then press the ESC key, the valve is placed at the required location.

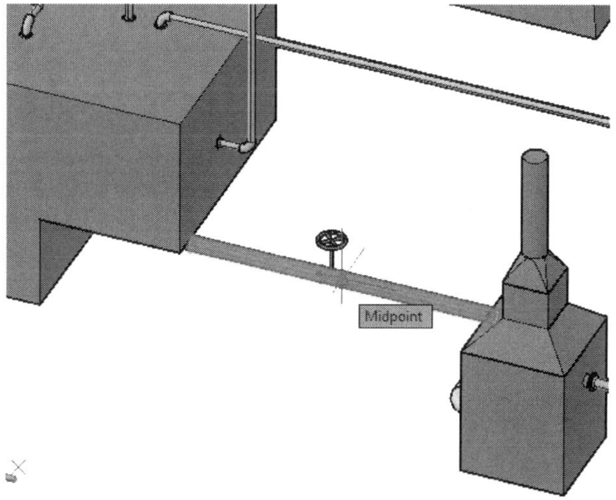

Figure P-104 *Placing Gate Valve on the pipe*

Again, choose **Gate Valve, FL, RF, 150** from the **Valve** area and follow the same procedure to place it on the midpoints of the pipes numbered **3004** and **3006**. Figure P-105 shows the valves placed at the required locations.

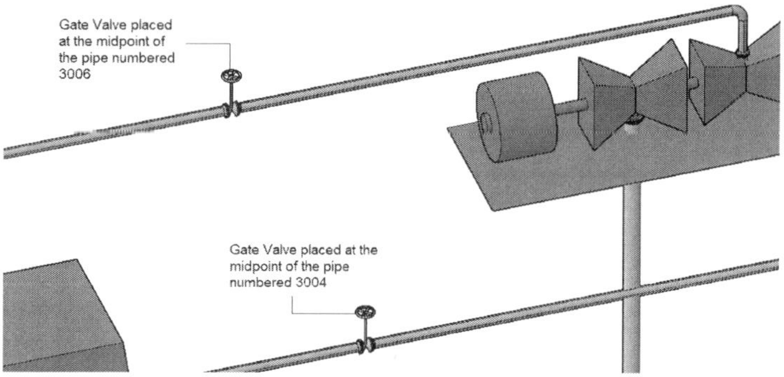

Figure P-105 *Gate Valves placed on pipes numbered **3004** and **3006***

Next, you need to place control valves on the same pipelines (numbered **3004** and **3006**). To do so, choose **Control Valve, Ball, WF, RF, 150** from the **Valve** area in the **Dynamic Pipe Spec** tab of the **TOOL PALETTES**; you are prompted to specify the insertion point. Specify the insertion point for both the control valves at some distance from the previously placed gate valves. Refer to Figure P-106 for the location of control valves to be placed on the pipelines.

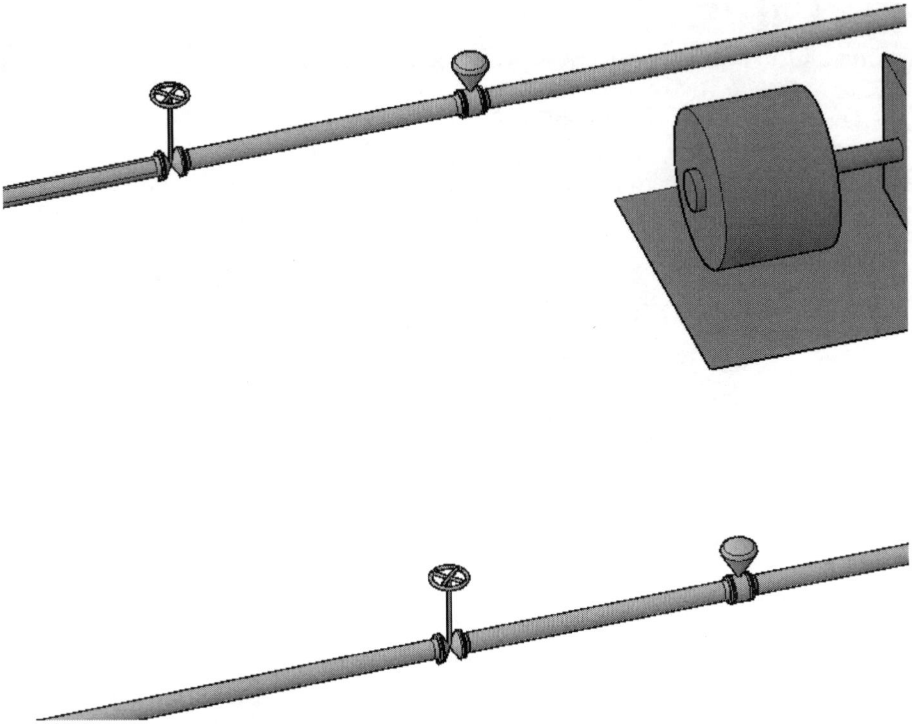

Figure P-106** Control Valves placed on pipes numbered **3004** and **3006

ADDING STRUCTURES AND SUPPORTS

Now, you need to add structures and supports to the equipment which are above ground level. First, you will add a structure that will hold the Turbine and Generator unit. To do so, invoke the **RECTANGLE** tool; you are prompted to specify the first corner point. Enter **52', -128', 3'5"** at the command prompt, you are prompted to specify the other corner point. Enter @**44', -15'** at the command prompt; a rectangle is created at the specified location. Next, choose the **Plate** tool from the **Parts** panel of the **Structure** tab; the **Create Plate/Grate** dialog box is displayed. Refer to Figure P-107 for the values to be entered in the **Create Plate/Grate** dialog box.

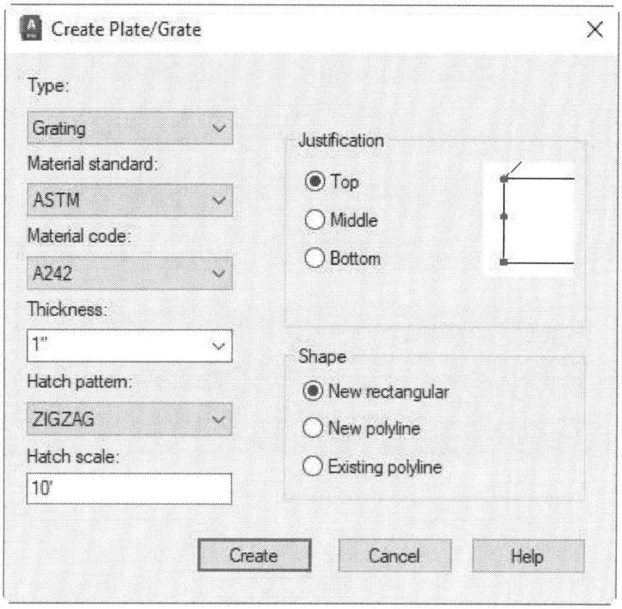

Figure P-107 *Values to be entered in the* ***Create Plate/Grate*** *dialog box*

After entering all the values in the **Create Plate/Grate** dialog box, choose the **Create** button in it; you are prompted to specify the first corner point of the grate. Specify the first point of the rectangle created earlier as the first corner point; you are prompted to specify the other corner point. Specify the diagonally opposite corner point of the rectangle as the second corner point; a grate will be created as shown in Figure P-108.

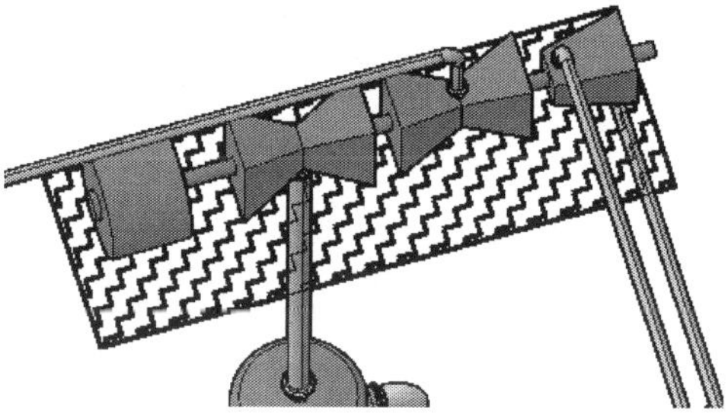

Figure P-108 ***Grate*** *created below the Turbine and Generator unit*

Next, you need to add structural members to the grate. To do so, choose the **Line Model** option from the **Line Model** drop-down list in the **Parts** panel of the **Structure** tab. Next, choose the **Member** tool from the **Parts** panel; the **Specify start point of structural member or [Line Settings]:** prompt is displayed at the command prompt. Choose the **Settings** option from the command prompt; the **Member Settings** dialog box is displayed. Next, choose **W** and **W16x40** from the **Shape type** and **Shape size** list boxes, respectively and then choose the **OK** button to exit the **Member Settings** dialog box; you are prompted again to specify the start point of the structural member. Refer to Figure P-109 for the start point of the structural member. After specifying the start point, you are prompted to specify the endpoint of the structural member. Move the cursor vertically down and enter **70'** at the command prompt; the structural member is placed at the required location. Next, choose the **Shape Model** option from the **Line Model** drop-down list in the **Parts** panel of the **Structure** tab; you will notice that the structural member is not properly aligned to the Grate. To fix it, select the structural member and right-click; a shortcut menu is displayed. Next, choose the **Edit Structure** option from the shortcut menu; the **Edit Member** dialog box is displayed. In this dialog box, choose the top right justification point in the **Orientation** window and then choose the **OK** button; the structural member is properly aligned to the Grate. Similarly, add the structural members to the remaining corners of the Grate. After adding the structural members to the Grate, it should look similar to the one shown in Figure P-110.

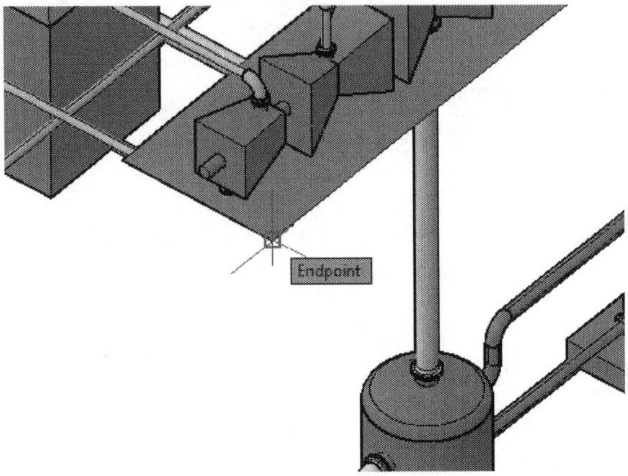

Figure P-109 *Start point of the structural member*

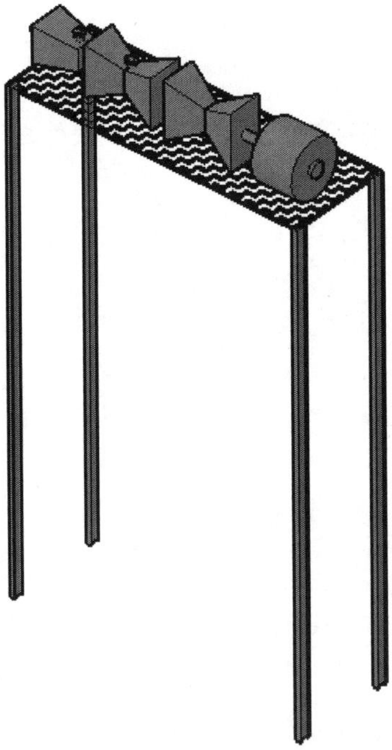

Figure P-110 *Grate after adding structural members*

Next, you need to add Grate and structural members to support the Condenser. To do so, type **RECTANGLE** at the command prompt and press ENTER; you are prompted to specify the first corner point. Enter **77',-140',-46'** at the command prompt; you are prompted to specify the other corner point. Enter @**-14',12'** at the command prompt; a rectangle is created at the specified location. Next, choose the **Plate** tool from the **Parts** panel of the **Structure** tab; the **Create Plate/Grate** dialog box is displayed. Refer to Figure P-107 for the values to be entered in this dialog box. After entering all the values in the **Create Plate/Grate** dialog box, choose the **Create** button in it; you are prompted to specify the first corner point of the Grate. Specify the first point of the rectangle created earlier as the first corner point; you are prompted to specify the other corner point. Specify the diagonally opposite corner point of the rectangle as the second corner point; a grate is created as shown in Figure P-111.

Figure P-111 ***Grate*** *created below the Condenser unit*

Next, follow the same procedure explained earlier to add structural members to the Grate. The member length in this case is **20'8"** (Vertically Downwards). After adding structural members to the Grate, it should look similar to the one shown in Figure P-112.

Note

The values to be entered in the ***Create Plate/Grate*** *and* ***Member Settings*** *dialog boxes will remain the same as mentioned before for all the gratings and structural members to be added further.*

Figure P-112 *Grate after adding structural members*

Similarly, add Gratings and structural members to the Vertical Inline Pump, Deaerator, and the Boiler Drum. The values required to do so are given below.

For Vertical Inline Pump
First corner point of rectangle: **37'6",-128'6",-45'**
Other corner point of the rectangle: **@11',-11'**
Member length: **21'8"** (Vertically Downwards)

For Deaerator
First corner point of rectangle: **108',-194',-15'**
Other corner point of the rectangle: **@18',-18'**
Member length: **51'8"** (Vertically Downwards)

For Boiler Drum
First corner point of rectangle: **-61',-176',-14'**
Other corner point of the rectangle: **@10',-10'**
Member length: **52'8"** (Vertically Downwards)

After adding the gratings and structural members to the Vertical Inline Pump, the Deaerator, and the Boiler Drum, they should look similar to the ones shown in Figures P-113, P-114, and P-115, respectively.

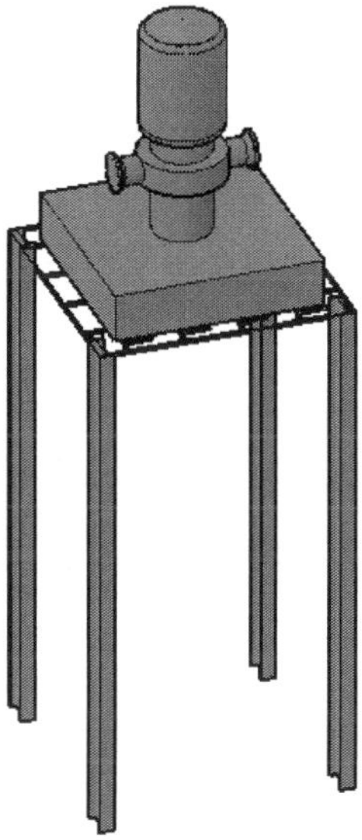

Figure P-113 *Vertical Inline Pump after adding grate and structural members*

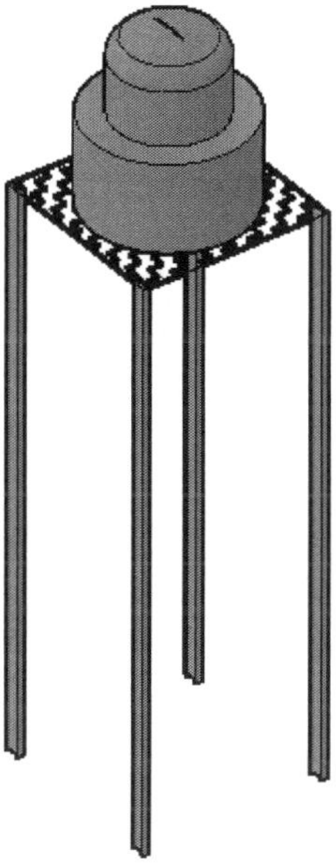

Figure P-114 *Deaerator after adding grate and structural members*

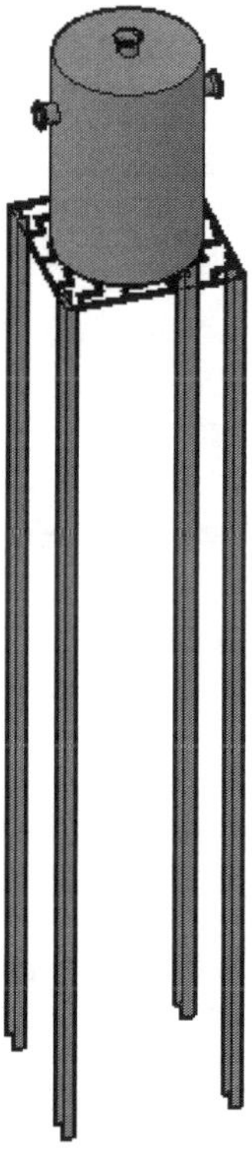

Figure P-115 *Boiler Drum after adding grate and structural members*

Next, you need to add support to the Cooling Tower. To do so, type **RECTANGLE** at the command prompt and press enter; you are prompted to specify the first corner point. Enter **-11',-117',-45'** at the command prompt; you are prompted to specify the other corner point. Enter @**32'6",-33'** at the command prompt; a rectangle is created at the specified location. Next, type **EXTRUDE** at the command prompt; you are prompted to select the objects to extrude. Select the rectangle created earlier and press ENTER; you are prompted to specify the height of extrusion. Enter **-21'8"** at the command prompt; a box is placed below the Cooling Tower.

Next, choose the **Attach Equipment** tool from the **Equipment** panel of the **Home** tab; you are prompted to select a single equipment item. Select the Cooling Tower; you are prompted to select objects. Select the box created earlier and press ENTER; it will get attached with the Cooling Tower.

After adding support to the Cooling Tower, it should look like the one shown in Figure P-116.

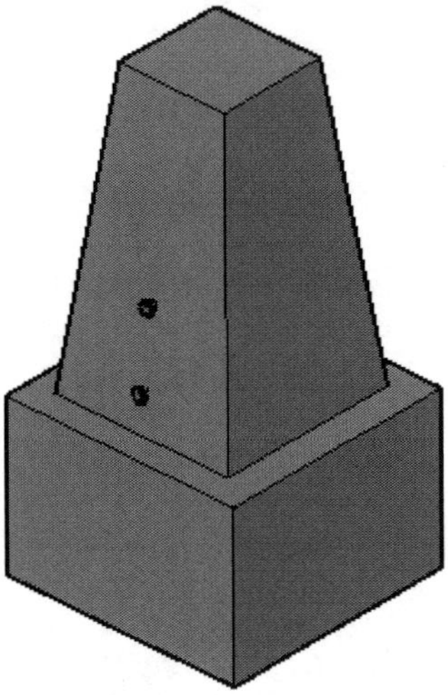

Figure P-116 *Cooling Tower after adding support*

ADDING LADDERS AND RAILINGS

After adding gratings and structural members to the equipment, you need to add ladders to them. First, you will add ladder to the Turbine and Generator unit. To do so, choose the **Line Model** option from the **Line Model** drop-down list in the **Parts** panel of the **Structure** tab. Also, make sure that **UCS** is set to **World**. Next, type **LINE** at the command prompt; you are prompted to specify the start point of the line. Snap the cursor to the mid point of the edge of the grating and click to specify the first point, refer to Figure P-117; you are prompted to specify the next point (second point). Next, drag the cursor vertically downwards (-ve Z-direction) and enter **70'** in the command prompt; you are prompted to specify the next point (third point). Again, drag the cursor in +ve Y-direction and enter **6"** at the command prompt. The line drawn should be similar to one shown in Figure P-117.

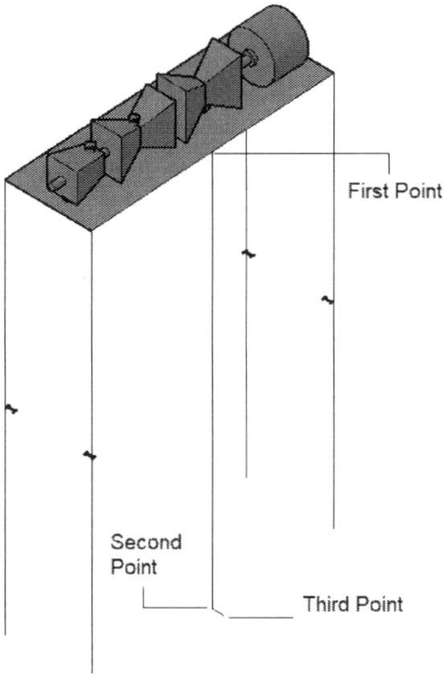

Figure P-117 Line added to the grating

Next, choose the **Ladder Settings** tool from the **Structural Settings** drop-down of the **Parts** panel of the **Structure** tab; the **Ladder Settings** dialog box is displayed with the **Ladder** tab chosen. Figures P-118 and P-119 show the values to be entered in the **Ladder Settings** dialog box when the **Ladder** and **Cage** tabs are chosen, respectively.

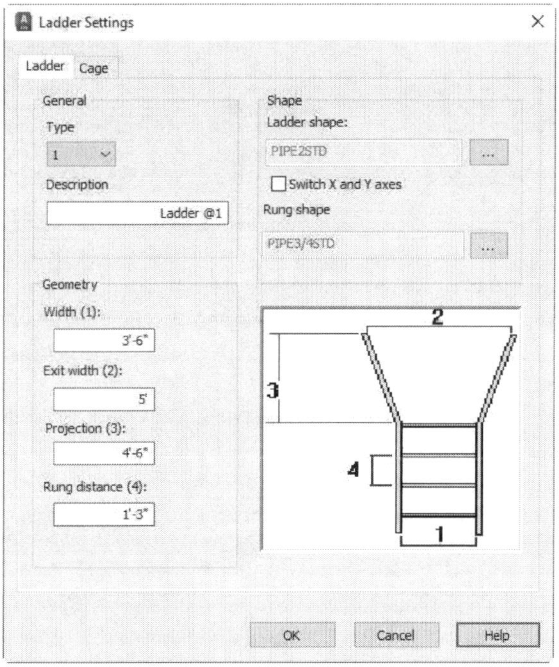

*Figure P-118 Values to be entered in the **Ladder** tab of the **Ladder Settings** dialog box*

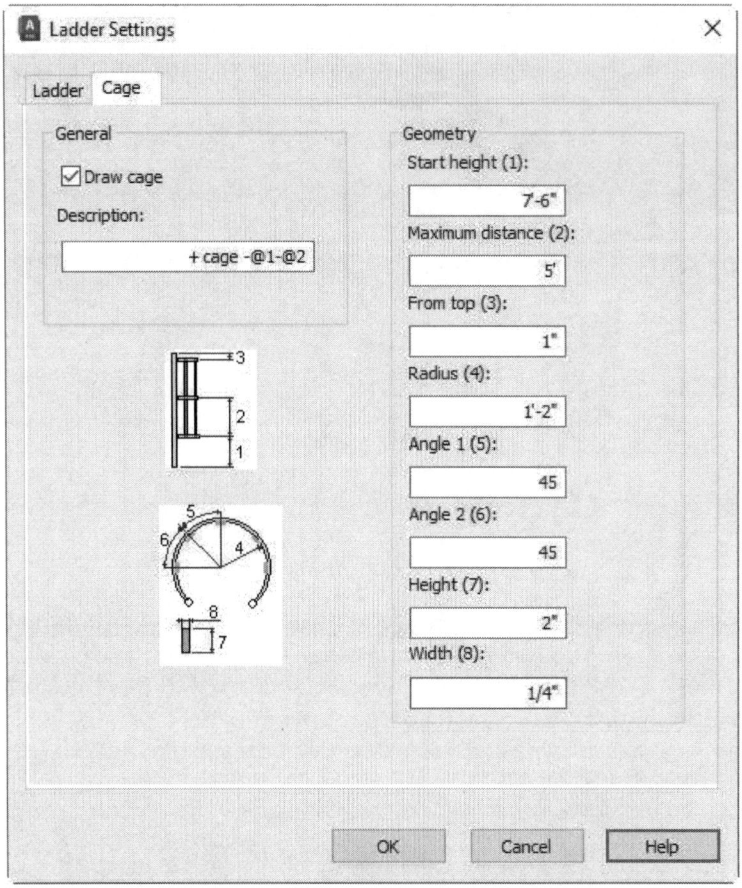

***Figure P-119** Values to be entered in the **Cage** tab of the **Ladder Settings** dialog box*

After entering all the values in both the tabs, choose the **OK** button to exit the **Ladder Settings** dialog box. Next, choose the **Ladder** tool from the **Parts** panel of the **Structure** tab; you are prompted to specify the start point of the ladder. Click on the first point of the line drawn earlier, refer to Figure P-117; you are prompted to specify the end point of the ladder. Click on the second point of the line, refer to Figure P-117; you are prompted to specify the directional distance point. Click on the third point of the line, refer to Figure P-117; a ladder is placed at the specified location. Next, choose the **Shape Model** option from the **Line Model** drop-down list of the **Parts** panel. The ladder placed should look similar to the one shown in Figure P-120.

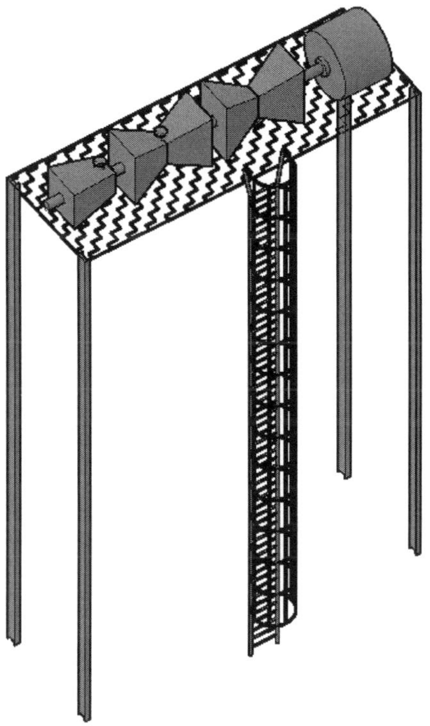

Figure P-120 *Ladder added to the Turbine and Generator unit*

Now, you need to add Railing to the Grating created for Turbine and Generator unit. To do so, choose the **Line Model** option from the **Line Model** drop-down list in the **Parts** panel of the **Structure** tab. Next, choose the **Railing Settings** tool from the Structural Settings drop-down of the **Parts** panel of the **Structure** tab; the **Railing Settings** dialog box is displayed. Refer to the Figure P-121 for the values to be entered in the **Railing Settings** dialog box. After entering all the values, choose the **OK** button to exit the **Railing Settings** dialog box. Next, choose the **Railing** tool from the **Parts** panel; you are prompted to specify the start point of the railing. Specify the start point of the railing, refer to Figure P-122; you are prompted to specify the end point. Click on the second, third, fourth, fifth, and sixth points successively; a railing is added to the grating. Next, choose the **Shape Model** option from the **Line Model** drop-down list of the **Parts** panel. After adding the railing, the Turbine and Generator unit should look similar to the one shown in Figure P-123.

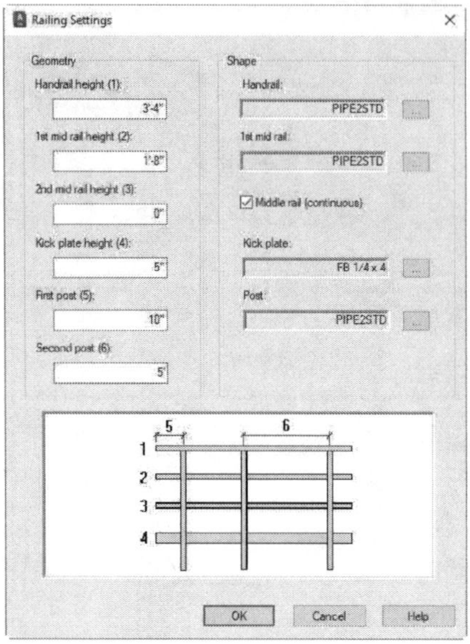

Figure P-121 *Values to be entered in the* ***Railing Settings*** *dailog box*

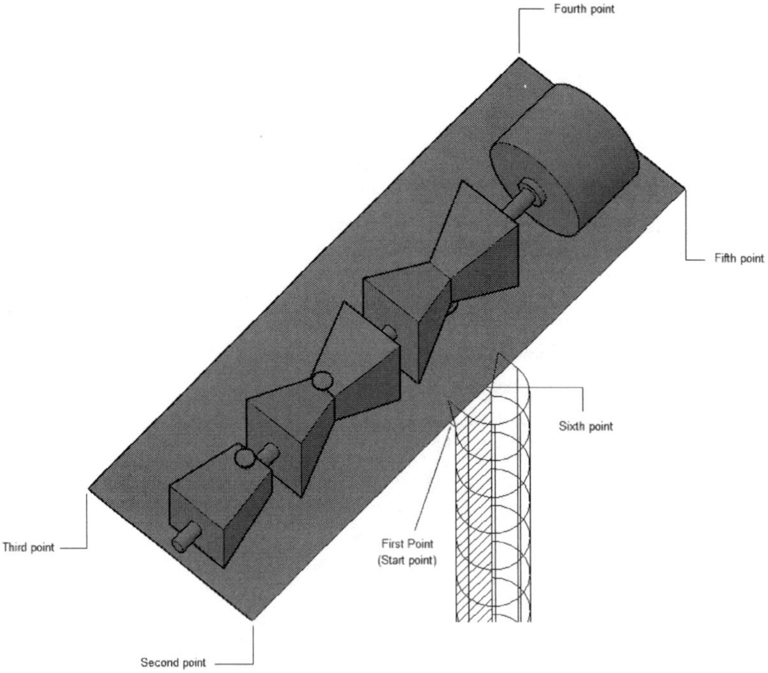

Figure P-122 *Points to be selected while adding the Railing*

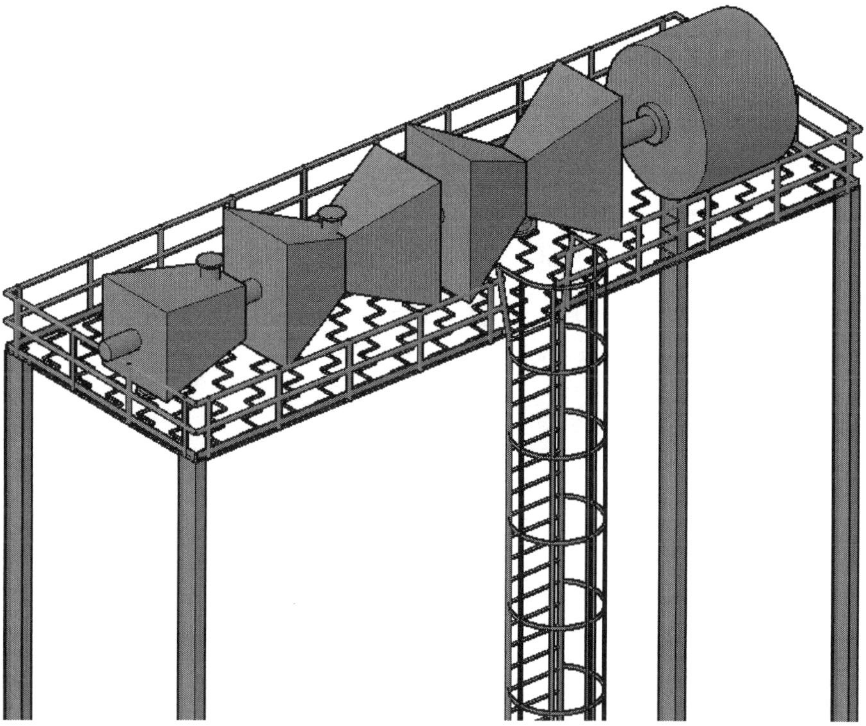

Figure P-123 *Railing added to the Turbine and Generator unit*

Similarly, add ladders and railings to the Condenser, Vertical Inline Pump, Deaerator, and the Boiler Drum.

Note

*The values to be entered in the **Ladder Settings** and **Railing Settings** dialog boxes will remain same for all the ladders to be added to the remaining equipment mentioned above. Also, the directional distance point value will remain same **(6")** while adding ladders.*

Tip

The directional distance point will always lie away from the equipment to which the ladder is to be added.

After adding ladder and railing to the Condenser, it should look similar to the one shown in Figure P-124. Note that the ladder to be added to the Condenser should start from the midpoint of the edge of the grate facing the High Pressure Heater and should be upto the ground level.

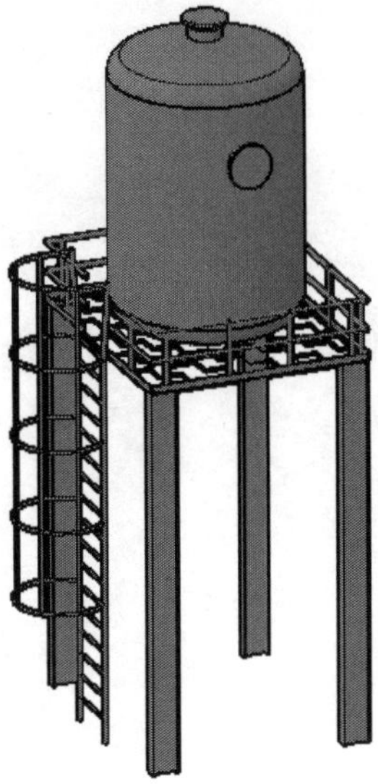

Figure P-124 *Condenser after adding ladder and railing*

After adding ladder and railing to the Vertical Inline Pump, it should look similar to the one shown in Figure P-125. Note that the ladder to be added to the Vertical Inline Pump should start from the midpoint of the edge of the grate facing the High Pressure Heater and should be upto the ground level.

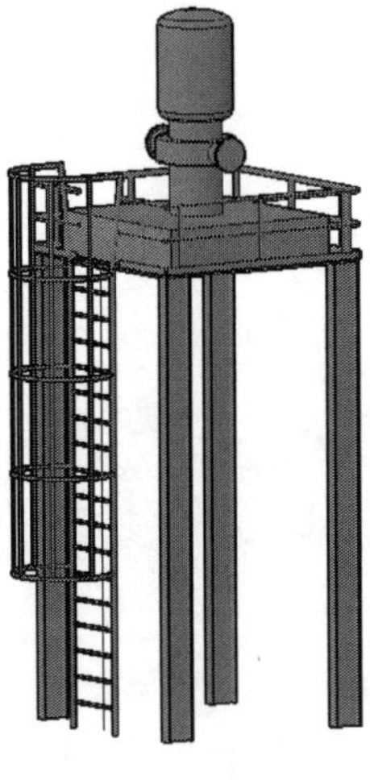

Figure P-125 *Vertical Inline Pump after adding ladder and railing*

After adding ladder and railing to the Deaerator, it should look similar to the one shown in Figure P-126. Note that the ladder to be added to the Deaerator should start from the midpoint of the edge of the grate which does not face any of the equipment and should be upto the ground level.

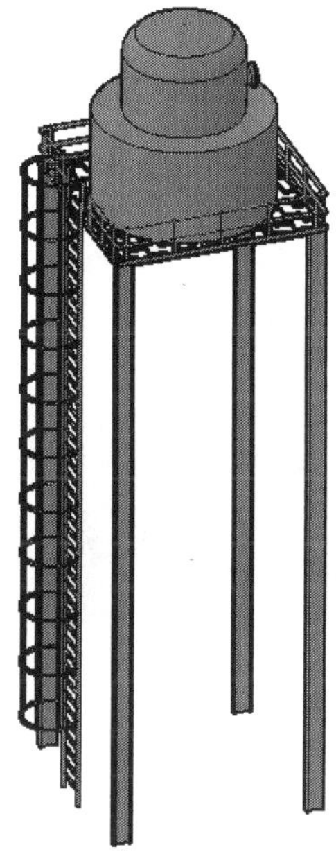

Figure P-126 *Deaerator after adding ladder and railing*

After adding ladder and railing to the Boiler Drum, it should look similar to the one shown in Figure P-127. Note that the ladder to be added to the Boiler Drum should start from the midpoint of the edge of the grate facing the Cooling Tower and should be upto the ground level.

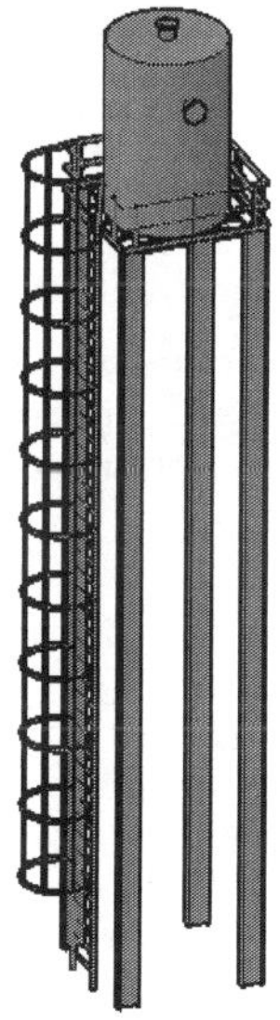

Figure P-127 *Boiler Drum after adding ladder and railing*

Figure P-128 shows the final model of the Thermal Power Plant.

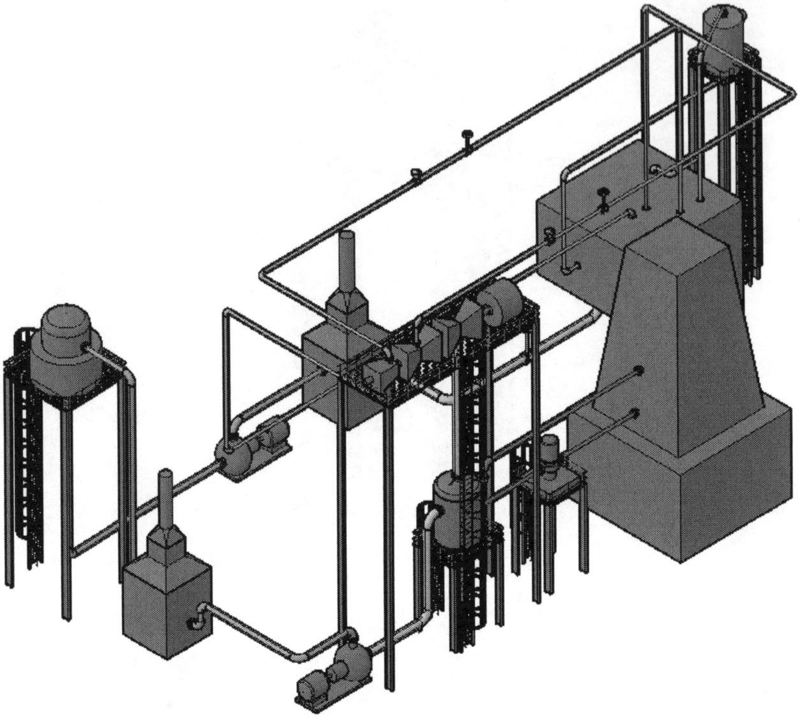

Figure P-128 *Final model of the Thermal Power Plant*

CREATING ISOMETRIC DRAWINGS

In this section, you will create isometric drawings for the pipelines numbered 1007,3004, and 3006.

Creating Quick Isometric Drawing for Line Number 1007

Now, you need to create the Quick Isometric drawing of the pipe numbered 1007 which connects the High Pressure Heater and the Boiler. Zoom into the pipeline connecting the High Pressure Heater and the Boiler and hover the cursor on the pipe, the line number of pipe is displayed, refer to Figure P-129.

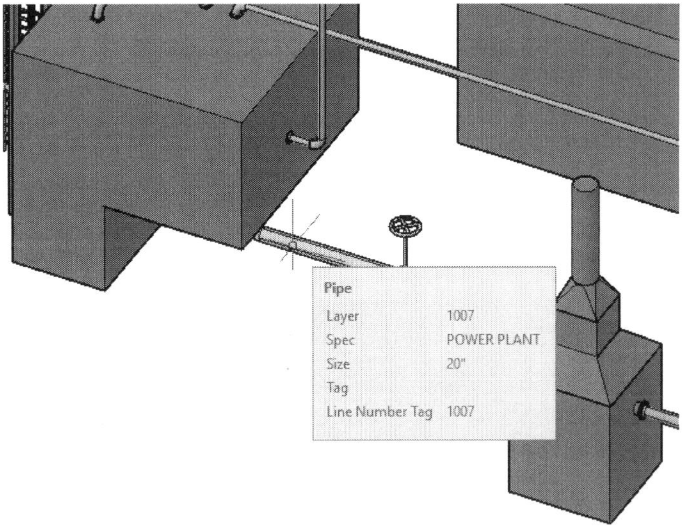

Figure P-129 *Checking the line number of the pipe*

Next, choose the **Quick Iso** tool from the **Iso Creation** panel of the **Isos** tab; you get a prompt as **Select components to ISO or select by [Line number]**. Choose the **Line number** option from the command prompt; the **Create Quick Iso** dialog box is displayed, refer to Figure P-130.

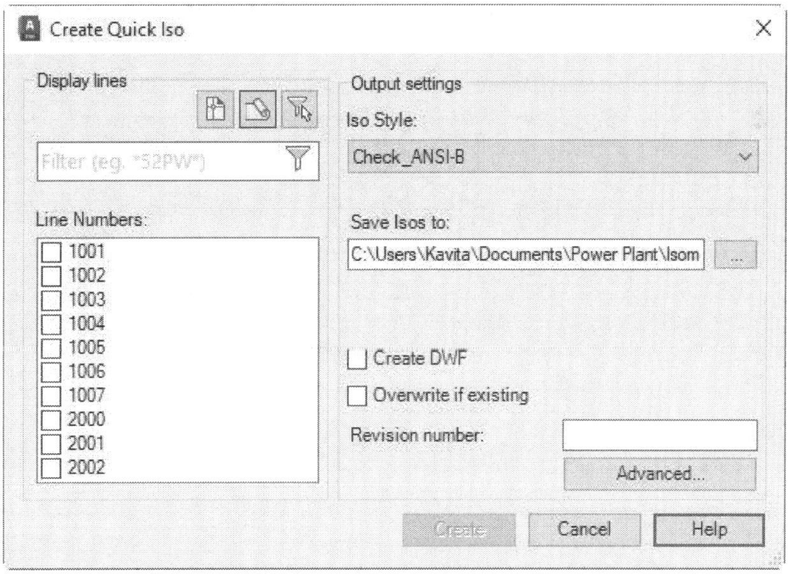

Figure P-130 *The* ***Create Quick Iso*** *dialog box*

In this dialog box, select the check box next to the line number **1007** in the **Line Numbers** list box. Next, select the **Check_ANSI-B** option from the **Iso Style** drop-down list in the **Output Settings** area and then choose the Create button; the isometric drawing creation will start. When the drawing creation is complete, a balloon is displayed at the right end of the Status Bar, refer to Figure P-131.

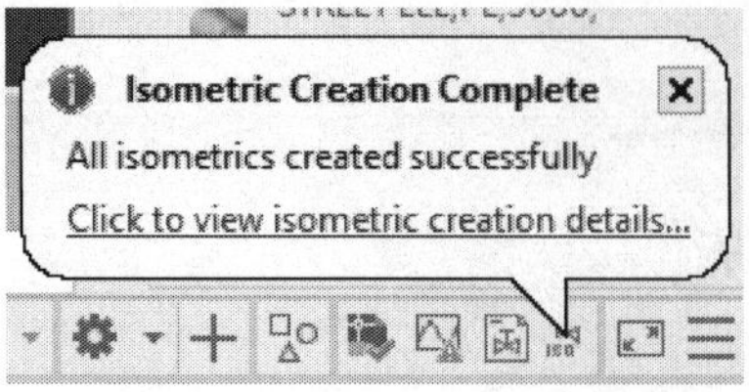

Figure P-131 Balloon displayed in the Status Bar

Next, click on the **Click to view isometric creation details** link displayed in the balloon; the **Isometric Creation Results** dialog box is displayed, refer to Figure P-132.

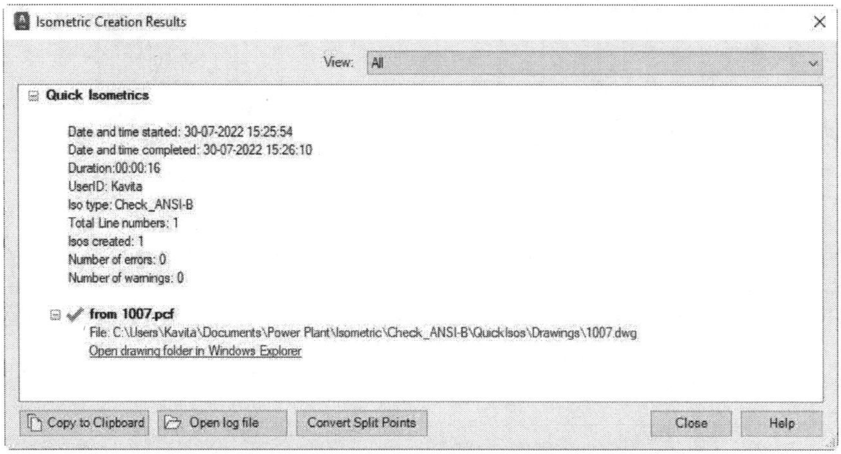

*Figure P-132 The **Isometric Creation Results** dialog box*

In this dialog box, click on the file path of the isometric drawing, refer to Figure P-132; the isometric drawing is displayed, as shown in Figure P-133.

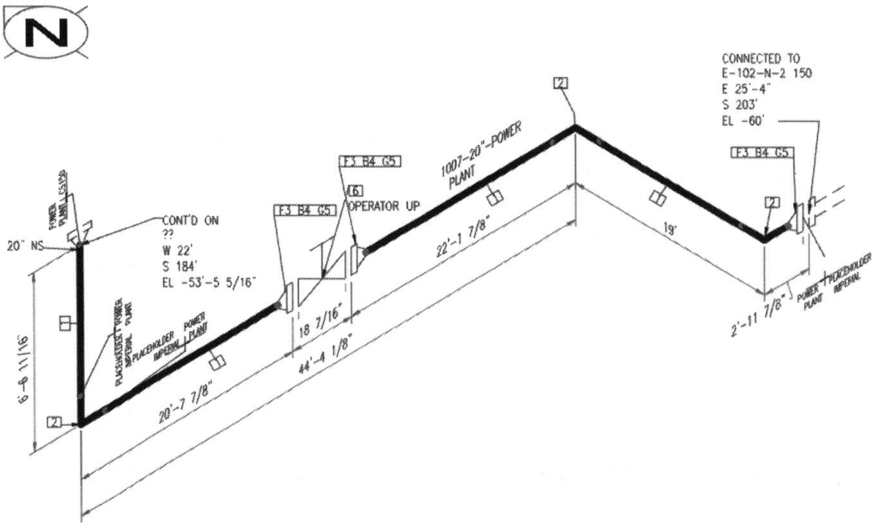

*Figure P-133 Quick isometric drawing for line number **1007***

Along with the isometric drawing, the bill of materials will also be displayed, refer to Figure P-134.

BILL OF MATERIALS			
ID	QTY	ND	DESCRIPTION
1	55'	20"	PIPE, SEAMLESS, 40, PE, ASTM A106
2	3	20"	PH IMPERIAL ELBOW 90.0°
3	3	20"	FLANGE WN, 150 LB, RF, ASME B16.5
4	72	1.1/4"X8 1/4"	PH IMPERIAL STUD BOLT
5	3	20"	PH IMPERIAL GASKET
6	1	20"	GATE VALVE, CONDUIT, 150 LB, RF, ASME B16.10

Figure P-134 *Bill of Materials for pipe number* ***1007***

Note

Quick Isometric drawings are created only to check the piping before creating a production isometric drawing. Moreover, these drawings are not added to the project and are available for current session only.

Creating Production Isometric Drawing for Line Number 3004

Now, you need to create the Production Isometric drawing of the pipe numbered **3004** which connects the Boiler and the High Pressure Turbine. Zoom into the pipeline connecting the Boiler and the High Pressure Turbine and hover the cursor on the pipe, the line number of the pipe is displayed, refer to Figure P-135.

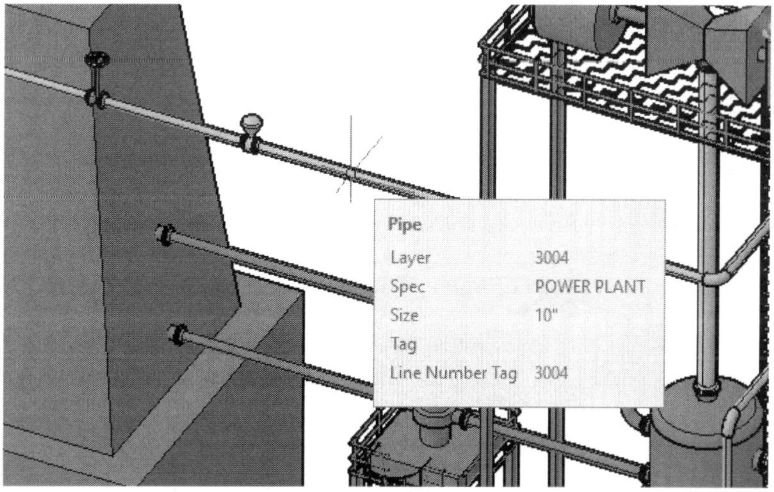

Figure P-135 *Checking the line number of the pipe*

Next, choose the **Production Iso** tool from the **Iso Creation** panel of the **Isos** tab; the **Create Production Iso** dialog box is displayed, refer to Figure P-136.

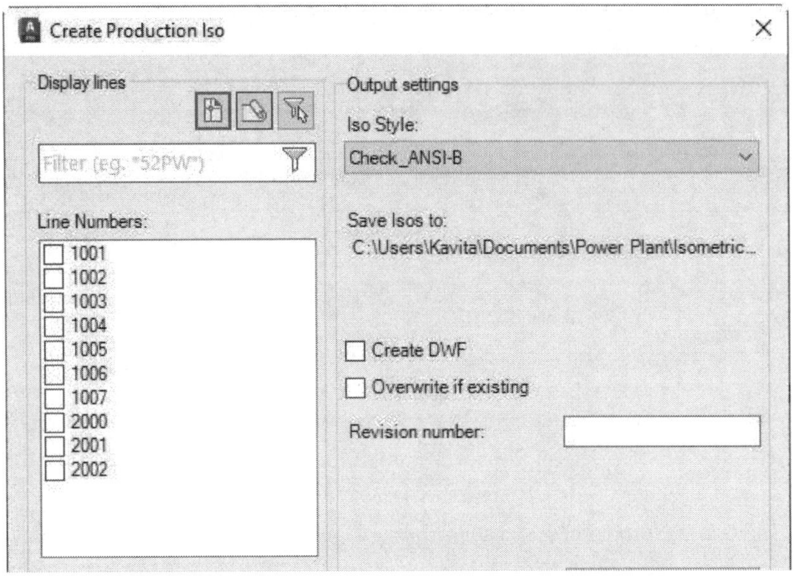

***Figure P-136** The **Create Production Iso** dialog box*

In this dialog box, select the check box next to the line number **3004** in the **Line Numbers** list box. Next, select the **Final_ANSI-B** option from the **Iso Style** drop-down list in the **Output settings** area and then choose the **Create** button; the isometric drawing creation will start. When the drawing creation is complete, a balloon is displayed at the right end of the Status Bar. Click on the **Click to view isometric creation details** link displayed in the balloon; the **Isometric Creation Results** dialog box is displayed, refer to Figure P-137.

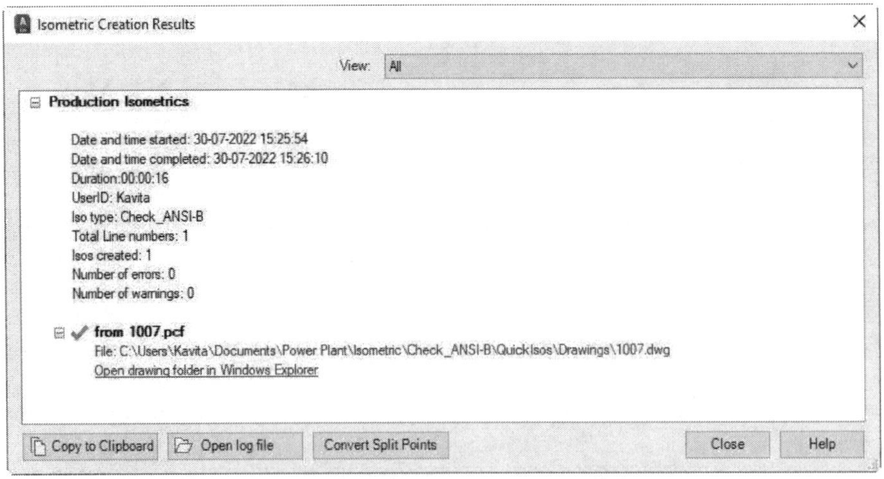

***Figure P-137** The **Isometric Creation Results** dialog box*

In this dialog box, click on the file path of the isometric drawing, refer to Figure P-137; the isometric drawing is displayed, as shown in Figure P-138.

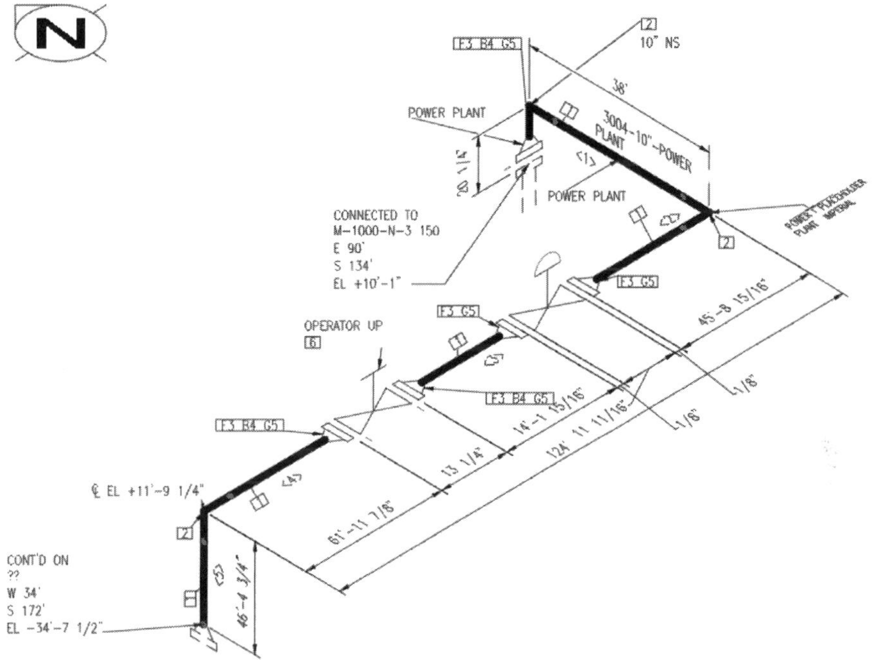

Figure P-138 *Production Isometric drawing for pipe number* ***3004***

Figures P-139 and P-140 show the Bill of Materials and Cut Piece List for pipe numbered **3004**.

BILL OF MATERIALS

ID	QTY	ND	SCH/ CLASS	DESCRIPTION
1	198'-3"	10"	100	PIPE, SEAMLESS, 40, PE, ASTM A106
2	3	10"		PH IMPERIAL ELBOW 90.0°
3	5	10"	150	FLANGE WN, 150 LB, RF, ASME B16.5
4	48	1"X6 1/4"		PH IMPERIAL STUD BOLT
5	5	10"		PH IMPERIAL GASKET
6	1	10"	150	GATE VALVE, CONDUIT, 150 LB, RF, ASME B16.10

Figure P-139 *Bill of Materials for pipe number* ***3004***

CUT PIECE LIST				
ID	LENGTH	ND	END1	END2
1	35'-3 3/4"	10"	UNIVERSAL_ET	UNIVERSAL_ET
2	44'-0 13/16"	10"	UNIVERSAL_ET	BEVEL
3	13'-6"	10"	BEVEL	BEVEL
4	60'-3 3/4"	10"	BEVEL	UNIVERSAL_ET
5	45'-0 11/16"	10"	UNIVERSAL_ET	SQUARE CUT

Figure P-140 *Cut Piece List for pipe number* ***3004***

Creating Production Isometric Drawing for Line Number 3006

Follow the same procedure to create production isometric drawing for pipe numbered **3006**. Note that for this pipeline, the production isometric drawing will split into two drawing files: **3006-1** and **3006-2**. Figure P-141 shows the Production Isometric drawing **3006-1**.

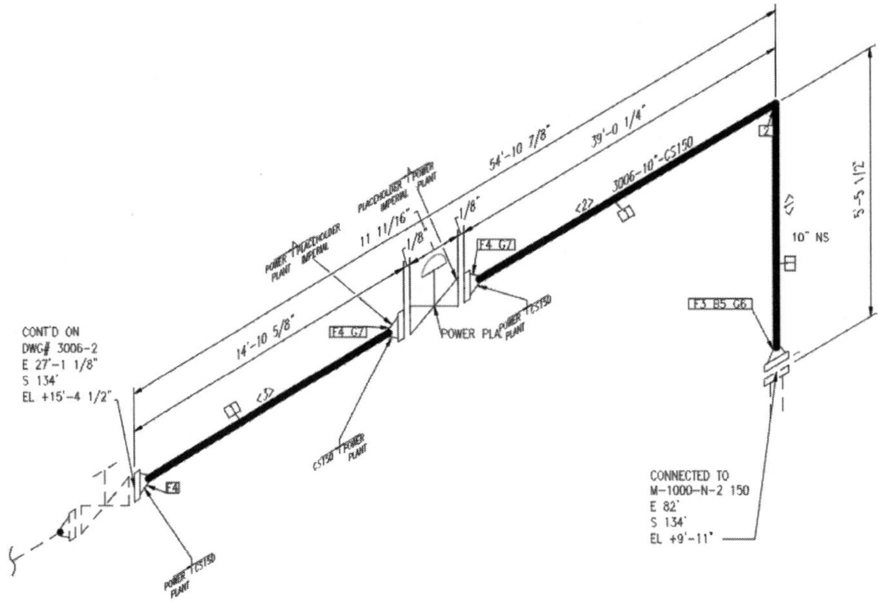

Figure P-141 *Production Isometric drawing* ***3006-1***

Figures P-142 and P-143 show the Bill of Materials and Cut Piece List for **3006-1** isometric drawing.

BILL OF MATERIALS

ID	QTY	ND	SCH/ CLASS	DESCRIPTION
1	55'-7"	10"		PIPE, SEAMLESS, PL, ASME B36.10
2	1	10"		ELL 90 LR, BV, ASME B16.9
3	1	10"	150	FLANGE WN, RF, 150 LB, ASME B16.5
4	3	10"	150	FLANGE WN, 150 LB, RF, ASME B16.5
5	12	7/8"X4 1/2"	150	BOLT SET, RF, 150 LB, STUD BOLT
6	1	10"	150	GASKET, SWG, 1/8" THK, RF, 150 LB, ASME B16.20, CS/PTFE
7	2	10"		PH IMPERIAL GASKET

Figure P-142 *Bill of Materials for* ***3006-1*** *isometric drawing*

CUT PIECE LIST

ID	LENGTH	ND	END1	END2
1	3'-10 7/16"	10"	BEVEL	BEVEL
2	37'-5 1/4"	10"	BEVEL	BEVEL
3	14'-2 11/16"	10"	BEVEL	BEVEL

Figure P-143 *Cut Piece List for* ***3006-1*** *isometric drawing*

Figure P-144 shows the production isometric drawing **3006-2**. Figures P-145 and P-146 show the Bill of Materials and Cut Piece List for **3006-2** isometric drawing.

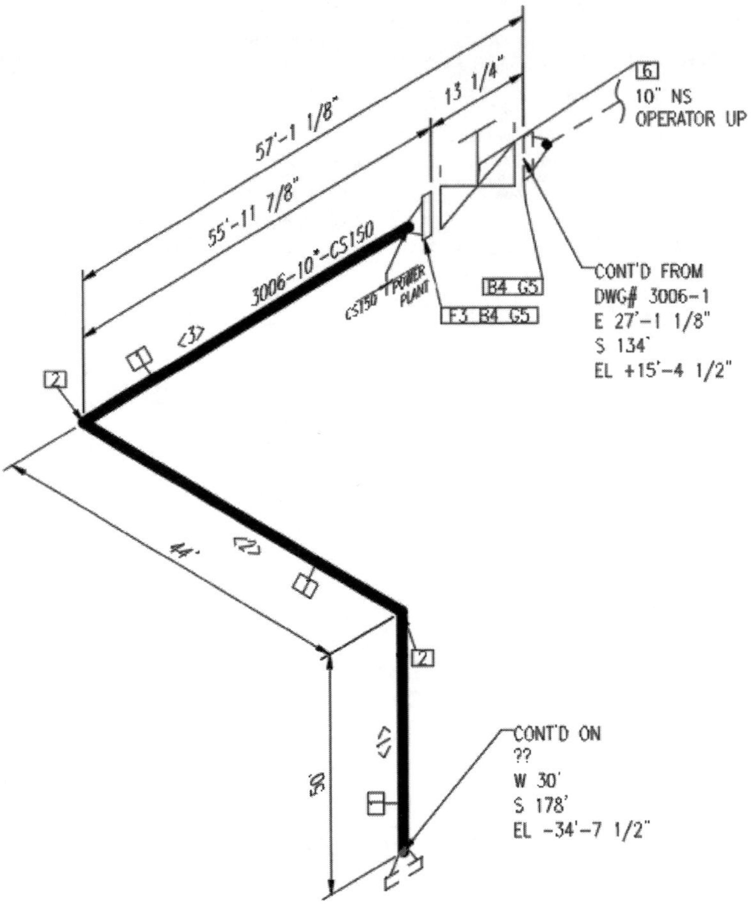

Figure P-144 *Production Isometric drawing* ***3006-2***

BILL OF MATERIALS

ID	QTY	ND	SCH/ CLASS	DESCRIPTION
1	144'-8"	10"		PIPE, SEAMLESS, PL, ASME B36.10
2	2	10"		ELL 90 LR, BV, ASME B16.9
3	1	10"	150	FLANGE WN, 150 LB, RF, ASME B16.5
4	32	1"X6 1/4"		PH IMPERIAL STUD BOLT
5	2	10"		PH IMPERIAL GASKET
6	1	10"	150	GATE VALVE, CONDUIT, 150 LB, RF, ASME B16.10

Figure P-145 *Bill of Materials for* ***3006-2*** *isometric drawing*

CUT PIECE LIST				
ID	LENGTH	ND	END1	END2
1	48'-9"	10"	BEVEL	SQUARE CUT
2	41'-6"	10"	BEVEL	BEVEL
3	54'-4 7/8"	10"	BEVEL	BEVEL

Figure P-146 *Cut Piece List for **3006-2** isometric drawing*

Next, save and close all the Isometric drawings.

CREATING ORTHOGRAPHIC DRAWINGS

Now, you need to create the Orthographic drawings for the plant model. To do so, choose the **Orthographic DWG** tab from the **PROJECT MANAGER**; the **Orthographic Drawings** node is displayed in the **Orthos** area. Click on the **Orthographic Drawings** node; it is highlighted. Next, right-click on it; a shortcut menu is displayed. Choose the **New Drawing** option from the shortcut menu; the **New DWG** dialog box is displayed. Next, enter **Ortho** in the **name** edit box and choose the **OK** button; a new drawing file is created and the **Ortho View** tab is added to the **Ribbon**.

Note

*While creating the Orthographic Views of the plant model, make sure that the orientation of the model is set to **NE Isometric**.*

First, you will create the Top view of the plant model. To do so, choose the **New View** tool from the **Ortho Views** panel of the **Ortho View** tab; the **Select Reference Models** dialog box is displayed, refer to Figure P-147.

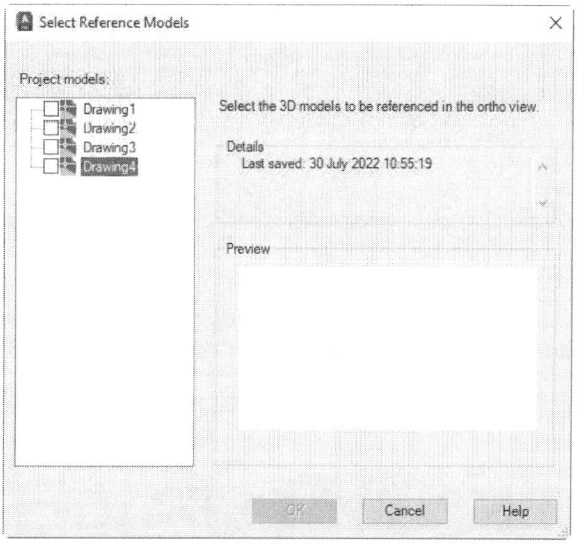

Figure P-147 *The **Select Reference Model** dialog box*

In this dialog box, select the check box next to the **P3D** model and then choose the **OK** button; the **Ortho Editor** tab is activated and the plant model is enclosed in the Ortho Cube, refer to Figure P-148.

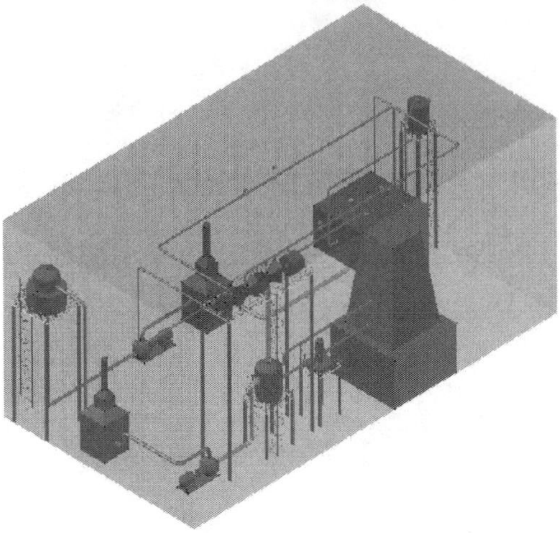

Figure P-148 *Plant model enclosed in the Ortho Cube*

Next, enter **0.006** as the scale factor in the **Scale** edit box of the **Output Size** panel of the **Ortho Editor** tab. Also, make sure that the **Top** option is selected from the drop-down list in the **Ortho Cube** panel of the **Ortho Editor** tab. Next, choose the **OK** button from the **Create** panel of the **Ortho Editor** tab; you are prompted to specify the insertion point of the viewport. Place the view on the top right of the drawing sheet; the **Ortho Generation** window is displayed and the Top view of the plant model is placed in the drawing sheet once the ortho generation is completed. Figure P-149 shows the top view of the plant model.

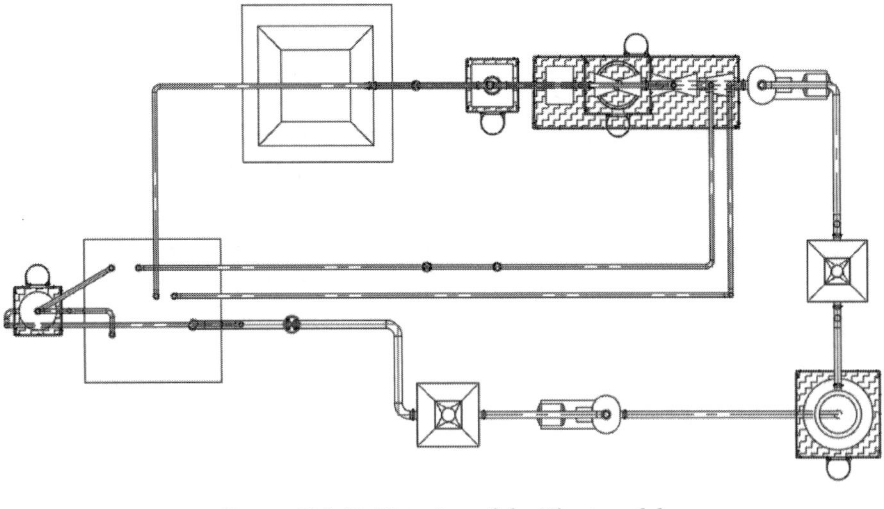

Figure P-149 *Top view of the Plant model*

Next, choose the **Adjacent View** tool from the **Ortho Views** panel of the **Ortho View** tab; you are prompted to select an orthographic view to create an adjacent view. Select the view boundary of the Top view placed earlier; the **Create an Adjacent View** dialog box is displayed, refer to Figure P-150.

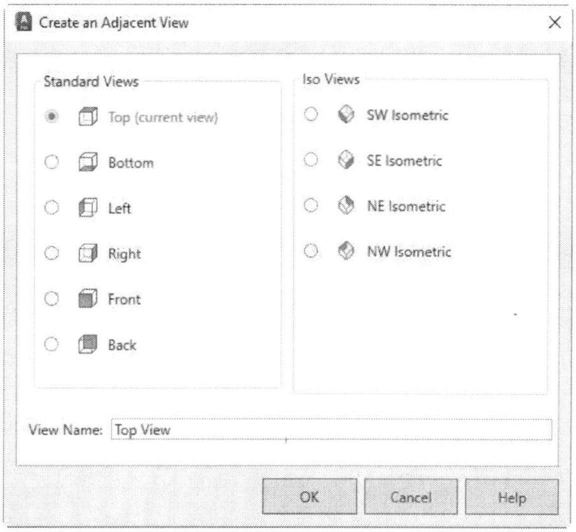

Figure P-150 *The **Create an Adjacent View** dialog box*

In this dialog box, select the **Front** radio button from the **Standard Views** area and then choose the **OK** button; you are prompted to specify the insertion point of the viewport. Place the front view below the previously created top view and then press ESC to exit the **Adjacent View** tool. Figure P-151 shows the front view of the plant model.

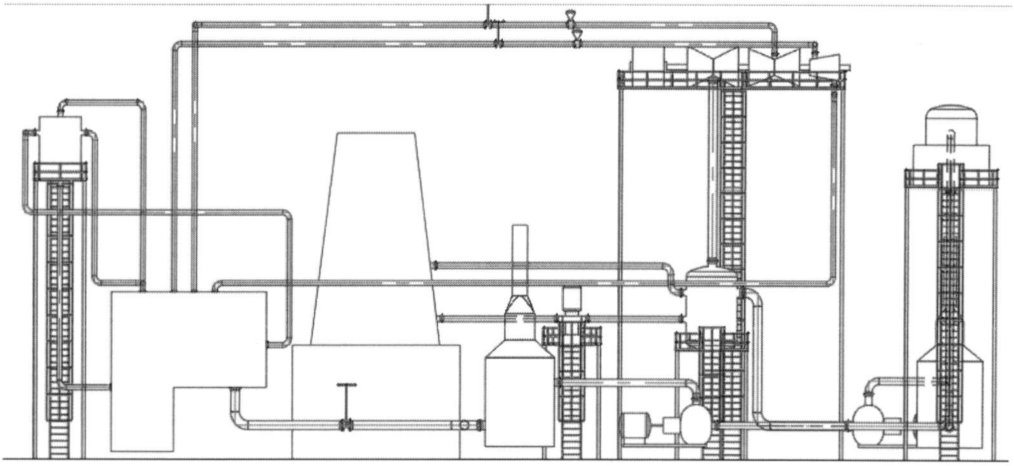

Figure P-151 *Front view of the plant model*

Again, choose the **Adjacent View** tool from the **Ortho Views** panel of the **Ortho View** tab; you are prompted to select an orthographic view to create an adjacent view. Select the view boundary of the front view; the **Create an Adjacent View** dialog box is displayed.

In this dialog box, select the **Left** radio button from the **Standard Views** area and then choose the **OK** button; you are prompted to specify the insertion point of the viewport. Place the left view at the left of the previously created front view and then press ESC to exit the **Adjacent View** tool. Figure P-152 shows the Left view of the plant model.

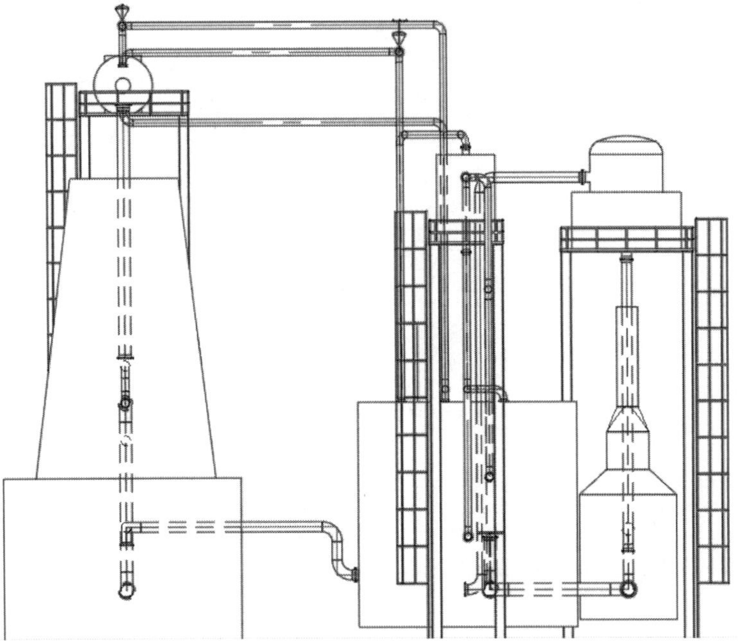

Figure P-152 *Left view of the plant model*

Next, choose the **Adjacent View** tool from the **Ortho Views** panel of the **Ortho View** tab; you are prompted to select an orthographic view to create an adjacent view. Select the view boundary of the top view placed earlier; the **Create an Adjacent View** dialog box is displayed.

In this dialog box, select the **NE Isometric** radio button from the **Iso Views** area and then choose the **OK** button; you are prompted as **Specify insertion point of the viewport or [Scale/Rotate/Existing]**. Choose the **Scale** option from the command prompt; you are prompted to specify the scale. Type **0.0035** at the command prompt and press ENTER. Next, place the view at top left of the drawing sheet. Figure P-153 shows the **NE Isometric** view of the plant model.

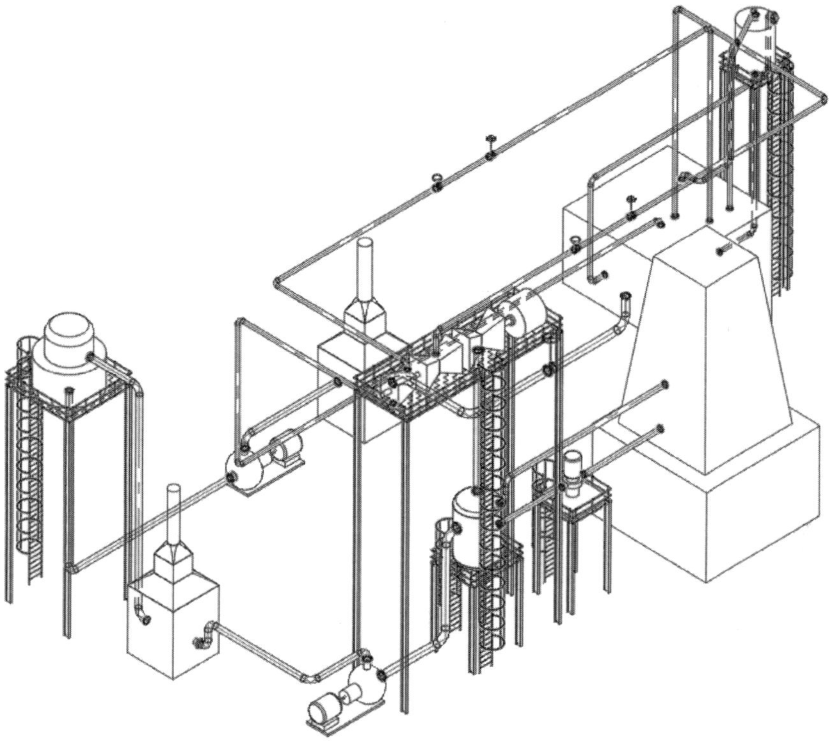

Figure P-153 NE Isometric *view of the plant model*

After placing all the views in the drawing sheet, it should look similar to the one shown in Figure P-154.

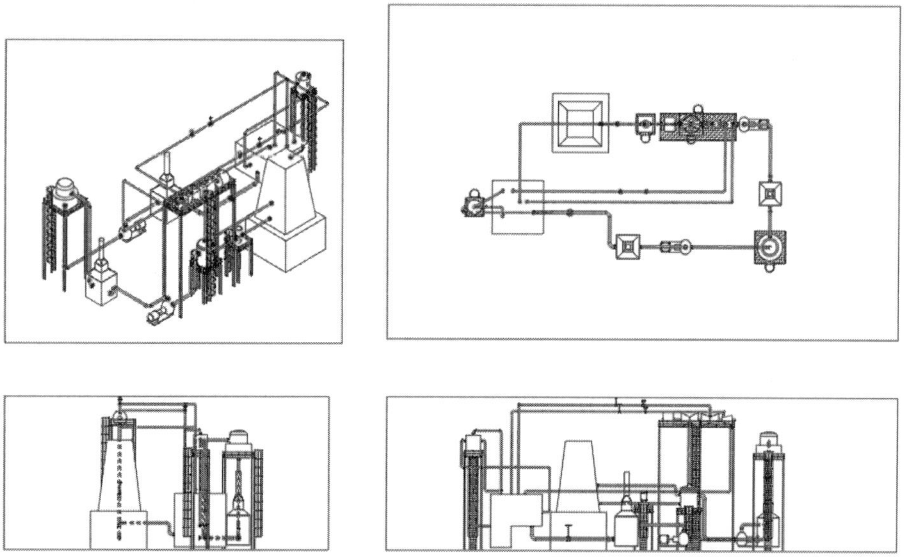

Figure P-154 *Drawing sheet after placing views*

Adding Annotations and Dimensions

First, you will add annotations to the components. To do so, choose the **Ortho Annotate** tool from the **Annotation** panel of the **Ortho View** tab; you are prompted to select a component to annotate. Select the Low Pressure Heater in the top view; you are prompted to specify the annotation style. Press ENTER; the annotation is attached to the cursor and you are prompted to specify the annotation position. Place the annotation at the right of the Low Pressure Heater, refer to Figure P-155.

Next, follow the same procedure to annotate the other equipment in the Top view. Figure P-155 shows the top view of the plant model after adding annotations to the equipment.

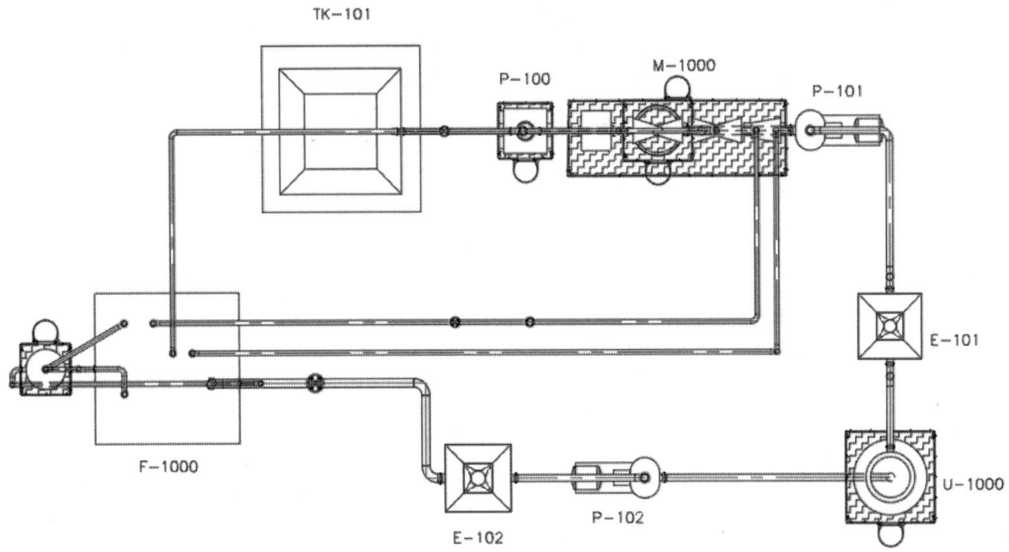

***Figure P-155** Top view after adding annotations*

Now, you will add dimensions to the top view of the plant model. To do so, choose the **Linear** tool from the **Dimension** drop-down of the **Dimensions** panel of the **Ortho View** tab; you are prompted to specify the first line origin. Select the center point of the Deaerator; you are prompted to specify the second extension line origin. Select the center point of the exit nozzle of the Centrifugal Pump 2; the dimension is attached to the cursor. Place the dimension into the drawing area, refer to Figure P-156 for location of the dimension.

Follow the same procedure to add other dimensions to the top view of the plant model. Figure P-156 shows the top view of the model after adding dimensions.

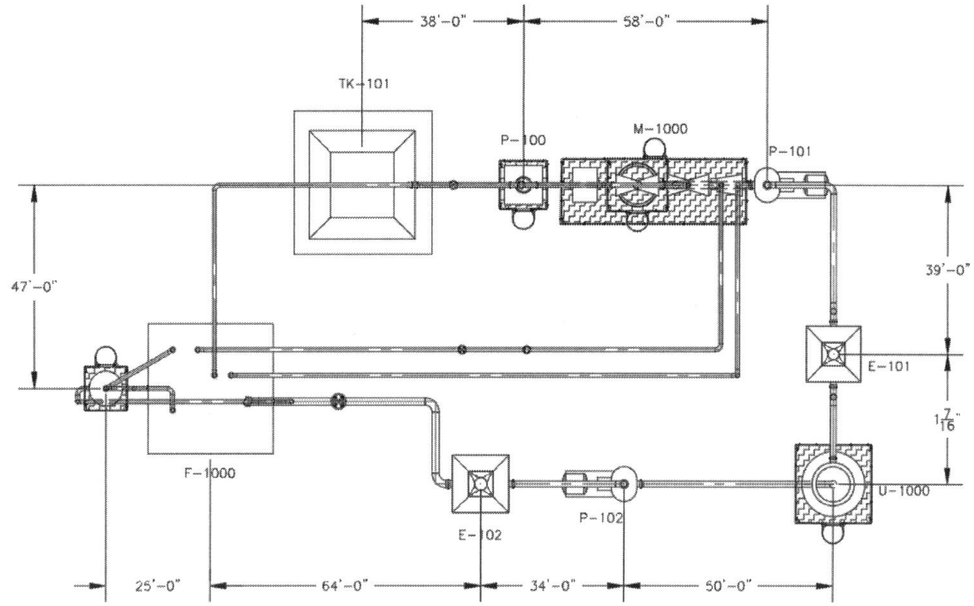

Figure P-156 Top view after adding dimensions

Next, save and close the orthographic drawing.

CREATING REPORTS AND MANAGING DATA

In this section, you will manage and create reports of the Thermal Power Plant model. To do so, choose the **Data Manager** button from the **Project** panel of the **Home** tab; the **DATA MANAGER** is displayed, refer to Figure P-157.

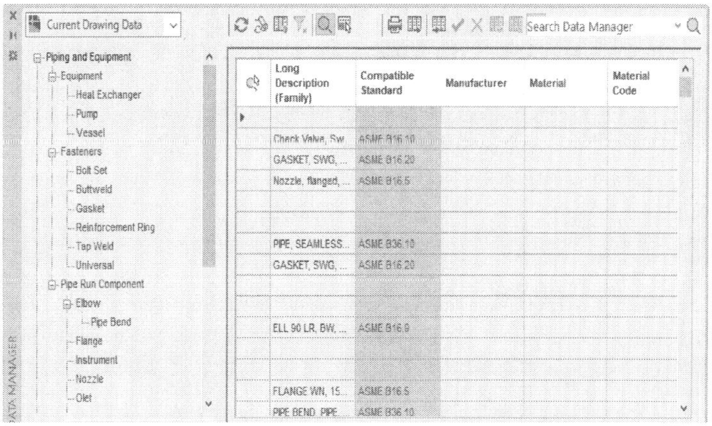

*Figure P-157 The **DATA MANAGER***

In the **DATA MANAGER**, select the **Plant 3D Project Data** option from the drop-down list located at the top-left corner; the Plant 3D project data is displayed. Select the **Equipment** node from the Class tree; the data related to equipment is displayed in the Data table.

Filtering and Exporting the Data

Now, you need to filter and export the data related to a particular equipment. Right-click on **Mechanical Drivers** in the **Long Description(Size)** column of the Data table and choose the **Filter By Selection** option from the shortcut menu displayed, refer to Figure P-158; the data related to **Mechanical Drivers** is displayed in the Data table.

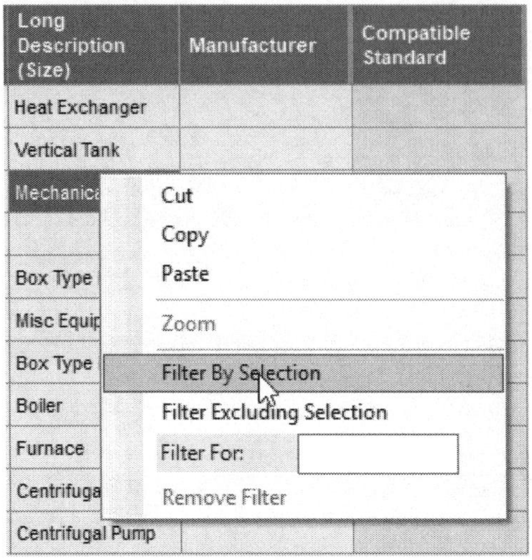

*Figure P-158 Choosing the **Filter By Selection** option*

Next, choose the **Export** button from the **DATA MANAGER** toolbar; the **Export Data** dialog box is displayed, refer to Figure P-159.

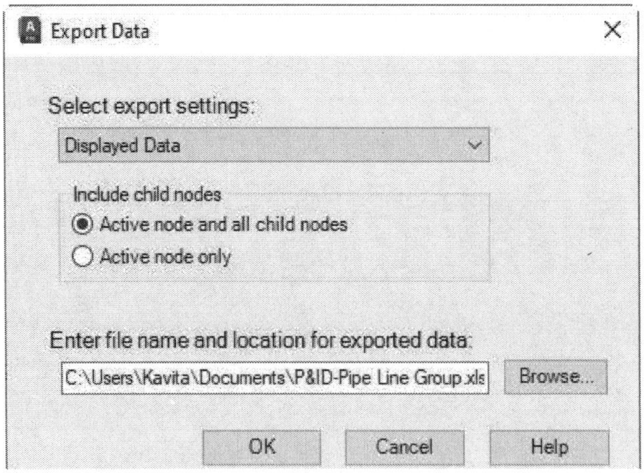

*Figure P-159 The **Export Data** dialog box*

In this dialog box, select the **Active node only** radio button from the **Include child nodes** area and then choose the **Browse** button; the **Export To** dialog box is displayed, refer to Figure P-160.

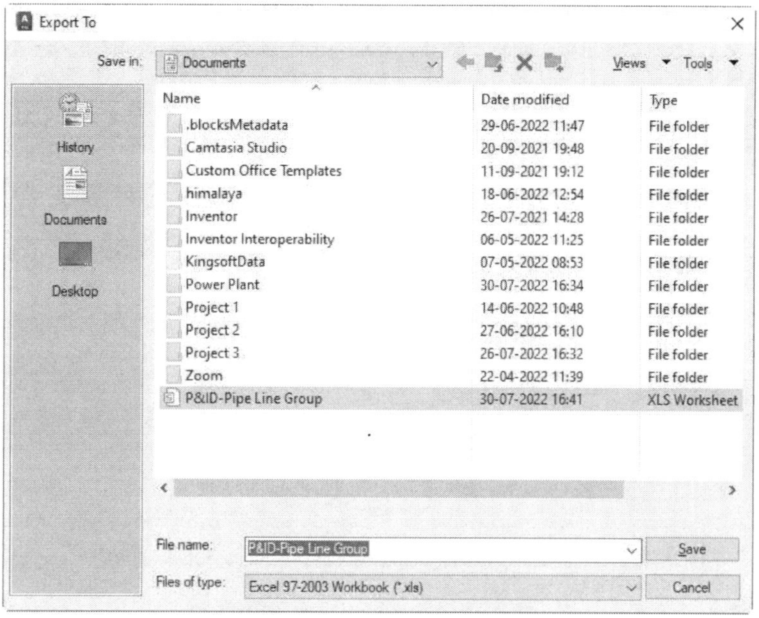

Figure P-160 *The **Export To** dialog box*

In this dialog box, select the **Excel Workbook (*.xlsx)** option from the **Files of type** drop-down list and specify the file location as C:\Users\User_name\Documents\POWER PLANT. Next, choose the **Save** button to exit the **Export To** dialog box; the **Export Data** dialog box is displayed again. Choose the **OK** button to exit the **Export Data** dialog box. The data is exported to the excel file and it is saved at the specified location.

Modifying the Data

Browse to the location C:\Users\User_name\Documents\POWER PLANT and open the **POWER PLANT- Equipment.xlsx** file; a spreadsheet is displayed, refer to Figure P-161.

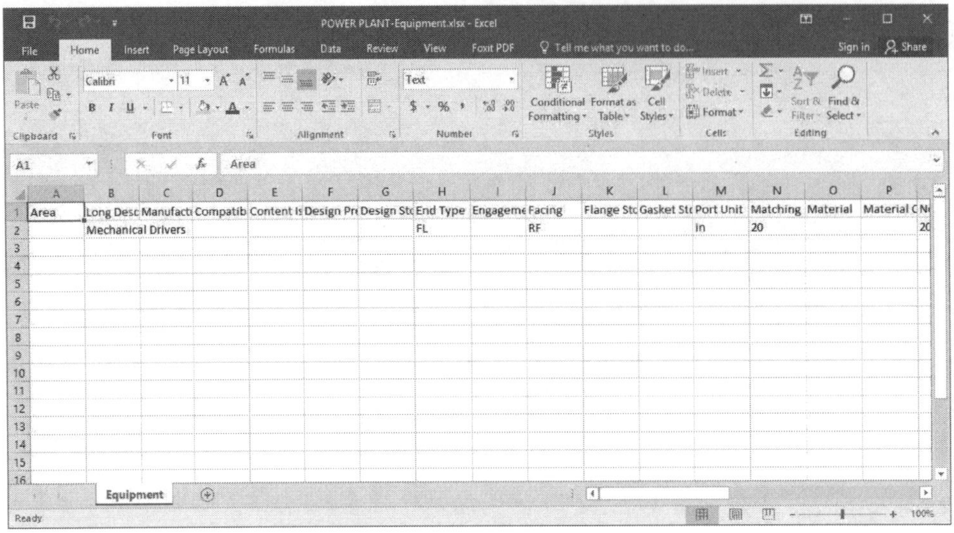

Figure P-161 Excel Sheet displaying Equipment Data

In the spreadsheet, change the **Long Description (Size)** from **Mechanical Drivers** to **Turbine**. Figures P-162 and P-163 show the **Long Description (Size)** column before and after editing, respectively.

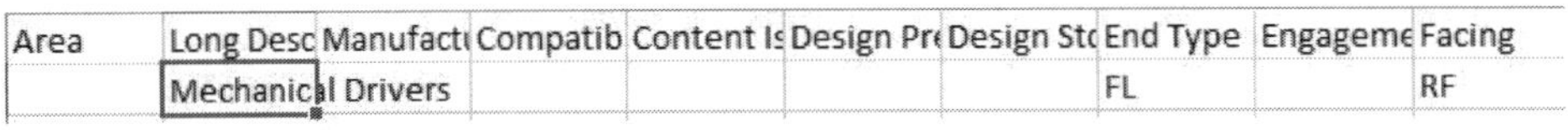

***Figure P-162 Long Description (Size)** before editing*

***Figure P-163 Long Description (Size)** after editing*

Next, save the excel file after making all the modifications in it and close it.

Importing the Data

Now, you will import the modified data into the DATA MANAGER. To do so, switch to AutoCAD Plant 3D and open the **DATA MANAGER** if it is closed. In the **DATA MANAGER**, make sure that **Plant 3D Project Data** is selected in the drop-down list located at the top left corner and equipment data is displayed in the Data table. Next, choose the **Import** button from the **DATA MANAGER** toolbar, refer to Figure P-164; the **AutoCAD Plant** message box is displayed, refer to Figure P-165.

***Figure P-164** Choosing the **Import** button*

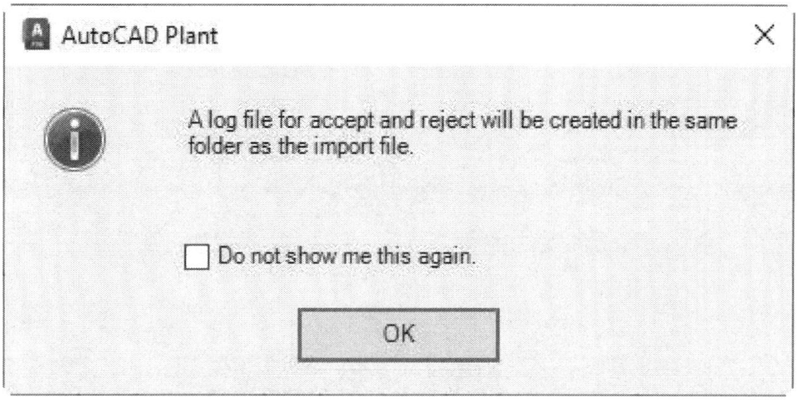

Figure P-165 *The* ***AutoCAD Plant*** *message box*

Next, choose the **OK** button from the **AutoCAD Plant** message box; the **Import From** dialog box is displayed, refer to Figure P-166.

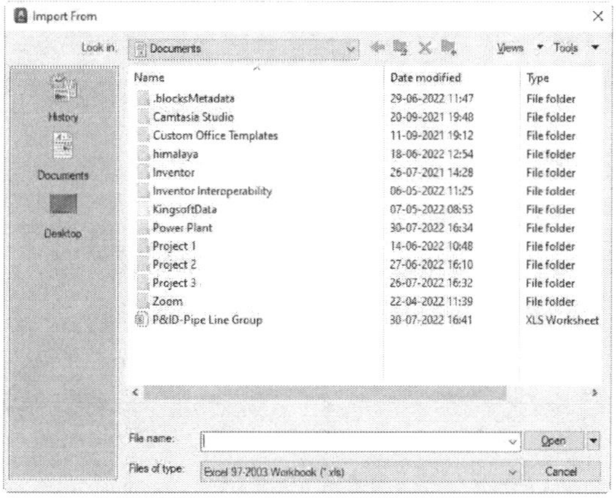

Figure P-166 *The* ***Import From*** *dialog box*

Next, browse to the location C:\Users\User_name\Documents\POWER PLANT and open the **POWER PLANT- Equipment.xlsx** file; the **Import Data** dialog box is displayed, refer to Figure P-167.

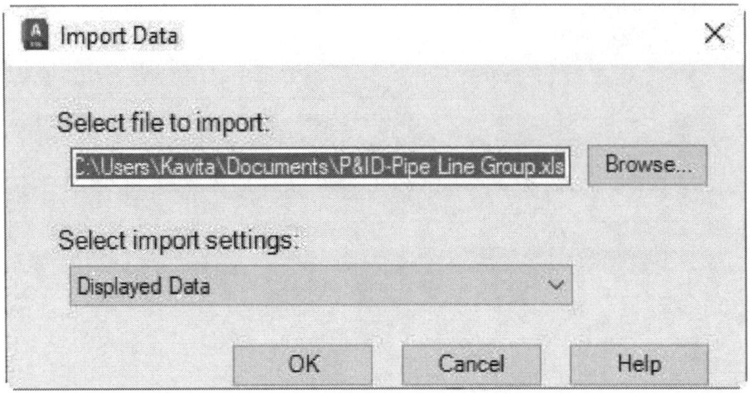

Figure P-167 *The* ***Import Data*** *dialog box*

Next, choose the **OK** button from the **Import Data** dialog box; the data in the excel file is imported into the **DATA MANAGER**, refer to Figure P-168.

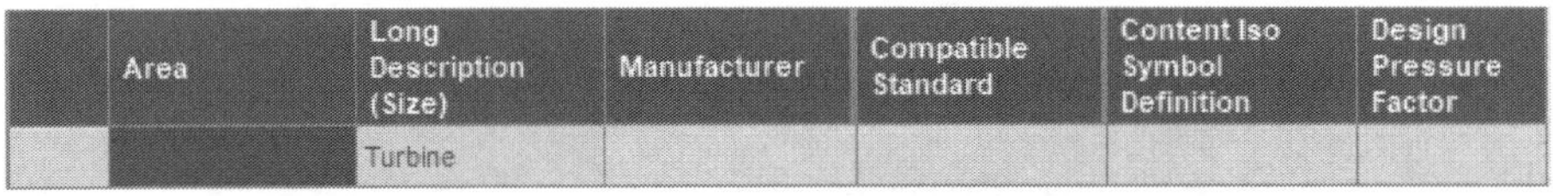

	Area	Long Description (Size)	Manufacturer	Compatible Standard	Content Iso Symbol Definition	Design Pressure Factor
		Turbine				

Figure P-168 *The Data Table after importing the data*

After importing the data into the **DATA MANAGER**, choose the **Accept All** button from the **DATA MANAGER** toolbar to accept the changes made to the data, refer to Figure P-169, and then exit the **DATA MANAGER**.

Figure P-169 *Choosing the* ***Accept All*** *button*

Next, choose the **Save** button from the **Application** menu and then the **Close** button to close the file.

Using Autodesk AutoCAD Plant Report Creator and Generating Reports

Now, you will be using **Autodesk AutoCAD Plant Report Creator** to generate reports. To do so, double-click on the shortcut icon of **AutoCAD Plant Report Creator 2023 - English** available on your desktop; the **Autodesk AutoCAD Plant 3D Report Creator** will be displayed, refer to Figure P-170.

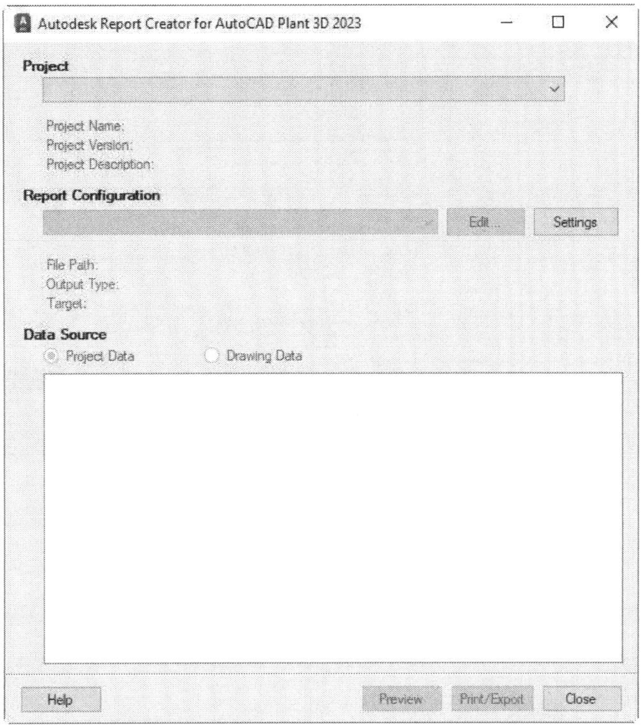

Figure P-170** The **Autodesk AutoCAD Plant 3D Report Creator

In **Autodesk AutoCAD Plant Report Creator**, select the **Open** option from the **Project** drop-down list; the **Open** dialog box will be displayed. Next, browse to the location C:\Users\User_name\Documents\POWER PLANT and double-click on the **Project.xml** file; the data sources related to the **POWER PLANT** project will be displayed in the **Autodesk AutoCAD Plant Report Creator**. Select the **3D Parts** option from the **Report Configuration** drop-down list and select the **Project Data** radio button from the **Data Source** area, if not already selected. Next, choose the **Print/Export** button from **Autodesk AutoCAD Plant Report Creator**; the **PDF Export Options** dialog box will be displayed. In this dialog box, enter the values in different text boxes as shown in Figure P-171 and then choose the **OK** button; the report will be exported in the PDF file format and the **Export results** window will be displayed, refer to Figure P-172.

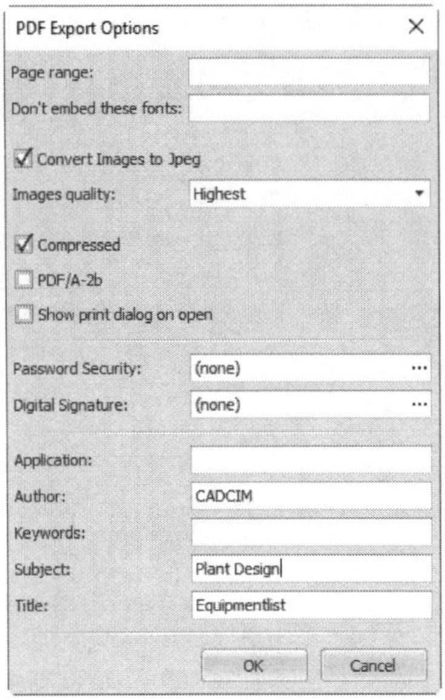

Figure P-171 *The* ***PDF Export Options*** *dialog box*

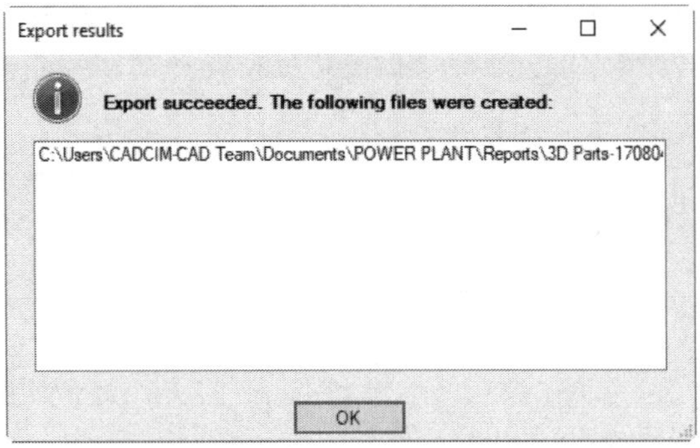

Figure P-172 *The* ***Export Results*** *window*

In this window, double-click on the file path; the PDF file containing the 3D Parts report will be displayed, refer to Figures P-173 and P-174.

Bill of Material

Autodesk

Project: POWER PLANT

Note: Fixed-length pipes are not included in pipes.

Quantity	Unit	Description	ND	Standard	Schedule	Material	PN	Angle
Type: Pipe, Seamless								
200'-2"	in	Pipe, Seamless, PL, ASME B36.10	10 in	ASME B36.10				
Type: PIPE, SEAMLESS								
604'-8"	in	PIPE, SEAMLESS, 40, 10" ND, PE, ASTM A106	10 in	ASTM A106	100	A106		
83'-10"	in	PIPE, SEAMLESS, 40, 12" ND, PE, ASTM A106	12 in	ASTM A106	100	A106		
211'-10"	in	PIPE, SEAMLESS, 40, 16" ND, PE, ASTM A106	16 in	ASTM A106	100	A106		
122'-10"	in	PIPE, SEAMLESS, 40, 20" ND, PE, ASTM A106	20 in	ASTM A106	100	A106		
Type: Elbow 90.0°								
19		PH IMPERIAL Elbow 90.0° ND 10"	10 in					
2		PH IMPERIAL Elbow 90.0° ND 12"	12 in					
8		PH IMPERIAL Elbow 90.0° ND 16"	16 in					
5		PH IMPERIAL Elbow 90.0° ND 20"	20 in					
Type: ELL 90 LR								
3		ELL 90 LR, BV, ASME B16.9	10 in	ASME B16.9				
Type: FLANGE WN								
23		FLANGE WN, RF, 150 LB, ASME B16.5	10 in	ASME B16.5			150	
6		FLANGE WN, 12" ND, 150 LB, RF, ASME B16.5	12 in	ASME B16.5			150	
8		FLANGE WN, 16" ND, 150 LB, RF, ASME B16.5	16 in	ASME B16.5			150	
9		FLANGE WN, RF, 150 LB, ASME B16.5	20 in	ASME B16.5			150	
Type: Bolt set								
8		Bolt set, RF, 150 LB, Stud Bolt	10 in				150	
1		Bolt set, RF, 150 LB, Stud Bolt	20 in				150	
Type: Stud Bolt								
10		PH IMPERIAL Stud Bolt 1"x6 1/4" Lg w/2 Hex. Nut 1"	10 in					
6		PH IMPERIAL Stud Bolt 1 1/8"x6 3/4" Lg w/2 Hex. Nut 1 1/8"	12 in					
8		PH IMPERIAL Stud Bolt 1 1/4"x7 1/2" Lg w/2 Hex. Nut 1 1/4"	16 in					
7		PH IMPERIAL Stud Bolt 1 1/4"x8 1/4" Lg w/2 Hex. Nut 1 1/4"	20 in					

Figure P-173 *Page 1 of the 3D Parts report in PDF file*

Quantity	Unit Description	ND	Standard	Schedule	Material	PN	Angle
Type: Gasket							
14	PH IMPERIAL Gasket ND 10"	10 in					
6	PH IMPERIAL Gasket ND 12"	12 in					
8	PH IMPERIAL Gasket ND 16"	16 in					
7	PH IMPERIAL Gasket ND 20"	20 in					
Type: Gasket, SWG							
8	GASKET, SWG, 10" ND, 1/8" THK, 150 LB, RF, ASME B16.20	10 in	ASME B16.20			150	
1	GASKET, SWG, 20" ND, 1/8" THK, 150 LB, RF, ASME B16.20	20 in	ASME B16.20			150	
Type: Control Valve, Ball							
2	CONTROL VALVE, BALL, 10" ND, 150 LB, RF, ISA 75.08.02, 11 69/100" LG	10 in	ISA 75.08.02			150	
Type: Gate Valve							
2	GATE VALVE, CONDUIT, 10" ND, 150 LB, RF, ASME B16.10, 13" LG	10 in	ASME B16.10			150	
1	GATE VALVE, CONDUIT, 20" ND, 150 LB, RF, ASME B16.10, 18" LG	20 in	ASME B16.10			150	

Figure P-174 *Page 2 of the 3D Parts report in PDF file*

After viewing the report, close the PDF file. Next, choose the **OK** button from the **Export results** window and close **Autodesk AutoCAD Plant Report Creator**.

Index

Symbols

3D Piping Workspace 1-13,1-14
3D Model Selection tool 9-4

A

All Hidden Lines tool 9-4
Annotation Settings 8-10
Application Menu 1-15
Attach Equipment tool 4-9
Interface Screen of AutoCAD Plant 3D 1-2
Autodesk Connection point 6-23
Autodesk App Store 1-25
Alert message box 1-23

B

Backup Files 1-19
Base AutoCAD Object Errors 2-22
Bends 6-14
Bill of materials 8-15, 9-8
Block 2-22
Block Based Graphics 5-17
BOM 9-9
Box Type Heater 4-5
Branch Table Editor 5-2
Branch Table Legends 5-19

C

Cage 3-10
Catalog Browser 5-14
Catalog Editor 5-13
Change Pipe Elevation grip 7-16
Check Isometric 8-2
Class Tree 10-2
Compass panel 6-8
Connection Port Properties Area 5-14
Control Valve Browser dialog box 2-18
Convert Supports tool 7-14
Create Equipment dialog box 4-5,4-2
Create Equipment tool 4-18
Create New Component dialog box 5-15
Create Quick Iso dialog box 8-2
Custom Parts Builder palette 7-5
Cutback Elbow 6-12
Cut Back Member tool 3-13, 3-21
Cut Pipe Symbols button 9-5

D

Data Manager 1-6
Data Manager Toolbar 10-2
Data table 10-3
Detach Equipment tool 4-10
Detach Supports tool 7-15
Dimensional Settings 8-11
Drawing Recovery Manager 1-20
Drawing Recovery message box 1-20
Dummy Leg 7-10
Dynamic Input Mode 6-4, 6-22

E

Edit Valve Operator 5-12
Editing Grids 3-3
Edit View tool 9-8
Equipment 4-2
Equipment drop-down 4-2
EXIT 1-24
Export button 10-9
Export PCF dialog box 8-7
Equipment in a P&ID 2-8
Export Report Data dialog box 10-9

F

Fittings 2-20
Flip Part grip 7-15
Flow Arrow 8-7
Footings 3-12

G

Grid 3-2
Grouping Lines 2-27

H

Hanger 7-10
Heater 4-5
Heat Exchanger 4-4
HELP 1-24
Hide Selected tool 3-16

I

Import button 10-9
Import Data dialog box 10-8
In-Canvas Viewport Controls 1-9
Inline Flip Part grip 7-16
Instrumentation Lines 2-18
Instruments 2-18
Iso 8-4, 8-9
Iso message tool 8-6
Isometric 8-2, 8-8
Isometric drawings 8-2
Isos 1-13
Inventor Models 4-9

L

Ladder 3-9, 3-10
Line 3-3
Line Number Selector 6-4
Line to Pipe tool 6-6
Locate in 3D model tool 9-8
Locking 8-8
Long description (family) style 5-11
Long description (size) style 5-11

M

Manage Xrefs icon 1-9
Matchlines button 9-5
Members 3-4
Menu Bar 1-17
Modify Equipment tool 4-7

N

NEWPROJECT command 2-2
Nozzles 4-3

O

Off page Connectors 2-20
Offset Connect 6-16
O-let Branch 6-19
OPEN command 1-21
Ortho 1-14, 9-2
OrthoCube 9-4
Ortho Editor contextual tab 9-2
Ortho Generation message box 9-6
Orthographic 8-2
Orthographic Drawings 9-2
Orthographic DWG tab 9-2
Outline Model 3-3

P

Piping 1-13
Pan tool 1-7
Parametric Graphics 5-16
Part Use Priority 5-7
PCF to Iso tool 8-8
P&ID Line List palette 7-3
P&ID Project Data 10-4
P&ID Validation Settings dialog box 7-18
Pipe Component File 8-8
Pipe Gap Tool 9-8
Adding Pipe Lines 2-9
Pipe Spec Viewer 6-2
Pipe Supports Spec tab 7-9
Piping Component Editor 5-13
Piping Component Properties Area 5-14
Plant 3D Project Data 10-2
Plant 3D Shape Browser dialog box 7-5
PLANTPARTCONVERT command 5-17
Plate/Grate 3-11
Production isometric drawing 8-5

PROJECT MANAGER 1-5, 2-2
Project Setup tool 2-30, 8-9
PROPERTIES Palette 1-10
Pump 4-4
Pipe Supports 7-8
Project Reports 10-8
Pipe Support dialog box 7-8, 7-9

Q

Quick Iso tool 8-2
QUIT command 1-24

R

Railing 3-7
Reinforcing Pad 6-22
Report Creator 10-10
Restore Member tool 3-13
Roll Elbow 6-13
Rotate Part grip 7-14
Route line 6-5
Route Pipe tool 6-11
Route a Pipe 6-3
Route a Pipe using a P&ID 6-6

S

Schematic Line panel 2-26
Select File dialog box 1-22
Select Reference Models dialog box 9-3
Select Size and Spec dialog box 7-4
Set Elevation edit box 6-14
Shape Model mode 3-3
ShowMotion tool 1-8
Size Mismatch Errors 2-23
Sloped and Offset piping 8-13
Spec Editor 5-2
Spec Editor Layout and Settings dialog box 5-9
Spec Mismatch Errors 2-23
Spec Sheet 5-4
Spec Viewer button 6-2
Stairs 3-6
Status Bar 1-9
SteeringWheels 1-7
Structural Members 3-3, 3-12
Structure tab 1-13
Structure Explode tool 3-15
Stub-In Branch 6-20
Stub-In Branch at an Offset 6-20
Stub-in Branch at a Precise Location 6-21
Substitute Part grip 7-15
Substituting Components 2-28

T

Tags to a Line 2-10
Tee Branch 6-18
Themes 8-9
Title Block and Display Properties 8-14
Title Block Setup contextual tab 8-14
Toggle Compass 6-8
Toggle Slope button 6-16
TOOL PALETTES - P&ID PIP 2-9
Tray Settings dialog box 1-8
Trim Member tool 3-14

U

Unresolved Off Page Connectors Errors 2-24
Update Annotation tool 9-7

V

VALIDATECONFIG command 7-18
Validate the drawing 2-21
Validate a 3D model 7-18
Validation Summary palette 7-18
Vessel 4-3
ViewCube 1-7
Visibility Options 3-15
Valves and Fittings 7-2
Valves 2-17
Valves and Fittings Using a P&ID 7-3

Z

Zoom tools 1-7

Other Publications by CADCIM Technologies

The following is the list of some of the publications by CADCIM Technologies. Please visit *www.cadcim.com* for the complete listing.

AutoCAD Textbooks

- AutoCAD 2023: A Problem-Solving Approach, Basic and Intermediate, 29th Edition
- Advanced AutoCAD 2023: A Problem-Solving Approach (3D and Advanced), 26th Edition
- AutoCAD 2022: A Problem-Solving Approach, Basic and Intermediate, 28th Edition
- AutoCAD 2021: A Problem-Solving Approach, Basic and Intermediate, 27th Edition
- AutoCAD 2020: A Problem-Solving Approach, Basic and Intermediate, 26th Edition
- AutoCAD 2019: A Problem-Solving Approach, Basic and Intermediate, 25th Edition
- AutoCAD 2018: A Problem-Solving Approach, Basic and Intermediate, 24th Edition
- Advanced AutoCAD 2021: A Problem-Solving Approach (3D and Advanced), 25th Edition

Autodesk Inventor Textbooks

- Autodesk Inventor Professional 2023 for Designers, 23rd Edition
- Autodesk Inventor Professional 2022 for Designers, 22nd Edition
- Autodesk Inventor Professional 2021 for Designers, 21st Edition
- Autodesk Inventor Professional 2020 for Designers, 20th Edition

AutoCAD MEP Textbooks

- AutoCAD MEP 2020 for Designers, 5th Edition
- AutoCAD MEP 2018 for Designers, 4th Edition
- AutoCAD MEP 2016 for Designers, 3rd Edition

AutoCAD Plant 3D Textbooks

- AutoCAD Plant 3D 2021 for Designers, 6th Edition
- AutoCAD Plant 3D 2020 for Designers, 5th Edition
- AutoCAD Plant 3D 2018 for Designers, 4th Edition

Autodesk Fusion 360 Textbooks

- Autodesk Fusion 360: A Tutorial Approach, 3rd Edition
- Autodesk Fusion 360: A Tutorial Approach, 2nd Edition

Solid Edge Textbooks

- Solid Edge 2022 for Designers, 19th Edition
- Solid Edge 2021 for Designers, 18th Edition
- Solid Edge 2020 for Designers, 17th Edition

NX Textbooks

- Siemens NX 2021 for Designers,14th Edition
- Siemens NX 2020 for Designers,13th Edition
- Siemens NX 2019 for Designers,17th Edition

NX Mold Textbook
- Mold Design Using NX 11.0: A Tutorial Approach

NX Nastran Textbook
- NX Nastran 9.0 for Designers

SOLIDWORKS Textbooks
- SOLIDWORKS 2022 for Designers, 20th Edition
- SOLIDWORKS 2021 for Designers, 19th Edition
- SOLIDWORKS 2020 for Designers, 18th Edition
- Learning SOLIDWORKS 2019: A Project Based Approach

SOLIDWORKS Simulation Textbooks
- SOLIDWORKS Simulation 2018: A Tutorial Approach
- SOLIDWORKS Simulation 2016: A Tutorial Approach

CATIA Textbooks
- CATIA V5-6R2021 for Designers, 19th Edition
- CATIA V5-6R2020 for Designers, 18th Edition
- CATIA V5-6R2019 for Designers, 17th Edition

Creo Parametric Textbooks
- Creo Parametric 8.0 for Designers, 8th Edition
- Creo Parametric 7.0 for Designers, 7th Edition

ANSYS Textbooks
- ANSYS Workbench 2021 R1: A Tutorial Approach
- ANSYS Workbench 2019 R2: A Tutorial Approach

Creo Direct Textbook
- Creo Direct 2.0 and Beyond for Designers

Autodesk Alias Textbooks
- Learning Autodesk Alias Design 2016, 5th Edition
- Learning Autodesk Alias Design 2015, 4th Edition

AutoCAD LT Textbooks
- AutoCAD LT 2020 for Designers, 13th Edition
- AutoCAD LT 2017 for Designers, 12th Edition

EdgeCAM Textbooks
- EdgeCAM 11.0 for Manufacturers
- EdgeCAM 10.0 for Manufacturers

Autodesk Revit MEP Textbooks
- Exploring Autodesk Revit 2022 for MEP, 8th Edition
- Exploring Autodesk Revit 2021 for MEP, 7th Edition

AutoCAD Civil 3D Textbooks
- Exploring AutoCAD Civil 3D 2023, 12th Edition
- Exploring AutoCAD Civil 3D 2022, 11th Edition
- Exploring AutoCAD Civil 3D 2020, 10th Edition

AutoCAD Map 3D Textbooks
- Exploring AutoCAD Map 3D 2022, 9th Edition
- Exploring AutoCAD Map 3D 2018, 8th Edition
- Exploring AutoCAD Map 3D 2017, 7th Edition

RISA-3D Textbook
- Exploring RISA-3D 14.0

Autodesk Navisworks Textbooks
- Exploring Autodesk Navisworks 2022, 9th Edition
- Exploring Autodesk Navisworks 2021, 8th Edition
- Exploring Autodesk Navisworks 2020, 7th Edition

AutoCAD Raster Design Textbooks
- Exploring AutoCAD Raster Design 2017
- Exploring AutoCAD Raster Design 2016

Bentley STAAD.Pro Textbooks
- Exploring Bentley STAAD.Pro CONNECT Edition, 5th Edition
- Exploring Bentley STAAD.Pro CONNECT Edition, 4th Edition
- Exploring Bentley STAAD.Pro (CONNECT) Edition

Autodesk 3ds Max Design Textbooks
- Autodesk 3ds Max Design 2015: A Tutorial Approach
- Autodesk 3ds Max Design 2014: A Tutorial Approach

Autodesk 3ds Max Textbooks
- Autodesk 3ds Max 2022 for Beginners: A Tutorial Approach, 22nd Edition
- Autodesk 3ds Max 2021: A Comprehensive Guide, 21st Edition
- Autodesk 3ds Max 2020: A Comprehensive Guide, 20th Edition

Autodesk Maya Textbooks
- Autodesk Maya 2023: A Comprehensive Guide, 14th Edition
- Autodesk Maya 2022: A Comprehensive Guide, 13th Edition
- Autodesk Maya 2020: A Comprehensive Guide, 12th Edition

Pixologic ZBrush Textbooks
- Pixologic ZBrush 2022: A Comprehensive Guide, 8th Edition
- Pixologic ZBrush 2021: A Comprehensive Guide, 7th Edition
- Pixologic ZBrush 2020: A Comprehensive Guide, 6th Edition

Fusion Textbooks
- Blackmagic Design Fusion 7 Studio: A Tutorial Approach, 3rd Edition
- The eyeon Fusion 6.3: A Tutorial Approach

Flash Textbooks
- Adobe Flash Professional CC 2015: A Tutorial Approach, 3rd Edition
- Adobe Flash Professional CC: A Tutorial Approach

Computer Programming Textbooks
- Introducing PHP 7/MySQL
- Introduction to C++ programming, 2nd Edition
- Learning Oracle 12c - A PL/SQL Approach

MAXON CINEMA 4D Textbooks
- MAXON CINEMA 4D S24 Studio: A Tutorial Approach, 8th Edition
- MAXON CINEMA 4D R20 Studio: A Tutorial Approach, 7th Edition
- MAXON CINEMA 4D R19 Studio: A Tutorial Approach, 6th Edition
- MAXON CINEMA 4D R18 Studio: A Tutorial Approach, 5th Edition

Oracle Primavera Textbooks
- Exploring Oracle Primavera P6 Professional 18, 3rd Edition
- Exploring Oracle Primavera P6 v8.4

Coming Soon from CADCIM Technologies
- Flow Simulation Using SOLIDWORKS 2021

Made in the USA
Monee, IL
06 May 2026

49640598R00249